SOCIOLOGICAL THEORY IN THE CLASSICAL ERA

EDITION 2

KEY TITLES OF RELATED INTEREST FROM SAGE AND PINE FORGE PRESS

SOCIOLOGICAL THEORY IN THE CLASSICAL ERA

TEXT AND READINGS

EDITION 2

LAURA DESFOR EDLES | SCOTT APPELROUTH

California State University, Northridge

PINE FORGE PRESS
An Imprint of SAGE Publications, Inc.
Los Angeles • London • New Delhi • Singapore • Washington DC

For information:

Pine Forge Press
An Imprint of SAGE Publications, Inc.
2455 Teller Road
Thousand Oaks, California 91320
E-mail: order@sagepub.com

SAGE Publications India Pvt. Ltd.
B 1/I 1 Mohan Cooperative Industrial Area
Mathura Road, New Delhi 110 044
India

SAGE Publications Ltd.
1 Oliver's Yard
55 City Road
London EC1Y 1SP
United Kingdom

SAGE Publications Asia-Pacific Pte. Ltd.
33 Pekin Street #02-01
Far East Square
Singapore 048763

Printed in the United States of America

Library of Congress Cataloging-in-Publication Data

Edles, Laura Desfor.
 Sociological theory in the classical era: text and readings / Laura Desfor Edles, Scott Appelrouth.—2nd ed.
   p. cm.
 Includes bibliographical references and index.
 ISBN 978-1-4129-7564-3 (pbk.)
  1. Sociology—History. 2. Sociology—Philosophy. 3. Sociologists—Biography.
 I. Appelrouth, Scott, 1965- II. Title.

HM461.E35 2010
301.01—dc22   2009019881

This book is printed on acid-free paper.

13 10 9 8 7 6 5

Acquisitions Editor:	Jerry Westby
Associate Editor:	Leah Mori
Editorial Assistant:	Eve Oettinger
Production Editor:	Laureen Gleason
Copy Editor:	Alison Hope
Typesetter:	C&M Digitals (P) Ltd.
Proofreader:	Jennifer Gritt
Indexer:	Mary Mortensen
Cover Designer:	Gail Buschman
Marketing Manager:	Jennifer Reed Banando

BRIEF CONTENTS

DETAILED CONTENTS

LIST OF ILLUSTRATIONS AND PHOTOS

LIST OF FIGURES AND TABLES

ABOUT THE AUTHORS

Laura Desfor Edles (Ph.D., University of California, Los Angeles, 1990) is Associate Professor of Sociology at California State University, Northridge. She is the author of *Symbol and Ritual in the New Spain: The Transition to Democracy after Franco* (1998) and *Cultural Sociology in Practice* (2002), as well as various articles on culture, theory, race/ethnicity, and social movements.

Scott Appelrouth (Ph.D., New York University, 2000) is Associate Professor at California State University, Northridge. His interests include sociological theory, cultural sociology, and social movements. He has taught classical and contemporary theory at both the graduate and undergraduate levels, and has published several articles in research- and teaching-oriented journals. His current research focuses on the controversies over jazz during the 1920s and rap during the 1980s.

PREFACE

Every semester, we begin our sociological theory courses by telling students that we love sociological theory, and that one of our goals is to get each and every one of them to love theory too. This challenge we set for ourselves makes teaching sociological theory exciting. If you teach "sexy" topics like the sociology of drugs, crime, or sex, students come into class expecting to be titillated. By contrast, when you teach sociological theory, students tend to come into class expecting the course to be abstract, dry, and absolutely irrelevant to their lives. The fun in teaching sociological theory is in proving students wrong. The thrill in teaching sociological theory is in getting students to see that sociological theory is absolutely central to their everyday lives—and *fascinating* as well. What a reward it is to have students who adamantly insisted that they "hated" theory at the beginning of the semester "converted" into theorists by the end!

In teaching sociological theory, we use original texts. We rely on original texts in part because every time we read these works we derive new meaning from them. Core sociological works tend to become "core" precisely for this reason. However, using original readings requires that the professor spend lots of time and energy explaining issues and material that is unexplained or taken for granted by the theorist. This book was born of this process—teaching from original works and explaining them to our students. Hence, this book includes the original readings we use in our courses, as well as our interpretation and explanation of them.

Thus, this book is distinct in that it is both a reader *and* a text. It is unlike existing readers in several ways, however. First and foremost, this book is not just a collection of seemingly disconnected readings. Rather, in this book we provide an overarching theoretical framework with which to understand, compare, and contrast these selections. In our experience, this overarching theoretical framework is essential in explaining the relevance and excitement of sociological theory.

In addition, we discuss the social and intellectual milieu in which the selections were written, as well as their contemporary relevance. Thus, we connect these seemingly disparate works not only theoretically, but also via concrete applications to today's world.

Finally, this book is unique in that we provide a variety of visuals and pedagogical devices—historical and contemporary photographs, and diagrams and charts illuminating core theoretical concepts and comparing specific ideas—to enhance student understanding. Our thinking is, why should only introductory level textbooks have visual images and pedagogical aids? Most everyone, not just the youngest audiences, enjoys—and learns best from—visuals.

The second edition of this book is distinct in that it includes even more elements, such as

- more "Significant Other" boxes,
- new original readings,
- more discussion questions,
- more photos, and
- instructor resources, such as test questions and PowerPoint slides.

As is often the case in book projects, this turned out to be a much bigger and thornier project than either of us first imagined. And, in the process of writing this book, we have accrued many intellectual and social debts. First, we especially thank Jerry Westby of SAGE/Pine Forge Press, for helping us get this project started. Jerry literally walked into our offices at California State University, Northridge, and turned what had been a nebulous, long-standing idea into a concrete plan. Diana Axelsen, who oversaw the first edition of this book through its final stages of production made several critical suggestions regarding the layout of the book that we continue to appreciate. In the production of this second edition, we are grateful to Production Editor Laureen Gleason, Associate Editor Lindsay Dutro, Editorial Assistant Eve Oettinger, and Copy Editor Alison Hope, all of whom made the process of production extraordinarily smooth. We thank them for their conscientiousness and hard work. At California State University, Northridge, we are especially indebted to our Dean of the College of Social and Behavioral Sciences, Stella Theodoulou, for supporting this undertaking. We are particularly grateful for having received College grants that allowed us to complete the second edition of this book.

We thank the following reviewers for their comments:

For the First Edition

Cynthia Anderson
University of Iowa

Jeralynn Cossman
Mississippi State University

Lara Foley
University of Tulsa

Paul Gingrich
University of Regina

Leslie Irvine
University of Colorado

Doyle McCarthy
Fordham University

Martha A. Myers
University of Georgia

Riad Nasser
Farleigh Dickinson University

Paul Paolucci
Eastern Kentucky University

Chris Ponticelli
University of South Florida

Larry Ridener
Pfeiffer University

Chaim Waxman
Rutgers University

For the Second Edition

James J. Dowd
University of Georgia

Alison Faupel
Emory University

Greg Fulkerson
SUNY Oneonta

Gesine Hearn
Idaho State University

Jacques Henry
University of Louisiana at Lafayette

Gabe Ignatow
University of North Texas

David Levine
Florida Atlantic University

E. Dianne Mosley
Texas Southern University

Finally, we both want to thank our families—Amie, Alex, and Julia; and Mike, Benny, and Ellie—for supporting us while we spent so much time and energy on this project.

1 INTRODUCTION

Key Concepts

■■ Theory
■■ Order
  ❑ Collective/Individual
■■ Action
  ❑ Rational/Nonrational
■■ Enlightenment
■■ Counter-Enlightenment

"But I'm not *a serpent, I tell you!" said Alice. "I'm a—I'm a—"*

"Well! What are *you?" said the Pigeon. "I can see you're trying to invent something!"*

"I—I'm a little girl," said Alice, rather doubtfully, as she remembered the number of changes she had gone through that day.

"A likely story indeed!" said the Pigeon, in a tone of the deepest contempt. "I've seen a good many little girls in my time, but never one *with such a neck as that! No, no! You're a serpent; and there's no use denying it. I suppose you'll be telling me next that you never tasted an egg!"*

"I have *tasted eggs, certainly," said Alice, who was a very truthful child; "but little girls eat eggs quite as much as serpents do, you know."*

"I don't believe it," said the Pigeon; "but if they do, why, then they're a kind of serpent: that's all I can say."

—Lewis Carroll, *Alice's Adventures in Wonderland* (1865/1960:54)

1

In the passage above, the Pigeon had a theory—Alice is a serpent because she has a long neck and eats eggs. Alice, however, had a different theory—she was a little girl. It was not the "facts" that were disputed in the above passage, however. Alice freely admitted she had a long neck and ate eggs. So why did Alice and the Pigeon come to such different conclusions? Why didn't the facts "speak for themselves"?

Alice and the Pigeon both *interpreted* the question (What *is* Alice?) using the categories, concepts, and assumptions with which each was familiar. It was these unarticulated concepts, assumptions, and categories that led the Pigeon and Alice to have such different conclusions.

Likewise, social life can be perplexing and complex. It is hard enough to know "the facts," let alone to know *why* things are as they seem. In this regard, theory is vital to making sense of social life because it holds assorted observations and facts together (as it did for Alice and the Pigeon). Facts make sense only because we interpret them using preexisting categories and assumptions, that is, "theories." The point is that even so-called facts are based on implicit assumptions and unacknowledged presuppositions. Whether or not we are consciously aware of them, our everyday life is filled with theories as we seek to understand the world around us. The importance of formal sociological theorizing is that it makes assumptions and categories explicit, hence makes them open to examination, scrutiny, and reformulation.

To be sure, some students find classical sociological theory as befuddling as Alice found her conversation with the Pigeon. Some students find it difficult to understand and interpret what classical theorists are saying. Indeed, some students wonder why they have to read works written more than a century ago, or why they have to study sociological theory at all. After all, they maintain, classical sociological theory is abstract and dry and has "nothing to do with my life." So why not just study contemporary theory (or, better yet, just examine empirical "reality"), and leave the old, classical theories behind?

In this book, we seek to demonstrate the continuing relevance of classical sociological theory. We argue that the theorists whose work you will read in this book are vital: first, because they helped chart the course of the discipline of sociology from its inception until the present time, and second, because their concepts and theories still permeate contemporary concerns. Sociologists still seek to explain such critical issues as the nature of capitalism, the basis of social solidarity or cohesion, the role of authority in social life, the benefits and dangers posed by modern bureaucracies, the dynamics of gender and racial oppression, and the nature of the "self," to name but a few. Classical sociological theory provides a pivotal conceptual base with which to explore today's world. To be sure, this world is more complex than it was a century ago, or for that matter, than it has been throughout most of human history, during which time individuals lived in small bands as hunter-gatherers. With agricultural and later industrial advances, however, societies grew increasingly complex. The growing complexity, in turn, led to questions about what is distinctively "modern" about contemporary life. Sociology was born as a way of thinking about just such questions; today, we face similar questions about the "postmodern" world. The concepts and ideas introduced by classical theorists enable us to ponder the causes and consequences of the incredible rate and breadth of change.

The purpose of this book is to provide students not only with core classical sociological readings, but also with a framework for comprehending them. In this introductory chapter, we discuss (1) *what* sociological theory is, (2) *why* it is important for students to read the original works of the "core" figures in sociology, (3) *who* these "core" theorists are, and (4) *how* students can develop a more critical and gratifying understanding of some of the most important ideas advanced by these theorists. To this end, we introduce a metatheoretical framework that enables students to navigate, compare, and contrast the theorists' central ideas as well as to contemplate *any* social issue within our own increasingly complex world.

Theory is a system of generalized statements or propositions about phenomena. There are two additional features, however, that together distinguish scientific theories from other idea systems such as those found in religion or philosophy. Scientific theories

1. explain and predict the phenomena in question, and

2. produce testable and thus falsifiable hypotheses.

Universal laws are intended to explain and predict events occurring in the natural or physical world. For instance, Isaac Newton established three laws of motion. The first law, the law of inertia, states that objects in motion will remain in motion and objects at rest will remain at rest, unless acted on by another force. In its explanation and predictions regarding the movement of objects, this law extends beyond the boundaries of time and space. For their part, sociologists seek to develop or refine general statements about some aspect of *social* life. For example, a long-standing (although not uncontested) sociological theory predicts that as a society becomes more modern, the salience of religion will decline. Similar to Newton's law of inertia, the secularization theory, as it is called, is not restricted in its scope to any one time period or population. Instead, it is an abstract proposition that can be tested in any society once the key concepts making up the theory—"modern" and "religion"—are defined, and once observable measures are specified.

Thus, sociological theories share certain characteristics with theories developed in other branches of science. However, there are significant differences between social and other scientific theories (i.e., theories in the social sciences as opposed to the natural sciences) as well. First, sociological theories tend to be more evaluative and critical than theories in the natural sciences. Sociological theories are often rooted in implicit moral assumptions, which contrast with traditional notions of scientific objectivity. In other words, it is often supposed the pursuit of scientific knowledge should be free from value judgments or moral assessments, that the first and foremost concern of science is to uncover what *is*, not what *ought* to be. Indeed, such objectivity is often cast as a defining feature of science, one that separates it from other forms of knowledge based on tradition, religion, or philosophy. But sociologists tend to be interested not only in understanding the workings of society, but also in realizing a more just or equitable social order. As you will see, the work of the core classical theorists is shaped in important respects by their own moral sensibilities regarding the condition of modern societies and what the future may bring. Thus, sociological theorizing at times falls short of the "ideal" science practiced more closely (though still imperfectly) by "hard" sciences like physics, biology, or chemistry. For some observers, this failure to conform consistently to the ideals of either science or philosophy is a primary reason for the discipline's troublesome identity crisis and "ugly duckling" status within the academic world. For others, it represents the opportunity to develop a unique understanding of social life.

A second difference between sociological theories and those found in other scientific disciplines stems from the nature of their respective subjects. Societies are always in the process of change, while the changes themselves can be spurred by any number of causes including internal conflicts, wars with other countries, scientific or technological advances, or through the expansion of economic markets that in turn spread foreign cultures and goods. As a result, it is more difficult to fashion universal laws to explain societal dynamics. Moreover, we must also bear in mind that humans, unlike other animals or naturally occurring elements in the physical world, are motivated to act by a complex array of social and psychological forces. Our behaviors are not the product of any one principle; instead, they can be driven by self-interest, altruism, loyalty, passion, tradition, or habit, to name but a few factors. From these remarks, you can see the difficulties inherent in developing universal laws of societal development and individual behavior, despite our earlier example of the secularization theory as well as other efforts to forge such laws.

These two aspects of sociological theory (the significance of moral assumptions and the nature of the subject matter) are responsible, in part, for the form in which much sociological theory is written. While some theorists construct formal propositions or laws to explain and predict social events and individual actions, more often theories are developed through storylike narratives. Thus, few of the original readings included in this volume will contain explicitly stated propositions. One of the intellectual challenges you will face in studying the selections is to uncover the general propositions embedded in the texts. Regardless of the style in which they are presented, however, the theories (or narratives) you will explore in this text answer the most central social questions, while revealing taken-for-granted truths and encouraging you to examine who you are and where we, as a society, are headed.

▪▪ WHY READ ORIGINAL WORKS?

Some professors agree with students that original works are just too hard to decipher. These professors use secondary textbooks that interpret and simplify the ideas of core theorists. Their argument is that you simply cannot capture students' attention using original works; students must be engaged in order to understand, and secondary texts ultimately lead to a better grasp of the covered theories.

However, there is an important problem with reading only interpretations of original works: The secondary and original texts are not the same. Secondary texts do not simply translate what the theorist wrote into simpler terms; rather, in order to simplify, they must revise what an author has said.

The problems that can arise from even the most faithfully produced interpretations can be illustrated by the "telephone game." Recall that childhood game where you and your friends sit in a circle. One person thinks of a message and whispers it to the next person, who passes the message on to the next person, until the last person in the circle announces the message aloud. Usually, everyone roars with laughter because the message at the end typically is nothing like the one circulated at the beginning. This is because the message inadvertently is misinterpreted and changed as it goes around.

In the telephone game, the goal is to repeat exactly what has been said to you. Yet, misinterpretations and modifications are commonplace. Consider now a secondary text in which the goal is not to restate exactly what originally was written, but to take the original source and make it "easier" to understand. While this process of simplification perhaps allows you to understand the secondary text, you are at least one step removed from what the original author actually wrote.[1] At the same time, you have no way of knowing what was written in the original works. Moreover, when you start thinking and writing about the material presented in the secondary reading, you are not one, but *two* steps removed from the original text. If the object of a course in classical sociological theory is to grapple with the ideas that preoccupied the core figures of the field—the ideas and analyses that would come

---

[1]Further complicating the matter is that many of the original works that make up the core of sociological theory were written in a language other than English. Language translation is itself an imperfect exercise.

to shape the direction of sociology for more than a century—then studying original works must be a cornerstone of the course.

To this end, we provide excerpts from the original writings of those we consider to be sociology's core classical theorists. If students are to understand Karl Marx's writings, they must read *Marx*, and not a simplified interpretation of his ideas. They must learn to study for themselves what the initiators of sociology have said about some of the most fundamental social issues, the relevance of which is timeless.

Yet, we also provide in this book a secondary interpretation of the theorists' overall frameworks and the selected readings. Our intent is to provide a guide (albeit simplified) for understanding the original works. The secondary interpretation will help you navigate the different writing styles often resulting from the particular historical, contextual, and geographical locations in which the theorists were rooted.

WHO ARE SOCIOLOGY'S CORE THEORISTS?

Our conviction that students should read the core classical sociological theorists raises an important question: Who are the core theorists? After all, the discipline of sociology has been influenced by dozens of philosophers and social thinkers. Given this fact, is it right to hold up a handful of scholars as *the* core theorists of sociology? Doesn't this lead to the canonization of a few "dead, white, European men"?

In our view, the answer is yes, it is right (or at least not wrong) to cast a select group of intellectuals as the core writers in the discipline; and yes, this is, to an extent, the canonization of a few dead, white, European men. On the other hand, it is these thinkers from whom later social theorists (who are not all dead, white, European, or male) primarily have drawn for inspiration and insight. To better understand our rationale for including some theorists while excluding others, it is important first to briefly consider the historical context that set the stage for the development of sociology as a discipline.

The Enlightenment

Many of the seeds for what would become sociology were first planted in the **Enlightenment**, a period of remarkable intellectual development that occurred in Europe during the late seventeenth and early eighteenth centuries. During the Enlightenment, a number of long-standing ideas and beliefs on social life were turned upside down. The development of civil society (open spaces of debate relatively free from government control) and the rapid pace of the modern world enabled a critical mass of literate citizens to think about the economic, political, and cultural conditions that shaped society. Before this period, explanations of the conditions of existence were so taken for granted that there was no institutionalized discipline examining them (Lemert 1993; Seidman 1994). Enlightenment intellectuals advocated rule by rational, impersonal laws and the end to arbitrary, despotic governments. They sought to define the rights and responsibilities of free citizens. In so doing, Enlighteners called into question the authority of kings whose rule was justified by divine right.

However, the Enlightenment was not so much a fixed set of ideas as it was a new attitude, a new method of thought. One of the most important aspects of this new attitude was an emphasis on *reason*. Central to this new attitude was questioning and reexamining received ideas and values.

The Enlightenment emphasis on reason was part and parcel of the rise of science. Scientific thought had begun to emerge in the fifteenth century through the efforts of astronomers and physicists such as Copernicus, Galileo, and Newton. Enlightenment intellectuals developed an approach to the world based on methodical observations. Rather than see the universe as divinely created and hierarchically ordered, Enlighteners insisted that the universe was a mechanical system composed of matter in motion that obeyed natural laws. Moreover, they argued that these laws could be uncovered by means of science and empirical research. In advocating the triumph of reasoned investigation over faith, Enlightenment intellectuals rebuked existing knowledge as fraught with prejudice and mindless tradition (Seidman 1994:20–21). Not surprisingly, such views were dangerous, because they challenged the authority of religious beliefs and those charged with advancing them. Indeed, some Enlighteners were tortured and imprisoned, or their work was burned for being heretical.

The rise of science and empiricism would give birth to sociology in the mid-nineteenth century. The central idea behind the emerging discipline was that society could be the subject of scientific examination in the same manner as biological organisms or the physical properties of material objects. Indeed, the French intellectual **Auguste Comte** (1798–1857), who coined the term "sociology" in 1839, also used the term "social physics" to refer to this new discipline and his organic conceptualization of society (see Significant Others box on p. 97). The term "social physics" reflects the Enlightenment view that the discipline of sociology parallels other natural sciences. Comte argued that, like natural scientists, sociologists should uncover, rationally and scientifically, the laws of the social world.[2] For Enlighteners, the main difference between scientific knowledge and either theological explanation or mere conjecture is that scientific knowledge can be tested. Thus, for Comte, the new science of society—sociology—involved (1) the analysis of the central elements and functions of social systems, using (2) concrete historical and comparative methods in order to (3) establish testable generalizations about them (Fletcher 1966:14).[3]

However, it was the French theorist **Émile Durkheim** (1858–1917), discussed in Chapter 3, who arguably was most instrumental in laying the groundwork for the emerging discipline of sociology. Durkheim emphasized that, while the primary domain of psychology is to understand processes internal to the individual (e.g., personality or instincts), the primary domain of sociology is "social facts": that is, conditions and circumstances external to the individual that, nevertheless, determine that individual's course of action. As a scientist, Durkheim advocated a systematic and methodical examination of social facts and their impact on individuals.

Interestingly, sociology reflects a complex mix of Enlightenment and **counter-Enlightenment** ideas (Seidman 1994). In the late eighteenth century, a conservative reaction to the Enlightenment took place. Under the influence of Jean-Jacques Rousseau (1712–1778), the unabashed embrace of rationality, technology, and progress was challenged. Against the emphasis on reason, counter-Enlighteners highlighted the significance of nonrational factors, such as tradition, emotions, ritual, and ceremony. Most importantly, counter-Enlighteners were concerned that the accelerating pace of industrialization and urbanization and the growing pervasiveness of bureaucratization were producing profoundly disorganizing effects. In one of his most important works, *The Social Contract* (1762), Rousseau argued that in order to have a free and equal society, there must be a genuine social contract in which everyone participates in creating laws for the good of

---

[2]Physics is often considered the most scientific and rational of all the natural sciences because it focuses on the basic elements of matter and energy and their interactions.

[3]Of course, the scientists of the Enlightenment were not uninfluenced by subjectivity or morality. Rather, as Seidman (1994:30–31) points out, paradoxically, the Enlighteners sacralized science, progress and reason; they deified the creators of science such as Galileo and Newton, and fervently believed that "science" could resolve all social problems and restore social order, which is itself a type of "faith."

society. Thus, rather than being oppressed by impersonal bureaucracy and laws imposed from above, people would willingly obey the laws because they had helped make them. Rousseau also challenged the age of reason, echoing Blaise Pascal's view that the heart has reasons that reason does not know. When left to themselves, our rational faculties leave us lifeless and cold, uncertain and unsure (see McMahon 2001:35).

In a parallel way, as you will see in Chapter 3, Durkheim was interested in both objective or external social facts and the more subjective elements of society, such as feelings of solidarity or commitment to a moral code. Akin to Rousseau, Durkheim believed that it was these subjective elements that ultimately held societies together. Similarly, **Karl Marx** (1818–1883), who is another of sociology's core figures (though he saw himself as an economist and social critic), fashioned an economic philosophy that was at once rooted in science and humanist prophecy. As you will see in Chapter 2, Marx analyzed not only the economic dynamics of capitalism, but also the social and moral problems inherent to the capitalist system. Additionally, as you will see in Chapter 4, another of sociology's core theorists, **Max Weber** (1864–1920), combined a methodical, scientific approach with a concern about both the material conditions and idea systems of modern societies.

Economic and Political Revolutions

Thus far, we have discussed how the discipline of sociology emerged within a specific intellectual environment. But of course, the Enlightenment and counter-Enlightenment were both the cause and the effect of a host of political and social developments, which also affected the newly emerging discipline of sociology. Tremendous economic, political, and religious transformations had been taking place in western Europe since the sixteenth century. The new discipline of sociology sought to explain scientifically both the causes and the effects of such extraordinary social change.

One of the most important of these changes was the Industrial Revolution, a period of enormous change that began in England in the eighteenth century. The term "Industrial Revolution" refers to the application of power-driven machinery to agriculture, transportation, and manufacturing. Although industrialization began in remote times and continues today, this process completely transformed Europe in the eighteenth century. It turned Europe from a predominantly agricultural to a predominantly industrial society. It not only radically altered how goods were produced and distributed, but galvanized the system of capitalism as well.

Specifically, large numbers of people left farms and agricultural work to become wage earners in factories in the rapidly growing cities. Indeed, although most of the world's population was rural before the Industrial Revolution, by the mid-nineteenth century, half of the population of England lived in the cities, and by the end of the nineteenth century, so did half of the population of Europe. Moreover, while there were scarcely cities in Europe with a population of 100,000 in 1800, there were more than 150 cities that size a century later. At the same time, factories were transformed by a long series of technological changes. Ever-more efficient machines were adopted, and tasks were routinized. Thus, for instance, with the introduction of the power loom in the textile industry, an unskilled worker could produce three and a half times as much as could the best handloom weaver.

However, this rise in efficiency came at a tremendous human cost. Mechanized production reduced both the number of jobs available and the technical skills needed for work in the factory. A few profited enormously, but most worked long hours for low wages. Accidents were frequent and often quite serious. Workers were harshly punished and their wages were docked for the slightest mistakes. Women and children worked alongside men in noisy, unsafe conditions. Most factories were dirty, poorly ventilated and lit, and dangerous.

As you will read in Chapter 2, Karl Marx was particularly concerned about the economic changes and disorganizing social effects that followed in the wake of the Industrial

Revolution. Marx not only wrote articles and books on the harsh conditions faced by workers under capitalism, but also was a political activist who helped organize revolutionary labor movements to provoke broad social change.

As you will read in Chapter 4, Max Weber also explored the profound social transformations taking place in European society in the eighteenth and nineteenth centuries. Akin to Marx, Weber was concerned about the social consequences wrought by such profound structural change. However, in contrast to Marx, Weber argued that it was not only economic structures (e.g., capitalism), but also organizational structures—most importantly bureaucracies—that profoundly affected social relations. Indeed, in one of the most famous metaphors in all of sociology, Weber compared modern society to an "iron cage." Even more important, in contrast to Marx, Weber also examined the particular systems of meaning, or ideas, that both induced and resulted from such profound structural change.

The eighteenth century was a time of not only tremendous economic, but also political transformation. One of the most significant political events of that time was the French Revolution, which shook France between 1787 and 1799 and toppled the *ancien régime*, or old rule. Inspired in large part by Rousseau's *Social Contract* (1762), the basic principle of the French Revolution as contained in its primary manifesto, "La Déclaration des Droits de l'Homme et du Citoyen" ("The Declaration of Rights of Man and of the Citizen"), was that "all men are born and remain free and equal in rights." The French revolutionaries called for "liberty, fraternity, and equality." They sought to substitute reason for tradition, and equal rights for privilege. Because the revolutionaries sought to rebuild government from the bottom up, the French Revolution stimulated profound political rethinking about the nature of government from its inception, and set the stage for democratic uprisings throughout Europe.

However, the French Revolution sparked a bloody aftermath, making it clear that even democratic revolutions involve tremendous social disruption and that heinous deeds can be done in the name of freedom. During the Reign of Terror led by Maximilien Robespierre, radical democrats rounded up and executed anyone—whether on the left or right of the political spectrum—suspected of being opposed to the revolution. In the months between September 1793 (when Robespierre took power) and July 1794 (when Robespierre was overthrown), revolutionary zealots arrested about 300,000 people, executed some 17,000, and imprisoned thousands more. It was during this radical period of the Republic that the guillotine, adopted as an efficient and merciful method of execution, became the symbol of the Terror.

The Ins and Outs of Classical Canons

Thus far, we have argued that the central figures at the heart of classical sociological theory all sought to explain the extraordinary economic, political, and social transformations taking place in Europe in the late nineteenth century. Yet, concerns about the nature of social bonds and how these bonds can be maintained in the face of extant social change existed long before the eighteenth century and in many places, not only in western Europe. Indeed, in the late fourteenth century, Abdel Rahman Ibn-Khaldun (1332–1406), born in Tunis, Tunisia, in North Africa, thought and wrote extensively on subjects that have much in common with contemporary sociology (Martindale 1981:134–36; Ritzer 2000:10). And long before the fourteenth century, Plato (ca. 428–ca. 347 BC), Aristotle (384–322 BC), and Thucydides (ca. 460–ca. 400 BC) wrote about the nature of war, the origins of the family and the state, and the relationship between religion and the government—topics that have since become central to sociology (Seidman 1994:19). Aristotle, for example, emphasized that human beings were naturally political animals—*zoon politikon* (Martin 1999:157). He sought to identify the essence that made a stone a stone or a society a society (Ashe 1999:89). For that matter, well before Aristotle's time, Confucius (551–479 BC) developed a theory for understanding Chinese society. Akin to Aristotle, Confucius maintained that government is the center of people's lives and that all other considerations derive from it. According to Confucius, a good government must be concerned with three things: sufficient food, a sufficient army, and the confidence of the people (Jaspers 1957/1962:47).

These premodern thinkers are better understood as philosophers, however, and not as sociologists. Both Aristotle and Confucius were less concerned with explaining social dynamics than with prescribing a perfected, moral social world. As a result, their ideas are guided less by a scientific pursuit of knowledge than by an ideological commitment to a specific set of values. Moreover, in contrast to modern sociologists, premodern thinkers tended to see the universe as a static, hierarchical order in which all beings, human and otherwise, have a more or less fixed and proper place and purpose, and they sought to identify the "natural" moral structure of the universe (Seidman 1994:19).

Our key point here is that, while the ideas of Marx, Weber, and Durkheim are today at the heart of the classical sociological theoretical canon, this does not mean that they are inherently better or more original than those of other intellectuals who wrote before or after them. Rather, it is to say that, for specific historical, social, and cultural as well as intellectual reasons, their works have helped define the discipline of sociology, and that sociologists refine, rework, and challenge their ideas to this day.

For that matter, Marx, Weber, and Durkheim have not always been considered the core theorists in sociology. On the contrary, until 1940, Weber and Durkheim were not especially adulated by American sociologists (Bierstedt 1981). Until that time, discussions of their work were largely absent from texts. Marx was not included in the canon until the 1960s. Meanwhile, even a cursory look at mid-century sociological theory textbooks reveals an array of important "core figures," including Sumner, Sorokin, Sorel, Pareto, Le Play, Ammon, Veblen, de Tocqueville, Cooley, Spencer, Tönnies, and Martineau. Although an extended discussion of all of these theorists is outside the scope of this volume, we provide a brief look at some of these scholars in the Significant Others boxes of the chapters that follow.

In the second half of this book, we focus on several writers who for social or cultural reasons were underappreciated as sociologists in their day. **Charlotte Perkins Gilman** (1860–1935), for example, was well known as a writer and radical feminist in her time, but not as a sociologist (Degler 1966:vii). It was not until the 1960s that there was a formalized sociological area called "feminist theory." Gilman sought to explain the basis of gender inequality in modern industrial society. She explored the fundamental questions that would become the heart of feminist social theory some 50 years later, when writers such as Simone de Beauvoir and Betty Friedan popularized these same concerns.

Georg Simmel (1858–1918), a German sociologist, wrote works that would later become pivotal in sociology, though his career was consistently stymied both because of the unusual breadth and content of his work, and because of his Jewish background.[4] Simmel sought to uncover the basic *forms* of social interaction, such as "exchange," "conflict," and "domination," that take place between individuals. Above all, Simmel underscored the contradictions of modern life. For instance, he emphasized how individuals strive both to conform to social groups and, at the same time, to distinguish themselves from others. Simmel's provocative work is gaining more and more relevance in today's world in which contradictions and ironies abound.

---

[4]Durkheim was also Jewish (indeed, he was the son of a rabbi), but anti-Semitism did not significantly impede Durkheim's career. In fact, it was Durkheim's eloquent article, "Individualism and Intellectuals" (1898) on the Dreyfus affair (a political scandal that emerged after a Jewish staff officer named Captain Alfred Dreyfus was erroneously court-martialed for selling secrets to the German Embassy in Paris) that shot him to prominence and eventually brought Durkheim his first academic appointment in Paris. In sum, German anti-Semitism was much more harmful to Georg Simmel than French anti-Semitism was to Durkheim.

While anti-Semitism prevented Simmel from receiving his full due, and sexism impeded Gilman (as well as other women scholars) from achieving hers, the forces of racism in the United States forestalled the sociological career of the African American intellectual **W. E. B. Du Bois** (1868–1963). Not surprisingly, it was this very racism that would become Du Bois's most pressing scholarly concern. Du Bois sought to develop a sociological theory about the interpenetration of race and class in America at a time when most sociologists ignored or glossed over the issue of racism. Although underappreciated in his day, Du Bois's insights are at the heart of contemporary sociological theories of race relations.

We conclude this book with the work of the social philosopher **George Herbert Mead** (1863–1931). Mead laid the foundation for symbolic interactionism, which has been one of the major perspectives in sociological theory since the middle of the twentieth century. Mead challenged prevailing psychological theories about the mind by highlighting the social basis of thinking and communication. Mead's provocative work on the emergent, symbolic dimensions of human interaction continue to shape virtually all social, psychological, and symbolic interactionist research today.

▓▓ How Can We Navigate Sociological Theory?

Thus far, we have (1) explained the imperativeness of sociological theory, (2) argued that students should read original theoretical works, and (3) discussed the theorists who we consider to be at the heart of classical sociological theory. Now we come to the fourth question: How can we best navigate the wide range of ideas that these theorists bring to the fore? To this end, in this section we explain the metatheoretical framework or "map" that we use in this book to explore and compare and contrast the work of each theorist.

The Questions of "Order" and "Action"

Our framework revolves around two central questions that social theorists and philosophers have grappled with since well before the establishment of sociology as an institutionalized discipline: the questions of *order* and *action* (Alexander 1987). Indeed, these two questions have been a cornerstone in social thought at least since the time of the ancient Greek philosophers. The first question (illustrated in Figure 1.1) is that of order. It asks what accounts for the patterns or predictability of behavior that lead us to experience social life as routine. Or, expressed somewhat differently, how do we explain the fact that social life is not random, chaotic, or disconnected, but instead demonstrates the existence of an ordered social universe? The second question (illustrated in Figure 1.2) is that of action. It considers the factors that motivate individuals or groups to act. The question of action, then, turns our attention to the forces held to be responsible for steering individual or group behavior in a particular direction.

Similar to how the north-south, east-west coordinates allow you to orient yourself to the details on a street map, our analytical map is anchored by four coordinates that assist you in navigating the details of the theories presented in this volume. In this case, the coordinates situate the answers to the two questions. Thus, to the question of order, one answer is that the patterns of social life are the product of structural arrangements or historical conditions that confront individuals or groups. As such, preexisting social arrangements produce the apparent orderliness of social life because individuals and groups are pursuing trajectories that, in a sense, are not of their own making. Society is thus pictured as an overarching system that works *down* on individuals and groups to determine the shape of the social order. Society is understood as a reality *sui generis* that operates according to its own logic distinct from the will of individuals. This orientation has assumed many different names—macro, holistic, objectivist, structuralist, and the label we use here, **collective** (or **collectivist**).

Figure 1.1 Basic Theoretical Continuum as to the Nature of Social Order

Figure 1.2 Basic Theoretical Continuum as to the Nature of Social Action

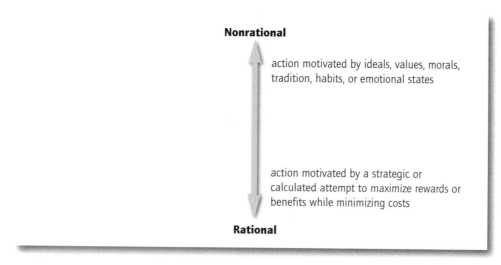

By contrast, the other answer to the question of order is that social order is a product of ongoing interactions between individuals and groups. Here, it is individuals and groups creating, re-creating, or altering the social order that works *up* to produce society. This position grants more autonomy to actors, because they are seen as relatively free to reproduce the patterns and routines of social life (i.e., the social order) or transform them. Over time, this orientation has earned several names as well—micro, elementarism, subjectivist, and the label we adopt here, **individual** (or **individualist**). (See Figure 1.1.)

Turning to the question of action, we again find two answers, labeled here **nonrational** and **rational**.[5] Specifically, if the motivation for action is primarily nonrational, the individual takes his bearings from subjective ideals, symbolic codes, values, morals, norms, traditions, the quest for meaning, unconscious desires, or emotional states, or a

---

[5]The terms "rational" and "nonrational" are problematic in that they have a commonsensical usage at odds with how theorists use these terms. By "rational" we do not mean "good and smart" and by "nonrational" we do not mean irrational, nonsensical, or stupid (Alexander 1987:11). Despite these problems, however, we continue to use the terms "rational" and "nonrational" because (although it is outside the scope of this discussion) the semantic alternatives (subjectivist, idealist, internal, etc.) are even more problematic.

combination of these. While the nonrationalist orientation is relatively broad in capturing a number of motivating forces, the rationalist orientation is far less encompassing. It contends that individual and group actions are motivated primarily by the attempt to maximize rewards while minimizing costs. Here, individuals and groups are viewed essentially as calculating and strategic as they seek to achieve the "selfish" goal of improving their position. Here, actors are seen as taking their bearings from the external conditions in which they find themselves rather than internal ideals.

Intersecting the two questions and their answers, we can create a four-celled map on which we are able to plot the basic theoretical orientation of the social thinkers featured in this book (see Figure 1.3). The four cells are identified as collective-nonrational, collective-rational, individual-nonrational, and individual-rational. We cannot overemphasize that these four coordinates are " ideal types"; theorists and theories are never "pure," i.e., situated completely in one cell. Implicitly or explicitly, or both, theorists inevitably incorporate more than one orientation in their work. These coordinates (or cells in the table) are best understood as endpoints to a continuum on which theories typically occupy a position

Figure 1.3 Theorists' Basic Orientation

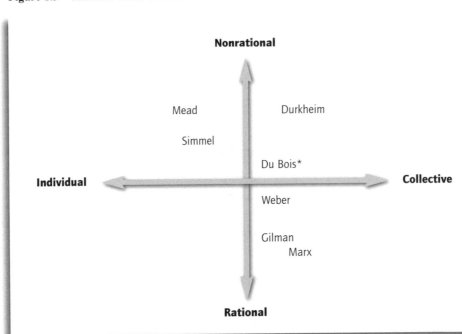

NOTE: This diagram reflects the basic theoretical orientation of each thinker. However, every theorist in this volume is far more nuanced and multidimensional than this simple figure lets on. The point is not to fix each theorist in a predetermined box, but rather to provide a means for illuminating and discussing each theorist's orientation relative to one another and within their various works.

*Our placement of Du Bois on the nonrational side of the continuum reflects the excerpts in this volume that were chosen because of their theoretical significance. In our view, it is his understanding of racial consciousness that constitutes his single most important theoretical contribution. However, he continually underscored the intertwined, structural underpinnings of race and class that, in the latter part of his life, led him to adopt a predominantly rationalist, Marxist-inspired orientation. In our view, however, Du Bois's later work has more empirical than theoretical significance.

somewhere between the extremes. Multidimensionality and ambiguity are reflected in our maps by the lack of fixed points.

In addition, it is important to note that this map is something *you* apply to the theories under consideration. Although each theorist addresses the questions of order and action, they generally did not use these terms in their writing. For that matter, their approaches to order and action tend to be implicit rather than explicit in their work. Thus, at times you will have to read between the lines to determine a theorist's position on these fundamental questions. While this may pose some challenges, it also expands your opportunities for learning.

Consequently, not everyone views each theorist in exactly the same light. Moreover, even within one major work a theorist may draw from both ends of the continuum. In each chapter, we will discuss the ambiguities and alternative interpretations within the body of work of each theorist. Nevertheless, these maps enable you to (1) recognize the general tendencies that exist within each theorist's body of work, and (2) compare and contrast (and argue about) thinkers' general theoretical orientations. (For further examples as to the flexibility of this framework, see the discussion questions at the end of the chapter.)

Put another way, when navigating the forest of theory, individual theorists are like trees. Our analytic map is a tool or device for locating the trees within the forest so you can enter and leave having developed a better sense of direction or, in this case, having learned far more than might otherwise have been the case. By enabling you to compare theorists' positions on two crucial issues, their work is likely to be seen less as a collection of separate, unrelated ideas. Bear in mind, however, that the map is only a tool. Its simplicity does not capture the complexities of the theories or of social life itself.

In sum, it is essential to remember that this four-cell table is an analytical device that helps us understand and compare and contrast theorists better, but it does not mirror or reflect reality. The production and reproduction of the social world is never a function of either individuals or social structures, but rather a complex combination of both. So too, motivation is never completely rational or completely nonrational. To demonstrate this point as well as how our analytical map on action and order works in general, we turn to a very simple example.

Consider this question: Why do people stop at red traffic lights? First, in terms of action, the answer to this question resides on a continuum with rational and nonrational orientations serving as the endpoints. On the one hand, you might say people stop at red traffic lights because it is in their best interest to avoid getting a ticket or having an accident. This answer reflects a *rationalist* response; it demonstrates that rationalist motivations involve the individual taking her bearings from outside herself. (See Table 1.1.) The action (stopping at the red light) proceeds primarily in light of external conditions (e.g., a police officer that could ticket you, oncoming cars that could hit you).

A *nonrationalist* answer to his question is that people stop at red traffic lights because they believe it is good and right to follow the law. Here, the individual takes his bearings from morals or values from within himself, rather than from external conditions (e.g., oncoming cars). Interestingly, if this moral or normative imperative is the only motivation for action, the individual will stop at the traffic light even if there is no police car or oncoming cars in sight. By contrast, if one's only motivation for action is rationalist and there are absolutely no visible dangers (i.e., no police officers or other cars in sight and hence no possibility of getting a ticket or having an accident), the driver will *not* stop at the red light: instead, she will go.

Another *nonrationalist* answer to the question "Why do people stop at red traffic lights?" involves "habits." (See Table 1.1.) By definition, habits are relatively unconscious: that is, we do not think about them. They come "automatically" not from strategic calculations or

Table 1.1 Why Do People Stop at Red Traffic Lights? Basic Approaches to Order and Action

ORDER

		Individual	Collective
ACTION	**Nonrational**	Value fidelity: Individual believes it is good and right to follow the law. Habit: Individual stops without thinking.	Hegemonic moral order: Society teaches it is wrong to disobey the law. "Red" means "stop" and "green" means "go" in hegemonic symbolic system.
	Rational	Instrumentality: Individual doesn't want to get a traffic ticket. Individual doesn't want to get into an accident.	Hegemonic legal structure: Society punishes those who break the law.

external circumstances, but from within; that is why they are typically considered *nonrational*. Interestingly, habits may or may not have their roots in morality. Some habits are "folkways" or routinized ways people do things in a particular society (e.g., paying your bills by mail rather than in person, driving on the right side of the road), while other habits are attached to sacred values (e.g., putting your hand over your heart when you salute the flag). Getting back to our example, say you are driving in your car on a deserted road at 2:00 in the morning and you automatically stop at a red traffic light out of habit. Your friend riding with you might say, "Why are you stopping? There's not a car in sight." If your action were motivated simply from habit and not a moral imperative to follow the law, you might say, "Hey you're right!" and drive through the red light.

Of course, actions often have—indeed, they usually have—both rational *and* nonrational dimensions. For instance, in this last example, you might have interpreted your friend's question, "Why are you stopping? There's not a car in sight," to mean, "Don't be a goody-goody—let's go!" In other words, you may have succumbed to peer pressure even though you knew it was wrong to do so. If such was the case, you may have wittingly or unwittingly believed your ego, or your sense of self, was on the line. Thus, it was not so much that rational trumped nonrational motivation as it was that you acted out of the external pressure from your friend and internal pressure to do the "cool" thing and be the particular type of person you want to be. If such were the case, your action is a complex combination of conditions both outside and within yourself.

Indeed, a basic premise of this book is that because social life is extremely complex, a complete social theory must account for multiple sources of action and levels of social order. Theorists must be able to account for the wide variety of components (e.g., individual predispositions, personality and emotions, social and symbolic structures) constitutive of this world. Thus, for instance, our rationalist response to the question as to why people stop at red traffic lights—that people stop simply because they don't want to get a ticket or get into an accident—is, in fact, incomplete. It is undercut by a series of unacknowledged nonrational motivations. There is a whole host of information that undergirds the very ability of an individual to make this choice. For example, before one can even begin to make the decision as to whether to stop for the red light, one must know that normally (and legally) "red" *means* "stop" and "green" *means* "go." That we know and take for granted that "red" means "stop" and "green" means "go" and then consciously think about and decide to override that cultural knowledge (and norm) indicates that even at our most rationalist moments we are still using the tools of a largely taken-for-granted, symbolic, or nonrational realm (see Table 1.1).

Now let's turn to the issue of *order*.

If we say "people stop at red lights because they don't want to get a ticket," this can be said to reflect a collectivist approach to order if we are emphasizing there is a coercive state apparatus (e.g., the law, police) that hems in behavior. If such is the case, we are emphasizing that external social structures precede and shape individual choice.

If we say "people stop because they believe it is good and right to follow the law," we might be taking a collectivist approach to order as well. Here we assume individuals are socialized to obey the law. We emphasize that specific social or collective morals and norms are internalized by individuals and reproduced in their everyday behavior. Similarly, if we emphasize it is only because of the preexisting symbolic code in which "red" means "stop" and "green" means "go" that individuals can decide what to do, then we would be taking a collectivist approach. These various versions of order and action are illustrated in Table 1.1.

On the other hand, that people stop at red traffic lights because they don't want to get into an accident or get a ticket also might reflect an *individualist* approach to order, if the assumption is that the individual determines his action using his own free will, and that from this the traffic system is born. Another important individualist albeit nonrationalist answer to this question emphasizes the role of emotions. For instance, one might fear getting a ticket, and—to the extent the fear comes from within the individual rather than from the actual external circumstances—we can say this fear represents a *nonrational* motivating force at the level of the individual.

In this book, you will see that the core sociological theorists hold a wide variety of views on the action/order continuum even within their own work. Overall, however, each theorist can be said to have a basic or general theoretical orientation. For instance, Marx was interested above all in the collectivist and rationalist conditions behind and within order and action, while Durkheim, especially in his later work, was most interested in the collectivist and nonrationalist realms. Thus, juxtaposing Figure 1.3 and Table 1.1, you can see that if we were to resurrect Marx and Durkheim from their graves and ask them the hypothetical question, "Why do people stop at red traffic lights?" Marx would be more likely to emphasize the rationalist motivation behind this act ("they seek to avoid getting a ticket"), while Durkheim would be more likely to emphasize the nonrational motivation ("they consider it the 'right' thing to do"). Both, however, would emphasize that these seemingly individualist acts are actually rooted in collective social and cultural structures (i.e., it is the *law* with its coercive and moral force that undergirds individual behavior). Meanwhile, at the more individualist end of the continuum, Mead would probably emphasize the immediate ideational process in which individuals interpret the meanings for and consequences of each possible action. (Note, though, that obviously each of these theorists is far more complex and multidimensional than this simple example lets on.)

Of course, the purpose of this book is not to examine the work of core sociological theorists in order to figure out how they might answer a hypothetical question about red traffic lights. Rather, the purpose of this book is to examine the central issues these theorists themselves raise and to analyze the particular theoretical stance they take as they explore these concerns. It is to this task that we now turn.

Discussion Questions

1. Explain the difference between "primary" and "secondary" theoretical sources. What are the advantages and disadvantages of reading each type of work?

2. The metatheoretical framework we introduce in this chapter is useful not only for navigating classical sociological theory, but also for thinking about virtually *any* social issue. Using Table 1.1 as a reference, devise your own question, and then give hypothetical answers that reflect the four different basic theoretical orientations: individual/rational, individual/nonrational, collective/rational, and collective/nonrational. For instance, why do 16-year-olds stay in (or drop out of) high school? Why might a man or woman stay in a situation of domestic violence? What are possible explanations for gender inequality? What are possible causal explanations for the Holocaust? What are the various arguments for and against affirmative action? What are the central arguments for and against capital punishment? Why are you reading this book?

3. Numerous works of fiction speak to the social conditions that early sociologists were examining. For instance, Charles Dickens's *Hard Times* (1854) portrays the hardships of the Industrial Revolution, while Victor Hugo's *Les Miserables* addresses the political and social dynamics of the French Revolution. Read either of these works (or watch the movies or play), and discuss the tremendous social changes they highlight.

2 KARL MARX (1818–1883)

Key Concepts

- Class
- Bourgeoisie
- Proletariat
- Forces and relations of production
- Capital
- Surplus value
- Alienation
- Labor theory of value
- Exploitation
- Class consciousness

The history of all hitherto existing society is the history of class struggles.

(Marx and Engels 1848/1978:473)

Have you ever worked at a job that left you feeling empty inside? Perhaps you have worked as a telemarketer, reading a script and selling a product that, in all likelihood, you had never seen or used. Or perhaps you have worked in a fast-food restaurant, or in a large factory or corporation. Sometimes we have jobs that make us feel like we are "just a number," that even though we do our job, we might be easily replaced. This is precisely the type of situation that greatly concerned Karl Marx. Marx sought to explain the nature of the capitalist economies that came to the fore in western Europe in the eighteenth and nineteenth centuries. He maintained that the economic deficiencies and social injustices inherent to capitalism would ultimately lead to the breakdown of capitalist societies. Yet Marx was not an academic writing in an "ivory tower": he was an activist, a revolutionary committed to the overthrow of capitalism. And as you will see shortly, Marx paid a personal price for his revolutionary activities.

Though Marx's prediction that capitalism would be replaced by communism has not come true (some would say, "not yet"), his critique of capitalism continues to resonate with contemporary society. His discussions regarding the concentration of wealth, the growth of monopoly capitalism, business's unscrupulous pursuit of profit (demonstrated, for instance, by the recent scandals surrounding WorldCom, Enron, Countrywide, Bear Stearns, Merrill Lynch, and Bernard Madoff, to name but a few), the relationship between government economic policy and the interests of the capitalist class, and the alienation experienced in the workplace all speak to concerns that affect almost everyone, even today. Indeed, who has not felt at one time or another that his job was solely a means to an end—a paycheck, money—instead of a forum for fulfilling his aspirations or cultivating his talents? Who has not felt as though she were an expendable "commodity," a means or tool in the production of a good or the provision of a service where even her emotions must be manufactured for the sake of the job? Clearly, Marx's ideas are as relevant today as they were more than a century ago.

A Biographical Sketch

Karl Marx was born on May 5, 1818, in Trier, a commercial city in southwestern Germany's Rhineland.[1] Descended from a line of rabbis on both sides of his family, Marx's father, Heinrich, was a secularly educated lawyer. Though Heinrich did not actively practice Judaism, he was subject to anti-Semitism. With France's ceding of the Rhineland to Prussia after the defeat of Napoleon, Jews living in the region were faced with a repeal of the civil rights granted under French rule. In order to keep his legal practice, Heinrich converted to Lutheranism in 1817. As a result, Karl was afforded the comforts of a middle-class home.

Following in his father's footsteps, Marx pursued a secular education. He enrolled as a law student at the University of Bonn in 1835, then transferred the following year to the University of Berlin. In addition to studying law, Marx devoted himself to the study of history and philosophy. While in Berlin, Marx also joined the Young Hegelians, a group of radical thinkers who developed a powerful critique of the philosophy of **Georg W. F. Hegel** (1770–1831), the dominant German intellectual figure of the day and one of the most influential thinkers of the nineteenth century. Marx constructed the basis of his theoretical system, historical materialism, by inverting Hegel's philosophy of social change. (See pp. 31–33 for a brief sketch of Hegel's philosophy and its relation to Marx's theory.)

In 1841, Marx earned a doctorate in philosophy from the University of Jena. However, his ambitions for an academic career ended when the Berlin ministry of education blacklisted him for his radical views.[2] Having established little in the way of career prospects during his student years, Marx accepted an offer to write for the *Rheinische Zeitung*, a liberal newspaper published in Cologne.

Marx soon worked his way up to become editor of the newspaper. Writing on the social conditions in Prussia, Marx criticized the government's treatment of the poor and exposed the harsh conditions of peasants working in the Moselle wine-producing region. However, Marx's condemnation of the authorities brought on the censors, and he was forced to resign his post.

Soon after, Marx married his childhood love, Jenny Von Westphalen, the daughter of a Prussian baron. The two moved to Paris in the fall of 1843. At the time, Paris was the center of European intellectual and political movements. While there, Marx became acquainted with a

---

[1]Prussia was a former kingdom in eastern Europe established in 1701 that included present-day Germany and Poland. It was dissolved following World War II.

[2]Marx's mentor and colleague, Bruno Bauer, had promised him a faculty position at the University of Bonn. But when Bauer was dismissed from the university for advocating leftist, antireligious views, Marx was effectively shut off from pursuing an academic career.

number of leading socialist writers and revolutionaries. Of particular importance to his intellectual development were the works of the French philosopher **Henri de Saint-Simon** (1760–1825) and his followers. Saint-Simon's ideas led to the creation of Christian Socialism, a movement that sought to organize modern industrial society according to the social principles espoused by Christianity. In their efforts to counter the exploitation and egoistic competition that accompany industrial capitalism, Saint-Simonians advocated that industry and commerce be guided according to an ethic of brotherhood and cooperation. By instituting common ownership of society's productive forces and an end to rights of inheritance, they believed that the powers of science and industry could be marshaled to create a more just society free from poverty.

Marx also studied the work of the seminal political economists **Adam Smith** (1723–1790) and **David Ricardo** (1772–1823). Smith's book *An Inquiry into the Nature and Causes of the Wealth of Nations* (1776) represents the first systematic examination of the relationship between government policy and a nation's economic growth. As such, it played a central role in defining the field of political economy. (See p. 22 for summary remarks on Smith's views.) For his part, Ricardo, building on Smith's earlier works, would further refine the study of economics. He wrote on a number of subjects, including the condition of wages, the source of value, taxation, and the production and distribution of goods. A leading economist in his day, Ricardo's writings were influential in shaping England's economic policies. It was from his critique of these writers that Marx would develop his humanist philosophy and economic theories.

During his time in Paris, Marx also began what would become a lifelong collaboration and friendship with **Friedrich Engels**, whom he met while serving as editor of the *Zeitung*. Marx's stay in France was short-lived, however, and again it was his journalism that sparked the ire of government authorities. In January 1845, he was expelled from the country at the request of the Prussian government for his antiroyalist articles. Unable to return to his home country (Prussia), Marx renounced his Prussian citizenship and settled in Brussels, where he lived with his family until 1848. In Brussels, Marx extended his ties to revolutionary working-class movements through associations with members of the League of the Just and the Communist League. Moreover, it was while living in Brussels that Marx and Engels produced two of their most important early works, *The German Ideology* (see below) and *The Communist Manifesto* (see below). In 1848, workers and peasants began staging revolts throughout much of Europe. As the revolution spread, Marx and Engels left Brussels and headed for Cologne to serve as coeditors of the radical *Neue Rheinische Zeitung*, a paper devoted to furthering the revolutionary cause. For his part in the protests, Marx was charged with inciting rebellion and defaming the Prussian royal family. Though acquitted, Marx was forced to leave the country. He returned to Paris, but soon was pressured by the French government to leave the country as well, so Marx and his family moved to London in 1849.

In London, Marx turned his attention more fully to the study of economics. Spending some 60 hours per week in the British Museum, Marx produced a number of important works, including *Capital* (see below), considered a masterpiece critique of capitalist economic principles and their human costs. Marx also continued his political activism.

From 1851–62, he was a regular contributor to the *New York Daily Tribune*, writing on such issues as political upheavals in France, the Civil War in the United States, Britain's colonization of India, and the hidden causes of war.[3] In 1864, Marx helped found and direct the *International Working Men's Association*, a socialist movement committed to ending the inequities and alienation or "loss of self" experienced under capitalism. The *International*

---

[3] A number of articles attributed to Marx were actually written by Engels, whose assistance allowed Marx to continue to collect a wage from the newspaper. Engels, whose father owned textile mills in Germany and England (that he would later inherit), also provided Marx with financial support throughout his years in London. The depth of Engels's devotion even led him to support an out-of-wedlock child fathered by Marx.

had branches across the European continent and the United States, and Marx's popular writing and activism gave him an international audience for his ideas.

Yet, the revolutionary workers' movements were floundering. In 1876, the *International* disintegrated and Marx was barely able to support himself and his wife as they struggled against failing health. Jenny died on December 2, 1881, and Marx himself died on March 14, 1883.

INTELLECTUAL INFLUENCES AND CORE IDEAS

The revolutionary spirit that inflamed Marx's work cannot be understood outside the backdrop of the sweeping economic and social changes occurring during this period. By the middle of the nineteenth century, the industrial revolution that began in Britain 100 years earlier was spreading throughout western Europe. Technological advances in transportation, communication, and manufacturing spurred an explosion in commercial markets for goods. The result was the birth of modern capitalism and the rise of middle-class owners of capital, or the **bourgeoisie**, to economic and political power. In the wake of these changes came a radical reorganization of both work and domestic life. With the rapid expansion of industry, agricultural work declined, forcing families to move from rural areas to the growing urban centers. It would not take long for the size of the manufacturing labor force to rival and then surpass the numbers working in agriculture.

Nowhere were the disorganizing effects of the industrial revolution and the growth of capitalism more readily apparent than in Manchester, England. In the first half of the nineteenth century Manchester's population exploded by 1,000 percent as it rapidly became a major industrial city.[4] The excessive rate of population growth meant that families had to live in makeshift housing without heat or light and in dismal sanitary conditions that fueled the spread of disease. The conditions in the mechanized factories were no better. The factories were poorly ventilated and lit and often dangerous, and factory owners disciplined workers to the monotonous rhythms of mass production. A 70-hour workweek was not uncommon for men and women, and children as young as six often worked as much as 50 hours a week. Yet, the wages earned by laborers left families on the brink of beggary. The appalling living and work standards led Engels to describe Manchester as "Hell upon Earth."

It was in reaction to such dire economic and social conditions that Marx sought to forge a theoretical model intended not only to *interpret* the world, but also to *change* it. In doing so, he centered his analysis on economic **classes**. For Marx, classes are groups of individuals who share a common position in relation to the means or **forces of production**. These refer to the raw materials, technology, machines, factories, and land that are necessary in the production of goods. Each class is distinguished by what it owns with regard to the means of production. Marx argued, "wage labourers, capitalists and landowners constitute [the] three big classes of modern society based upon the capitalist mode of production." Thus, under capitalism, there are "the owners merely of labour-power, owners of capital, and landowners, whose respective sources of income are wages, profit, and ground-rent" (Marx 1867/1978:441).[5]

---

[4]Manchester was also the site of Engels's urban ethnography, *The Condition of the Working Class in England*, and the location of one of his family's textile mills. It was Engels's work that early on helped to crystallize Marx's conception of the proletariat as the revolutionary force in modern industrial society.

[5]Marx was not entirely consistent when discussing the number and types of classes that compose capitalist societies. Most often, however, he described such societies as consisting of two antagonistic classes: the bourgeoisie and the proletariat.

Photo 2.1 Sordid Factory Conditions: A Young Girl Working as a Spinner in a U.S. Textile Mill, Circa 1910

Photo 2.2 Sadly, for some factory workers, little has changed over the past century. Here, 16-year-old girls are assembling Keds sneakers at the Kunshan Sun Hwa Footwear Company, in China. The girls apply the toxic glue with their bare hands. At the end of the day, they must line up and leave single file. The factory is surrounded by a 15-foot wall topped with barbed wire.

Private ownership of the means of production leads to class relations based on domination and subordination. While wage earners are free to quit or refuse a particular job, they nevertheless must sell their labor power to someone in the capitalist class in order to live. This is because laborers have only their ability to work to exchange for money that can then be used to purchase the goods necessary for their survival. However, the amount of wages paid is far exceeded by the profits reaped by those who control the productive forces. As a result, classes are pitted against each other in a struggle to control the means of production, the distribution of resources, and profits.

For Marx, this class struggle is the catalyst for social change and the prime mover of history. This is because any mode of production based on private property (e.g., slavery, feudalism, capitalism) bears the seeds of its own destruction by igniting ongoing economic conflicts that inevitably will sweep away existing social arrangements and give birth to new classes of oppressors and the oppressed. Indeed, as Marx states in one of the most famous passages in *The Communist Manifesto*, "The history of all hitherto existing society is the history of class struggles" (Marx and Engels 1848/1978:473; see below).

Marx developed his theory in reaction to laissez-faire capitalism, an economic system based on individual competition for markets. It emerged out of the destruction of feudalism, in which peasant agricultural production was based on subsistence standards in the service to lords, and the collapse of merchant and craft guilds, where all aspects of commerce and industry were tightly controlled by monopolistic professional organizations. The basic premise behind this form of capitalism, as outlined by Adam Smith, is that any and all should be free to enter and compete in the marketplace of goods and services. Under the guiding force of the "invisible hand," the best products at the lowest prices will prevail, and a "universal opulence [will] extend itself to the lowest ranks of the people" (Smith 1776/1990:6). Without the interference of regulations that artificially distort supply and demand and disturb the "natural" adjusting of prices, the economy will be controlled by those in the best position to dictate its course of development: consumers and producers. Exchanges between buyers and sellers are rooted not in appeals to the others' "humanity but to their self-love . . . [by showing] them that it is for their own advantage to do for him what he requires of them" (ibid.:8). The potentially destructive drive for selfishly bettering one's lot is checked, however, by a rationally controlled competition for markets that discourages deceptive business practices, because whatever gains a seller can win through illicit means will be nullified as soon as the "market" uncovers them. According to Smith, a "system of perfect liberty" is thus created that both generates greater wealth for all and promotes the general well-being of society.

Marx shared much of Smith's analysis of economics. For instance, both viewed history as unfolding through evolutionary stages in economic organization and understood the central role of governments to be protecting the privilege of the wealthy through upholding the right to private property. Nevertheless, important differences separate the two theories. Most notable is Marx's insistence that, far from establishing a system of perfect liberty, private ownership of the means of production necessarily leads to the alienation of workers. They sell not only their labor power, but also their souls. They have no control over the product they are producing, while their work is devoid of any redeeming human qualities (see below). Although capitalism produces self-betterment for owners of capital, it necessarily prevents workers from realizing their essential human capacity to engage in creative labor.

Indeed, in highly mechanized factories, a worker's task might be so mundane and repetitive (e.g., "insert bolt A into widget B") that she seems to become part of the machine itself. For example, a student once said she worked in a job in which she had a scanner attached to her arm. Her job was simply to stand by a conveyer belt in which boxes of various sizes came by. She stuck her arm out and "read" the boxes with her scanner arm. Her individual human potential was completely irrelevant to her job. She was just a "cog in a wheel" of mechanization. Marx maintained that when human actions are no different from those of a machine, the individual is dehumanized.

Moreover, according to Marx, capitalism is inherently exploitative. It is the labor power of workers that produces the products to be sold by the owners of businesses. Workers mine the raw materials, tend to the machines, and assemble the products. Yet, it is the owner who takes for himself the profits generated by the sale of goods. Meanwhile, workers' wages hover around subsistence levels, allowing them to purchase only the necessities—sold at a profit by capitalists to ensure their return to work the next day. One of Marx's near contemporaries, Thorstein Veblen (1857–1929), an American sociologist and economist, held a similar view on the nature of the relationship between owners and workers. (See the Significant Others box that follows.)

From the point of view of the business owner, capitalism is a "dog-eat-dog" system in which business owners must always watch the "bottom line" in order to compete for market dominance. Business owners can never rest on their laurels—because someone can always come along and create either a better or newer product, or the same product at a lower price. Thus, a business owner must constantly think strategically and work to improve her product or reduce her costs, or both. Cutting costs can increase a business owner's profit either directly (as she keeps more money for herself) or indirectly (by enabling the business owner to lower the price and sell more of her products).

While competition between capitalists may lead to greater levels of productivity, it also results in a concentration of wealth into fewer and fewer hands. One of the basic truths of capitalism is that it takes money to make money, and the more money a business owner has at his disposal, the more ability he has to generate profit-making schemes. For

Photo 2.3   Many of Charlie Chaplin's silent films during the 1920s and 1930s offered a comedic—and quite critical—look at the industrial order. Here, in a scene from *Modern Times* (1936), Chaplin is literally a "cog in a machine."

instance, a wealthy capitalist might temporarily underprice his product (i.e., sell it below the cost of its production) in order to force his competitors out of business. Once the competition is eliminated and a monopoly is established, the product can be priced as high as the market will bear.

Significant Others

Thorstein Veblen (1857–1929): The Leisure Class and Conspicuous Consumption

While Karl Marx's ideas would remain largely on the periphery of sociology until the 1960s, his ideas, nevertheless, inspired a legion of scholars even before his death. One early student of Marx's theories was Thorstein Veblen. Veblen was born in Wisconsin, the son of Norwegian immigrants. His parents, like so many others of that time and place, were poor tenant farmers who came to America seeking to better their lives. Fortunately, after a number of years of hardship and thrift, the Veblens were able to attain a modest lifestyle working as family farmers. Thorstein's humble upbringing, however, contrasted sharply with the vast fortunes being reaped by America's robber barons, who ruthlessly dominated the nation's budding industrial economy.

Veblen's cognizance of the nation's gross inequities of wealth found expression in his writings, most notably *The Theory of the Leisure Class* (1899) and *The Theory of Business Enterprise* (1904). As a sociologist and economist, Veblen, in his scholarly analyses, did not pretend to value the neutrality often associated with scientific endeavors. Instead, his work presents a highly critical picture of modern capitalism and the well-to-do, the "leisure class," who benefit most from the economic system built on "waste." Though the efficiency of mechanized production is capable of creating a surplus of goods that could in turn provide a decent standard of living for all, Veblen argued that "parasitic" business leaders "sabotaged" the industrial system in their quest for personal profit.

Though Veblen by no means embraced Marxist models of society and social change in their entirety, his work nevertheless contains important parallels with some of Marx's key ideas. For instance, his assertion that the state of a society's technological development forms the foundation for its "schemes of thought" bears a pronounced resemblance to Marx's distinction between the economic base and superstructure. Additionally, Veblen's analysis of the modern-day conflict between "business" (those who make money) and "industry" (those who make "things") recalls Marx's own two-class model of capitalist society and its attendant moral critique of the exploitation of workers and the clash between the forces and relations of production. However, it was his twin notions of "conspicuous consumption" and "conspicuous leisure" that would come to have the greatest impact on sociology. Veblen here calls our attention to the "waste" of both money and time that individuals of all social classes engage in as a means for improving their self-esteem and elevating their status in the community. Whether it's purchasing expensive cars or clothes when inexpensive brands will suffice, or dedicating oneself to learning the finer points of golf or dining etiquette, such practices signal an underlying competitive attempt to best others and secure one's position in the status order.

Figure 2.1 Marx's Model of Social Change: The Communist Revolution

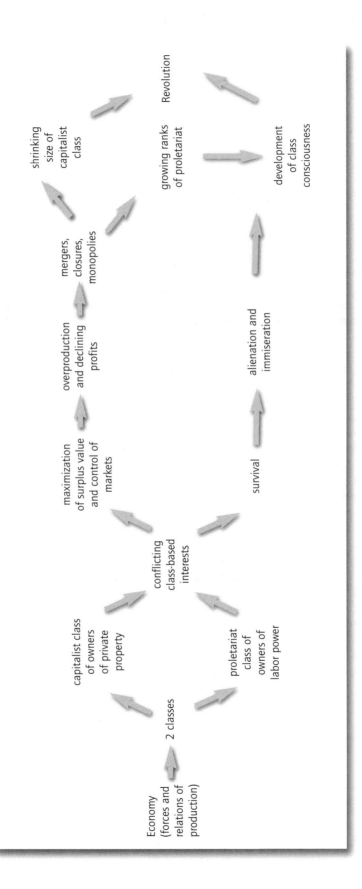

The business owners who are unable to compete successfully for a share of the market find themselves joining the swelling ranks of propertyless wage earners: the **proletariat**. This adds to the revolutionary potential of the working-class movement in two ways. First, the proletariat is transformed into an overwhelming majority of the population, making its class interests an irresistible force for change. Second, as Marx points out, the former capitalists bring with them a level of education not possessed by the typical wage laborer. This breeds political consequences as the former members of the bourgeoisie translate their economic resentment into a radicalization of the proletariat by educating the workers with regard to both the nature of capitalist accumulation and the workers' essential role in overthrowing the system of their oppression.

This was precisely the purpose of Marx's political activities: He sought to generate **class consciousness**—an awareness on the part of the working class of its common relationship to the means of production. Marx believed that this awareness was a vital key for sparking a revolution that would create a "dictatorship of the proletariat," transforming it from a wage-earning, propertyless mass into the ruling class. Unlike all previous class-based revolutions, however, this one would be fought in the interests of a vast majority of the population and not for the benefit of a few as the particular class interests of the proletariat had come to represent the universal interests of humanity. The epoch of capitalism was a necessary stage in this evolution—and the last historical period rooted in class conflict (see Figure 2.1, page 25). Capitalism, with its unleashing of immense economic productivity, had created the capital and technology needed to sustain a communist society, the final stage of history.

Using the power of the state to abolish private ownership of the means of production, the proletariat would wrest control of society's productive forces from the hands of the bourgeoisie and create a centralized, socialist economy. Socialism, however, would be but a temporary phase. Without private ownership of the means of production, society would no longer be divided along class lines; without antagonistic class interests, the social conditions that produce conflict, exploitation, and alienation would no longer exist. The disappearance of classes and class conflict would render obsolete the state whose primary charge is to secure the right to private property. Finally, without class conflict—the fuel that ignites social change—the dialectical progression of human history comes to a utopian end. With the production of goods controlled collectively and not by private business elites, individuals would be free to cultivate their natural talents and actualize their full potential.[6] (You will read more about this below, in the excerpt from *The Communist Manifesto.*)

As indicated previously, this evolutionary type of thinking was typical of Enlightenment intellectuals. Today, however, many consider Marx's "end of prehistory" vision of communism as the least viable part of his theory. While the internal contradictions of capitalism are real, they have been checked by a number of practices, including ongoing government intervention in the economy, the continued expansion of markets (i.e., Western-dominated globalization), and cost-saving advances in production and organizational technologies.

⠿ MARX'S THEORETICAL ORIENTATION

In terms of our metatheoretical framework, Figure 2.2 illustrates how Marx's work is predominantly collectivist and rationalist in orientation. Of course, as discussed previously, the action/order dimensions are intended to serve as heuristic devices. Certainly, there are

---

[6]By no means have modern communist societies—for instance, the former Soviet Union, China, and North Vietnam—resembled the type of free and creative society envisioned by Marx.

Figure 2.2   Marx's Basic Theoretical Orientation

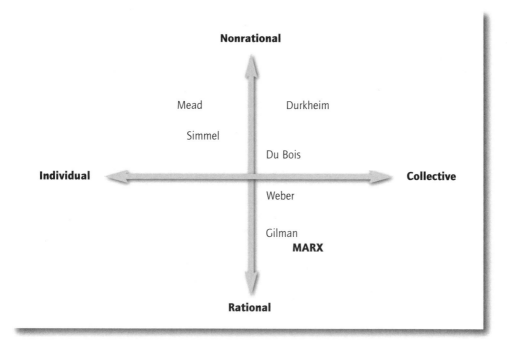

elements of Marx's theory that do not fit neatly into this particular "box." Nevertheless, Marx pursued themes that, taken as a whole, underscored his vision of a social order shaped by broad historical transitions and classes of actors (collectivist) pitted against one another in a struggle to realize their economic interests (rationalist).

Regarding the question of order, Marx saw human societies as evolving toward an ultimate, utopian end—a process spurred by *class* conflict. It is the struggle to control the forces of production and the distribution of resources and profits they create that leads classes—not individuals—to become the prime movers of history.

Of course, one might counter that it is *individuals* who "make history." Is it not individuals who make up classes, join labor unions, manage factories, merge corporations, and devise industry strategies? Though this is perhaps true on one level, throughout his work Marx emphasized the structural parameters that inhibit and shape individual decisions and actions. On this point Marx stated in one of his most famous passages, although "men make their own history . . . they do not make it just as they please; they do not make it under circumstances chosen by themselves, but under circumstances directly found, given and transmitted from the past" (Marx 1852/1978:595).

The circumstance of greatest import in this regard is that individuals are born into societies where the **forces and relations of production** that make up "material life"—classes and property relations—are already established independent of their will. From this existing economic base is born a "superstructure" or "the social, political, and intellectual life processes in general" (Marx 1859/1978:4). The superstructure, in short, consists of everything noneconomic in nature such as a society's legal, political, and educational systems, as well as its stock of commonsense knowledge. As a result, an individual's very consciousness—how she views the world, develops aspirations, and defines her interests—is not determined by the individual's own subjectivity. Instead, ideas about the world and one's place in it are structured by, or built into, the *objective* class position an individual occupies.

And while there are capitalists and laborers who seemingly do not pursue their antagonistic class interests, such exceptions to the rule do not disprove it: "It is not the consciousness of men that determines their being, but, on the contrary, their social being that determines their consciousness" (ibid.).

In terms of the motivation for action, Marx's work is primarily rationalist. This tendency is most clearly reflected in his emphasis on class-based *interests*. According to Marx, humans are separated from other animals due to our innate need to realize our full potential through engaging in creative labor. It is through freely developed "conscious life-activity" (Marx 1844/1978:76) that women and men are able to develop their "true" selves and forge meaningful relationships with others. It is in the process of production and in the objects that result that women and men realize themselves and their significance in a world that they create. (The corruption of the link between labor and self-realization by capitalism is addressed most fully in the selection "Alienated Labour," below.)

Because self-fulfillment is derived through labor, it is in the individual's interest to control the production process that is so vital to meeting this most basic of human needs. The crucial arena of this struggle is the network of economic relationships: the manufacturing, distribution, and sale of goods, and the selling of one's labor. Even if individuals are unaware of their true class interests, they will still be moved by them. For recall that interests are a reflection of one's objective position in relation to the process of production; they are not spawned by one's subjective disposition. The essential point here is that Marx's model *presupposes* that our actions are driven by our attempts to maximize our interests. (See Figure 2.3.) Of course, whether or not we are truly as rationalistic as Marx maintains is a point of great theoretical debate.

Figure 2.3 Marx's Core Concepts

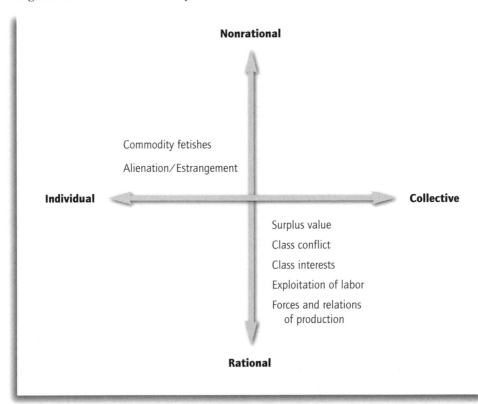

Antonio Gramsci (1891–1937): Hegemony and the Ruling Ideas

Antonio Gramsci was an Italian philosopher, journalist, and political activist who spent much of his adult life ardently supporting the revolutionary cause of the working class. His foray into politics began in earnest in 1915 when he became a member of the Italian Socialist Party (PSI) and published critical essays in the Party's official paper *L'Avanti*. In 1919, he cofounded the periodical, *The New Order: A Weekly Review of Socialist Culture*. Covering political events across Europe, the United States, and the Soviet Union, the paper was widely influential among Italy's radical Left. After an internal split within the PSI in 1921, Gramsci became a prominent member of the Italian Communist Party (PCI), serving first in the Party's central committee and then as a delegate to the Communist International in Moscow. He would go on to be elected to the Party's Chamber of Deputies. and later rise to the position of General Secretary.

Gramsci would pay a heavy cost for his political activism. His sympathies with the Bolshevik revolution and its leaders, and his alliance with his country's workers' movements, made him an enemy of Italy's newly formed fascist government. In 1926, Gramsci was arrested for his political activities, and was sentenced to 20 years in prison. He would serve only 11 years in prison, however, before dying of a brain hemorrhage in April 1937.

Despite the harsh conditions of his imprisonment and his fragile health, Gramsci produced 29 notebooks—some 3,000 pages—of political and philosophical analysis. The notebooks were smuggled out, but none was published until several years after the end of World War II. It would be another 20 years before the notebooks were compiled and published in English, under the title *The Prison Notebooks*. The notebooks reveal one of Gramsci's central concerns: to explain why Europe's working-class failed to spearhead a socialist revolution, and how, in Italy and elsewhere, it could act against its own class interests by supporting a fascist regime. In addressing these issues, Gramsci confronted an oft-noted weakness in Marx's historical materialism: the role of ideas in preventing or advancing revolutionary change. Asserting, "the ruling ideas are the ideas of the ruling class," Marx portended that the proletariat, with its numbers increasing, would come to recognize its class interests and unite to overthrow the bourgeoisie and the conditions of alienation and exploitation that serve the capitalists' narrow ambitions for profit. Yet, despite the fact that the material or economic conditions were ripe for a revolutionary movement across much of Europe, no successful challenge to the ruling powers was mounted.

To account for the lack of revolutionary foment on the part of the working class, Gramsci emphasized the role of ideas in establishing "hegemony," or domination, over subaltern classes. For Gramsci, the bourgeoisie maintained its dominance not primarily through force or coercion, but through the willing, "spontaneous" consent of the ruled. This consent was the outgrowth of the proletariat adopting as its own the values, beliefs, and attitudes that serve the interests of the ruling class. In other words, the working class is socialized (particularly through the educational system) into accepting a bourgeois ideology as an unquestioned or commonsense view of the world and their place in it. As a result, the working class aligns itself with the status quo, thus granting legitimacy to social and economic arrangements that perpetuate their own exploitation.

(Continued)

(Continued)

Recognizing that economic crises alone could not spark a socialist revolution, Gramsci was convinced that in order for the proletariat to unmask the real sources of its oppression and generate a unified, popular revolt, it must first develop its own "organic" consciousness, or counter hegemony. This counter hegemony would articulate the real interests and needs of the masses. Moreover, he insisted that this counter ideology must originate from within the masses; to be effective in provoking revolutionary change, it cannot be imposed on them by bourgeois "traditional" intellectuals who remain detached from the everyday realities of working-class life. Declaring, "all men are intellectuals," Gramsci sought to encourage the development of "organic" intellectuals from within the ranks of the working class through his political journalism and active participation in the workers' movement. Such individuals are intellectuals not in the sense of their profession or social function, but in terms of their "directing the ideas and aspirations of the class to which they organically belong" (Gramsci 1971:3). In this way, the factory worker and truck driver, the financial accountant and government bureaucrat, are all potential intellectuals. Indeed, the intellectuals most capable of contributing to progressive social change were not those of the "traditional" or professional type—writers, artists, scientists, philosophers—but rather those who engage in "praxis," connecting theoretical insights to an active attempt to fashion a more just society. For Gramsci, this was the "new intellectual" drawn from the working class:

> In the modern world, technical education, closely bound to industrial labour even at the most primitive and unqualified level, must form the basis of the new type of intellectual. The mode of being of the new intellectual can no longer consist in eloquence, which is an exterior and momentary mover of feelings and passions, but in active participation in practical life, as constructor, organiser, "permanent persuader" and not just a simple orator. . . . One of the most important characteristics of any group that is developing towards dominance is its struggle to assimilate and to conquer "ideologically" the traditional intellectuals, but this assimilation and conquest is made quicker and more efficacious the more the group in question succeeds in simultaneously elaborating its own organic intellectuals. (Gramsci 1971:10)

Readings

Marx's writings included here are divided into four sections. The first section centers on his "materialist conception of history," developed in reaction to the works of the German idealist philosopher Georg W. F. Hegel (1770–1831). The second section offers his critique of the human costs of capitalism. The third section contains Marx's call for the inevitable communist revolution that will usher in the "end of prehistory" and, with it, the end of alienation, private property, and oppressive government. In the fourth set of readings, we move from Marx's prophecy of emancipation to his theory of economics. Here you will read his analyses of the sources of value and the nature of commodities.

The final reading offers Engels's analysis of the subordinate role of women within the family, forms of class conflict, and the rise of the state as an alienating power for containing class antagonisms and preserving the dominance of property holders.

Introduction to *The German Ideology*

Written in 1845–46, *The German Ideology* presents the most detailed account of Marx's theory of history. In it, Marx set out to reformulate the work of the eminent German philosopher Georg W. F. Hegel. In contrast to previous philosophers who focused on explaining the roots of stability in the physical and social worlds (i.e., why things seemingly stayed the same), Hegel saw change as the motor of history. For Hegel, change was driven by a dialectical process in which a given state of being or idea contains within it the seeds of an opposing state of being or opposing idea. The resolution of the conflict produces yet a new state of being or idea. This synthesis, in turn, forms the basis of a new contradiction, thus continuing the process of change.

As an example, consider the division of gender roles. Traditionally, women and the roles they perform have been devalued relative to the positions occupied by men. Thus, the notion that a "woman's place is in the home" serves as a justification for male dominance in economic and political affairs. Out of the ideas that sustained the oppression of women was born the opposing view that women are in fact superior to men. From this vantage, it is argued, for instance, that women's "innate" compassion or empathy better qualifies them for positions of leadership compared to the "innate" aggression said to characterize men. From the clash of these opposing ideas, a synthesis or state of being has evolved in which neither women nor men are considered superior to the other, but instead are viewed as equals. Thus, women have entered into roles formerly reserved for men, while men have begun to perform more traditionally "feminine" tasks.

Is society thus faced with a never-ending challenging of ideas as one "truth" replaces another in the evolution of history? Hegel's answer is a definitive "no." He expressed, instead, a belief in the ultimate perfectibility of the consciousness of humankind. Such perfection occurs through the progressive realization of "Absolute Idea" as revealed by God. In other words, every idea (thesis) is a distorted expression of an all-embracing "Spirit" or "Mind" (God) that produces an opposite idea (antithesis). The two contradictory ideas are unified to form a synthesis that in turn becomes the basis for a new idea (thesis). Progress and history itself come to an end as the contradictions between our ideas about reality and the "Truth" of reality as designed by God are finally resolved. In arguing that the evolution of human history proceeds purposively according to an immanent or predestined design, Hegel offers a teleological vision shared with both Christian theology and Enlightenment philosophy. (As you will read below, Marx, too, fashioned a teleological theory, but one that casts communism as the end toward which history progresses.)

If this seems abstract, it is because it is! Perhaps we can clarify Hegel's dialectic idealism a bit further. The essence of reality lies in thought or ideas because it is only in and through the concepts that order our experiences that experiences, as such, are known. Reality is a product of our conceptual categories or consciousness and thus has no existence independent of our own construction of it. As our ideas or knowledge changes, so does reality. The stages of history or reality are then defined by progressive stages in the negation of the prevailing conceptual ordering of experience. The utopian aspect of this development is found in the assertion that humankind's knowledge will reach the perfected state of "Pure Reason" or "Absolute Idea" in which freedom takes the form of self-knowledge.

In contending that history is marked by a distortion of "Truth" or "Pure Reason," it follows that our consciousness is alienated from Spirit (God). The condition of alienation thus stems from a religiously grounded misunderstanding of reality. At its core, this misunderstanding comes from the failure to recognize that man and Spirit are one. Instead, man exists as an "unhappy soul," placing in God all that is good and righteous, while seeing in himself only that which is base and sinful. God becomes an alien, all-knowing, powerful force separated from ignorant, powerless man. Yet, as consciousness evolves through the historical dialectic, it advances closer to utopia in the form of an absolute self-knowledge that recognizes that reality is a product of the human spirit and not an alien force. No longer plagued by the irrationality that comes from a distorted view of the essence of mankind, man, in unity with Mind, can order the world in a rational way.[1]

The German Ideology reflects both Marx's indebtedness to and break from Hegel's philosophy. On the one hand, akin to Hegel, Marx depicts the unfolding of history as a progressive, dialectical process that culminates in a utopia of freedom and self-realization. In other words, like Hegel, Marx argues that each successive period in societal evolution is a necessary consequence of the preceding stage; and Marx projects a millennial significance onto the process itself, claiming that social development ends in a "necessary" utopia free of conflict and exploitation.

However, Marx breaks decisively from Hegel by insisting that it is *material* existence—not consciousness—that fuels historical change. Thus, Marx sought to take Hegel's idealism, which had the evolution of history "standing on its head," and "turn it right side up" in order to discover the real basis of the progression of human societies. Theoretically, this inversion is of utmost significance because it reflects a shift from a nonrationalist to a rationalist theoretical orientation.

The German Ideology is a pivotal writing because it offers the fullest treatment of Marx's materialist conception of history. It is in Marx's theory of historical materialism that we find one of his most important philosophical contributions, namely his conviction that ideas or interests have no existence independent of physical reality. In numerous passages, you will see Marx's rejection of Hegel's notion that ideas determine experience in favor of the materialist view that experience determines ideas. For instance, Marx asserts, "Consciousness can never be anything else than conscious existence, and the existence of men is their actual life-process" (Marx and Engels 1846/1978:154). And again, "Life is not determined by consciousness, but consciousness by life" (ibid.:155). In short, Marx argues that the essence of individuals, what they truly are and how they see the world, is determined by their material, economic conditions—"both with *what* they produce and with *how* they produce"—in which they live out their very existence (ibid.:150; emphasis in original).

Moreover, to argue that experience determines consciousness yields a radical conclusion: "The ideas of the ruling class are in every epoch the ruling ideas" (ibid.:172). In other words, Marx maintains that the dominant economic class controls not only a society's means of material production, but the production of ideas as well. To illustrate this point, consider, for instance, the *idea* of individual equality. From where did it spring? The notion of equality is by no means universal. Not only do some contemporary societies reject the concept of equality, but even those societies that do guarantee such rights (the United States, France, England, to name but a few) have not always done so. How are we then to account for the development of this principle? The answer, in short, lies in the development of capitalism.

---

[1]Hegel's notion of alienation would play a central role in Marx's work. Marx, however, argued that alienation was not a consequence of distorted consciousness but, rather, that it resulted from the material conditions of production. Marx takes up this issue in his essay "Alienated Labour," excerpted below.

As an economic system, capitalism is based on the notion of "freedom"—workers are "free" to find work or to quit their job. Entrepreneurs are "free" to open or close their businesses. In order for competitive capitalism to develop to its fullest productive capacities, individuals must be able to move, work, and invest their capital freely. This ability is expressed through the idea of individual equality. Thus, the concept of equality is born out of the capitalist mode of production and the nature of the social relationships it demands. It is an idea advanced by the bourgeoisie to sanction individualism that, in turn, justifies and sustains the economic conditions in which they themselves are the dominant force. In short, it serves the economic and political interests of the ruling class.

From *The German Ideology* (1845–1846)

Karl Marx and Friedrich Engels

The premises from which we begin are not arbitrary ones, not dogmas, but real premises from which abstraction can only be made in the imagination. They are the real individuals, their activity and the material conditions under which they live, both those which they find already existing and those produced by their activity. The premises can thus be verified in a purely empirical way.

The first premise of all human history is, of course, the existence of living human individuals. Thus the first fact to be established is the physical organisation of these individuals and their consequent relation to the rest of nature. Of course, we cannot here go either into the actual physical nature of man, or into the natural conditions in which man finds himself—geological, orohydrographical, climatic and so on. The writing of history must always set out from these natural bases and their modification in the course of history through the action of men.

Men can be distinguished from animals by consciousness, by religion or anything else you like. They themselves begin to distinguish themselves from animals as soon as they begin to produce their means of subsistence, a step which is conditioned by their physical organisation. By producing their means of subsistence men are indirectly producing their actual material life.

The way in which men produce their means of subsistence depends first of all on the nature of the actual means of subsistence they find in existence and have to reproduce. This mode of production must not be considered simply as being the reproduction of the physical existence of the individuals. Rather it is a definite form of activity of these individuals, a definite form of expressing their life, a definite *mode of life* on their part. As individuals express their life, so they are. What they are, therefore, coincides with their production, both with *what* they produce and with how they produce. The nature of individuals thus depends on the material conditions determining their production. . . .

The relations of different nations among themselves depend upon the extent to which each has developed its productive forces, the division of labour and internal intercourse. This statement is generally recognised. But not only the relation of one nation to others, but also the whole internal structure of the nation itself depends on the stage of development reached by its production and its internal and external intercourse. How far the productive forces of a nation are developed is shown most manifestly by the degree to which the division of labour has been carried. Each new productive force, insofar as it is not merely a quantitative of productive forces already known (for instance the bringing into cultivation of fresh land), causes a further development of the division of labour.

The division of labour inside a nation leads at first to the separation of industrial and commercial from agricultural labour, and hence to the separation of *town* and *country* and to the conflict

SOURCE: Excerpts from "The German Ideology: Part I," translated by Robert Tucker, from *The Marx-Engels Reader,* Second Edition, by Karl Marx and Friedrich Engels, edited by Robert C. Tucker. Copyright © 1978, 1972 by W.W. Norton & Company, Inc. Used by permission of W.W. Norton & Company, Inc.

of their interests. Its further development leads to the separation of commercial from industrial labour. At the same time through the division of labour inside these various branches there develop various divisions among the individuals co-operating in definite kinds of labour. The relative position of these individual groups is determined by the methods employed in agriculture, industry and commerce (patriarchalism, slavery, estates, classes). These same conditions are to be seen (given a more developed intercourse) in the relations of different nations to one another.

The various stages of development in the division of labour are just so many different forms of ownership, i.e., the existing stage in the division of labour determines also the relations of individuals to one another with reference to the material, instrument, and product of labour.

The first form of ownership is tribal [*Stammeigentum*] ownership. It corresponds to the undeveloped stage of production, at which a people lives by hunting and fishing, by the rearing of beasts or, in the highest stage, agriculture. In the latter case it pre-supposes a great mass of uncultivated stretches of land. The division of labour is at this stage still very elementary and is confined to a further extension of the natural division of labour existing in the family. The social structure is, therefore, limited to an extension of the family; patriarchal family chieftains, below them the members of the tribe, finally slaves. The slavery latent in the family only develops gradually with the increase of population, the growth of wants, and with the extension of external relations, both of war and of barter.

The second form is the ancient communal and State ownership which proceeds especially from the union of several tribes into a city by agreement or by conquest, and which is still accompanied by slavery. Beside communal ownership we already find movable, and later also immovable, private property developing, but as an abnormal form subordinate to communal ownership. The citizens hold power over their labouring slaves only in their community, and on this account alone, therefore, they are bound to the form of communal ownership. It is the communal private property which compels the active citizens to remain in this spontaneously derived form of association over against their

slaves. For this reason the whole structure of society based on this communal ownership, and with it the power of the people, decays in the same measure as, in particular, immovable private property evolves. The division of labour is already more developed. We already find the antagonism of town and country; later the antagonism between those states which represent town interests and those which represent country interests, and inside the towns themselves the antagonism between industry and maritime commerce. The class relation between citizens and slaves is now completely developed. . . .

The third form of ownership is feudal or estate property. If antiquity started out from the town and its little territory, the Middle Ages started out from the *country*. This different starting-point was determined by the sparseness of the population at that time, which was scattered over a large area and which received no large increase from the conquerors. In contrast to Greece and Rome, feudal development at the outset, therefore, extends over a much wider territory, prepared by the Roman conquests and the spread of agriculture at first associated with them. The last centuries of the declining Roman Empire and its conquest by the barbarians destroyed a number of productive forces; agriculture had declined, industry had decayed for want of a market, trade had died out or been violently suspended, the rural and urban population had decreased. From these conditions and the mode of organisation of the conquest determined by them, feudal property developed under the influence of the Germanic military constitution. Like tribal and communal ownership, it is based again on a community; but the directly producing class standing over against it is not, as in the case of the ancient community, the slaves, but the enserfed small peasantry. As soon as feudalism is fully developed, there also arises antagonism to the towns. The hierarchical structure of landownership, and the armed bodies of retainers associated with it, gave the nobility power over the serfs. This feudal organisation was, just as much as the ancient communal ownership, an association against a subjected producing class; but the form of association and the relation to the direct producers were different because of the different conditions of production.

This feudal system of landownership had its counterpart in the *towns* in the shape of corporative property, the feudal organisation of trades. Here property consisted chiefly in the labour of each individual person. The necessity for association against the organised robber nobility, the need for communal covered markets in an age when the industrialist was at the same time a merchant, the growing competition of the escaped serfs swarming into the rising towns, the feudal structure of the whole country: these combined to bring about the *guilds*. The gradually accumulated small capital of individual craftsmen and their stable numbers, as against the growing population, evolved the relation of journeyman and apprentice, which brought into being in the towns a hierarchy similar to that in the country.

Thus the chief form of property during the feudal epoch consisted on the one hand of landed property with serf labour chained to it, and on the other of the labour of the individual with small capital commanding the labour of journeymen. The organisation of both was determined by the restricted conditions of production—the small-scale and primitive cultivation of the land, and the craft type of industry. There was little division of labour in the heyday of feudalism. Each country bore in itself the antithesis of town and country; the division into estates was certainly strongly marked; but apart from the differentiation of princes, nobility, clergy and peasants in the country, and masters, journeymen, apprentices and soon also the rabble of casual labourers in the towns, no division of importance took place. In agriculture it was rendered difficult by the strip-system, beside which the cottage industry of the peasants themselves emerged. In industry there was no division of labour at all in the individual trades themselves, and very little between them. The separation of industry and commerce was found already in existence in older towns; in the newer it only developed later, when the towns entered into mutual relations. . . .

The fact is, therefore, that definite individuals who are productively active in a definite way enter into these definite social and political relations. Empirical observation must in each separate instance bring out empirically, and without any mystification and speculation, the connection of the social and political structure with production. The social structure and the State are continually evolving out of the life process of definite individuals, but of individuals, not as they may appear in their own or other people's imagination, but as they *really* are; i.e., as they operate, produce materially, and hence as they work under definite material limits, presuppositions and conditions independent of their will.

The production of ideas, of conceptions, of consciousness, is at first directly interwoven with the material activity and the material intercourse of men, the language of real life. Conceiving, thinking, the mental intercourse of men, appear at this stage as the direct efflux of their material behaviour. The same applies to mental production as expressed in the language of politics, laws, morality, religion, metaphysics, etc., of a people. Men are the producers of their conceptions, ideas, etc.—real, active men, as they are conditioned by a definite development of their productive forces and of the intercourse corresponding to these, up to its furthest forms. Consciousness can never be anything else than conscious existence, and the existence of men is their actual life-process. If in all ideology men and their circumstances appear upside-down as in a *camera obscura,* this phenomenon arises just as much from their historical life-process as the inversion of objects on the retina does from their physical life-process.

In direct contrast to German philosophy which descends from heaven to earth, here we ascend from earth to heaven. That is to say, we do not set out from what men say, imagine, conceive, nor from men as narrated, thought of, imagined, conceived, in order to arrive at men in the flesh. We set out from real, active men, and on the basis of their real life-process we demonstrate the development of the ideological reflexes and echoes of this life-process. The phantoms formed in the human brain are also, necessarily, sublimates of their material life-process, which is empirically verifiable and bound to material premises. Morality, religion, metaphysics, all the rest of ideology and their corresponding forms of consciousness, thus no longer retain the semblance of independence. They have no history, no development; but men, developing their material production and their material intercourse, alter, along with this their real existence, their thinking and the products of their thinking. Life is not determined by consciousness, but consciousness by life. In the first method of approach the starting-point is consciousness taken as the living individual; in the second method, which conforms to real life, it is the real living individuals themselves, and consciousness is considered solely as *their* consciousness. . . .

The production of life, both of one's own in labour and of fresh life in procreation, now appears as a double relationship: on the one hand as a natural, on the other as a social relationship. By social we understand the co-operation of several individuals, no matter under what conditions, in what manner and to what end. It follows from this that a certain mode of production, or industrial stage, is always combined with a certain mode of co-operation, or social stage, and this mode of co-operation is itself a "productive force." Further, that the multitude of productive forces accessible to men determines the nature of society, hence, that the "history of humanity" must always be studied and treated in relation to the history of industry and exchange. . . . Thus it is quite obvious from the start that there exists a materialistic connection of men with one another, which is determined by their needs and their mode of production, and which is as old as men themselves. This connection is ever taking on new forms, and thus presents a "history" independently of the existence of any political or religious nonsense which would especially hold men together.

Only now, after having considered four moments, four aspects of the primary historical relationships, do we find that man also possesses "consciousness";[i] but, even so, not inherent, not "pure" consciousness. From the start the "spirit" is afflicted with the curse of being "burdened" with matter, which here makes its appearance in the form of agitated layers of air, sounds, in short, of language. Language is as old as consciousness, language is practical consciousness that exists also for other men, and for that reason alone it really exists for me personally as well; language, like consciousness, only arises from the need, the necessity, of intercourse with other men. Where there exists a relationship, it exists for me: the animal does not enter into *"relations"* with anything, it does not enter into any relation at all. For the animal, its relation to others does not exist as a relation. Consciousness is, therefore, from the very beginning a social product, and remains so as long as men exist at all. Consciousness is at first, of course, merely consciousness concerning the *immediate* sensuous environment and consciousness of the limited connection with other persons and things outside the individual who is growing self-conscious. At the same time it is consciousness of nature, which first appears to men as a completely alien, all-powerful and unassailable force, with which men's relations are purely animal and by which they are overawed like beasts; it is thus a purely animal consciousness of nature (natural religion).

We see here immediately: this natural religion or this particular relation of men to nature is determined by the form of society and vice versa. Here, as everywhere, the identity of nature and man appears in such a way that the restricted relation of men to nature determines their restricted relation to one another, and their restricted relation to one another determines men's restricted relation to nature, just because nature is as yet hardly modified historically; and, on the other hand, man's consciousness of the necessity of associating with the individuals around him is the beginning of the consciousness that he is living in society at all. This beginning is as animal as social life itself at this stage. It is mere herd-consciousness, and at this point man is only distinguished from sheep by the fact that with him consciousness takes the place of instinct or that his instinct is a conscious one. This sheep-like or tribal consciousness receives its further development and extension through increased productivity, the increase of needs, and, what is fundamental to both of these, the

---

[i]Marginal note by Marx: "Men have history because they must *produce* their life, and because they must produce it moreover in a *certain* way: this is determined by their physical organisation: their consciousness is determined in just the same way."

increase of population. With these there develops the division of labour, which was originally nothing but the division of labour in the sexual act, then that division of labour which develops spontaneously or "naturally" by virtue of natural predisposition (e.g., physical strength), needs, accidents, etc., etc. Division of labour only becomes truly such from the moment when a division of material and mental labour appears.[ii] From this moment onwards consciousness *can* really flatter itself that it is something other than consciousness of existing practice, that it *really* represents something without representing something real; from now on consciousness is in a position to emancipate itself from the world and to proceed to the formation of "pure" theory, theology, philosophy, ethics, etc. But even if this theory, theology, philosophy, ethics, etc., comes into contradiction with the existing relations, this can only occur because existing social relations have come into contradiction with existing forces of production. . . .

With the division of labour, in which all these contradictions are implicit, and which in its turn is based on the natural division of labour in the family and the separation of society into individual families opposed to one another, is given simultaneously the *distribution,* and indeed the *unequal* distribution, both quantitative and qualitative, of labour and its products, hence property: the nucleus, the first form, of which lies in the family, where wife and children are the slaves of the husband. This latent slavery in the family, though still very crude, is the first property, but even at this early stage it corresponds perfectly to the definition of modern economists who call it the power of disposing of the labour-power of others. Division of labour and private property are, moreover, identical expressions: in the one the same thing is affirmed with reference to activity as is affirmed in the other with reference to the product of the activity.

Further, the division of labour implies the contradiction between the interest of the separate individual or the individual family and the communal interest of all individuals who have intercourse with one another. And indeed, this communal interest does not exist merely in the imagination, as the "general interest," but first of all in reality, as the mutual interdependence of the individuals among whom the labour is divided. And finally, the division of labour offers us the first example of how, as long as man remains in natural society, that is, as long as a cleavage exists between the particular and the common interest, as long, therefore, as activity is not voluntarily, but naturally, divided, man's own deed becomes an alien power opposed to him, which enslaves him instead of being controlled by him. For as soon as the distribution of labour comes into being, each man has a particular, exclusive sphere of activity, which is forced upon him and from which he cannot escape. He is a hunter, a fisherman, a shepherd, or a critical critic, and must remain so if he does not want to lose his means of livelihood: while in communist society, where nobody has one exclusive sphere of activity but each can become accomplished in any branch he wishes, society regulates the general production and thus makes it possible for me to do one thing today and another tomorrow, to hunt in the morning, fish in the afternoon, rear cattle in the evening, criticise after dinner, just as I have a mind, without ever becoming hunter, fisherman, shepherd or critic. This fixation of social activity, this consolidation of what we ourselves produce into an objective power above us, growing out of our control, thwarting our expectations, bringing to naught our calculations, is one of the chief factors in historical development up till now.

And out of this very contradiction between the interest of the individual and that of the community the latter takes an independent form as the *State,* divorced from the real interests of individual and community, and at the same time as an illusory communal life, always based, however, on the real ties existing in every family and tribal conglomeration—such as flesh and blood, language, division of labour on a larger scale, and other interests—and especially, as we shall enlarge upon later, on the classes, already determined by the division of labour, which in every such mass of men separate out, and of which one dominates all the others. It follows from this that all struggles within the State, the struggle between democracy, aristocracy, and

---

[ii]Marginal note by Marx: "The first form of ideologists, *priests,* is concurrent."

monarchy, the struggle for the franchise, etc., etc., are merely the illusory forms in which the real struggles of the different classes are fought out among one another.... Further, it follows that every class which is struggling for mastery, even when its domination, as is the case with the proletariat, postulates the abolition of the old form of society in its entirety and of domination itself, must first conquer for itself political power in order to represent its interest in turn as the general interest, which in the first moment it is forced to do. Just because individuals seek only their particular interest, which for them does not coincide with their communal interest (in fact the general is the illusory form of communal life), the latter will be imposed on them as an interest "alien" to them, and "independent" of them, as in its turn a particular, peculiar "general" interest; or they themselves must remain within this discord, as in democracy. On the other hand, too, the *practical* struggle of these particular interests, which constantly *really* run counter to the communal and illusory communal interests, makes *practical* intervention and control necessary through the illusory "general" interest in the form of the State. The social power, i.e., the multiplied productive force, which arises through the co-operation of different individuals as it is determined by the division of labour, appears to these individuals, since their co-operation is not voluntary but has come about naturally, not as their own united power, but as an alien force existing outside them, of the origin and goal of which they are ignorant, which they thus cannot control, which on the contrary passes through a peculiar series of phases and stages independent of the will and the action of man, nay even being the prime governor of these.

This *"estrangement"* (to use a term which will be comprehensible to the philosophers) can, of course, only be abolished given two *practical* premises. For it to become an "intolerable" power, i.e., a power against which men make a revolution, it must necessarily have rendered the great mass of humanity "propertyless," and produced, at the same time, the contradiction of an existing world of wealth and culture, both of which conditions presuppose a great increase in productive power, a high degree of its development. And, on the other hand, this development of productive forces (which itself implies the actual empirical existence of men in their *world-historical,* instead of local, being) is an absolutely necessary practical premise because without it *want* is merely made general, and with *destitution* the struggle for necessities and all the old filthy business would necessarily be reproduced; and furthermore, because only with this universal development of productive forces is a universal intercourse between men established, which produces in all nations simultaneously the phenomenon of the "propertyless" mass (universal competition), makes each nation dependent on the revolutions of the others, and finally has put *world-historical,* empirically universal individuals in place of local ones. Without this, (1) communism could only exist as a local event; (2) the *forces* of intercourse themselves could not have developed as *universal,* hence intolerable powers: they would have remained home-bred conditions surrounded by superstition; and (3) each extension of intercourse would abolish local communism. Empirically, communism is only possible as the act of the dominant peoples "all at once" and simultaneously, which presupposes the universal development of productive forces and the world intercourse bound up with communism. How otherwise could for instance property have had a history at all, have taken on different forms, and landed property, for example, according to the different premises given, have proceeded in France from parcellation to centralisation in the hands of a few, in England from centralisation in the hands of a few to parcellation, as is actually the case today? Or how does it happen that trade, which after all is nothing more than the exchange of products of various individuals and countries, rules the whole world through the relation of supply and demand—a relation which, as an English economist says, hovers over the earth like the fate of the ancients, and with invisible hand allots fortune and misfortune to men, sets up empires and overthrows empires, causes nations to rise and to disappear—while with the abolition of the basis of private property, with the communistic regulation of production (and, implicit in this, the destruction of the alien relation between men and what they themselves produce), the power of the relation of supply and demand is dissolved into nothing, and men get exchange, production, the mode of their mutual relation, under their own control again?

Communism is for us not a *state of affairs* which is to be established, an *ideal* to which reality [will] have to adjust itself. We call communism the *real* movement which abolishes the present state of things. The conditions of this movement result from the premises now in existence. Moreover, the mass of *propertyless* workers—the utterly precarious position of labour-power on a mass scale cut off from capital or from even a limited satisfaction and, therefore, no longer merely temporarily deprived of work itself as a secure source of life—presupposes the *world market* through competition. The proletariat can thus only exist *world-historically,* just as communism, its activity, can only have a "world-historical" existence. World-historical existence of individuals, i.e., existence of individuals which is directly linked up with world history.

The form of intercourse determined by the existing productive forces at all previous historical stages, and in its turn determining these, is *civil society.* The latter, as is clear from what we have said above, has as its premises and basis the simple family and the multiple, the so-called tribe, and the more precise determinants of this society are enumerated in our remarks above. Already here we see how this civil society is the true source and theatre of all history, and how absurd is the conception of history held hitherto, which neglects the real relationships and confines itself to high-sounding dramas of princes and states.

Civil society embraces the whole material intercourse of individuals within a definite stage of the development of productive forces. It embraces the whole commercial and industrial life of a given stage and, insofar, transcends the State and the nation, though, on the other hand again, it must assert itself in its foreign relations as nationality, and inwardly must organise itself as State. The term "civil society" [*bürgerliche Gesellschaft*][iii] emerged in the eighteenth century, when property relationships had already extricated themselves from the ancient and medieval communal society. Civil society as such only develops with the bourgeoisie; the social organisation evolving directly out of production and commerce, which in all ages forms the basis of the State and of the rest of the idealistic superstructure, has, however, always been designated by the same name. . . .

This conception of history depends on our ability to expound the real process of production, starting out from the material production of itself, and to comprehend the form of intercourse connected with this and created by this mode of production (i.e., civil society in its various stages), as the basis of all history; and to show it in its action as State, to explain all the different theoretical products and forms of consciousness, religion, philosophy, ethics, etc., etc., and trace their origins and growth from that basis; by which means, of course, the whole thing can be depicted in its totality (and therefore, too, the reciprocal action of these various sides on one another). It has not, like the idealistic view of history, in every period to look for a category, but remains constantly on the real *ground* of history; it does not explain practice from the idea but explains the formation of ideas from material practice; and accordingly it comes to the conclusion that all forms and products of consciousness cannot be dissolved by mental criticism, by resolution into "self-consciousness" or transformation into "apparitions," "spectres," "fancies," etc., but only by the practical overthrow of the actual social relations which gave rise to this idealistic humbug; that not criticism but revolution is the driving force of history, also of religion, of philosophy and all other types of theory. It shows that history does not end by being resolved into "self-consciousness" as "spirit of the spirit," but that in it at each stage there is found a material result: a sum of productive forces, a historically created relation of individuals to nature and to one another, which is handed down to each generation from its predecessor; a mass of productive forces, capital funds and conditions, which on the one hand, is indeed modified by the new generation, but also on the other prescribes for it its conditions of life and gives it a definite development, a special character. It shows that circumstances make men just as much as men make circumstances. This sum of productive forces, capital funds and social forms of intercourse, which every individual and generation finds in existence as something given, is the real basis of what the philosophers have conceived as "substance" and "essence of man," and what they have deified and attacked: a real basis which is not in the least disturbed, in its effect and influence on the development of men, by the

---

[iii]*Burgerliche Gesellschaft* can mean either "bourgeois society" or "civil society."

fact that these philosophers revolt against it as "self-consciousness" and the "Unique." These conditions of life, which different generations find in existence, decide also whether or not the periodically recurring revolutionary convulsion will be strong enough to overthrow the basis of the entire existing system. And if these material elements of a complete revolution are not present (namely, on the one hand the existing productive forces, on the other the formation of a revolutionary mass, which revolts not only against separate conditions of society up till then, but against the very "production of life" till then, the "total activity" on which it was based), then, as far as practical development is concerned, it is absolutely immaterial whether the idea of this revolution has been expressed a hundred times already, as the history of communism proves. . . .

The ideas of the ruling class are in every epoch the ruling ideas: i.e., the class which is the ruling *material* force of society, is at the same time its ruling *intellectual* force. The class which has the means of material production at its disposal, has control at the same time over the means of mental production, so that thereby, generally speaking, the ideas of those who lack the means of mental production are subject to it. The ruling ideas are nothing more than the ideal expression of the dominant material relationships, the dominant material relationships grasped as ideas; hence of the relationships which make the one class the ruling one, therefore, the ideas of its dominance. The individuals composing the ruling class possess among other things consciousness, and therefore think. Insofar, therefore, as they rule as a class and determine the extent and compass of an epoch, it is self-evident that they do this in its whole range, hence among other things rule also as thinkers, as producers of ideas, and regulate the production and distribution of the ideas of their age: thus their ideas are the ruling ideas of the epoch. For instance, in an age and in a country where royal power, aristocracy and bourgeoisie are contending for mastery and where, therefore, mastery is shared, the doctrine of the separation of powers proves to be the dominant idea and is expressed as an "eternal law."

The division of labour, which we have already seen above as one of the chief forces of history up till now, manifests itself also in the ruling class as the division of mental and material labour, so that

inside this class one part appears as the thinkers of the class (its active, conceptive ideologists, who make the perfecting of the illusion of the class about itself their chief source of livelihood), while the others' attitude to these ideas and illusions is more passive and receptive because they are in reality the active members of this class and have less time to make up illusions and ideas about themselves. Within this class this cleavage can even develop into a certain opposition and hostility between the two parts, which, however, in the case of a practical collision, in which the class itself is endangered, automatically comes to nothing, in which case there also vanishes the semblance that the ruling ideas were not the ideas of the ruling class and had a power distinct from the power of this class. The existence of revolutionary ideas in a particular period presupposes the existence of a revolutionary class; about the premises for the latter sufficient has already been said above.

If now in considering the course of history we detach the ideas of the ruling class from the ruling class itself and attribute to them an independent existence, if we confine ourselves to saying that these or those ideas were dominant at a given time, without bothering ourselves about the conditions of production and the producers of these ideas, if we thus ignore the individuals and world conditions which are the source of the ideas, we can say, for instance, that during the time that the aristocracy was dominant, the concepts honour, loyalty, etc., were dominant, during the dominance of the bourgeoisie the concepts freedom, equality, etc. The ruling class itself on the whole imagines this to be so. This conception of history, which is common to all historians, particularly since the eighteenth century, will necessarily come up against the phenomenon that increasingly abstract ideas hold sway, i.e., ideas which increasingly take on the form of universality. For each new class which puts itself in the place of one ruling before it, is compelled, merely in order to carry through its aim, to represent its interest as the common interest of all the members of society, that is, expressed in ideal form: it has to give its ideas the form of universality, and represent them as the only rational, universally valid ones. The class making a revolution appears from the very start, if only because it is opposed to a *class,* not as a class but as the representative of the whole of society; it appears as the whole mass of society confronting the one ruling class.[iv] It can do this because, to start with, its interest really is more connected with the common interest of all other non-ruling classes, because under the pressure of hitherto existing conditions its interest has not yet been able to develop as the particular interest of a particular class. Its victory, therefore, benefits also many individuals of the other classes which are not winning a dominant position, but only insofar as it now puts these individuals in a position to raise themselves into the ruling class. When the French bourgeoisie overthrew the power of the aristocracy, it thereby made it possible for many proletarians to raise themselves above the proletariat, but only insofar as they became bourgeois. Every new class, therefore, achieves its hegemony only on a broader basis than that of the class ruling previously, whereas the opposition of the non-ruling class against the new ruling class later develops all the more sharply and profoundly. Both these things determine the fact that the struggle to be waged against this new ruling class, in its turn, aims at a more decided and radical negation of the previous conditions of society than could all previous classes which sought to rule.

This whole semblance, that the rule of a certain class is only the rule of certain ideas, comes to a natural end, of course, as soon as class rule in general ceases to be the form in which society is organised, that is to say, as soon as it is no longer necessary to represent a particular interest as general or the "general interest" as ruling

Introduction to *Economic and Philosophic Manuscripts of 1844*

In the essay "Alienated Labour" (taken from the *Economic and Philosophic Manuscripts of 1844*), Marx examines the condition of **alienation** or estrangement. For Marx, alienation is inherent in capitalism, because the process of production and the results of our labor confront us as a dominating power. It stems not from religiously rooted errors of consciousness, as Hegel argued, but from the material conditions in which we apply our essential productive capacities. For, contrary to Hegel's assertion, God does not create man and his ideas. Instead, it is man who creates the idea of God.

How is it that alienation is a necessary feature of capitalism? For the wage earner, work is alienating because it serves solely to provide the means (i.e., money) for maintaining her physical existence. Instead of labor representing an end in itself—an activity that expresses our capacity to shape our lives and our relationships with others—private ownership of the means of production reduces the role of the worker to that of a cog in a machine. The worker is an expendable object that performs routinized tasks. Put in another way, for Marx, working just for money—and not for the creative potential of labor itself—is akin to selling your soul.

The wage earner has little, if any, control over the production *process*. The types of materials or machines to be used, how to divide the necessary tasks, and the rate at which goods are to be manufactured are all determined by the owner of the factory or business. The worker is thus subject to the demands of the production process; it confronts her as an alienating power that controls her labor. Because the worker is alienated in her role as producer, she can only be but alienated from that which the process of her labor produces. In turn, the *product* opposes the worker as an object over which she has no control. The questions of where and how it is sold and how much to charge are determined by the capitalist. More profoundly, the worker is dependent on the object for her very existence. It is only for her labor expended in producing the object that she earns a wage and is thus able to survive. If the

---

[iv]Marginal note by Marx: "Universality corresponds to (1) the class versus the estate, (2) the competition, world-wide intercourse, etc., (3) the great numerical strength of the ruling class, (4) the illusion of the *common* interests (in the beginning this illusion is true), (5) the delusion of the ideologists and the division of labour."

object disappears—when the factory closes or technology renders the worker's labor obsolete—through no fault of her own she is left clinging to survival.

Because the worker is alienated from the process of production as well as the product of his labor, he becomes inescapably alienated from *himself.* The wage earner spends two-thirds of his waking hours engaged in a meaningless activity, save its providing him with the means of subsistence. Torn away from the object of his labor, he is unable to realize the essence of his creative nature or "species being" through his work. Finally, the worker is alienated from the rest of humanity, and becomes just another commodity to be bought and sold. To himself and others he is more like an animal or a machine than a human. Tragically, Marx asserts that the worker is free only in the performance of his "animal functions—eating, drinking, procreating . . . and in his human functions [labor] he no longer feels himself to be anything but an animal" (Marx 1844/1978:74).

In "The Power of Money in Bourgeois Society" (also taken from the *Economic and Philosophic Manuscripts*), Marx extends his critique of capitalist production to money itself. Here he describes how the possessor of money can be transformed into anything money can buy; how one's individuality is determined not by his own characteristics or capacities, but by the power of money to transform what he wants to be into what he *is.* Money is a medium capable of being exchanged not only for a specific good or service, but also for traits such as beauty, talent, or honesty. It is not simply something that we earn, spend, or save—rather, it *does* things, it makes us who and what we are. Money is "the alienating *ability of mankind"* (Marx 1844/1978:104, emphasis in the original) that bonds us to life itself and to our relationships with others, not through our innate qualities, but through what we have the power to buy.

Significantly, this concern with the subjective consequences of the capitalist system reflects a nonrationalist dimension to Marx's argument that contrasts with his overall rationalist theoretical orientation. In "Alienated Labour," Marx does not focus on the nature of class interests and the struggle to realize them (though it certainly would be in our interest to reform, if not abolish, the productive arrangements he describes). Rather, he describes a "way of being," a sensibility imposed on workers and capitalists alike by the properties inherent to capitalism. Indeed, the nonrationalist logic of this essay is highlighted further by the fact that Marx is constructing a moral critique as much as a scientific argument concerning the degradation wreaked by capitalism.

—— From *Economic and Philosophic Manuscripts of 1844* ——

Karl Marx

ALIENATED LABOUR

We have proceeded from the premises of political economy. We have accepted its language and its laws. We presupposed private property, the separation of labour, capital and land, and of wages, profit of capital and rent of land—likewise division of labour, competition, the concept of exchange-value, etc. On the basis of political economy itself, in its own words, we have shown that the worker sinks to the level of a commodity and becomes indeed the most wretched of commodities; that the wretchedness of the worker is in inverse proportion to the power and magnitude of his production; that the necessary result of competition is the accumulation of capital in a few hands, and thus the restoration of monopoly in a more terrible form; that finally the distinction between capitalist and land-rentier, like that between the tiller of the soil and the factory-worker, disappears and that the whole of society must fall apart into the two classes—the property-*owners* and the propertyless *workers.* . . .

Now, therefore, we have to grasp the essential connection between private property, avarice, and the separation of labour, capital and landed property; between exchange and competition, value and the devaluation of men, monopoly and competition, etc.; the connection between this whole estrangement and the *money*-system.

Do not let us go back to a fictitious primordial condition as the political economist does, when he tries to explain. Such a primordial condition explains nothing. He merely pushes the question away into a grey nebulous distance. He assumes in the form of fact, of an event, what he is supposed to deduce—namely, the necessary relationship between two things—between, for example, division of labour and exchange. Theology in the same way explains the origin of evil by the fall of man: that is, it assumes as a fact, in historical form, what has to be explained.

We proceed from an *actual* economic fact.

The worker becomes all the poorer the more wealth he produces, the more his production increases in power and range. The worker becomes an ever cheaper commodity the more commodities he creates. With the *increasing value* of the world of things proceeds in direct proportion the *devaluation* of the world of men. Labour produces not only commodities; it produces itself and the worker as a *commodity*—and does so in the proportion in which it produces commodities generally.

This fact expresses merely that the object which labour produces—labour's product—confronts it as *something alien,* as a *power independent* of the producer. The product of labour is labour which has been congealed in an object, which has become material: it is the *objectification* of labour. Labour's realization is its objectification. In the conditions dealt with by political economy this realization of labour appears as *loss of reality* for the workers; objectification as *loss of the object* and *object-bondage*; appropriation as *estrangement,* as *alienation.*

So much does labour's realization appear as loss of reality that the worker loses reality to the point of starving to death. So much does objectification appear as loss of the object that the worker is robbed of the objects most necessary not only for his life but for his work. Indeed, labour itself becomes an object which he can get hold of only with the greatest effort and with the most irregular interruptions. So much does the appropriation of the object appear as estrangement that the more objects the worker produces the fewer can he possess and the more he falls under the dominion of his product, capital.

All these consequences are contained in the definition that the worker is related to the *product of his labour* as to an *alien* object. For on this premise it is clear that the more the worker spends himself, the more powerful the alien objective world becomes which he creates over-against himself, the poorer he himself—his inner world—becomes, the less belongs to him as his own. It is the same in religion. The more man puts into God, the less he retains in himself. The worker puts his life into the object; but now his life no longer belongs to him but to the object. Hence, the greater this activity, the greater is the worker's lack of objects. Whatever the product of his labour is, he is not. Therefore the greater this product, the less is he himself. The *alienation* of the worker in his product means not only that his labour becomes an object, an *external* existence, but that it exists *outside him,* independently, as something alien to him, and that it becomes a power of its own confronting him; it means that the life which he has conferred on the object confronts him as something hostile and alien.

Let us now look more closely at the *objectification,* at the production of the worker; and therein at the *estrangement,* the *loss* of the object, his product.

The worker can create nothing without *nature,* without the *sensuous external world.* It is the material on which his labor is manifested, in which it is active, from which and by means of which it produces.

But just as nature provides labor with the *means of life* in the sense that labour cannot *live* without objects on which to operate, on the other hand, it also provides the *means of life* in the more restricted sense—i.e., the means for the physical subsistence of the *worker* himself.

Thus the more the worker by his labour *appropriates* the external world, sensuous nature, the more he deprives himself of *means of life* in the double respect: first, that the sensuous external world more and more ceases to be an object belonging to his labour—to be his labour's *means of life*; and secondly, that it more and more ceases to be *means of life* in the immediate sense, means for the physical subsistence of the worker.

Thus in this double respect the worker becomes a slave of his object, first, in that he receives an *object of labour*, i.e., in that he receives *work*; and secondly, in that he receives *means of subsistence*. Therefore, it enables him to exist, first, as a *worker;* and, second, as a *physical subject.* The extremity of this bondage is that it is only as a *worker* that he continues to maintain himself as a *physical subject,* and that it is only as a *physical subject* that he is a worker.

(The laws of political economy express the estrangement of the worker in his object thus: the more the worker produces, the less he has to consume; the more values he creates, the more valueless, the more unworthy he becomes; the better formed his product, the more deformed becomes the worker; the more civilized his object, the more barbarous becomes the worker; the mightier labour becomes, the more powerless becomes the worker; the more ingenious labour becomes, the duller becomes the worker and the more he becomes nature's bondsman.)

Political economy conceals the estrangement inherent in the nature of labour by not considering the direct relationship between the worker (labour) and production. It is true that labour produces for the rich wonderful things—but for the worker it produces privation. It produces palaces—but for the worker, hovels. It produces beauty—but for the worker, deformity. It replaces labour by machines—but some of the workers it throws back to a barbarous type of labour, and the other workers it turns into machines. It produces intelligence—but for the worker idiocy, cretinism.

The direct relationship of labour to its produce is the relationship of the worker to the objects of his production. The relationship of the man of means to the objects of production and to production itself is only a *consequence* of this first relationship—and confirms it. We shall consider this other aspect later.

When we ask, then, what is the essential relationship of labour we are asking about the relationship of the *worker* to production.

Till now we have been considering the estrangement, the alienation of the worker only in one of its aspects, i.e., the worker's *relationship to the products of his labour.* But the estrangement is manifested not only in the result but in the *act of production*—within the *producing activity* itself. How would the worker come to face the product of his activity as a stranger, were it not that in the very act of production he was estranging himself from himself? The product is after all but the summary of the activity of production. If then the product of labour is alienation, production itself must be active alienation, the alienation of activity, the activity of alienation. In the estrangement of the object of labour is merely summarized the estrangement, the alienation, in the activity of labour itself.

What, then, constitutes the alienation of labour?

First, the fact that labour is *external* to the worker, i.e., it does not belong to his essential being; that in his work, therefore, he does not affirm himself but denies himself, does not feel content but unhappy, does not develop freely his physical and mental energy but mortifies his body and ruins his mind. The worker therefore only feels himself outside his work, and in his work feels outside himself. He is at home when he is not working, and when he is working he is not at home. His labour is therefore not voluntary, but coerced; it is *forced labour.* It is therefore not the satisfaction of a need; it is merely a *means* to satisfy needs external to it. Its alien character emerges clearly in the fact that as soon as no physical or other compulsion exists, labour is shunned like the plague. External labour, labour in which man alienates himself, is a labour of self-sacrifice, of mortification. Lastly, the external character of labour for the worker appears in the fact that it is not his own, but someone else's, that it does not belong to him, that in it he belongs, not to himself, but to another. Just as in religion the spontaneous activity of the human imagination, of the human brain and the human heart, operates independently of the individual—that is, operates on him as an alien, divine or diabolical activity—in the same way the worker's activity is not his spontaneous activity. It belongs to another; it is the loss of his self.

As a result, therefore, man (the worker) no longer feels himself to be freely active in any but his animal functions—eating, drinking, procreating, or at most in his dwelling and in dressing-up, etc.; and in his human functions he no longer feels himself to be anything but an animal. What is animal becomes human and what is human becomes animal.

Certainly eating, drinking, procreating, etc., are also genuinely human functions. But in the abstraction which separates them from the sphere of all other human activity and turns them into sole and ultimate ends, they are animal.

We have considered the act of estranging practical human activity, labour, in two of its aspects. (1) The relation of the worker to the *product of labour* as an alien object exercising power over him. This relation is at the same time the relation to the sensuous external world, to the objects of nature as an alien world antagonistically opposed to him. (2) The relation of labour to the *act of production* within the *labour* process. This relation is the relation of the worker to his own activity as an alien activity not belonging to him; it is activity as suffering, strength as weakness, begetting as emasculating, the worker's *own* physical and mental energy, his personal life or what is life other than activity—as an activity which is turned against him, neither depends on nor belongs to him. Here we have *self-estrangement,* as we had previously the estrangement of the *thing.*

We have yet a third aspect of *estranged labour* to deduce from the two already considered.

Man is a species being, not only because in practice and in theory he adopts the species as his object (his own as well as those of other things), but—and this is only another way of expressing it—but also because he treats himself as the actual, living species; because he treats himself as a *universal* and therefore a free being.

The life of the species, both in man and in animals, consists physically in the fact that man (like the animal) lives on inorganic nature; and the more universal man is compared with an animal, the more universal is the sphere of inorganic nature on which he lives. Just as plants, animals, stones, the air, light, etc., constitute a part of human consciousness in the realm of theory, partly as objects of natural science, partly as objects of art—his spiritual inorganic nature, spiritual nourishment which he must first prepare to make it palatable and digestible—so too in the realm of practice they constitute a part of human life and human activity. Physically man lives only on these products of nature, whether they appear in the form of food, heating, clothes, a dwelling, or whatever it may be. The universality of man is in practice manifested precisely in the universality which makes all nature his *inorganic* body—both inasmuch as nature is (1) his direct means of life, and (2) the material, the object, and the instrument of his life-activity. Nature is man's *inorganic body*—nature, that is, in so far as it is not itself the human body. Man *lives* on nature—means that nature is his *body,* with which he must remain in continuous intercourse if he is not to die. That man's physical and spiritual life is linked to nature means simply that nature is linked to itself, for man is a part of nature.

In estranging from man (1) nature, and (2) himself, his own active functions, his life-activity, estranged labour estranges the *species* from man. It turns for him the *life of the species* into a means of individual life. First it estranges the life of the species and individual life, and secondly it makes individual life in its abstract form the purpose of the life of the species, likewise in its abstract and estranged form.

For in the first place labour, *life-activity, productive life* itself, appears to man merely as a *means* of satisfying a need—the need to maintain the physical existence. Yet the productive life is the life of the species. It is life-engendering life. The whole character of a species—its species character—is contained in the character of its life-activity; and free, conscious activity is man's species character. Life itself appears only as *a means to life.*

The animal is immediately identical with its life-activity. It does not distinguish itself from it. It is its *life-activity.* Man makes his life-activity itself the object of his will and of his consciousness. He has conscious life-activity. It is not a determination with which he directly merges. Conscious life-activity directly distinguishes man from animal life-activity. It is just because of this that he is a species being. Or it is only because he is a species being that he is a Conscious Being, i.e., that his own life is an object for him. Only because of that is his activity free activity. Estranged labour reverses this relationship, so that it is just because man is a conscious being that he makes his life-activity, his *essential* being, a mere means to his *existence.*

In creating an *objective world* by his practical activity, in *working-up* inorganic nature, man proves himself a conscious species being, i.e., as a being that treats the species as its own essential being, or that treats itself as a species being. Admittedly animals also produce. They build themselves nests, dwellings, like the bees, beavers, ants, etc. But an animal only produces what it immediately needs for itself or its young. It produces one-sidedly, whilst man produces universally. It produces only under the dominion of immediate physical need, whilst man produces even when he is free from physical need and only truly produces in freedom therefrom. An animal produces only itself, whilst man reproduces the whole of nature. An animal's product belongs immediately to its physical body, whilst man freely confronts his product. An animal forms things in accordance with the standard and the need of the species to which it belongs, whilst man knows how to produce in accordance with the standard of every species, and knows how to apply everywhere the inherent standard to the object. Man therefore also forms things in accordance with the laws of beauty.

It is just in the working-up of the objective world, therefore, that man first really proves himself to be a *species being.* This production is his active species life. Through and because of this production, nature appears as *his* work and his reality. The object of labour is, therefore, the *objectification of man's species life:* for he duplicates himself not only, as in consciousness, intellectually, but also actively, in reality, and therefore he contemplates himself in a world that he has created. In tearing away from man the object of his production, therefore, estranged labour tears from him his *species life,* his real species objectivity, and transforms his advantage

over animals into the disadvantage that his inorganic body, nature, is taken from him.

Similarly, in degrading spontaneous activity, free activity, to a means, estranged labour makes man's species life a means to his physical existence.

The consciousness which man has of his species is thus transformed by estrangement in such a way that the species life becomes for him a means.

Estranged labour turns thus:

(3) *Man's species being,* both nature and his spiritual species property, into a being *alien* to him, into a *means* to his *individual existence.* It estranges man's own body from him, as it does external nature and his spiritual essence, his *human* being.

(4) An immediate consequence of the fact that man is estranged from the product of his labour, from his life-activity, from his species being is the *estrangement of man* from *man.* If a man is confronted by himself, he is confronted by the *other* man. What applies to a man's relation to his work, to the product of his labour and to himself, also holds of a man's relation to the other man, and to the other man's labour and object of labour.

In fact, the proposition that man's species nature is estranged from him means that one man is estranged from the other, as each of them is from man's essential nature.

The estrangement of man, and in fact every relationship in which man stands to himself, is first realized and expressed in the relationship in which a man stands to other men.

Hence within the relationship of estranged labour each man views the other in accordance with the standard and the position in which he finds himself as a worker.

We took our departure from a fact of political economy—the estrangement of the worker and his production. We have formulated the concept of this fact—*estranged, alienated* labour. We have analysed this concept—hence analysing merely a fact of political economy.

Let us now see, further, how in real life the concept of estranged, alienated labour must express and present itself.

If the product of labour is alien to me, if it confronts me as an alien power, to whom, then, does it belong?

If my own activity does not belong to me, if it is an alien, a coerced activity, to whom, then, does it belong?

To a being *other* than me.

Who is this being?

The *gods?* To be sure, in the earliest times the principal production (for example, the building of temples, etc., in Egypt, India and Mexico) appears to be in the service of the gods, and the product belongs to the gods. However, the gods on their own were never the lords of labour. No more was *nature.* And what a contradiction it would be if, the more man subjugated nature by his labour and the more the miracles of the gods were rendered superfluous by the miracles of industry, the more man were to renounce the joy of production and the enjoyment of the produce in favour of these powers.

The *alien* being, to whom labour and the produce of labour belongs, in whose service labour is done and for whose benefit the produce of labour is provided, can only be *man* himself.

If the product of labour does not belong to the worker, if it confronts him as an alien power, this can only be because it belongs to some *other man than the worker.* If the worker's activity is a torment to him, to another it must be *delight* and his life's joy. Not the gods, not nature, but only man himself can be this alien power over man.

We must bear in mind the above-stated proposition that man's relation to himself only becomes *objective* and *real* for him through his relation to the other man. Thus, if the product of his labour, his labour *objectified,* is for him an *alien,* hostile, powerful object independent of him, then his position towards it is such that someone else is master of this object, someone who is alien, hostile, powerful, and independent of him. If his own activity is to him an unfree activity, then he is treating it as activity performed in the service, under the dominion, the coercion and the yoke of another man.

Every self-estrangement of man from himself and from nature appears in the relation in which he places himself and nature to men other than and differentiated from himself. For this reason religious self-estrangement necessarily appears in the relationship of the layman to the priest, or again to a mediator, etc., since we are here dealing with the intellectual world. In the real practical world self-estrangement can only become manifest through the real practical relationship to other men. The medium through which estrangement takes place is itself *practical.* Thus through estranged labour man not only engenders his relationship to the object and to the act of production as to powers that are alien and hostile to him; he also engenders the relationship in which other men stand to his production and to his product, and the relationship in which he stands to these other men. Just as he begets his own production as the loss of his reality, as his punishment; just as he begets his own product as a loss, as a product not belonging to him; so he begets the dominion of the one who does not produce over production and over the product. Just as he estranges from himself his own activity, so he confers to the stranger activity which is not his own.

Till now we have only considered this relationship from the standpoint of the worker and later we shall be considering it also from the standpoint of the non-worker.

Through *estranged, alienated labour,* then, the worker produces the relationship to this labour of a man alien to labour and standing outside it. The relationship of the worker to labour engenders the relation to it of the capitalist, or whatever one chooses to call the master of labour. *Private property* is thus the product, the result, the necessary consequence, of *alienated labour,* of the external relation of the worker to nature and to himself.

Private property thus results by analysis from the concept of *alienated labour*—i.e., of *alienated man,* of estranged labour, of estranged life, of *estranged* man.

True, it is a result of the *movement of private property* that we have obtained the concept of *alienated labour (of alienated life)* from political economy. But on analysis of this concept it becomes clear that though private property appears to be the source, the cause of alienated labour, it is really its consequence, just as the gods *in the beginning* are not the cause but the effect of man's intellectual confusion. Later this relationship becomes reciprocal.

Only at the very culmination of the development of private property does this, its secret, re-emerge, namely, that on the one hand it is the *product* of alienated labour, and that secondly it is the *means* by which labour alienates itself, the *realization of this alienation.*

This exposition immediately sheds light on various hitherto unsolved conflicts.

(1) Political economy starts from labour as the real soul of production; yet to labour it gives nothing, and to private property everything. From this contradiction Proudhon has concluded in favour of labour and against private property. We understand, however, that this apparent contradiction is the contradiction of *estranged labour* with itself, and that political economy has merely formulated the laws of estranged labour.

We also understand, therefore, that *wages* and *private property* are identical: where the product, the object of labour pays for labour itself, the wage is but a necessary consequence of labour's estrangement, for after all in the wage of labour, labour does not appear as an end in itself but as the servant of the wage. We shall develop this point later, and meanwhile will only deduce some conclusions.

A *forcing-up of wages* (disregarding all other difficulties, including the fact that it would only be by force, too, that the higher wages, being an anomaly, could be maintained) would therefore be nothing but *better payment for the slave,* and would not conquer either for the worker or for labour their human status and dignity.

Indeed, even the *equality of wages* demanded by Proudhon only transforms the relationship of the present-day worker to his labour into the relationship of all men to labour. Society is then conceived as an abstract capitalist.

Wages are a direct consequence of estranged labour, and estranged labour is the direct cause of private property. The downfall of the one aspect must therefore mean the downfall of the other.

(2) From the relationship of estranged labour to private property it further follows that the emancipation of society from private property, etc., from servitude, is expressed in the *political* form of the *emancipation of the workers;* not that *their* emancipation alone was at stake but because the emancipation of the workers contains universal human emancipation—and it contains this, because the whole of human servitude is involved in the relation of the worker to production, and every relation of servitude is but a modification and consequence of this relation. . . .

THE POWER OF MONEY IN BOURGEOIS SOCIETY

If man's *feelings,* passions, etc., are not merely anthropological phenomena in the [narrower] sense, but truly *ontological* affirmations of essential being (of nature), and if they are only really affirmed because their *object* exists for them as an object of *sense,* then it is clear:

(1) That they have by no means merely one mode of affirmation, but rather that the distinctive character of their existence, of their life, is constituted by the distinctive mode of their affirmation. In what manner the object exists for them, is the characteristic mode of their *gratification.*

(2) Whenever the sensuous affirmation is the direct annulment of the object in its independent form (as in eating, drinking, working up of the object, etc.), this is the affirmation of the object.

(3) In so far as man, and hence also his feeling, etc., are *human,* the affirmation of the object by another is likewise his own enjoyment.

(4) Only through developed industry—i.e., through the medium of private property—does the ontological essence of human passion come to be both in its totality and in its humanity; the science of man is therefore itself a product of man's establishment of himself by practical activity.

(5) The meaning of private property—liberated from its estrangement—is the *existence of essential objects* for man, both as objects of enjoyment and as objects of activity.

By possessing the *property* of buying everything, by possessing the property of appropriating all objects, *money* is thus the *object* of eminent possession. The universality of its *property* is the omnipotence of its being. It therefore functions as the almighty being. Money is the *pimp* between man's need and the object, between his life and his means of life. But that which mediates *my* life for me, also

mediates the existence of other people *for me.*
For me it is the *other* person.

"What, man! confound it, hands and feet
And head and backside, all are yours!
And what we take while life is sweet,
Is that to be declared not ours?
Six stallions, say, I can afford.
Is not their strength my property?
I tear along, a sporting lord,
As if their legs belonged to me."

(Mephistopheles, in *Faust*)[i]

Shakespeare in *Timon of Athens:*

"Gold? Yellow, glittering, precious gold?
No, Gods, I am no idle votarist! . . .
Thus much of this will make black white,
    foul fair,
Wrong right, base noble, old young, coward
    valiant.
. . . Why, this
Will lug your priests and servants from your
    sides,
Pluck stout men's pillows from below their
    heads:
This yellow *slave*
Will knit and break religions, bless the accursed;
Make the hoar leprosy adored, place thieves
And give them title, knee and approbation
With senators on the bench: This is it
That makes the wappen'd widow wed again;
She, whom the spital-house and ulcerous sores
Would cast the gorge at, this embalms and
    spices
To the April day again. . . . Damned earth,
Thou common whore of mankind, that putt'st
    odds
Among the rout of nations."[ii]

And also later:

"O thou sweet king-killer, and dear divorce
Twixt natural son and sire! thou bright defiler

Of Hymen's purest bed! thou valiant Mars!
Thou ever young, fresh, loved and delicate
    wooer,
Whose blush doth thaw the consecrated snow
That lies on Dian's lap! Thou *visible God!*
That solder'st *close impossibilities,*
And mak'st them kiss! That speak'st with
    every tongue,
To every purpose! O thou touch of hearts!
Think thy slave man rebels, and by thy virtue
Set them into confounding odds, that beasts
May have the world in empire!"[iii]

Shakespeare excellently depicts the real nature of *money.* To understand him, let us begin, first of all, by expounding the passage from Goethe.

That which is for me through the medium of *money*—that for which I can pay (i.e., which money can buy)—that am I, the possessor of the money. The extent of the power of money is the extent of my power. Money's properties are my properties and essential powers—the properties and powers of its possessor. Thus, what I *am* and *am capable* of is by no means determined by my individuality. I am ugly, but I can buy for myself the most *beautiful* of women. Therefore I am not *ugly,* for the effect of *ugliness*—its deterrent power—is nullified by money. I, in my character as an individual, am *lame,* but money furnishes me with twenty-four feet. Therefore I am not lame. I am bad, dishonest, unscrupulous, stupid; but money is honoured, and therefore so is its possessor. Money is the supreme good, therefore its possessor is good. Money, besides, saves me the trouble of being dishonest: I am therefore presumed honest. I am *stupid,* but money is the *real mind* of all things and how then should its possessor be stupid? Besides, he can buy talented people for himself, and is he who has power over the talented not more talented than the talented? Do not I, who thanks to money am capable of *all* that the human heart longs for, possess all human capacities? Does not my money therefore transform all my incapacities into their contrary?

[i]Goethe, *Faust,* (Part I–Faust's Study, III), translated by Philip Wayne (Penguin, 1949), p. 91.

[ii]Shakespeare, *Timon of Athens,* Act 4, Scene 3. Marx quotes the Schlegel-Tieck German translation. (Marx's emphasis.)

[iii]Ibid.

If *money* is the bond binding me to *human* life, binding society to me, binding me and nature and man, is not money the bond of all *bonds?* Can it not dissolve and bind all ties? Is it not, therefore, the universal *agent of divorce?* It is the true *agent of divorce* as well as the true *binding agent*—the [universal][iv] *galvano-chemical* power of Society.

Shakespeare stresses especially two properties of money:

(1) It is the visible divinity—the transformation of all human and natural properties into their contraries, the universal confounding and overturning of things: it makes brothers of impossibilities.

(2) It is the common whore, the common pimp of people and nations.

The overturning and confounding of all human and natural qualities, the fraternization of impossibilities—the *divine* power of money—lies in its *character* as men's estranged, alienating and self-disposing *species-nature*. Money is the alienated *ability of mankind.*

That which I am unable to do as a *man,* and of which therefore all my individual essential powers are incapable, I am able to do by means of *money.* Money thus turns each of these powers into something which in itself it is not—turns it, that is, into its *contrary.*

If I long for a particular dish or want to take the mail-coach because I am not strong enough to go by foot, money fetches me the dish and the mail-coach: that is, it converts my wishes from something in the realm of imagination, translates them from their meditated, imagined or willed existence into their *sensuous, actual* existence—from imagination to life, from imagined being into real being. In effecting this mediation, money is the *truly creative* power.

No doubt *demand* also exists for him who has no money, but his demand is a mere thing of the imagination without effect or existence for me, for a third party, for the others, and which therefore remains for me *unreal* and *objectless.* The difference between effective demand based on money and ineffective demand based on my need, my passion, my wish, etc., is the difference

between being and *thinking,* between the imagined which *exists* merely within me and the imagined as it is for me outside me as a *real object.*

If I have no money for travel, I have no *need*—that is, no real and self-realizing need—to travel. If I have the *vocation* for study but *no* money for it, I have no vocation for study—that is, no *effective,* no *true* vocation. On the other hand, if I have really *no* vocation for study but have the will *and* the money for it, I have an *effective* vocation for it. Being the external, common *medium* and *faculty* for turning an *image* into *reality* and *reality* into a mere *image* (a faculty not springing from man as man or from human society as society), *money* transforms the *real essential powers of man and nature* into what are merely abstract conceits and therefore *imperfections*—into tormenting chimeras—just as it transforms *real imperfections and chimeras*—essential powers which are really impotent, which exist only in the imagination of the individual—into *real powers* and *faculties.*

In the light of this characteristic alone, money is thus the general overturning of *individualities* which turns them into their contrary and adds contradictory attributes to their attributes.

Money, then, appears as this *overturning* power both against the individual and against the bonds of society, etc., which claim to be *essences* in themselves. It transforms fidelity into infidelity, love into hate, hate into love, virtue into vice, vice into virtue, servant into master, master into servant, idiocy into intelligence and intelligence into idiocy.

Since money, as the existing and active concept of value, confounds and exchanges all things, it is the general *confounding* and *compounding* of all things—the world upside-down—the confounding and compounding of all natural and human qualities.

He who can buy bravery is brave, though a coward. As money is not exchanged for any one specific quality, for any one specific thing, or for any particular human essential power, but for the entire objective world of man and nature, from the standpoint of its possessor it therefore serves to exchange every property for every

---

[iv]An end of the page is torn out of the manuscript [Trans.].

other, even contradictory, property and object: it is the fraternization of impossibilities. It makes contradictions embrace.

Assume *man* to be *man* and his relationship to the world to be a human one: then you can exchange love only for love, trust for trust, etc. If you want to enjoy art, you must be an artistically cultivated person; if you want to exercise influence over other people, you must be a person with a stimulating and encouraging effect on other people. Every one of your relations to man and to nature must be a *specific expression,* corresponding to the object of your will, of your *real individual* life. If you love without evoking love in return—that is, if your loving as loving does not produce reciprocal love; if through a *living expression* of yourself as a loving person you do not make yourself a *loved person,* then your love is impotent—a misfortune.

Introduction to *The Communist Manifesto*

In 1847, the Communist League, an association formed by radical workers in 1836, commissioned Marx and Engels to write a political tract outlining the organization's program. The result was the now-famous *Communist Manifesto* (also called *The Manifesto of the Communist Party*), which you will read below. In contrast to other readings in this volume, the *Manifesto* is a deliberately adversarial work intended to inspire allegiance to the movement's cause. Though it had only modest impact at the time of its publication in 1848, shortly afterward workers and peasants staged revolts throughout much of Europe including France, Germany, and Italy.

Notwithstanding its origins as a political tract, *The Communist Manifesto* is of great theoretical significance. In it, you will again encounter Marx's theory of historical materialism and his inversion of Hegel's idealism. You will also see Marx's commitment to the Enlightenment belief in the perfectibility of humanity, which in his view will be realized through an inevitable communist revolution. The *Manifesto* also describes the economic processes that led to the ascendancy of the capitalist class and that eventually will produce to its own "grave-diggers"—a class-conscious proletariat.

Indeed, much of the *Manifesto* is a "scientific prophecy" detailing the downfall of the capitalist class and the rise of the proletariat. As such, it represents a penetrating theory of social change. The eventual collapse of capitalism will occur much in the way as previous economic systems: the social **relations of production** (how productive activity is organized and the laws governing property ownership) will become a "fetter" or obstacle to the continued development of the means of production (i.e., machinery, technology). The result is an "epidemic of overproduction" (Marx and Engels 1848/1978:478) in which the bourgeoisie "chokes" on the overabundance of goods produced by ever-increasing industrial efficiency. The final crisis of capitalism is thus a necessary consequence of the technological progress that was itself spurred by the capitalist class's private ownership of the means of production and the goods produced.

As an example of this process, consider the recent debates on music file sharing over the Internet. Though by no means spelling the doom of capitalism, the controversy nevertheless highlights the contradictions that arise between the forces and relations of production. Technology—the forces of production—advances more quickly than changes in the laws governing the relations of production—that is, ownership of property. Computer and communication technologies have been developed that enable a virtually infinite number of users to simultaneously share data stored on their hard drives. To avail themselves of this capability, users must first connect to a central terminal that serves as a temporary holding station. Napster was such a central terminal for sharing music files. The company itself did not own or control the files; it simply provided a conduit for the individuals who did. However, because the company provided a singular, "tangible" site for the free exchange of music, lawyers for the record companies were able to successfully argue that *Napster*—and not the individual users—circumvented copyright regulations despite the fact that it did not "steal" the files for its own use.

This advancement in computer technology undermines current laws governing copyright ownership and the rights that accompany proprietorship. Internet users can download and thus possess music with unparalleled ease and speed without having to pay for it, making it impossible for the owners of the copyrights to fully control the distribution of their property. (While cassette tapes and albums have been copied for decades, the convenience with which such "pirated" versions are made and the scope of their distribution pales in comparison to that of file sharing on the Internet.) This, of course, runs completely counter to a legal cornerstone of capitalism, namely, that owners must be compensated for the use of their "private" property. However, the social relations of production—the laws of ownership—do not prevent *individuals* from sharing the products that they have purchased and thus rightfully own, and someone at some point purchased the music that now is stored on their hard drive. However, to the extent that current laws are enforced or rewritten in an effort to combat the infringement of property rights, the social relations of production become a "fetter" to the full development of advances in technology.

Returning to our theoretical discussion of the dynamics of capitalism, capitalists must forever seek to eliminate their competitors, create new markets, destroy some of their products, or cut back their productive capacity in order to minimize the oversupply of goods that results from increasingly sophisticated means of production. If production is reduced, however, capitalists, in turn, will be forced to reduce their work force and, with it, their source of profit as well as the size of the market able to purchase their goods. Yet, the bourgeoisie is confronted not only with these economic realities of capitalism, but also with political consequences, as competition creates an obstacle to class unity and to the ability to implement coherent economic policies that will ensure its dominance. And so the cycle continues.

Meanwhile, factory conditions themselves facilitate the development of a revolutionary class consciousness through which workers come to realize the true source of their alienation and the possibility of breaking free from the chains of their enslavement. Placed side by side in their performance of tedious, monotonous tasks, the physical settings of factories increase the contact between the workers, making it easier to communicate and spread allegiance to the proletariat's cause. Urging "WORKING MEN OF ALL COUNTRIES, UNITE!" Marx warns that the Communists

> openly declare that their ends can be attained only by the forcible overthrow of all existing social conditions. Let the ruling class tremble at a Communistic revolution. The proletarians have nothing to lose but their chains. They have a world to win. (Marx and Engels 1848/1978:500)

Yet, the question remains: Why would the establishment of a communist economy create a more humane society? At the risk of oversimplifying the matter, the communist utopia hinges on the abolition of private property. Marx maintains that once the means of production becomes collectively owned, exploitation of the worker is no longer possible. This is because the surplus value (i.e., profit) produced by the worker is not appropriated or siphoned off by an individual owner. Instead, it is distributed among the workers themselves. Alienation is also ended because the worker, now a part owner of the enterprise, is able to direct the production process and maintain control over the products she creates. In turn, the worker is no longer estranged from herself and the species being. Finally, the competition for profit that characterizes bourgeois capitalism is brought to a close and, with it, recurring economic crises. Periods of "boom or bust" and their accompanying disruptions to employment are replaced by a more stable form of economic planning that produces according to the needs of the population and not the whims of an unpredictable market. "In place of the old bourgeois society, with its classes and class antagonisms, we shall have an association, in which the free development of each is the condition for the free development of all" (Marx and Engels 1848/1978:491).

The economic crisis currently unfolding is a textbook example of the continuing relevance and prescience of Marx's ideas. Today's crisis is America's worst since the Great Depression. A record high of nearly 14 million workers are unemployed; those still employed are left with no choice but to accept steep cuts in pay and benefits. The United States, however, is by no means alone in experiencing the dramatic downturn. Just as Marx predicted, the spread of capitalism has ensured that the ever-worsening economic conditions cannot be confined to any one country's borders, but necessarily must reach across the entire globe. With sales of commodities plummeting worldwide, capitalists are beginning to "choke" on their supplies as warehouses are filling to capacity with unshipped goods. To compensate, stores are slashing prices—and losing profits—in order to sell their products. But what of the workers, the proletariat? Millions of people are looking for jobs, struggling to meet their basic needs (as are many millions of the employed). Nevertheless, production across all sectors of the economy is slowing to a virtual halt, but not because the machines are broken or somehow malfunctioning, or because there are not enough skilled laborers available to carry out the required tasks. Production has been stopped artificially by capitalists, and they must do so in order to prevent glutting the market with their goods while preserving whatever profits they are still able to earn. The relations of production—private ownership and its accompanying drive for private profit—have become a fetter to the forces of production, despite the fact that millions are living in increasingly desperate conditions.

While the causes of the current crisis are complex, many analysts have pointed to the dominant role played by the bundling of individual home loans into mortgage-backed securities that were then sold to investors. When the housing bubble that made investment in these financial instruments profitable burst, banks and investment companies around the world were left holding assets with rapidly declining values. However, the very corporations who invented and sold this new form of security are unable to root out the problems caused by these "troubled assets" because the originally bundled securities have been rebundled and traded so frequently that it has become impossible to determine the value of the securities as well as who actually owns a specific asset. Capitalists, like a "sorcerer who is no longer able to control the powers of the nether world whom he has called upon by his spells" (Marx and Engels 1848/1978:478), created a financial instrument that they are incapable of controlling and that has metastasized to the point where it threatens the stability of the global capitalist economy.

To stem the tide of the fallout, governments are intensifying their intervention in their respective economies. In the United States, intervention to this point has taken the form of giving billions of taxpayer dollars to the very financial institutions that are largely responsible for creating the crisis with little oversight or accountability for how the funds will be used. And should the government decide to use public funds to purchase the troubled assets from the banks and investment companies, it will be impossible for taxpayers to know whether or not they are paying a fair price for them, because the value of the assets cannot be determined. At the same time, the government has provided comparatively little funds for the increasingly distressed auto industry—one of the few remaining manufacturing industries in the country—prompting some observers to claim that the government is concerned only with the well-being of Wall Street and not Main Street. In rescuing the "moneyed interests" while letting drown those blue-collar workers who *make* things, a ring of truth is sounded in Marx's assertion, "The executive of the modern State is but a committee for managing the common affairs of the whole bourgeoisie" (Marx and Engels 1848/1978:475). Yet, to avoid a complete economic collapse, the capitalists and the state have no choice but to appeal to the public—the proletariat—"to ask for its help, and thus drag it into the political arena" (ibid.:481), in turn supplying it with a political and intellectual education that will later be used as a weapon against them.

From *The Communist Manifesto* (1848)

Karl Marx and Friedrich Engels

A spectre is haunting Europe—the spectre of communism. All the powers of old Europe have entered into a holy alliance to exorcise this spectre: Pope and Tsar, Metternich and Guizot, French Radicals and German police-spies.

Where is the party in opposition that has not been decried as communistic by its opponents in power? Where the opposition that has not hurled back the branding reproach of Communism, against the more advanced opposition parties, as well as against its reactionary adversaries?

Two things result from this fact.

I. Communism is already acknowledged by all European powers to be itself a power.

II. It is high time that Communists should openly, in the face of the whole world, publish their views, their aims, their tendencies, and meet this nursery tale of the spectre of communism with a manifesto of the party itself.

To this end, Communists of various nationalities have assembled in London, and sketched the following manifesto, to be published in the English, French, German, Italian, Flemish and Danish languages.

BOURGEOIS AND PROLETARIANS[i]

The history of all hitherto existing society[ii] is the history of class struggles.

Freeman and slave, patrician and plebeian, lord and serf, guild-master[iii] and journeyman, in a word, oppressor and oppressed, stood in constant opposition to one another, carried on an uninterrupted, now hidden, now open fight, a fight that each time ended, either in a revolutionary re-constitution of society at large, or in the common ruin of the contending classes.

In the earlier epochs of history, we find almost everywhere a complicated arrangement of society into various orders, a manifold gradation of social rank. In ancient Rome we have patricians, knights, plebeians, slaves; in the Middle Ages, feudal lords, vassals, guild-masters, journeymen, apprentices, serfs; in almost all of these classes, again, subordinate gradations.

The modern bourgeois society that has sprouted from the ruins of feudal society has not done away with class antagonisms. It has but established new classes, new conditions of oppression, new forms of struggle in place of the old ones.

SOURCE: Marx/Engels Internet Archive.

[i]By bourgeoisie is meant the class of modern Capitalists, owners of the means of social production and employers of wage-labour. By proletariat, the class of modern wage-labourers who, having no means of production of their own, are reduced to selling their labour-power in order to live. [*Engels, English edition of 1888*]

[ii]That is, all *written* history. In 1847, the pre-history of society, the social organisation existing previous to recorded history, was all but unknown. Since then, Haxthausen discovered common ownership of land in Russia, Maurer proved it to be the social foundation from which all Teutonic races started in history, and by and by village communities were found to be, or to have been the primitive form of society everywhere from India to Ireland. The inner organisation of this primitive Communistic society was laid bare, in its typical form, by Morgan's crowning discovery of the true nature of the *gens* and its relation to the *tribe*. With the dissolution of these primaeval communities society begins to be differentiated into separate and finally antagonistic classes. I have attempted to retrace this process of dissolution in: "Der Ursprung der Familie, des Privateigenthums und des Staats" [*The Origin of the Family, Private Property and the State*], 2nd edition, Stuttgart 1886. [*Engels, English edition of 1888*]

[iii]Guild-master, that is, a full member of a guild, a master within, not a head of a guild. [*Engels, English edition of 1888*]

Our epoch, the epoch of the bourgeoisie, possesses, however, this distinct feature: it has simplified class antagonisms: Society as a whole is more and more splitting up into two great hostile camps, into two great classes directly facing each other: bourgeoisie and proletariat.

From the serfs of the Middle Ages sprang the chartered burghers of the earliest towns. From these burgesses the first elements of the bourgeoisie were developed.

The discovery of America, the rounding of the Cape, opened up fresh ground for the rising bourgeoisie. The East-Indian and Chinese markets, the colonisation of America, trade with the colonies, the increase in the means of exchange and in commodities generally, gave to commerce, to navigation, to industry, an impulse never before known, and thereby, to the revolutionary element in the tottering feudal society, a rapid development.

The feudal system of industry, in which industrial production was monopolized by closed guilds, now no longer suffices for the growing wants of the new markets. The manufacturing system took its place. The guild-masters were pushed aside by the manufacturing middle class; division of labor between the different corporate guilds vanished in the face of division of labor in each single workshop.

Meantime the markets kept ever growing, the demand ever rising. Even manufacturers no longer sufficed. Thereupon, steam and machinery revolutionized industrial production. The place of manufacture was taken by the giant, Modern Industry, the place of the industrial middle class, by industrial millionaires, the leaders of the whole industrial armies, the modern bourgeois.

Modern industry has established the world-market, for which the discovery of America paved the way. This market has given an immense development to commere, to navigation, to communication by land. This development has, in turn, reacted on the extension of industry; and in proportion as industry, commerce, navigation, railways extended, in the same proportion the bourgeoisie developed, increased its capital, and pushed into the background every class handed down from the Middle Ages.

We see, therefore, how the modern bourgeoisie is itself the product of a long course of development, of a series of revolutions in the modes of production and of exchange.

Each step in the development of the bourgeoisie was accompanied by a corresponding political advance in that class. An oppressed class under the sway of the feudal nobility, an armed and self-governing association in the medieval commune;[iv] here independent urban republic (as in Italy and Germany), there taxable "third estate" of the monarchy (as in France), afterward, in the period of manufacturing proper, serving either the semi-feudal or the absolute monarchy as a counterpoise against the nobility, and, in fact, cornerstone of the great monarchies in general, the bourgeoisie has at last, since the establishment of Modern Industry and of the world-market, conquered for itself, in the modern representative state, exclusive political sway. The executive of the modern state is but a committee for managing the common affairs of the whole bourgeoisie.

The bourgeoisie, historically, has played a most revolutionary part.

The bourgeoisie, wherever it has got the upper hand, has put an end to all feudal, patriarchal, idyllic relations. It has pitilessly torn asunder the motley feudal ties that bound man to his "natural superiors," and has left no other nexus between man and man than naked self-interest, than callous "cash payment." It has drowned out the most heavenly ecstasies of religious fervour, of chivalrous enthusiasm, of philistine sentimentalism, in the icy water of egotistical calculation. It has resolved personal worth into exchange value, and in place of the numberless indefeasible chartered freedoms, has set up that single, unconscionable freedom—Free Trade. In one word, for exploitation, veiled by religious and political illusions, it has substituted naked, shameless, direct, brutal exploitation.

[iv] "Commune" was the name taken, in France, by the nascent towns even before they had conquered from their feudal lords and masters local self-government and political rights as the "Third Estate." Generally speaking, for the economical development of the bourgeoisie, England is here taken as the typical country; for its political development, France. [*Engels, English edition of 1888*] This was the name given their urban communities by the townsmen of Italy and France, after they had purchased or wrested their initial rights of self-government from their feudal lords. [*Engels, German edition of 1890*]

The bourgeoisie has stripped of its halo every occupation hitherto honored and looked up to with reverent awe. It has converted the physician, the lawyer, the priest, the poet, the man of science, into its paid wage-laborers.

The bourgeoisie has torn away from the family its sentimental veil, and has reduced the family relation into a mere money relation.

The bourgeoisie has disclosed how it came to pass that the brutal display of vigour in the Middle Ages, which reactionaries so much admire, found its fitting complement in the most slothful indolence. It has been the first to show what man's activity can bring about. It has accomplished wonders far surpassing Egyptian pyramids, Roman aqueducts, and Gothic cathedrals; it has conducted expeditions that put in the shade all former exoduses of nations and crusades.

The bourgeoisie cannot exist without constantly revolutionizing the instruments of production, and thereby the relations of production, and with them the whole relations of society. Conservation of the old modes of production in unaltered form, was, on the contrary, the first condition of existence for all earlier industrial classes. Constant revolutionizing of production, uninterrupted disturbance of all social conditions, everlasting uncertainty and agitation distinguish the bourgeois epoch from all earlier ones. All fixed, fast-frozen relations, with their train of ancient and venerable prejudices and opinions, are swept away, all new-formed ones become antiquated before they can ossify. All that is solid melts into air, all that is holy is profaned, and man is at last compelled to face with sober senses, his real condition of life, and his relations with his kind.

The need of a constantly expanding market for its products chases the bourgeoisie over the entire surface of the globe. It must nestle everywhere, settle everywhere, establish connections everywhere.

The bourgeoisie has through its exploitation of the world-market given a cosmopolitan character to production and consumption in every country. To the great chagrin of reactionaries, it has drawn from under the feet of industry the national ground on which it stood. All old-established national industries have been destroyed or are daily being destroyed. They are dislodged by new industries, whose introduction becomes a life and death question for all civilized nations, by industries that no longer work up indigenous raw material, but raw material drawn from the remotest zones; industries whose products are consumed, not only at home, but in every quarter of the globe. In place of the old wants, satisfied by the production of the country, we find new wants, requiring for their satisfaction the products of distant lands and climes. In place of the old local and national seclusion and self-sufficiency, we have intercourse in every direction, universal inter-dependence of nations. And as in material, so also in intellectual production. The intellectual creations of individual nations become common property. National one-sidedness and narrow-mindedness become more and more impossible, and from the numerous national and local literatures, there arises a world literature.

The bourgeoisie, by the rapid improvement of all instruments of production, by the immensely facilitated means of communication, draws all, even the most barbarian, nations into civilization. The cheap prices of commodities are the heavy artillery with which it batters down all Chinese walls, with which it forces the barbarians' intensely obstinate hatred of foreigners to capitulate. It compels all nations, on pain of extinction, to adopt the bourgeois mode of production; it compels them to introduce what it calls civilization into their midst, *i.e.,* to become bourgeois themselves. In one word, it creates a world after its own image.

The bourgeoisie has subjected the country to the rule of the towns. It has created enormous cities, has greatly increased the urban population as compared with the rural, and has thus rescued a considerable part of the population from the idiocy of rural life. Just as it has made the country dependent on the towns, so it has made barbarian and semi-barbarian countries dependent on the civilized ones, nations of peasants on nations of bourgeois, the East on the West.

The bourgeoisie keeps more and more doing away with the scattered state of the population, of the means of production, and of property. It has agglomerated population, centralized means of production, and has concentrated property in a few hands. The necessary consequence of this was political centralization. Independent, or but loosely connected provinces, with separate interests, laws, governments and systems of taxation, became lumped together into one nation, with one government, one code of laws, one national class-interest, one frontier and one customs-tariff.

The bourgeoisie, during its rule of scarce one hundred years, has created more massive and more colossal productive forces than have all preceding generations together. Subjection of nature's forces to man, machinery, application of chemistry to industry and agriculture, steam-navigation, railways, electric telegraphs, clearing of whole continents for cultivation, canalization of rivers, whole populations conjured out of the ground—what earlier century had even a presentiment that such productive forces slumbered in the lap of social labor?

We see then: the means of production and of exchange, on whose foundation the bourgeoisie built itself up, were generated in feudal society. At a certain stage in the development of these means of production and of exchange, the conditions under which feudal society produced and exchanged, the feudal organization of agriculture and manufacturing industry, in one word, the feudal relations of property became no longer compatible with the already developed productive forces; they became so many fetters. They had to be burst asunder; they were burst asunder.

Into their place stepped free competition, accompanied by a social and political constitution adapted in it, and the economic and political sway of the bourgeois class.

A similar movement is going on before our own eyes. Modern bourgeois society with its relations of production, of exchange and of property, a society that has conjured up such gigantic means of production and of exchange, is like the sorcerer, who is no longer able to control the powers of the nether world whom he has called up by his spells. For many a decade past the history of industry and commerce is but the history of the revolt of modern productive forces against modern conditions of production, against the property relations that are the conditions for the existence of the bourgeois and of its rule. It is enough to mention the commercial crises that, by their periodical return, put the existence of the entire bourgeois society on its trial, each time more threateningly. In these crises a great part not only of the existing products, but also of the previously created productive forces, are periodically destroyed. In these crises there breaks out an epidemic that, in all earlier epochs, would have seemed an absurdity—the epidemic of over-production. Society suddenly finds itself put back into a state of momentary barbarism; it appears as if a famine, a universal war of devastation had cut off the supply of every means of subsistence; industry and commerce seem to be destroyed. And why? Because there is too much civilization, too much means of subsistence, too much industry, too much commerce. The productive forces at the disposal of society no longer tend to further the development of the conditions of bourgeois property; on the contrary, they have become too powerful for these conditions, by which they are fettered, and so soon as they overcome these fetters, they bring disorder into the whole of bourgeois society, endanger the existence of bourgeois property. The conditions of bourgeois society are too narrow to comprise the wealth created by them. And how does the bourgeoisie get over these crises? On the one hand by enforced destruction of a mass of productive forces; on the other, by the conquest of new markets, and by the more thorough exploitation of the old ones. That is to say, by paving the way for more extensive and more destructive crises, and by diminishing the means whereby crises are prevented.

The weapons with which the bourgeoisie felled feudalism to the ground are now turned against the bourgeoisie itself.

But not only has the bourgeoisie forged the weapons that bring death to itself; it has also called into existence the men who are to wield those weapons—the modern working class—the proletarians.

In proportion as the bourgeoisie, *i.e.,* capital, is developed, in the same proportion is the proletariat, the modern working class, developed—a class of laborers, who live only so long as they find work, and who find work only so long as their labor increases capital. These laborers, who must sell themselves piece-meal, are a commodity, like every other article of commerce, and are consequently exposed to all the vicissitudes of competition, to all the fluctuations of the market.

Owing to the extensive use of machinery and to the division of labor, the work of the proletarians has lost all individual character, and consequently, all charm for the workman. He becomes an appendage of the machine, and it is only the most simple, most monotonous, and most easily acquired knack, that is required of him. Hence, the cost of production of a workman is restricted, almost entirely, to the means of subsistence that he requires for maintenance, and for the propagation of his race. But the price of a commodity, and therefore also of labor,[v] is equal to its cost of production. In proportion, therefore, as the repulsiveness of the work increases, the wage decreases. What is more, in proportion as the use of machinery and division of labor increases, in the same proportion the burden of toil also increases, whether by prolongation of the working hours, by the increase of the work exacted in a given time or by increased speed of machinery, etc.

Modern Industry has converted the little workshop of the patriarchal master into the great factory of the industrial capitalist. Masses of laborers, crowded into the factory, are organized like soldiers. As privates of the industrial army they are placed under the command of a perfect hierarchy of officers and sergeants. Not only are they slaves of the bourgeois class, and of the bourgeois state; they are daily and hourly enslaved by the machine, by the over-looker, and, above all, in the individual bourgeois manufacturer himself. The more openly this despotism proclaims gain to be its end and aim, the more petty, the more hateful and the more embittering it is.

The less the skill and exertion of strength implied in manual labor, in other words, the more modern industry becomes developed, the more is the labor of men superseded by that of women. Differences of age and sex have no longer any distinctive social validity for the working class. All are instruments of labor, more or less expensive to use, according to their age and sex.

No sooner is the exploitation of the laborer by the manufacturer, so far, at an end, that he receives his wages in cash, than he is set upon by the other portion of the bourgeoisie, the landlord, the shopkeeper, the pawnbroker, etc.

The lower strata of the middle class—the small tradespeople, shopkeepers, and retired tradesmen generally, the handicraftsmen and peasants—all these sink gradually into the proletariat, partly because their diminutive capital does not suffice for the scale on which Modern Industry is carried on, and is swamped in the competition with the large capitalists, partly because their specialized skill is rendered worthless by new methods of production. Thus the proletariat is recruited from all classes of the population.

The proletariat goes through various stages of development. With its birth begins its struggle with the bourgeoisie. At first the contest is carried on by individual laborers, then by the work of people of a factory, then by the operative of one trade, in one locality, against the individual bourgeois who directly exploits them. They direct their attacks not against the bourgeois condition of production, but against the instruments of production themselves; they destroy imported wares that compete with their labor, they smash to pieces machinery, they set factories ablaze, they seek to restore by force the vanished status of the workman of the Middle Ages.

At this stage the laborers still form an incoherent mass scattered over the whole country, and broken up by their mutual competition. If anywhere they unite to form more compact bodies, this is not yet the consequence of their own active union, but of the union of the bourgeoisie, which class, in order to attain its own political ends, is compelled to set the whole proletariat in motion, and is moreover yet, for a time, able to do so. At this stage, therefore, the proletarians do not fight their enemies, but the enemies of their enemies, the remnants of absolute monarchy, the landowners, the non-industrial bourgeois, the petty bourgeois. Thus the whole historical movement is concentrated in the hands of the bourgeoisie; every victory so obtained is a victory for the bourgeoisie.

But with the development of industry the proletariat not only increases in number; it becomes concentrated in greater masses, its strength grows, and it feels that strength more. The various interests and conditions of life within the ranks of the proletariat are more and more equalized, in proportion as machinery obliterates all

---

[v]Subsequently Marx pointed out that the worker sells not his labor but his labor power.

distinctions of labor, and nearly everywhere reduces wages to the same low level. The growing competition among the bourgeois, and the resulting commercial crises, make the wages of the workers ever more fluctuating. The increasing improvement of machinery, ever more rapidly developing, makes their livelihood more and more precarious; the collisions between individual workmen and individual bourgeois take more and more the character of collisions between two classes. Thereupon the workers begin to form combinations (trade unions) against the bourgeois; they club together in order to keep up the rate of wages; they found permanent associations in order to make provision beforehand for these occasional revolts. Here and there the contest breaks out into riots.

Now and then the workers are victorious, but only for a time. The real fruit of their battles lie, not in the immediate result, but in the ever-expanding union of the workers. This union is helped on by the improved means of communication that are created by Modern Industry and that place the workers of different localities in contact with one another. It was just this contact that was needed to centralize the numerous local struggles, all of the same character, into one national struggle between classes. But every class struggle is a political struggle. And that union, to attain which the burghers of the Middle Ages, with their miserable highways, required centuries, the modern proletarians, thanks to railways, achieve in a few years.

This organization of the proletarians into a class, and consequently into a political party, is continually being upset again by the competition between the workers themselves. But it ever rises up again, stronger, firmer, mightier. It compels legislative recognition of particular interests of the workers, by taking advantage of the divisions among the bourgeoisie itself. Thus the Ten-Hours Bill in England was carried.

Altogether collisions between the classes of the old society further, in many ways, the course of development of the proletariat. The bourgeoisie finds itself involved in a constant battle. At first with the aristocracy; later on, with those portions of the bourgeoisie itself, whose interests have become antagonistic to the progress of industry; at all time, with the bourgeoisie of foreign countries. In all these battles it sees itself compelled to appeal to the proletariat, to ask for its help, and thus, to drag it into the political arena. The bourgeoisie itself, therefore, supplies the proletariat with its own elements of political and general education, in other words, it furnishes the proletariat with weapons for fighting the bourgeoisie.

Further, as we have already seen, entire sections of the ruling class are, by the advance of industry, precipitated into the proletariat, or are at least threatened in their conditions of existence. These also supply the proletariat with fresh elements of enlightenment and progress.

Finally, in times when the class struggle nears the decisive hour, the progress of dissolution going on within the ruling class, in fact within the whole range of old society, assumes such a violent, glaring character, that a small section of the ruling class cuts itself adrift, and joins the revolutionary class, the class that holds the future in its hands. Just as, therefore, at an earlier period, a section of the nobility went over to the bourgeoisie, so now a portion of the bourgeoisie goes over to the proletariat, and in particular, a portion of the bourgeois ideologists, who have raised themselves to the level of comprehending theoretically the historical movement as a whole.

Of all the classes that stand face to face with the bourgeoisie today, the proletariat alone is a genuinely revolutionary class. The other classes decay and finally disappear in the face of Modern Industry; the proletariat is its special and essential product.

The lower middle class, the small manufacturer, the shopkeeper, the artisan, the peasant, all these fight against the bourgeoisie, to save from extinction their existence as fractions of the middle class. They are therefore not revolutionary, but conservative. Nay more, they are reactionary, for they try to roll back the wheel of history. If by chance they are revolutionary, they are so only in view of their impending transfer into the proletariat, they thus defend not their present, but their future interests, they desert their own standpoint to place themselves at that of the proletariat.

The "dangerous class," the social scum, that passively rotting mass thrown off by the lowest layers of the old society, may, here and there, be swept into the movement by a proletarian revolution; its conditions of life, however, prepare it far more for the part of a bribed tool of reactionary intrigue.

In the condition of the proletariat, those of old society at large are already virtually swamped. The proletarian is without property; his relation to his wife and children has no longer anything in common with the bourgeois family-relations; modern industry labor, modern subjection to capital, the same in England as in France, in America as in Germany, has stripped him of every trace of national character. Law, morality, religion, are to him so many bourgeois prejudices, behind which lurk in ambush just as many bourgeois interests.

All the preceding classes that got the upper hand, sought to fortify their already acquired status by subjecting society at large to their conditions of appropriation. The proletarians cannot become masters of the productive forces of society, except by abolishing their own previous mode of appropriation, and thereby also every other previous mode of appropriation. They have nothing of their own to secure and to fortify; their mission is to destroy all previous securities for, and insurances of, individual property.

All previous historical movements were movements of minorities, or in the interest of minorities. The proletarian movement is the self-conscious, independent movement of the immense majority, in the interest of the immense majority. The proletariat, the lowest stratum of our present society, cannot stir, cannot raise itself up, without the whole superincumbent strata of official society being sprung into the air.

Though not in substance, yet in form, the struggle of the proletariat with the bourgeoisie is at first a national struggle. The proletariat of each country must, of course, first of all settle matters with its own bourgeoisie.

In depicting the most general phases of the development of the proletariat, we traced the more or less veiled civil war, raging within existing society, up to the point where that war breaks out into open revolution, and where the violent overthrow of the bourgeoisie lays the foundation for the sway of the proletariat.

Hitherto, every form of society has been based, as we have already seen, on the antagonism of oppressing and oppressed classes. But in order to oppress a class, certain conditions must be assured to it under which it can, at least, continue its slavish existence. The serf, in the period of serfdom, raised himself to membership in the commune, just as the petty bourgeois, under the yoke of the feudal absolutism, managed to develop into a bourgeois. The modern laborer, on the contrary, instead of rising with the process of industry, sinks deeper and deeper below the conditions of existence of his own class. He becomes a pauper, and pauperism develops more rapidly than population and wealth. And here it becomes evident, that the bourgeoisie is unfit any longer to be the ruling class in society, and to impose its conditions of existence upon society as an over-riding law. It is unfit to rule because it is incompetent to assure an existence to its slave within his slavery, because it cannot help letting him sink into such a state, that it has to feed him, instead of being fed by him. Society can no longer live under this bourgeoisie, in other words, its existence is no longer compatible with society.

The essential conditions for the existence, and for the sway of the bourgeois class, is the formation and augmentation of capital; the condition for capital is wage-labor. Wage-labor rests exclusively on competition between the labourers. The advance of industry, whose involuntary promoter is the bourgeoisie, replaces the isolation of the laborers, due to competition, by the revolutionary combination, due to association. The development of Modern Industry, therefore, cuts from under its feet the very foundation on which the bourgeoisie produces and appropriates products. What the bourgeoisie, therefore, produces, above all, is its own grave-diggers. Its fall and the victory of the proletariat are equally inevitable.

PROLETARIANS AND COMMUNISTS

In what relation do the Communists stand to the proletarians as a whole?

The Communists do not form a separate party opposed to other working-class parties.

They have no interests separate and apart from those of the proletariat as a whole.

They do not set up any sectarian principles of their own, by which to shape and mold the proletarian movement.

The Communists are distinguished from the other working-class parties by this only: (1) In the national struggles of the proletarians of the different countries, they point out and bring to the front the common interests of the entire proletariat, independently of all nationality. (2) In the various stages of development which the struggle of the working class against the bourgeoisie has to pass through, they always and everywhere represent the interests of the movement as a whole.

The Communists, therefore, are on the one hand, practically, the most advanced and resolute section of the working-class parties of every country, that section which pushes forward all others; on the other hand, theoretically, they have over the great mass of the proletariat the advantage of clearly understanding the lines of march, the conditions, and the ultimate general results of the proletarian movement.

The immediate aim of the Communists is the same as that of all other proletarian parties: Formation of the proletariat into a class, overthrow of the bourgeois supremacy, conquest of political power by the proletariat.

The theoretical conclusions of the Communists are in no way based on ideas or principles that have been invented, or discovered, by this or that would-be universal reformer.

They merely express, in general terms, actual relations springing from an existing class struggle, from a historical movement going on under our very eyes. The abolition of existing property relations is not at all a distinctive feature of communism.

All property relations in the past have continually been subject to historical change consequent upon the change in historical conditions.

The French Revolution, for example, abolished feudal property in favor of bourgeois property.

The distinguishing feature of communism is not the abolition of property, but the abolition of bourgeois property generally, but modern bourgeois private property is the final and most complete expression of the system of producing and appropriating products, that is based on class antagonisms, on the exploitation of the many by the few.

In this sense, the theory of the Communists may be summed up in the single sentence: Abolition of private property.

We Communists have been reproached with the desire of abolishing the right of personally acquiring property as the fruit of a man's own labor,

which property is alleged to be the groundwork of all personal freedom, activity and independence.

Hard-won, self-acquired, self-earned property! Do you mean the property of petty artisan and of the small peasant, a form of property that preceded the bourgeois form? There is no need to abolish that; the development of industry has to a great extent already destroyed it, and is still destroying it daily.

Or do you mean the modern bourgeois private property?

But does wage-labour create any property for the laborer? Not a bit. It creates capital, i.e., that kind of property which exploits wage-labor, and which cannot increase except upon conditions of begetting a new supply of wage-labor for fresh exploitation. Property, in its present form, is based on the antagonism of capital and wage-labor. Let us examine both sides of this antagonism.

To be a capitalist, is to have not only a purely personal, but a social *status* in production. Capital is a collective product, and only by the united action of many members, nay, in the last resort, only by the united action of all members of society, can it be set in motion.

Capital is, therefore, not only personal; it is a social power.

When, therefore, capital is converted into common property, into the property of all members of society, personal property is not thereby transformed into social property. It is only the social character of the property that is changed. It loses its class-character.

Let us now take wage-labor.

The average price of wage-labor is the minimum wage, i.e., that quantum of the means of subsistence, which is absolutely requisite to keep the laborer in bare existence as a laborer. What, therefore, the wage-laborer appropriates by means of his labour, merely suffices to prolong and reproduce a bare existence. We by no means intend to abolish this personal appropriation of the products of labor, an appropriation that is made for the maintenance and reproduction of human life, and that leaves no surplus wherewith to command the labour of others. All that we want to do away with, is the miserable character of this appropriation, under which the laborer lives merely to increase capital, and is allowed to live only in so far as the interest of the ruling class requires it.

In bourgeois society, living labor is but a means to increase accumulated labor. In communist society, accumulated labor is but a means to widen, to enrich, to promote the existence of the laborer.

In bourgeois society, therefore, the past dominates the present; in communist society, the present dominates the past. In bourgeois society capital is independent and has individuality, while the living person is dependent and has no individuality.

And the abolition of this state of things is called by the bourgeois, abolition of individuality and freedom! And rightly so. The abolition of bourgeois individuality, bourgeois independence, and bourgeois freedom is undoubtedly aimed at.

By freedom is meant, under the present bourgeois conditions of production, free trade, free selling and buying.

But if selling and buying disappears, free selling and buying disappears also. This talk about free selling and buying, and all the other "brave words" of our bourgeois about freedom in general, have a meaning, if any, only in contrast with restricted selling and buying, with the fettered traders of the Middle Ages, but have no meaning when opposed to the communistic abolition of buying and selling, or the bourgeois conditions of production, and of the bourgeoisie itself.

You are horrified at our intending to do away with private property. But in your existing society, private property is already done away with for nine-tenths of the population; its existence for the few is solely due to its non-existence in the hands of those nine-tenths. You reproach us, therefore, with intending to do away with a form of property, the necessary condition for whose existence is the non-existence of any property for the immense majority of society.

In one word, you reproach us with intending to do away with your property. Precisely so; that is just what we intend.

From the moment when labor can no longer be converted into capital, money, or rent, into a social power capable of being monopolized, *i.e.,* from the moment when individual property can no longer be transformed into bourgeois property, into capital, from that moment, you say, individuality vanishes.

You must, therefore, confess that by "individual" you mean no other person than the bourgeois, than the middle-class owner of property. This person must, indeed, be swept out of the way, and made impossible.

Communism deprives no man of the power to appropriate the products of society; all that it does is to deprive him of the power to subjugate the labor of others by means of such appropriation.

It has been objected that upon the abolition of private property all work will cease, and universal laziness will overtake us.

According to this, bourgeois society ought long ago to have gone to the dogs through sheer idleness; for those who acquire anything, do not work. The whole of this objection is but another expression of the tautology: There can no longer be any wage-labor when there is no longer any capital.

All objections urged against the communistic mode of producing and appropriating material products have, in the same way, been urged against the communistic modes of producing and appropriating intellectual products. Just as, to the bourgeois, the disappearance of class property is the disappearance of production itself, so the disappearance of class culture is to him identical with the disappearance of all culture.

That culture, the loss of which he laments, is, for the enormous majority, a mere training to act as a machine.

But don't wrangle with us so long as you apply, to our intended abolition of bourgeois property, the standard of your bourgeois notions of freedom, culture, law, etc. Your very ideas are but the outgrowth of the conditions of your bourgeois production and bourgeois property, just as your jurisprudence is but the will of your class made into a law for all, a will, whose essential character and direction are determined by the economical conditions of existence of your class.

The selfish misconception that induces you to transform into eternal laws of nature and of reason, the social forms springing from your present mode of production and form of property—historical relations that rise and disappear in the progress of production—this misconception you share with every ruling class that has preceded you. What you see clearly in the case of ancient property, what you admit in the case of feudal property, you are of course forbidden to admit in the case of your own bourgeois form of property.

Abolition of the family! Even the most radical flare up at this infamous proposal of the Communists.

On what foundation is the present family, the bourgeois family, based? On capital, on private gain. In its completely developed form this family exists only among the bourgeoisie. But this state of things finds its complement in the practical absence of the family among proletarians, and in public prostitution.

The bourgeois family will vanish as a matter of course when its complement vanishes, and both will vanish with the vanishing of capital.

Do you charge us with wanting to stop the exploitation of children by their parents? To this crime we plead guilty.

But, you say, we destroy the most hallowed of relations, when we replace home education by social.

And your education! Is not that also social, and determined by the social conditions under which you educate, by the intervention, direct or indirect, of society, by means of schools, etc.? The Communists have not intended the intervention of society in education; they do but seek to alter the character of that intervention, and to rescue education from the influence of the ruling class.

The bourgeois clap-trap about the family and education, about the hallowed co-relation of parents and child, becomes all the more disgusting, the more, by the action of Modern Industry, all the family ties among the proletarians are torn asunder, and their children transformed into simple articles of commerce and instruments of labor.

But you Communists would introduce community of women, screams the bourgeoisie in chorus.

The bourgeois sees his wife a mere instrument of production. He hears that the instruments of production are to be exploited in common, and, naturally, can come to no other conclusion [than] that the lot of being common to all will likewise fall to the women.

He has not even a suspicion that the real point aimed at is to do away with the status of women as mere instruments of production.

For the rest, nothing is more ridiculous than the virtuous indignation of our bourgeois at the community of women which, they pretend, is to be openly and officially established by the Communists. The Communists have no need to introduce free love; it has existed almost from time immemorial.

Our bourgeois, not content with having wives and daughters of their proletarians at their disposal, not to speak of common prostitutes, take the greatest pleasure in seducing each other's wives.

Bourgeois marriage is in reality a system of wives in common and thus, at the most, what the Communists might possibly be reproached with, is that they desire to introduce, in substitution for a hypocritically concealed, an openly legalized system of free love. For the rest, it is self-evident that the abolition of the present system of production must bring with it the abolition of free love springing from that system, i.e., of prostitution both public and private.

The Communists are further reproached with desiring to abolish countries and nationality.

The working men have no country. We cannot take from them what they have not got. Since the proletariat must first of all acquire political supremacy, must rise to be the leading class of *the* nation, must constitute itself the nation, it is, so far, itself national, though not in the bourgeois sense of the word.

National differences and antagonism between peoples are daily more and more vanishing, owing to the development of the bourgeoisie, to freedom of commerce, to the world-market, to uniformity in the mode of production and in the conditions of life corresponding thereto.

The supremacy of the proletariat will cause them to vanish still faster. United action, of the leading civilized countries at least, is one of the first conditions for the emancipation of the proletariat.

In proportion as the exploitation of one individual by another will also be put an end to, the exploitation of one nation by another will also be put an end to. In proportion as the antagonism between classes within the nation vanishes, the hostility of one nation to another will come to an end.

The charges against communism made from a religious, a philosophical, and, generally, from an ideological standpoint, are not deserving of serious examination.

Does it require deep intuition to comprehend that man's ideas, views and conceptions, in one word, man's consciousness, changes with every change in the conditions of his material existence, in his social relations and in his social life?

What else does the history of ideas prove, than that intellectual production changes its character in proportion as material production is changed? The ruling ideas of each age have ever been the ideas of its ruling class.

When people speak of the ideas that revolutionize society, they do but express that fact, that within the old society, the elements of a new one have been created, and that the dissolution of the old ideas keeps even pace with the dissolution of the old conditions of existence.

When the ancient world was in its last throes, the ancient religions were overcome by Christianity. When Christian ideas succumbed in the eighteenth century to rationalist ideas, feudal society fought its death battle with the then revolutionary bourgeoisie. The ideas of religious liberty and freedom of conscience merely gave expression to the sway of free competition within the domain of knowledge.

"Undoubtedly," it will be said, "religious, moral, philosophical and juridicial ideas have been modified in the course of historical development. But religion, morality, philosophy, political science, and law, constantly survived this change."

"There are, besides, eternal truths, such as Freedom, Justice, etc., that are common to all states of society. But communism abolishes eternal truths, it abolishes all religion, and all morality, instead of constituting them on a new basis; it therefore acts in contradiction to all past historical experience."

What does this accusation reduce itself to? The history of all past society has consisted in the development of class antagonisms, antagonisms that assumed different forms at different epochs.

But whatever form they may have taken, one fact is common to all past ages, viz., the exploitation of one part of society by the other. No wonder, then, that the social consciousness of past ages, despite all the multiplicity and variety it displays, moves within certain common forms, or general ideas, which cannot completely vanish except with the total disappearance of class antagonisms.

The communist revolution is the most radical rupture with traditional property relations; no wonder that its development involved the most radical rupture with traditional ideas.

But let us have done with the bourgeois objections to communism.

We have seen above, that the first step in the revolution by the working class, is to raise the proletariat to the position of ruling class, to win the battle of democracy.

The proletariat will use its political supremacy to wrest, by degrees, all capital from the bourgeoisie, to centralize all instruments of production in the hands of the state, i.e., of the proletariat organized as the ruling class; and to increase the total productive forces as rapidly as possible.

Of course, in the beginning, this cannot be effected except by means of despotic inroads on the rights of property, and on the conditions of bourgeois production; by means of measures, therefore, which appear economically insufficient and untenable, but which, in the course of the movement, outstrip themselves, necessitate further inroads upon the old social order, and are unavoidable as a means of entirely revolutionizing the mode of production.

These measures will of course be different in different countries.

Nevertheless in most advanced countries, the following will be pretty generally applicable.

1. Abolition of property in land and application of all rents of land to public purposes.

2. A heavy progressive or graduated income tax.

3. Abolition of all right of inheritance.

4. Confiscation of the property of all emigrants and rebels.

5. Centralization of credit in the hands of the state, by means of a national bank with state capital and an exclusive monopoly.

6. Centralization of the means of communication and transport in the banks of the state.

7. Extension of factories and instruments of production owned by the state; the bringing into cultivation of waste-lands, and the improvement of the soil generally in accordance with a common plan.

8. Equal obligation of all to work. Establishment of industrial armies, especially for agriculture.

9. Combination of agriculture with manufacturing industries; gradual abolition of all the distinction between town and country, by a more equable distribution of the populace over the country.

10. Free education for all children in public schools. Abolition of children's factory labor in its present form. Combination of education with industrial production, etc.

When, in the course of development, class distinctions have disappeared, and all production has been concentrated in the hands of a vast association of the whole nation, the public power will lose its political character. Political power, properly so called, is merely the organised power of one class for oppressing another. If the proletariat during its contest with the bourgeoisie is compelled, by the force of circumstances, to organise itself as a class, if, by means of a revolution, it makes itself the ruling class, and, as such, sweeps away by force the old conditions of production, then it will, along with these conditions, have swept away the conditions for the existence of class antagonisms and of classes generally, and will thereby have abolished its own supremacy as a class.

In place of the old bourgeois society, with its classes and class antagonisms, we shall have an association, in which the free development of each is the condition for the free development of all.

▪▪
▪▪

Introduction to *Capital*

In this section, we turn to what many consider Marx's masterpiece of economic analysis: *Capital.* Here, we provide excerpts from two chapters: "Commodities" and "The General Formula for Capital."

In "Commodities," Marx explores the sources of "value" by asking what determines the worth or price of goods bought and sold on the market. In answering this question, Marx again borrowed from the work of Adam Smith to draw a distinction between "use-value" and "exchange-value." Use-value refers to the utility of a commodity or its ability to satisfy wants.[1] A commodity has use-value only if it is consumed or otherwise put to use. For instance, a one-legged stool cannot readily satisfy a person's desire to sit; therefore, it has no use-value for most individuals. The use-value of a commodity, however, does not determine its actual price; although the usefulness of a commodity may differ between individuals (maybe you really do prefer sitting on a one-legged stool), the cost of the good does not likewise change (we'll all pay the same price for it). Moreover, because use-value refers to the *qualities* of commodities—what they do—it cannot establish a *quantifiable* standard for measuring the price of goods. After all, how can one quantify and compare the usefulness of a light bulb with that of a fork?

Exchange-value, on the other hand, does express equivalencies—how much of a given commodity (e.g., corn) it takes to equal the value of another commodity (e.g., iron). Because exchange-value is derived from trade, it cannot be a property inherent in the commodity itself. Instead, it is dependent on what goods are being exchanged. For instance, one DVD player might be exchanged fairly for one guitar, two jackets, or three CD burners. Thus, a DVD player has not one, but many exchange-values. But if different quantities of different

---

[1]Marx explicitly excluded questions concerning the origins of "wants" as well as how commodities actually satisfied them. Some Marxist-inspired theorists, most notably those associated with the Frankfurt School, would later turn their attention to precisely such questions—that is, how the continued expansion of capitalism requires the production of "false" needs.

commodities can nevertheless be equal in exchange-value, then the value of the commodities must be determined by something else separate from yet common to the commodities themselves.

For Marx, this common "something else" is labor. In Marx's **labor theory of value** (which he appropriated from Adam Smith and David Ricardo), the value of an object is determined ultimately by the amount of labor time (hours, weeks, months, etc.) that it took to produce it. "Commodities, therefore, in which equal quantities of labour are embodied, or which can be produced in the same time, have the same value. . . . As values, all commodities are only definite masses of congealed labour-time" (Marx 1867/1978:306). By equating the value of goods with labor time, Marx not only outlined the economic principles that purportedly guide exchange, he also unmasked the root source of exploitation inherent in capitalist production.

In a capitalist economy, those who do not own the means of production have no choice but to sell their labor power in order to survive. The worker's labor power is thus treated as a commodity exchanged, in this case, for a wage. But at what rate is the worker paid? What determines the exchange-value of labor? Like all other commodities, the value of labor power is a function of the amount of labor time necessary to produce itself. In other words, the value of labor power is equivalent to the costs incurred by the worker for food, clothing, shelter, training, and other goods necessary to ensure both the survival of his family and his return to work the next day.

However, the length of the working day exceeds the time needed on the job in order for the worker to reproduce his labor power. Say, for instance, that in six hours of work a laborer is able to produce for the capitalist the equivalent value of what he needs in order to support his family and return to work. Because the worker's wage is equal to the value of the goods necessary for his family's survival, he is paid, in this case, for six hours worth of labor. Yet, the capitalist employs the worker for a longer duration, say 12 hours a day. During these additional six hours, the worker produces surplus value for the capitalist. **Surplus value** is the difference between what workers earn for their labor and the price or value of the goods that they produce. Surplus value is thus the source of the capitalist's profit: the capitalist pays the worker less than the value of what she actually produces. Human labor is thus the one commodity that is exchanged for its value while being capable of producing more than its value.

To illustrate this concept more clearly, consider a simplified example of a furniture manufacturing plant employing 100 workers. A worker paid $10.00 an hour to assemble tables would earn $400 for a 40-hour workweek. Annually, the worker would earn $20,800. This annual wage would barely keep a family of four out of poverty, to say nothing of attaining the "American Dream." On the other hand, let's assume the worker assembles 100 tables over the course of a year, each sold on the market for $300. The worker thus generates $30,000 for the owner of the plant. The nearly $10,000 difference between wages earned and money generated is appropriated by the capitalist both to reinvest in her business and to support her own family. While this may not seem like a significant difference, recall that the plant employs 100 workers, each of whose labor produces roughly $30,000 in sales. Now the owner is appropriating nearly $1 million in surplus value over the course of only one year, while the workers, whose labor produced the goods sold on the market for a profit, cling with their families to a near-poverty existence.

Additionally, private ownership of the means of the production allows the owner to control the production process and appropriate the products, thus enabling him to take this profit solely for himself. In turn, surplus value is also the source of the capitalists' **exploitation** of the worker because the worker gives more than is given in return without having any voice in this relationship of exchange.

In his effort to increase his profit and market share, the capitalist has two principal means at his disposal: increasing "absolute" or increasing "relative" surplus value. He can increase his *absolute surplus value* by extending the working day. The increase in hours on the job, in turn, increases the productivity of his workforce. With wages remaining constant, greater productivity yields higher profits for the capitalist. During Marx's time, 12- and 14-hour working days were not uncommon, and capitalists routinely opposed legislation aimed at reducing laborers' hours.

Capitalists can also increase their *relative surplus value*. This stems from increasing the productivity of labor by instituting timesaving procedures. With a decrease in the time and thus the cost of production, a capitalist is able to undersell his competitors and capture a larger share of the market. For instance, production efficiency can be improved as capitalists specialize their labor force by reorganizing workers and the allocation of tasks. Specialization simplifies a worker's role in the production process so that, rather than performing a variety of tasks, his contribution is reduced to one or two operations. Often this entails adopting an assembly line system of manufacturing such as Henry Ford did when he revolutionized the automobile industry in the early twentieth century. However, although specialization increases efficiency by enabling more products to be produced in less time, it also leads to the routinization of labor and the workers' loss of self-fulfillment.

Similarly, in their competition for markets, capitalists can turn to more-sophisticated machines and technology to enable laborers to produce more goods in less time. To the extent that mechanized production decreases the necessary labor time, surplus value is increased, along with the level of worker alienation and exploitation.

Although a machine may be able to run 24 hours a day (and does not need insurance or bathroom breaks), mechanized production has its costs. In the short run, it can lead to a reduction in profits, despite the higher volume of productivity, as machines take the place of workers who are the capitalists' source of surplus value. Increasing productivity as a means for selling commodities more cheaply than one's competitors sell also *compels* a capitalist to sell more products and dominate a larger share of the market. Without selling more commodities, the capitalist cannot offset the lower selling price and the expense of adopting more costly machines, to say nothing of turning a profit. Moreover, as the capitalist's competitors begin to make use of the new technology, she is forced to seek—and pay for—ever-newer and more-efficient machines, lest she suffer the very fate she intends to inflict on others.

The competition for markets and the need to increase productivity bear long-run costs, as well. Specialization and mechanization force more workers into unstable employment and a marginal existence. Needed to perform only the most monotonous of unskilled tasks, workers become easily replaceable and expendable. Indeed, "it is the absolute interest of every capitalist to press a given quantity of labour out of a smaller, rather than a greater number of labourers," because doing so increases their relative surplus value and accumulation of capital (Marx 1867/1978:425). As a result, an "industrial reserve army" of unemployed and underemployed laborers is created, the ranks of which swell as the employed segments of the proletariat are overworked. Thus, despite the increasing levels of productivity and growth in the amount of wealth controlled by the capitalists, the market for their products begins to shrink as a growing "relative surplus-population" of laborers is left unable to afford little more than the necessities for survival. At the same time, the increasing competition for jobs due to the expanding industrial reserve army combines with the marginalization of skills to decrease the wages of those fortunate enough to be employed. Meanwhile, competition between capitalists forever breeds greater specialization and mechanization, and all that follows in their wake. Recurring crises of overproduction and "boom or bust" are thus endemic to the capitalist system, while economic recessions and depressed wages become more severe.[2]

---

[2]Though Marx contended that the continuing expansion of the industrial reserve army operates as "a law of population peculiar to the capitalist mode of production" (Marx 1867/1978:423), it is clear that rising rates of unemployment are not inevitable, nor are fluctuations in rates of unemployment due entirely to changing levels of production. Instead, unemployment rates are as much a product of government policy as they are of general economic conditions. Nevertheless, a recent (2006) report issued by the International Labour Organization revealed that the number of people unemployed worldwide reached an all-time high of 191.8 million in 2005, an increase of 34.4 million (21 percent) since 1995. Additionally, of the more than 2.8 billion workers in the world, 1.4 billion earned less than $2 dollars per day.

In this chapter, Marx also reworks his earlier analysis of alienation in the form of the "fetishism of commodities." Recall that alienation, according to Marx, is a dehumanizing consequence of the worker's estrangement or separation from the means of production and the goods produced (see our discussion of "Alienated Labour" above). Similarly, commodity fetishism refers to the distorted relationship existing between individuals and the production and consumption of goods. However, in fetishizing commodities, Marx argues that we treat the goods we buy as if they have "magical" powers. We lose sight of the fact that *we* create commodities and, in doing so, grant them a power over us that in reality they do not hold.

Perhaps you can think of how products directed at our personal appearance are marketed. Advertisements for shampoos, lotions, deodorants, toothpastes, and the like routinely convey the message that interpersonal "success" is dependent on our using these products. Boy gets girl because he buys a specific brand of mouthwash. Girl gets boy because she uses a toothpaste that "whitens" her teeth. Likewise, driving a particular type of car or drinking a particular brand of soft drink or beer magically transforms us into the "type" of person who uses the products. In each instance, our accomplishments and failures are derived not from who we are as individuals, but magically from what we buy as consumers. As a result, our social interactions as well as our sense of self are mediated through or steered by products, not by our individual qualities. When we fetishize commodities, we relate to things, not people. (Compare Marx's argument here with the one made earlier in the excerpt from "The Power of Money in Bourgeois Society.")

Not only are commodities fetishized, but so too is the process of commodity production. When we blame machines for our dissatisfaction, we endow them with human qualities of conscious intent or will. In turn, we fail to recognize that it is the owner of the means of production who is responsible for transforming the production process, not the machines. Thus, if the introduction of new technology increases the speed of the labor process or alters how that process is organized among workers, fetishizing commodity production prevents laborers from holding capitalists accountable for their growing dissatisfaction. Instead, workers will assign the source of their increasing exploitation not to the capitalists who benefit from it, but to the new technology. This carries with it important political consequences, because the intrinsically social nature of the production process is veiled, making workers less able to effectively press their class-based interests for change. The Luddites were one such group of handicraft workers who in early nineteenth century England destroyed the textile machines that rendered their skilled labor obsolete, displacing them with cheap, unskilled laborers. Their protests were met with repressive government actions that included hangings and imprisonment in exile.

Finally, in "The General Formula for Capital," Marx describes the cycle or circulation of commodities peculiar to capitalism. Unlike other economic arrangements, production under capitalism is driven by the quest for increasing profits and capital for reinvestment, not toward simply fulfilling needs or wants established through tradition. Guiding the profit motive is a cycle of exchange Marx labeled "M-C-M." By definition, the capitalist enters into economic exchange already possessing **capital** (raw materials, machinery for production) or, more generally, money (M). Seeking to expand her business and profits, the capitalist converts her money into a commodity (C) by purchasing additional machinery, raw materials, or labor. The capitalist then uses these commodities to produce other commodities that are then sold for money (M). Hence, the meaning of the slogan, "It takes money to make money."

For the proletariat, the cycle of exchange takes an inverse path. Take a typical wage earner, for example. The worker enters into the labor market possessing only his labor power, which he sells as a commodity (C). His commodity, labor, is then exchanged for money (M) or a wage. The worker then takes the money and spends it on the commodities (C) necessary to his survival. The circulation of commodities here follows the pattern C-M-C. The worker sells his one commodity in order to purchase goods he does not otherwise possess. Such a pattern of exchange cannot generate a profit. Instead, it is a cycle of economic activity that provides solely for the satisfaction of basic needs and a subsistence level of existence. Moreover, this cycle must be repeated daily as the commodities bought by the worker—food, fuel, clothing, shelter—tied as they are to survival, are more or less immediately consumed or in need of continual replacement. Rent is paid not once, but monthly. Clothes are bought not once, but regularly, when worn out or outgrown.

From *Capital* (1867)

Karl Marx

COMMODITIES

The Two Factors of a Commodity: Use-Value and Value (The Substance of Value and the Magnitude of Value)

The wealth of those societies in which the capitalist mode of production prevails, presents itself as "an immense accumulation of commodities," its unit being a single commodity. Our investigation must therefore begin with the analysis of a commodity.

A commodity is, in the first place, an object outside us, a thing that by its properties satisfies human wants of some sort or another. The nature of such wants, whether, for instance, they spring from the stomach or from fancy, makes no difference. Neither are we here concerned to know how the object satisfies these wants, whether directly as means of subsistence, or indirectly as means of production.

Every useful thing, as iron, paper, etc., may be looked at from the two points of view of quality and quantity. It is an assemblage of many properties, and may therefore be of use in various ways. To discover the various uses of things is the work of history. So also is the establishment of socially recognized standards of measure for the quantities of these useful objects. The diversity of these measures has its origin partly in the diverse nature of the objects to be measured, partly in convention.

The utility of a thing makes it a use-value. But this utility is not a thing of air. Being limited by the physical properties of the commodity, it has no existence apart from that commodity. A commodity, such as iron, corn, or a diamond, is therefore, so far as it is a material thing, a use-value, something useful. This property of a commodity is independent of the amount of labour required to appropriate its useful qualities. When treating of use-value, we always assume to be dealing with definite quantities, such as dozens of watches, yards of linen, or tons of iron. The use-values of commodities furnish the material for a special study, that of the commercial knowledge of commodities.[i] Use-values become a reality only by use or consumption: they also constitute the substance of all wealth, whatever may be the social form of that wealth. In the form of society we are about to consider, they are, in addition, the material depositories of exchange-value.

Exchange-value, at first sight, presents itself as a quantitative relation, as the proportion in which values in use of one sort are exchanged for those of another sort, a relation constantly changing with time and place. Hence exchange-value appears to be something accidental and purely relative, and consequently an intrinsic value, *i.e.,* an exchange-value that is inseparably connected with, inherent in commodities, seems a contradiction in terms. Let us consider the matter a little more closely.

A given commodity, *e.g.,* a quarter of wheat is exchanged for x blacking, y silk, or z gold, etc.—in short, for other commodities in the most different proportions. Instead of one exchange-value, the wheat has, therefore, a great many.

SOURCE: Marx/Engels Internet Archive.

[i]In bourgeois societies the economic fictio juris prevails, that every one, as a buyer, possesses an encyclopaedic knowledge of commodities. *[Marx]*

But since x blacking, y silk, or z gold, etc., each represent the exchange-value of one quarter of wheat, x blacking, y silk, z gold, etc., must, as exchange-values, be replaceable by each other, or equal to each other. Therefore, first: the valid exchange-values of a given commodity express something equal; secondly, exchange-value, generally, is only the mode of expression, the phenomenal form, of something contained in it, yet distinguishable from it.

Let us take two commodities, *e.g.,* corn and iron. The proportions in which they are exchangeable, whatever those proportions may be, can always be represented by an equation in which a given quantity of corn is equated to some quantity of iron: *e.g.,* 1 quarter corn = x cwt. iron. What does this equation tell us? It tells us that in two different things—in 1 quarter of corn and x cwt. of iron, there exists in equal quantities something common to both. The two things must therefore be equal to a third, which in itself is neither the one nor the other. Each of them, so far as it is exchange-value, must therefore be reducible to this third.

A simple geometrical illustration will make this clear. In order to calculate and compare the areas of rectilinear figures, we decompose them into triangles. But the area of the triangle itself is expressed by something totally different from its visible figure, namely, by half the product of the base multiplied by the altitude. In the same way the exchange-values of commodities must be capable of being expressed in terms of something common to them all, of which thing they represent a greater or less quantity.

This common "something" cannot be either a geometrical, a chemical, or any other natural property of commodities. Such properties claim our attention only in so far as they affect the utility of those commodities, make them use-values. But the exchange of commodities is evidently an act characterised by a total abstraction from use-value. Then one use-value is just as good as another, provided only it be present in sufficient quantity. Or, as old Barbon says, "one sort of wares are as good as another, if the values be equal. There is no difference or distinction in things of equal value.... An hundred pounds' worth of lead or iron, is of as great value as one hundred pounds' worth of silver or gold." As use-values, commodities are, above all, of different qualities, but as exchange-values they are merely different quantities, and consequently do not contain an atom of use-value.

If then we leave out of consideration the use-value of commodities, they have only one common property left, that of being products of labour. But even the product of labour itself has undergone a change in our hands. If we make abstraction from its use-value, we make abstraction at the same time from the material elements and shapes that make the product a use-value; we see in it no longer a table, a house, yarn, or any other useful thing. Its existence as a material thing is put out of sight. Neither can it any longer be regarded as the product of the labour of the joiner, the mason, the spinner, or of any other definite kind of productive labour. Along with the useful qualities of the products themselves, we put out of sight both the useful character of the various kinds of labour embodied in them, and the concrete forms of that labour; there is nothing left but what is common to them all: all are reduced to one and the same sort of labour, human labour in the abstract.

Let us now consider the residue of each of these products; it consists of the same unsubstantial reality in each, a mere congelation of homogeneous human labour, of labour-power expended without regard to the mode of its expenditure. All that these things now tell us is, that human labour-power has been expended in their production, that human labour is embodied in them. When looked at as crystals of this social substance, common to them all, they are—Values.

We have seen that when commodities are exchanged, their exchange-value manifests itself as something totally independent of their use-value. But if we abstract from their use-value, there remains their Value as defined above. Therefore, the common substance that manifests itself in the exchange-value of commodities, whenever they are exchanged, is their value. The progress of our investigation will show that exchange-value is the only form in which the value of commodities can manifest itself or be expressed. For the present, however, we have to consider the nature of value independently of this, its form.

A use-value, or useful article, therefore, has value only because human labour in the abstract has been embodied or materialised in it.

How, then, is the magnitude of this value to be measured? Plainly, by the quantity of the value-creating substance, the labour, contained in the article. The quantity of labour, however, is measured by its duration, and labour-time in its turn finds its standard in weeks, days, and hours.

Some people might think that if the value of a commodity is determined by the quantity of labour spent on it, the more idle and unskilful the labourer, the more valuable would his commodity be, because more time would be required in its production. The labour, however, that forms the substance of value, is homogeneous human labour, expenditure of one uniform labour-power. The total labour-power of society, which is embodied in the sum total of the values of all commodities produced by that society, counts here as one homogeneous mass of human labour-power, composed though it be of innumerable individual units. Each of these units is the same as any other, so far as it has the character of the average labour-power of society, and takes effect as such; that is, so far as it requires for producing a commodity, no more time than is needed on an average, no more than is socially necessary. The labour-time socially necessary is that required to produce an article under the normal conditions of production, and with the average degree of skill and intensity prevalent at the time. The introduction of power-looms into England probably reduced by one-half the labour required to weave a given quantity of yarn into cloth. The hand-loom weavers, as a matter of fact, continued to require the same time as before; but for all that, the product of one hour of their labour represented after the change only half an hour's social labour, and consequently fell to one-half its former value.

We see then that that which determines the magnitude of the value of any article is the amount of labour socially necessary, or the labour-time socially necessary for its production. Each individual commodity, in this connexion, is to be considered as an average sample of its class. Commodities, therefore, in which equal quantities of labour are embodied, or which can be produced in the same time, have the same value. The value of one commodity is to the value of any other, as the labour-time necessary for the production of the one is to that necessary for the production of the other. "As

values, all commodities are only definite masses of congealed labour-time."

The value of a commodity would therefore remain constant, if the labour-time required for its production also remained constant. But the latter changes with every variation in the productiveness of labour. This productiveness is determined by various circumstances, amongst others, by the average amount of skill of the workmen, the state of science, and the degree of its practical application, the social organisation of production, the extent and capabilities of the means of production, and by physical conditions. For example, the same amount of labour in favourable seasons is embodied in 8 bushels of corn, and in unfavourable, only in four. The same labour extracts from rich mines more metal than from poor mines. Diamonds are of very rare occurrence on the earth's surface, and hence their discovery costs, on an average, a great deal of labour-time. Consequently much labour is represented in a small compass. Jacob doubts whether gold has ever been paid for at its full value. This applies still more to diamonds. According to Eschwege, the total produce of the Brazilian diamond mines for the eighty years, ending in 1823, had not realised the price of one-and-a-half years' average produce of the sugar and coffee plantations of the same country, although the diamonds cost much more labour, and therefore represented more value. With richer mines, the same quantity of labour would embody itself in more diamonds, and their value would fall. If we could succeed at a small expenditure of labour, in converting carbon into diamonds, their value might fall below that of bricks. In general, the greater the productiveness of labour, the less is the labour-time required for the production of an article, the less is the amount of labour crystallised in that article, and the less is its value; and *vice versa,* the less the productiveness of labour, the greater is the labour-time required for the production of an article, and the greater is its value. The value of a commodity, therefore, varies directly as the quantity, and inversely as the productiveness, of the labour incorporated in it.

A thing can be a use-value, without having value. This is the case whenever its utility to man is not due to labour. Such are air, virgin soil, natural meadows, etc. A thing can be useful, and the product of human labour, without being a commodity. Whoever directly satisfies his wants with the produce of his own labour, creates, indeed, use-values, but not commodities. In order to produce the latter, he must not only produce use-values, but use-values for others, social use-values. (And not only for others, without more. The medieval peasant produced quit-rent-corn for his feudal lord and tithe-corn for his parson. But neither the quit-rent-corn nor the tithe-corn became commodities by reason of the fact that they had been produced for others. To become a commodity a product must be transferred to another, whom it will serve as a use-value, by means of an exchange.)[ii] Lastly nothing can have value, without being an object of utility. If the thing is useless, so is the labour contained in it; the labour does not count as labour, and therefore creates no value. . . .

The Fetishism of Commodities and the Secret Thereof

A commodity appears, at first sight, a very trivial thing, and easily understood. Its analysis shows that it is, in reality, a very queer thing, abounding in metaphysical subtleties and theological niceties. So far as it is a value in use, there is nothing mysterious about it, whether we consider it from the point of view that by its properties it is capable of satisfying human wants, or from the point that those properties are the product of human labour. It is as clear as noon-day, that man, by his industry, changes the forms of the materials furnished by Nature, in such a way as to make them useful to him. The form of wood, for instance, is altered, by making a table out of it. Yet, for all that, the table continues to be that common, every-day thing, wood. But, so soon as it steps forth as a commodity, it is changed into something transcendent. It not only stands with its feet on the ground, but, in relation to all other commodities, it stands on its head, and evolves out of its wooden brain grotesque ideas, far more wonderful than "table-turning" ever was.

The mystical character of commodities does not originate, therefore, in their use-value. Just as little does it proceed from the nature of the determining factors of value. For, in the first place, however varied the useful kinds of labour, or productive activities, may be, it is a physiological fact, that they are functions of the human organism, and that each such function, whatever may be its nature or form, is essentially the expenditure of human brain, nerves, muscles, etc. Secondly, with regard to that which forms the groundwork for the quantitative determination of value, namely, the duration of that expenditure, or the quantity of labour, it is quite clear that there is a palpable difference between its quantity and quality. In all states of society, the labour-time that it costs to produce the means of subsistence, must necessarily be an object of interest to mankind, though not of equal interest in different stages of development. And lastly, from the moment that men in any way work for one another, their labour assumes a social form.

Whence, then, arises the enigmatical character of the product of labour, so soon as it assumes the form of commodities? Clearly from this form itself. The equality of all sorts of human labour is expressed objectively by their products all being equally values; the measure of the expenditure of labour-power by the duration of that expenditure, takes the form of the quantity of value of the products of labour; and finally, the mutual relations of the producers, within which the social character of their labour affirms itself, take the form of a social relation between the products.

A commodity is therefore a mysterious thing, simply because in it the social character of men's labour appears to them as an objective character stamped upon the product of that labour; because the relation of the producers to the sum total of their own labour is presented to them as a social relation, existing not between themselves, but between the products of their labour. This is the reason why the products of labour become commodities, social things whose qualities are at the same time perceptible

---

[ii] I am inserting the parenthesis because its omission has often given rise to the misunderstanding that every product that is consumed by someone other than its producer is considered in Marx a commodity. [Engels, 4th German edition]

and imperceptible by the senses. In the same way the light from an object is perceived by us not as the subjective excitation of our optic nerve, but as the objective form of something outside the eye itself. But, in the act of seeing, there is at all events, an actual passage of light from one thing to another, from the external object to the eye. There is a physical relation between physical things. But it is different with commodities. There, the existence of the things *quâ* commodities, and the value-relation between the products of labour which stamps them as commodities, have absolutely no connexion with their physical properties and with the material relations arising therefrom. There it is a definite social relation between men, that assumes, in their eyes, the fantastic form of a relation between things. In order, therefore, to find an analogy, we must have recourse to the mist-enveloped regions of the religious world. In that world the productions of the human brain appear as independent beings endowed with life, and entering into relation both with one another and the human race. So it is in the world of commodities with the products of men's hands. This I call the Fetishism which attaches itself to the products of labour, so soon as they are produced as commodities, and which is therefore inseparable from the production of commodities.

This Fetishism of commodities has its origin, as the foregoing analysis has already shown, in the peculiar social character of the labour that produces them.

As a general rule, articles of utility become commodities, only because they are products of the labour of private individuals or groups of individuals who carry on their work independently of each other. The sum total of the labour of all these private individuals forms the aggregate labour of society. Since the producers do not come into social contact with each other until they exchange their products, the specific social character of each producer's labour does not show itself except in the act of exchange. In other words, the labour of the individual asserts itself as a part of the labour of society, only by means of the relations which the act of exchange establishes directly between the products, and indirectly, through them, between the producers. To the latter, therefore, the relations connecting the labour of one individual with that of the rest appear, not as direct social relations between individuals at work, but as what they really are, material relations between persons and social relations between things. It is only by being exchanged that the products of labour acquire, as values, one uniform social status, distinct from their varied forms of existence as objects of utility. This division of a product into a useful thing and a value becomes practically important, only when exchange has acquired such an extension that useful articles are produced for the purpose of being exchanged, and their character as values has therefore to be taken into account, beforehand, during production. From this moment the labour of the individual producer acquires socially a two-fold character. On the one hand, it must, as a definite useful kind of labour, satisfy a definite social want, and thus hold its place as part and parcel of the collective labour of all, as a branch of a social division of labour that has sprung up spontaneously. On the other hand, it can satisfy the manifold wants of the individual producer himself, only in so far as the mutual exchangeability of all kinds of useful private labour is an established social fact, and therefore the private useful labour of each producer ranks on an equality with that of all others. The equalisation of the most different kinds of labour can be the result only of an abstraction from their inequalities, or of reducing them to their common denominator, viz., expenditure of human labour-power or human labour in the abstract. The two-fold social character of the labour of the individual appears to him, when reflected in his brain, only under those forms which are impressed upon that labour in everyday practice by the exchange of products. In this way, the character that his own labour possesses of being socially useful takes the form of the condition, that the product must be not only useful, but useful for others, and the social character that his particular labour has of being the equal of all other particular kinds of labour, takes the form that all the physically different articles that are the products of labour, have one common quality, viz., that of having value.

Hence, when we bring the products of our labour into relation with each other as values, it is not because we see in these articles the material receptacles of homogeneous human labour. Quite the contrary: whenever, by an exchange, we equate as values our different products, by that very act, we also equate, as human labour, the different kinds of labour expended upon them. We are not aware of this, nevertheless we do it. Value, therefore, does not stalk about with a label describing what it is. It is value, rather, that converts every product into a social hieroglyphic. Later on, we try to decipher the hieroglyphic, to get behind the secret of our own social products; for to stamp an object of utility as a value, is just as much a social product as language. The recent scientific discovery, that the products of labour, so far as they are values, are but material expressions of the human labour spent in their production, marks, indeed, an epoch in the history of the development of the human race, but, by no means, dissipates the mist through which the social character of labour appears to us to be an objective character of the products themselves. The fact, that in the particular form of production with which we are dealing, viz., the production of commodities, the specific social character of private labour carried on independently, consists in the equality of every kind of that labour, by virtue of its being human labour, which character, therefore, assumes in the product the form of value—this fact appears to the producers, notwithstanding the discovery above referred to, to be just as real and final, as the fact, that, after the discovery by science of the component gases of air, the atmosphere itself remained unaltered.

What, first of all, practically concerns producers when they make an exchange, is the question, how much of some other product they get for their own? in what proportions the products are exchangeable? When these proportions have, by custom, attained a certain stability, they appear to result from the nature of the products, so that, for instance, one ton of iron and two ounces of gold appear as naturally to be of equal value as a pound of gold and a pound of iron in spite of their different physical and chemical qualities appear to be of equal weight. The character of having value, when once impressed upon products, obtains fixity only by reason of their acting and re-acting upon each other as quantities of value. These quantities vary continually, independently of the will, foresight and action of the producers. To them, their own social action takes the form of the action of objects, which rule the producers instead of

being ruled by them. It requires a fully developed production of commodities before, from accumulated experience alone, the scientific conviction springs up, that all the different kinds of private labour, which are carried on independently of each other, and yet as spontaneously developed branches of the social division of labour, are continually being reduced to the quantitative proportions in which society requires them. And why? Because, in the midst of all the accidental and ever fluctuating exchange-relations between the products, the labour-time socially necessary for their production forcibly asserts itself like an over-riding law of Nature. The law of gravity thus asserts itself when a house falls about our ears. The determination of the magnitude of value by labour-time is therefore a secret, hidden under the apparent fluctuations in the relative values of commodities. Its discovery, while removing all appearance of mere accidentality from the determination of the magnitude of the values of products, yet in no way alters the mode in which that determination takes place.

Man's reflections on the forms of social life, and consequently, also, his scientific analysis of those forms, take a course directly opposite to that of their actual historical development. He begins, post festum, with the results of the process of development ready to hand before him. The characters that stamp products as commodities, and whose establishment is a necessary preliminary to the circulation of commodities, have already acquired the stability of natural, self-understood forms of social life, before man seeks to decipher, not their historical character, for in his eyes they are immutable, but their meaning. Consequently it was the analysis of the prices of commodities that alone led to the determination of the magnitude of value, and it was the common expression of all commodities in money that alone led to the establishment of their characters as values. It is, however, just this ultimate money-form of the world of commodities that actually conceals, instead of disclosing, the social character of private labour, and the social relations between the individual producers. When I state that coats or boots stand in a relation to linen, because it is the universal incarnation of abstract human labour, the absurdity of the statement is self-evident.

Nevertheless, when the producers of coats and boots compare those articles with linen, or, what is the same thing, with gold or silver, as the universal equivalent, they express the relation between their own private labour and the collective labour of society in the same absurd form.

The categories of bourgeois economy consist of such like forms. They are forms of thought expressing with social validity the conditions and relations of a definite, historically determined mode of production, viz., the production of commodities. The whole mystery of commodities, all the magic and necromancy that surrounds the products of labour as long as they take the form of commodities, vanishes therefore, so soon as we come to ther forms of production. . . .

The life-process of society, which is based on the process of material production, does not strip off its mystical veil until it is treated as production by freely associated men, and is consciously regulated by them in accordance with a settled plan. This, however, demands for society a certain material ground-work or set of conditions of existence which in their turn are the spontaneous product of a long and painful process of development.

Political Economy has indeed analysed, however incompletely, value and its magnitude, and has discovered what lies beneath these forms. But it has never once asked the question why labour is represented by the value of its product and labour-time by the magnitude of that value. These formulæ, which bear it stamped upon them in unmistakable letters that they belong to a state of society, in which the process of production has the mastery over man, instead of being controlled by him, such formulæ appear to the bourgeois intellect to be as much a self-evident necessity imposed by Nature as productive labour itself. Hence forms of social production that preceded the bourgeois form, are treated by the bourgeoisie in much the same way as the Fathers of the Church treated pre-Christian religions.

To what extent some economists are misled by the Fetishism inherent in commodities, or by the objective appearance of the social characteristics of labour, is shown, amongst other ways, by the dull and tedious quarrel over the part played by Nature in the formation of exchange-value. Since exchange-value is a definite social manner of expressing the amount of labour bestowed upon an object, Nature has no more to do with it, than it has in fixing the course of exchange.

The mode of production in which the product takes the form of a commodity, or is produced directly for exchange, is the most general and most embryonic form of bourgeois production. It therefore makes its appearance at an early date in history, though not in the same predominating and characteristic manner as now-a-days. Hence its Fetish character is comparatively easy to be seen through. But when we come to more concrete forms, even this appearance of simplicity vanishes. Whence arose the illusions of the monetary system? To it gold and silver, when serving as money, did not represent a social relation between producers but were natural objects with strange social properties. And modern economy, which looks down with such disdain on the monetary system, does not its superstition come out as clear as noon-day, whenever it treats of capital? How long is it since economy discarded the physiocratic illusion, that rents grow out of the soil and not out of society?

But not to anticipate, we will content ourselves with yet another example relating to the commodity-form. Could commodities themselves speak, they would say: Our use-value may be a thing that interests men. It is no part of us as objects. What, however, does belong to us as objects, is our value. Our natural intercourse as commodities proves it. In the eyes of each other we are nothing but exchange-values. Now listen how those commodities speak through the mouth of the economist. "Value"—(*i.e.,* exchange-value) "is a property of things, riches"—(*i.e.,* use-value) "of man. Value, in this sense, necessarily implies exchanges, riches do not." "Riches" (use-value) "are the attribute of men, value is the attribute of commodities. A man or a community is rich, a pearl or a diamond is valuable. . . . A pearl or a diamond is valuable" as a pearl or diamond. So far no chemist has ever discovered exchange-value either in a pearl or a diamond. The economic discoverers of this chemical element, who by-the-by lay special claim to critical acumen, find however that the use-value of objects belongs to them independently of their material properties, while their value, on the other hand, forms a part of them as objects. What confirms them in this view, is the peculiar circumstance that the use-value of objects is realised without exchange, by means of a direct relation between the objects and man, while, on the other hand, their value is realised only by exchange, that is, by means of a social process. Who fails here to call to mind our good friend, Dogberry, who informs neighbour Seacoal, that, "To be a well-favoured man is the gift of fortune; but reading and writing comes by Nature."

THE GENERAL FORMULA FOR CAPITAL

The circulation of commodities is the starting-point of capital. The production of commodities, their circulation, and that more developed form of their circulation called commerce, these form the historical ground-work from which it rises. The modern history of capital dates from the creation in the 16th century of a world-embracing commerce and a world-embracing market.

If we abstract from the material substance of the circulation of commodities, that is, from the exchange of the various use-values, and consider only the economic forms produced by this process of circulation, we find its final result to be money: this final product of the circulation of commodities is the first form in which capital appears.

As a matter of history, capital, as opposed to landed property, invariably takes the form at first of money; it appears as moneyed wealth, as the capital of the merchant and of the usurer. But we have no need to refer to the origin of capital in order to discover that the first form of appearance of capital is money. We can see it daily under our very eyes. All new capital, to commence with, comes on the stage, that is, on the market, whether of commodities, labour, or money, even in our days, in the shape of money that by a definite process has to be transformed into capital.

The first distinction we notice between money that is money only, and money that is capital, is nothing more than a difference in their form of circulation.

The simplest form of the circulation of commodities is C—M—C, the transformation of commodities into money, and the change of the money back again into commodities; or selling in order to buy. But alongside of this form we find another specifically different form:

M—C—M, the transformation of money into commodities, and the change of commodities back again into money; or buying in order to sell. Money that circulates in the latter manner is thereby transformed into, becomes capital, and is already potentially capital.

Now let us examine the circuit M—C—M a little closer. It consists, like the other, of two antithetical phases. In the first phase, M—C, or the purchase, the money is changed into a commodity. In the second phase, C—M, or the sale, the commodity is changed back again into money. The combination of these two phases constitutes the single movement whereby money is exchanged for a commodity, and the same commodity is again exchanged for money; whereby a commodity is bought in order to be sold, or, neglecting the distinction in form between buying and selling, whereby a commodity is bought with a commodity. The result, in which the phases of the process vanish, is the exchange of money for money, M—M. If I purchase 2,000 lbs. of cotton for £100, and resell the 2,000 lbs. of cotton for £110, I have, in fact, exchanged £100 for £110, money for money.

Now it is evident that the circuit M—C—M would be absurd and without meaning if the intention were to exchange by this means two equal sums of money, £100 for £100. The miser's plan would be far simpler and surer; he sticks to his £100 instead of exposing it to the dangers of circulation. And yet, whether the merchant who has paid £100 for his cotton sells it for £110, or lets it go for £100, or even £50, his money has, at all events, gone through a characteristic and original movement, quite different in kind from that which it goes through in the hands of the peasant who sells corn, and with the money thus set free buys clothes. We have therefore to examine first the distinguishing characteristics of the forms of the circuits M—C—M and C—M—C, and in doing this the real difference that underlies the mere difference of form will reveal itself.

Let us see, in the first place, what the two forms have in common.

Both circuits are resolvable into the same two antithetical phases, C—M, a sale, and M—C, a purchase. In each of these phases the same material elements—a commodity, and money, and the same economic dramatis personae, a buyer and a seller—confront one another. Each circuit is the unity of the same two antithetical phases, and in each case this unity is brought about by the intervention of three contracting parties, of whom one only sells, another only buys, while the third both buys and sells.

What, however, first and foremost distinguishes the circuit C—M—C from the circuit M—C—M, is the inverted order of succession of the two phases. The simple circulation of commodities begins with a sale and ends with a purchase, while the circulation of money as capital begins with a purchase and ends with a sale. In the one case both the starting-point and the goal are commodities, in the other they are money. In the first form the movement is brought about by the intervention of money, in the second by that of a commodity.

In the circulation C—M—C, the money is in the end converted into a commodity, that serves as a use-value; it is spent once for all. In the inverted form, M—C—M, on the contrary, the buyer lays out money in order that, as a seller, he may recover money. By the purchase of his commodity he throws money into circulation, in order to withdraw it again by the sale of the same commodity. He lets the money go, but only with the sly intention of getting it back again. The money, therefore, is not spent, it is merely advance.

In the circuit C—M—C, the same piece of money changes its place twice. The seller gets it from the buyer and pays it away to another seller. The complete circulation, which begins with the receipt, concludes with the payment, of money for commodities. It is the very contrary in the circuit M—C—M. Here it is not the piece of money that changes its place twice, but the commodity. The buyer takes it from the hands of the seller and passes it into the hands of another buyer. Just as in the simple circulation of commodities the double change of place of the same piece of money effects its passage from one hand into another, so here the double change of place of the same commodity brings about the reflux of the money to its point of departure.

Such reflux is not dependent on the commodity being sold for more than was paid for it. This circumstance influences only the amount of the money that comes back. The reflux itself takes place, so soon as the purchased commodity is resold, in other words, so soon as the circuit M—C—M is completed. We have here, therefore, a palpable difference between the circulation of money as capital, and its circulation as mere money.

The circuit C—M—C comes completely to an end, so soon as the money brought in by the sale of one commodity is abstracted again by the purchase of another.

If, nevertheless, there follow a reflux of money to its starting-point, this can only happen through a renewal or repetition of the operation. If I sell a quarter of corn of £3, and with this £3 buy clothes, the money, so far as I am concerned, is spent and done with. It belongs to the clothes merchant. If I now sell a second quarter of corn, money indeed flows back to me, not however as a sequel to the first transaction, but in consequence of its repetition. The money again leaves me, so soon as I complete this second transaction by a fresh purchase. Therefore, in the circuit C—M—C, the expenditure of money has nothing to do with its reflux. On the other hand, in M—C—M, the reflux of the money is conditioned by the very mode of its expenditure. Without this reflux, the operation fails, or the process is interrupted and incomplete, owing to the absence of its complementary and final phase, the sale.

The circuit C—M—C starts with one commodity, and finishes with another, which falls out of circulation and into consumption. Consumption, the satisfaction of wants, in one word, use-value, is its end and aim. The circuit M—C—M, on the contrary, commences with money and ends with money. Its leading motive, and the goal that attracts it, is therefore mere exchange-value.

In the simple circulation of commodities, the two extremes of the circuit have the same economic form. They are both commodities, and commodities of equal value. But they are also use-values differing in their qualities, as, for example, corn and clothes. The exchange of products, of the different materials in which the labour of society is embodied, forms here the basis of the movement. It is otherwise in the circulation M—C—M, which at first sight appears purposeless, because tautological. Both extremes have the same economic form. They are both money, and therefore are not qualitatively different use-values; for money is but the converted form of commodities, in which their particular use-values vanish. To exchange £100 for cotton, and then this same cotton again for £110, is merely is roundabout way of exchanging money for money, the same for the same, and appears to be an operation just as purposeless as it is absurd. One sum of money is distinguishable from another only by its amount. The character and tendency of the process M—C—M, is therefore not due to any qualitative difference between its extremes, both being money, but solely to their quantitative difference. More money is withdrawn from circulation at the finish than was thrown into it at the start. The cotton that was bought for £100 is perhaps resold for £100 £ +10 or £110. The exact form of this process is therefore M—C—M', where M'+ ∇M = M the original sum advanced, plus an increment. This increment or excess over the original value I call "surplus-value." The value originally advanced, therefore, not only remains intact while in circulation, but adds to itself a surplus-value or expands itself. It is this movement that converts it into capital.

Of course, it is also possible, that in C—M—C, the two extremes C—C, say corn and clothes, may represent different quantities of value. The farmer may sell his corn above its value, or may buy the clothes at less than their value. He may, on the other hand, "be done" by the clothes merchant. Yet, in the form of circulation now under consideration, such differences in value are purely accidental. The fact that the corn and the clothes are equivalents, does not deprive the process of all meaning, as it does in M—C—M. The equivalence of their values is rather a necessary condition to its normal course.

The repetition or renewal of the act of selling in order to buy, is kept within bounds by the very object it aims at, namely, consumption or the satisfaction of definite wants, an aim that lies altogether outside the sphere of circulation. But when we buy in order to sell, we, on the contrary, begin and end with the same thing, money,

exchange-value; and thereby the movement becomes interminable. No doubt, M becomes M + ∇M, £100 become £110. But when viewed in their qualitative aspect alone, £110 are the same as £100, namely money; and considered quantitatively, £110 is, like £100, a sum of definite and limited value. If now, the £110 be spent as money, they cease to play their part. They are no longer capital. Withdrawn from circulation, they become petrified into a hoard, and though they remained in that state till doomsday, not a single farthing would accrue to them. If, then, the expansion of value is once aimed at, there is just the same inducement to augment the value of the £110 as that of the £100; for both are but limited expressions for exchange-value, and therefore both have the same vocation to approach, by quantitative increase, as near as possible to absolute wealth. Momentarily, indeed, the value originally advanced, the £100, is distinguishable from the surplus-value of £10 that is annexed to it during circulation; but the distinction vanishes immediately. At the end of the process, we do not receive with one hand the original £100, and with the other, the surplus-value of £10. We simply get a value of £110, which is in exactly the same condition and fitness for commencing the expanding process, as the original £100 was. Money ends the movement only to begin it again.[iii] Therefore, the final result of every separate circuit, in which a purchase and consequent sale are completed, forms of itself the starting-point of a new circuit. The simple circulation of commodities—selling in order to buy—is a means of carrying out a purpose unconnected with circulation, namely, the appropriation of use-values, the satisfaction of wants. The circulation of money as capital is, on the contrary, an end in itself, for the expansion of value takes place only within this constantly renewed movement. The circulation of capital has therefore no limits.

As the conscious representative of this movement, the possessor of money becomes a capitalist. His person, or rather his pocket, is the point from which the money starts and to which it

---

[iii]"Capital is divisible . . . into the original capital and the profit, the increment to the capital . . . although in practice this profit is immediately turned into capital, and set in motion with the original." (F. Engels, "Umrisse zu einer Kritik der Nationalökonomie, in the "Deutsch-Französische Jahrbücher," edited by Arnold Ruge and Karl Marx." Paris, 1844, p. 99.) [Marx]

returns. The expansion of value, which is the objective basis or main-spring of the circulation M—C—M, becomes his subjective aim, and it is only in so far as the appropriation of ever more and more wealth in the abstract becomes the sole motive of his operations, that he functions as a capitalist, that is, as capital personified and endowed with consciousness and a will. Use-values must therefore never be looked upon as the real aim of the capitalist; neither must the profit on any single transaction. The restless never-ending process of profit- making alone is what he aims at. This boundless greed after riches, this passionate chase after exchange- value, is common to the capitalist and the miser; but while the miser is merely a capitalist gone mad, the capitalist is a rational miser. The never-ending augmentation of exchange-value, which the miser strives after, by seeking to save his money from circulation, is attained by the more acute capitalist, by constantly throwing it afresh into circulation.

The independent form, *i.e.,* the money-form, which the value of commodities assumes in the case of simple circulation, serves only one purpose, namely, their exchange, and vanishes in the final result of the movement. On the other hand, in the circulation M—C—M, both the money and the commodity represent only different modes of existence of value itself, the money its general mode, and the commodity its particular, or, so to say, disguised mode. It is constantly changing from one form to the other without thereby becoming lost, and thus assumes an automatically active character. If now we take in turn each of the two different forms which self-expanding value successively assumes in the course of its life, we then arrive at these two propositions: Capital is money: Capital is commodities. In truth, however, value is here the active factor in a process, in which, while constantly assuming the form in turn of money and commodities, it at the same time changes in magnitude, differentiates itself by throwing off surplus-value from itself; the original value, in other words, expands spontaneously. For the movement, in the course of which it adds surplus-value, is its own movement, its expansion, therefore, is automatic expansion. Because it is value, it has acquired the occult quality of being able to add value to itself. It brings forth living offspring, or, at the least, lays golden eggs.

Value, therefore, being the active factor in such a process, and assuming at one time the form of money, at another that of commodities, but through all these changes preserving itself and expanding, it requires some independent form, by means of which its identity may at any time be established. And this form it possesses only in the shape of money. It is under the form of money that value begins and ends, and begins again, every act of its own spontaneous generation. It began by being £100, it is now £110, and so on. But the money itself is only one of the two forms of value. Unless it takes the form of some commodity, it does not become capital. There is here no antagonism, as in the case of hoarding, between the money and commodities. The capitalist knows that all commodities, however scurvy they may look, or however badly they may smell, are in faith and in truth money, inwardly circumcised Jews, and what is more, a wonderful means whereby out of money to make more money.

In simple circulation, C—M—C, the value of commodities attained at the most a form independent of their use-values, i.e., the form of money; but that same value now in the circulation M—C—M, or the circulation of capital, suddenly presents itself as an independent substance, endowed with a motion of its own, passing through a life-process of its own, in which money and commodities are mere forms which it assumes and casts off in turn. Nay, more: instead of simply representing the relations of commodities, it enters now, so to say, into private relations with itself. It differentiates itself as original value from itself as surplus-value; as the father differentiates himself from himself quâ the son, yet both are one and of one age: for only by the surplus-value of £10 does the £100 originally advanced become capital, and so soon as this takes place, so soon as the son, and by the son, the father, is begotten, so soon does their difference vanish, and they again become one, £110.

Value therefore now becomes value in process, money in process, and, as such, capital. It comes out of circulation, enters into it again, preserves and multiplies itself within its circuit, comes back out of it with expanded bulk, and begins the same round ever afresh. M—M', money which begets money, such is the description of Capital from the mouths of its first interpreters, the Mercantilists.

Buying in order to sell, or, more accurately, buying in order to sell dearer, M—C—M', appears certainly to be a form peculiar to one kind of capital alone, namely merchants' capital. But industrial capital too is money, that is changed into commodities, and by the sale of these commodities, is re-converted into more money.

The events that take place outside the sphere of circulation, in the interval between the buying and selling, do not affect the form of this movement. Lastly, in the case of interest- bearing capital, the circulation M—C—M' appears abridged. We have its result without the intermediate stage, in the form M—M', "en style lapidaire" so to say, money that is worth more money, value that is greater than itself.

M—C—M' is therefore in reality the general formula of capital as it appears prima facie within the sphere of circulation.

Introduction to Friedrich Engels's *The Origin of the Family, Private Property and the State*

One year after Marx's death in 1883, Friedrich Engels published *The Origin of the Family, Private Property and the State*, a work in which he examined the relationship between the evolution of family structures and class-based societies, and the emergence of the state as an institution that governs in the interests of the economically powerful. Much of his argument is based on *Ancient Society*, a book written by the American anthropologist Lewis Morgan that was published in 1877. Marx himself had intended to produce a manuscript based on Morgan's work and Engels incorporated many of the notes Marx had prepared on the subject. Following their earlier collaborations, Engels rearticulates here their materialist conception of history whereby the organization of societies is determined by both the production of the means of existence and the reproduction of the species; that is, "by the stage of development of labor on the one hand and of the family on the other" (Engels 1884/1942:5).

Following Morgan's analysis, Engels argued that prehistoric societies had passed through two stages of development—savagery and barbarism. The period of savagery was characterized by communally organized hunting and gathering societies. All the members of a "gen" or clan participated in securing their survival free from any form of exploitation. In this natural, classless state, families were likewise communally organized through group marriages in which men had multiple wives, wives had multiple husbands, and the children belonged to all. This early stage of human existence disappeared as the gathering of nuts and fruits, the use of simple stone tools, and hunting with primitive bows and arrows gave way to more-sophisticated and dependable means of producing the necessities of life. With these changes, human societies transitioned to barbarism, a stage marked by the domestication and breeding of animals for food, the development of irrigation techniques for the cultivation of crops, and, later, iron plows for tilling large fields.

In conjunction with this evolution in production, clan organization changed as well, as group marriages were replaced by the "pairing family" consisting of one man, one woman, and their children. The advent of the pairing family effected a new division of labor in which the man took responsibility for obtaining food and, with it, ownership of the means of production. Thus, it was the husband who owned the cattle and the instruments of labor necessary for maintaining the family's physical existence. With their survival dependent on the man, the family was now organized under paternal power in which the wife and the children become subject to the rule and exploitation of the husband. The man's power was further consolidated through overturning "mother-right" lines of descent. Laws of inheritance would henceforth be assigned through the male, not the female. For Engels, the significance of this change cannot be overstated. Indeed, it represents, "the *world historical defeat* of the female sex. The man took command in the home also; the woman was degraded and reduced to servitude, she became the slave of his lust and a mere instrument for the production of children" (Engels 1884/1942:50, emphasis in the original).

Continual population growth and advances in production techniques and labor ushered in the transition from barbarism to civilization—the period of industry. The increase in production and wealth made possible by technological developments was accompanied by a parallel increase in the amount of work extracted from the members of the gens. To offset the new demands for labor, "the first great division of labor arose" and with it, "the first great cleavage of society into two classes: masters and slaves, exploiters and exploited" (Engels 1884/1942:147).

The transition from barbarism to civilization also marks the transition from the pairing family to the monogamous family. Like the pairing family that preceded it, the monogamous family is based on the unquestioned supremacy of the man. Yet, far from being based on love, monogamous marriages were "the first form of the family to be based, not on natural, but on economic conditions—on the victory of private property over primitive, natural communal property" (Engels 1884/1942:57). Monogamy was necessary to ensure that the wealth possessed by the man, made possible only by developments in the forces of production, would be inherited by his offspring alone. Thus, paternity needed to be ensured, making women, and only women, subject to the demands of monogamy, while men were free to engage in polygamous relations. Monogamous marriage stripped women of the freedom, honor, and respect naturally accorded during the "backward" period of barbarism. The "advances" of civilization in their stead have regulated the wife to "domestic slavery." Yet Engels maintained that under communism and its abolition of classes of every type, husbands would be stripped of their economic supremacy and with it their dominance within the family. Just as one class will no longer be able to subjugate another, the family will no longer be founded on "the subjugation of the one sex by the other" (ibid.:58).

It should be noted, however, that many anthropologists and historians contend that Engels's portrayal of the premodern family is ethnographically and historically inaccurate. Virtually all known human societies have been male dominant to some extent, though the extent and types of dominance vary significantly (Lane 1990:10). Even communes that intentionally seek to institute complete gender equality often end up reverting to a gendered division of labor (e.g., Rothman 1995). Nevertheless, what is remarkable about Engels's work is not only that he pointedly raises the issue of the subordination of women, but also that he traces it directly to material causes, explicitly maintaining that the alleged inferiority of women is anything but biological.

This materialist perspective is likewise evident in Engels's contention that increasing productivity and the development of classes brought changes not only to the organization of the family, but also to the organization of the broader social order. Societies existing during the periods of savagery and barbarism were based on kinship with members of a given clan tied together by blood. However, with growing populations and expanding trade and commerce, homogenous clans gave way to heterogeneous communities whose members were divided by differences in wealth, and thus into exploiting and exploited classes. The common concerns that had united kinship groups were submerged under the eruption of divisive class antagonisms. The clash of interests could not be contained by the "gentile constitution" that to this point had governed the communistic, tribal societies through long-standing, widely shared customs.

Such a society could only exist either in the continuous open fight of these classes against one another, or else under the rule of a third power, which, apparently standing above the warring classes, suppressed their open conflict and allowed the class struggle to be fought out at most in the economic field, in so-called legal form. The gentile constitution was finished. It had been shattered by the division of labor and its result, the cleavage of society into classes. It was replaced by the *state*. (Engels 1884/1942:154, emphasis in the original)

The state was now the decisive center of power within the civilized society, the boundaries of which were marked by territory, not blood. To secure its defense against foreign enemies and to maintain domestic order, the state compelled its citizens to pay taxes. Yet the instruments of protection—the armies, police forces, and prisons—served the interests not of society as a whole, but rather the interests of the dominant economic class. Thus, the state provided the ruling class with the political means for oppressing and exploiting the subordinate class and ensuring its continued economic dominance. The inevitable communist revolution, however, will bring an end to class struggle as its basis, private property, is abolished. In turn, the state, whose primary charge is to protect the property rights of the ruling class, will be rendered obsolete. "The society which organizes production anew on the basis of free and equal association of the producers will put the whole state machinery where it will then belong—into the museum of antiquities, next to the spinning wheel and the bronze ax" (Engels 1884/1942:158).

From *The Origin of the Family, Private Property and the State* (1884)

Friedrich Engels

THE FAMILY

The Pairing Family

[T]he history of the family in primitive times consists in the progressive narrowing of the circle, originally embracing the whole tribe, within which the two sexes have a common conjugal relation. The continuous exclusion, first of nearer, then of more and more remote relatives, and at last even of relatives by marriage, ends by making any kind of group marriage practically impossible. Finally, there remains only the single, still loosely linked pair, the molecule with whose dissolution marriage itself ceases. This in itself shows what a small part individual sex-love, in the modern sense of the word, played in the rise of monogamy. . . . Whereas in the earlier forms of the family men never lacked women, but, on the contrary, had too many rather than too few, women had now become scarce and highly sought after. Hence it is with the pairing marriage that there begins the capture and purchase of women—widespread symptoms, but no more than symptoms, of the much deeper change that had occurred. . . .

The pairing family, itself too weak and unstable to make an independent household necessary or even desirable, in no wise destroys the communistic household inherited from earlier times. Communistic housekeeping, however, means the supremacy of women in the house; just as the exclusive recognition of the female parent, owing to the impossibility of recognizing the male parent with certainty, means that the women—the mothers—are held in high respect. One of the most absurd notions taken over from eighteenth-century enlightenment is that in the beginning of society woman was the slave of man. Among all savages and all barbarians of the lower and middle stages, and to a certain extent of the upper stage also, the position of women is not only free, but honorable. . . .

The communistic household, in which most or all of the women belong to one and the same gens, while the men come from various gentes, is the material foundation of that supremacy of the women which was general in primitive times. . . . The division of labor between the two sexes is determined by quite other causes than by the position of woman in society. Among peoples

SOURCE: Marx/Engels Internet Archive.

where the women have to work far harder than we think suitable, there is often much more real respect for women than among our Europeans. The lady of civilization, surrounded by false homage and estranged from all real work, has an infinitely lower social position than the hardworking woman of barbarism, who was regarded among her people as a real lady . . . and who was also a lady in character. . . .

The first beginnings of the pairing family appear on the dividing line between savagery and barbarism. . . . The pairing family is the form characteristic of barbarism, as group marriage is characteristic of savagery and monogamy of civilization. . . . In the single pair the group was already reduced to its final unit, its two-atom molecule: one man and one woman. Natural selection, with its progressive exclusions from the marriage community, had accomplished its task; there was nothing more for it to do in this direction. Unless new, social forces came into play, there was no reason why a new form of family should arise from the single pair. But these new forces did come into play. . . .

Pairing marriage had brought a new element into the family. By the side of the natural mother of the child it placed its natural and attested father, with a better warrant of paternity, probably, than that of many a "father" today. According to the division of labor within the family at that time, it was the man's part to obtain food and the instruments of labor necessary for the purpose. He therefore also owned the instruments of labor, and in the event of husband and wife separating, he took them with him, just as she retained her household goods. Therefore, according to the social custom of the time, the man was also the owner of the new source of subsistence, the cattle, and later of the new instruments of labor, the slaves. But according to the custom of the same society, his children could not inherit from him. . . .

Thus, on the one hand, in proportion as wealth increased, it made the man's position in the family more important than the woman's, and on the other hand created an impulse to exploit this strengthened position in order to overthrow, in favor of his children, the traditional order of inheritance. This, however, was impossible so long as descent was reckoned according to mother-right. Mother-right, therefore, had to be overthrown, and overthrown it was. This was by no means so difficult as it looks to us today. For this revolution—one of the most decisive ever experienced by humanity—could take place without disturbing a single one of the living members of a gens. All could remain as they were. A simple decree sufficed that in the future the offspring of the male members should remain within the gens, but that of the female should be excluded by being transferred to the gens of their father. The reckoning of descent in the female line and the matriarchal law of inheritance were thereby overthrown, and the male line of descent and the paternal law of inheritance were substituted for them. As to how and when this revolution took place among civilized peoples, we have no knowledge. It falls entirely within prehistoric times. . . .

The overthrow of mother-right was the world historical defeat of the female sex. The man took command in the home also; the woman was degraded and reduced to servitude, she became the slave of his lust and a mere instrument for the production of children. . . .

The establishment of the exclusive supremacy of the man shows its effects first in the patriarchal family, which now emerges as an intermediate form. Its essential characteristic is not polygamy, of which more later, but "the organization of a number of persons, bond and free, into a family, under paternal power, for the purpose of holding lands, and for the care of flocks and herds" (Morgan [*Ancient Society*] 1877:474).

Its essential features are the incorporation of unfree persons, and paternal power; hence the perfect type of this form of family is the Roman. The original meaning of the word "family" (familia) is not that compound of sentimentality and domestic strife which forms the ideal of the present-day philistine; among the Romans it did not at first even refer to the married pair and their children, but only to the slaves. Famulus means domestic slave, and familia is the total number of slaves belonging to one man. . . . The term was invented by the Romans to denote a new social organism, whose head ruled over wife and children and a number of slaves, and was invested under Roman paternal power with rights of life and death over them all. . . .

Such a form of family shows the transition of the pairing family to monogamy. In order to make certain of the wife's fidelity and therefore of the paternity of the children, she is delivered over unconditionally into the power of the husband; if he kills her, he is only exercising his rights.

The Monogamous Family

It develops out of the pairing family . . . in the transitional period between the upper and middle stages of barbarism; its decisive victory is one of the signs that civilization is beginning. It is based on the supremacy of the man, the express purpose being to produce children of undisputed paternity; such paternity is demanded because these children are later to come into their father's property as his natural heirs. It is distinguished from pairing marriage by the much greater strength of the marriage tie, which can no longer be dissolved at either partner's wish. As a rule, it is now only the man who can dissolve it, and put away his wife. The right of conjugal infidelity also remains secured to him, at any rate by custom (the Code Napoleon explicitly accords it to the husband as long as he does not bring his concubine into the house), and as social life develops he exercises his right more and more; should the wife recall the old form of sexual life and attempt to revive it, she is punished more severely than ever. . . .

We meet this new form of the family in all its severity among the Greeks. While the position of the goddesses in their mythology, as Marx points out, brings before us an earlier period when the position of women was freer and more respected, in the heroic age we find the woman already being humiliated by the domination of the man and by competition from girl slaves. . . . In Homer young women are booty and are handed over to the pleasure of the conquerors, the handsomest being picked by the commanders in order of rank; the entire *Iliad*, it will be remembered, turns on the quarrel of Achilles and Agamemnon over one of these slaves. If a hero is of any importance, Homer also mentions the captive girl with whom he shares his tent and his bed. These girls were also taken back to Greece and brought under the same roof as the wife, as Cassandra was brought by Agamemnon in Aeschylus. . . . The legitimate wife was expected to put up with all this, but herself to remain strictly chaste and faithful. In the heroic age a Greek woman is, indeed, more respected than in the period of civilization, but to her husband she is after all nothing but the mother of his legitimate children and heirs, his chief housekeeper and the supervisor of his female slaves, whom he can and does take as concubines if he so fancies. It is the existence of slavery side by side with monogamy, the presence of young, beautiful slaves belonging unreservedly to the man, that stamps monogamy from the very beginning with its specific character of monogamy for the woman only, but not for the man. And that is the character it still has today. . . .

[Athenian] girls only learned spinning, weaving, and sewing, and at most a little reading and writing. They lived more or less behind locked doors and had no company except other women. The women's apartments formed a separate part of the house, on the upper floor or at the back, where men, especially strangers, could not easily enter, and to which the women retired when men visited the house. They never went out without being accompanied by a female slave; indoors they were kept under regular guard. . . . In Euripides a woman is called an oikourema, a thing (the word is neuter) for looking after the house, and, apart from her business of bearing children, that was all she was for the Athenian—his chief female domestic servant. The man had his athletics and his public business, from which women were barred; in addition, he often had female slaves at his disposal and during the most flourishing days of Athens an extensive system of prostitution which the state at least favored. . . .

This is the origin of monogamy as far as we can trace it back among the most civilized and highly developed people of antiquity. It was not in any way the fruit of individual sex-love, with which it had nothing whatever to do; marriages remained as before marriages of convenience. It was the first form of the family to be based, not on natural, but on economic conditions—on the victory of private property over primitive, natural communal property. The Greeks themselves put the matter quite frankly: the sole exclusive aims of monogamous marriage were to make the man supreme in the family, and to propagate, as the future heirs to his wealth, children indisputably his own. Otherwise, marriage was a burden, a duty which had to be performed, whether one liked it or not, to gods, state, and one's ancestors. . . .

Thus when monogamous marriage first makes its appearance in history, it is not as the reconciliation of man and woman, still less as the highest form of such a reconciliation. Quite the contrary. Monogamous marriage comes on the scene as the subjugation of the one sex by the other; it announces a struggle between the sexes unknown throughout the whole previous prehistoric period. In an old unpublished manuscript, written by Marx and myself in 1846, [the *German Ideology*] I find the words: "The first division of labor is that between man and woman for the propagation of children." And today I can add: The first class opposition that appears in history coincides with the development of the antagonism between man and woman in monogamous marriage, and the first class oppression coincides with that of the female sex by the male. Monogamous marriage was a great historical step forward; nevertheless, together with slavery and private wealth, it opens the period that has lasted until today in which every step forward is also relatively a step backward, in which prosperity and development for some is won through the misery and frustration of others. It is the cellular form of civilized society, in which the nature of the oppositions and contradictions fully active in that society can be already studied. . . .

Thus, wherever the monogamous family remains true to its historical origin and clearly reveals the antagonism between the man and the woman expressed in the man's exclusive supremacy, it exhibits in miniature the same oppositions and contradictions as those in which society has been moving, without power to resolve or overcome them, ever since it split into classes at the beginning of civilization. I am speaking here, of course, only of those cases of monogamous marriage where matrimonial life actually proceeds according to the original character of the whole institution, but where the wife rebels against the husband's supremacy. . . .

Nowadays there are two ways of concluding a bourgeois marriage. In Catholic countries the parents, as before, procure a suitable wife for their young bourgeois son, and the consequence is, of course, the fullest development of the contradiction inherent in monogamy: the husband abandons himself to hetaerism and the wife to adultery. Probably the only reason why the Catholic Church abolished divorce was because it had convinced itself that there is no more a cure for adultery than there is for death. In Protestant countries, on the other hand, the rule is that the son of a bourgeois family is allowed to choose a wife from his own class with more or less freedom; hence there may be a certain element of love in the marriage, as, indeed, in accordance with Protestant hypocrisy, is always assumed, for decency's sake. Here the husband's hetaerism is a more sleepy kind of business, and adultery by the wife is less the rule. But since, in every kind of marriage, people remain what they were before, and since the bourgeois of Protestant countries are mostly philistines, all that this Protestant monogamy achieves, taking the average of the best cases, is a conjugal partnership of leaden boredom, known as "domestic bliss." . . .

In both cases, however, the marriage is conditioned by the class position of the parties and is to that extent always a marriage of convenience. In both cases this marriage of convenience turns often enough into crassest prostitution—sometimes of both partners, but far more commonly of the woman, who only differs from the ordinary courtesan in that she does not let out her body on piece-work as a wage worker, but sells it once and for all into slavery. . . . Sex-love in the relationship with a woman becomes, and can only become, the real rule among the oppressed classes, which means today among the proletariat—whether this relation is officially sanctioned or not. But here all the foundations of typical monogamy are cleared away. Here there is no property, for the preservation and inheritance of which monogamy and male supremacy were established; hence there is no incentive to make this male supremacy effective. What is more, there are no means of making it so. Bourgeois law, which protects this supremacy, exists only for the possessing class and their dealings with the proletarians. The law costs money and, on account of the worker's poverty, it has no validity for his relation to his wife. Here quite other personal and social conditions decide. And now that large-scale industry has taken the wife out of the home onto the labor market and into the factory, and made her often the bread-winner of the family, no basis for any kind of male supremacy is left in the proletarian household— except, perhaps, for something of the brutality towards women that has spread since the introduction of monogamy. The proletarian family is therefore no longer monogamous in the strict sense, even where there is passionate love and firmest loyalty on both sides, and maybe all the blessings of religious and civil authority. Here, therefore, the eternal attendants of monogamy, hetaerism and adultery, play only an almost vanishing part. The wife has in fact regained the right to dissolve the marriage, and if two people cannot get on with one another, they prefer to separate. In short, proletarian marriage is monogamous in the etymological sense of the word, but not at all in its historical sense.

Our jurists, of course, find that progress in legislation is leaving women with no further ground of complaint. Modern civilized systems of law increasingly acknowledge, first, that for a marriage to be legal, it must be a contract freely entered into by both partners, and, second, that also in the married state both partners must stand on a common footing of equal rights and duties. If both these demands are consistently carried out, say the jurists, women have all they can ask.

This typically legalist method of argument is exactly the same as that which the radical republican bourgeois uses to put the proletarian in his place. The labor contract is to be freely entered into by both partners. But it is considered to have been freely entered into as soon as the law makes both parties equal on paper. The power conferred on the one party by the difference of class position, the pressure thereby brought to bear on the other party—the real economic position of both—that is not the law's business. Again, for the duration of the labor contract both parties are to have equal rights, in so far as one or the other does not expressly surrender them. That economic relations compel the worker to surrender even the last semblance of equal rights—here again, that is no concern of the law. . . .

As regards the legal equality of husband and wife in marriage, the position is no better. The legal inequality of the two partners, bequeathed to us from earlier social conditions, is not the cause but the effect of the economic oppression of the woman. In the old communistic household, which comprised many couples and their children, the task entrusted to the women of managing the household was as much a public and socially necessary industry as the procuring of food by the men. With the patriarchal family, and still more with the single monogamous family, a change came. Household management lost its public character. It no longer concerned society. It became a private service; the wife became the head servant, excluded from all participation in social production. Not until the coming of modern large-scale industry was the road to social production opened to her again— and then only to the proletarian wife. But it was opened in such a manner that, if she carries out her duties in the private service of her family, she remains excluded from public production and unable to earn; and if she wants to take part in public production and earn independently, she cannot carry out family duties. And the wife's position in the factory is the position of women in all branches of business, right up to medicine and the law. The modern individual family is founded on the open or concealed domestic slavery of the wife, and modern society is a mass composed of these individual families as its molecules.

In the great majority of cases today, at least in the possessing classes, the husband is obliged to earn a living and support his family, and that in itself gives him a position of supremacy, without any need for special legal titles and privileges. Within the family he is the bourgeois and the wife represents the proletariat. In the industrial world, the specific character of the economic oppression burdening the proletariat is visible in all its sharpness only when all special legal privileges of the capitalist class have been abolished and complete legal equality of both classes established. The democratic republic does not do away with the opposition of the two classes; on the contrary, it provides the clear field on which the fight can be fought out. And in the same way, the peculiar character of the supremacy of the husband over the wife in the modern family, the necessity of creating real social equality between them, and the way to do it, will only be seen in the clear light of day when both possess legally complete equality of rights. Then it will be plain that the first condition for the liberation of the wife is to bring the whole female sex back into public industry, and that this in turn demands the abolition of the monogamous family as the economic unit of society. . . .

We are now approaching a social revolution in which the economic foundations of monogamy as they have existed hitherto will disappear just as surely as those of its complement—prostitution. Monogamy arose from the concentration of considerable wealth in the hands of a single individual's man—and from the need to bequeath this wealth to the children of that man and of no other. For this purpose, the monogamy of the woman was required, not that of the man, so this monogamy of the woman did not in any way interfere with open or concealed polygamy on the part of the man. But by transforming by far the greater portion, at any rate, of permanent, heritable wealth—the means of production—into social property, the coming social revolution will reduce to a minimum all this anxiety about bequeathing and inheriting. Having arisen from economic causes, will monogamy then disappear when these causes disappear?

One might answer, not without reason: far from disappearing, it will, on the contrary, be realized completely. For with the transformation of the means of production into social property there will disappear also wage-labor, the proletariat, and therefore the necessity for a certain—statistically calculable—number of women to surrender themselves for money. Prostitution disappears; monogamy, instead of collapsing, at last becomes a reality—also for men.

In any case, therefore, the position of men will be very much altered. But the position of women, of all women, also undergoes significant change. With the transfer of the means of production into common ownership, the single family ceases to be the economic unit of society. Private housekeeping is transformed into a social industry. The care and education of the children becomes a public affair; society looks after all children alike, whether they are legitimate or not. This removes all the anxiety about the "consequences," which today is the most essential social—moral as well as economic—factor that prevents a girl from giving herself

completely to the man she loves. Will not that suffice to bring about the gradual growth of unconstrained sexual intercourse and with it a more tolerant public opinion in regard to a maiden's honor and a woman's shame? And, finally, have we not seen that in the modern world monogamy and prostitution are indeed contradictions, but inseparable contradictions, poles of the same state of society? Can prostitution disappear without dragging monogamy with it into the abyss? . . .

In the vast majority of cases . . . marriage remained, up to the close of the middle ages, what it had been from the start—a matter which was not decided by the partners. In the beginning, people were already born married—married to an entire group of the opposite sex. In the later forms of group marriage similar relations probably existed, but with the group continually contracting. In the pairing marriage it was customary for the mothers to settle the marriages of their children; here, too, the decisive considerations are the new ties of kinship, which are to give the young pair a stronger position in the gens and tribe. And when, with the preponderance of private over communal property and the interest in its bequeathal, father-right and monogamy gained supremacy, the dependence of marriages on economic considerations became complete. The form of marriage by purchase disappears, the actual practice is steadily extended until not only the woman but also the man acquires a price—not according to his personal qualities, but according to his property. That the mutual affection of the people concerned should be the one paramount reason for marriage, outweighing everything else, was and always had been absolutely unheard of in the practice of the ruling classes; that sort of thing only happened in romance—or among the oppressed classes, who did not count.

Such was the state of things encountered by capitalist production when it began to prepare itself, after the epoch of geographical discoveries, to win world power by world trade and manufacture. One would suppose that this manner of marriage exactly suited it, and so it did. And yet—there are no limits to the irony of history—capitalist production itself was to make the decisive breach in it. By changing all things into commodities, it dissolved all inherited and traditional relationships, and, in place of time-honored custom and historic right, it set up purchase and sale, "free" contract. And the English jurist, H. S. Maine, thought he had made a tremendous discovery when he said that our whole progress in comparison with former epochs consisted in the fact that we had passed "from status to contract," from inherited to freely contracted conditions—which, in so far as it is correct, was already in *The Communist Manifesto.*

But a contract requires people who can dispose freely of their persons, actions, and possessions, and meet each other on the footing of equal rights. To create these "free" and "equal" people was one of the main tasks of capitalist production. Even though at the start it was carried out only half-consciously, and under a religious disguise at that, from the time of the Lutheran and Calvinist Reformation the principle was established that man is only fully responsible for his actions when he acts with complete freedom of will, and that it is a moral duty to resist all coercion to an immoral act. But how did this fit in with the hitherto existing practice in the arrangement of marriages? Marriage, according to the bourgeois conception, was a contract, a legal transaction, and the most important one of all, because it disposed of two human beings, body and mind, for life. Formally, it is true, the contract at that time was entered into voluntarily: without the assent of the persons concerned, nothing could be done. But everyone knew only too well how this assent was obtained and who were the real contracting parties in the marriage. But if real freedom of decision was required for all other contracts, then why not for this? Had not the two young people to be coupled also the right to dispose freely of themselves, of their bodies and organs? Had not chivalry brought sex-love into fashion, and was not its proper bourgeois form, in contrast to chivalry's adulterous love, the love of husband and wife? And if it was the duty of married people to love each other, was it not equally the duty of lovers to marry each other and nobody else? Did not this right of the lovers stand higher than the right of parents, relations, and other traditional marriage-brokers and matchmakers? If the right of free, personal discrimination broke boldly into the Church and

religion, how should it halt before the intolerable claim of the older generation to dispose of the body, soul, property, happiness, and unhappiness of the younger generation? . . .

So it came about that the rising bourgeoisie, especially in Protestant countries, where existing conditions had been most severely shaken, increasingly recognized freedom of contract also in marriage, and carried it into effect in the manner described. Marriage remained class marriage, but within the class the partners were conceded a certain degree of freedom of choice. And on paper, in ethical theory and in poetic description, nothing was more immutably established than that every marriage is immoral which does not rest on mutual sexual love and really free agreement of husband and wife. In short, the love marriage was proclaimed as a human right, and indeed not only as a droit de l'homme, one of the rights of man, but also, for once in a way, as droit de la femme?," one of the rights of woman. . . .

And as sexual love is by its nature exclusive—although at present this exclusiveness is fully realized only in the woman—the marriage based on sexual love is by its nature individual marriage. We have seen how right Bachofen [Johann Jakob Bachofen (1815–1887), Swiss anthropologist and sociologist] was in regarding the advance from group marriage to individual marriage as primarily due to the women. Only the step from pairing marriage to monogamy can be put down to the credit of the men, and historically the essence of this was to make the position of the women worse and the infidelities of the men easier. If now the economic considerations also disappear which made women put up with the habitual infidelity of their husbands—concern for their own means of existence and still more for their children's future—then, according to all previous experience, the equality of woman thereby achieved will tend infinitely more to make men really monogamous than to make women polyandrous.

But what will quite certainly disappear from monogamy are all the features stamped upon it through its origin in property relations; these are, in the first place, supremacy of the man, and, secondly, indissolubility. The supremacy of the man in marriage is the simple consequence of his economic supremacy, and with the abolition of the latter will disappear of itself. The indissolubility of marriage is partly a consequence of the economic situation in which monogamy arose, partly tradition from the period when the connection between this economic situation and monogamy was not yet fully understood and was carried to extremes under a religious form. Today it is already broken through at a thousand points. If only the marriage based on love is moral, then also only the marriage in which love continues. But the intense emotion of individual sex-love varies very much in duration from one individual to another, especially among men, and if affection definitely comes to an end or is supplanted by a new passionate love, separation is a benefit for both partners as well as for society—only people will then be spared having to wade through the useless mire of a divorce case.

What we can now conjecture about the way in which sexual relations will be ordered after the impending overthrow of capitalist production is mainly of a negative character, limited for the most part to what will disappear. But what will there be new? That will be answered when a new generation has grown up: a generation of men who never in their lives have known what it is to buy a woman's surrender with money or any other social instrument of power; a generation of women who have never known what it is to give themselves to a man from any other considerations than real love, or to refuse to give themselves to their lover from fear of the economic consequences. When these people are in the world, they will care precious little what anybody today thinks they ought to do; they will make their own practice and their corresponding public opinion about the practice of each individual—and that will be the end of it. . . .

BARBARISM, CIVILIZATION, AND THE STATE

With the herds and the other new riches, a revolution came over the family. To procure the necessities of life had always been the business of the man; he produced and owned the means of doing so. The herds were the new means of producing these necessities; the taming of the animals in the first instance and their later tending were the man's work. To him, therefore, belonged the cattle, and to him the commodities and the slaves received in exchange for cattle. All the surplus which the acquisition of the necessities of life now yielded fell to the man; the woman shared in its enjoyment, but had no part in its ownership. . . . The division of labor within the family had regulated the division of property between the man and the woman. That division of labor had remained the same; and yet it now turned the previous domestic relation upside down, simply because the division of labor outside the family had changed. The same cause which had ensured to the woman her previous supremacy in the house—that her activity was confined to domestic labor—this same cause now ensured the man's supremacy in the house: the domestic labor of the woman no longer counted beside the acquisition of the necessities of life by the man; the latter was everything, the former an unimportant extra. We can already see from this that to emancipate woman and make her the equal of the man is and remains an impossibility so long as the woman is shut out from social productive labor and restricted to private domestic labor. The emancipation of woman will only be possible when woman can take part in production on a large, social scale, and domestic work no longer claims anything but an insignificant amount of her time. And only now has that become possible through modern large-scale industry, which does not merely permit of the employment of female labor over a wide range, but positively demands it, while it also tends towards ending private domestic labor by changing it more and more into a public industry. . . .

The next step leads us to the upper stage of barbarism. . . . The town, with its houses of stone or brick, encircled by stone walls, towers and ramparts, became the central seat of the tribe or the confederacy of tribes—an enormous architectural advance, but also a sign of growing danger and need for protection. Wealth increased rapidly, but as the wealth of individuals. The products of weaving, metal-work and the other handicrafts, which were becoming more and more differentiated, displayed growing variety and skill. In addition to corn, leguminous plants and fruit, agriculture now provided wine and oil, the preparation of which had been learned. Such manifold activities were no longer within the scope of one and the same individual; the second great division of labor took place: handicraft separated from agriculture. The continuous increase of production and simultaneously of the productivity of labor heightened the value of human labor-power. Slavery, which during the preceding period was still in its beginnings and sporadic, now becomes an essential constituent part of the social system; slaves no longer merely help with production—they are driven by dozens to work in the fields and the workshops. . . .

Civilization consolidates and intensifies all these existing divisions of labor, particularly by sharpening the opposition between town and country (the town may economically dominate the country, as in antiquity, or the country the town, as in the middle ages), and it adds a third division of labor, peculiar to itself and of decisive importance: it creates a class which no longer concerns itself with production, but only with the exchange of the products—the merchants. Hitherto whenever classes had begun to form, it had always been exclusively in the field of production; the persons engaged in production were separated into those who directed and those who executed, or else into large-scale and small-scale producers. Now for the first time a class appears which, without in any way participating in production, captures the direction of production as a whole and economically subjugates the producers; which makes itself into an indispensable middleman between any two producers and exploits them both. Under the pretext that they save the producers the trouble and risk of exchange, extend the sale of their products to distant markets and are therefore the most useful class of the population, a class of parasites comes into being, "genuine social ichneumons," who, as a reward for their actually very insignificant services, skim all the cream off production at home and abroad, rapidly amass enormous wealth and correspondingly social influence, and for that reason receive under civilization ever higher honors and ever greater control of production, until at last they also bring forth a product of their own—the periodical trade crises. . . .

Alongside wealth in commodities and slaves, alongside wealth in money, there now appeared wealth in land also. The individuals' rights of possession in the pieces of land originally allotted to them by gens or tribe had now become so established that the land was their hereditary property. . . . Full, free ownership of the land meant not only power, uncurtailed and unlimited, to possess the land; it meant also the power to alienate it. As long as the land belonged to the gens, no such power could exist. But when the new landed proprietor shook off once and for all the fetters laid upon him by the prior right of gens and tribe, he also cut the ties which had hitherto inseparably attached him to the land. Money, invented at the same time as private property in land, showed him what that meant. Land could now become a commodity; it could be sold and pledged. . . .

Confronted by the new forces in whose growth it had had no share, the gentile constitution was helpless. The necessary condition for its existence was that the members of a gens or at least of a tribe were settled together in the same territory and were its sole inhabitants. . . . Every territory now had a heterogeneous population belonging to the most varied gentes and tribes; everywhere slaves, protected persons and aliens lived side by side with citizens. The settled conditions of life which had only been achieved towards the end of the middle stage of barbarism were broken up by the repeated shifting and changing of residence under the pressure of trade, alteration of occupation and changes in the ownership of the land. The members of the gentile bodies could no longer meet to look after their common concerns. . . . In addition to the needs and interests with which the gentile bodies were intended and fitted to deal, the upheaval in productive relations and the resulting change in the social structure had given rise to new needs and interests, which were not only alien to the old gentile order, but ran directly counter to it at every point. The interests of the groups of handicraftsmen which had arisen with the division of labor, the special needs of the town as opposed to the country, called for new organs. But each of these groups was composed of people of the most diverse gentes, phratries, and tribes, and even included aliens. Such organs had therefore to be formed outside the gentile constitution, alongside of it, and hence in opposition to it. And this conflict of interests was at work within every gentile body, appearing in its most extreme form in the association of rich and poor, usurers and debtors, in the same gens and the same tribe. Further, there was the new mass of population outside the gentile bodies, which, as in Rome, was able to become a power in the land and at the same time was too numerous to be gradually absorbed into the kinship groups and tribes. In relation to this mass, the gentile bodies stood opposed as closed, privileged corporations; the primitive natural democracy had changed into a malign aristocracy. Lastly, the gentile constitution had grown out of a society which knew no internal contradictions, and it was only adapted to such a society. It possessed no means of coercion except public opinion. But here was a society which by all its economic conditions of life had been forced to split itself into freemen and slaves, into the exploiting rich and the exploited poor; a society which not only could never again reconcile these contradictions, but was compelled always to intensify them. Such a society could only exist either in the continuous open fight of these classes against one another, or else under the rule of a third power, which, apparently standing above the warring classes, suppressed their open conflict and allowed the class struggle to be fought out at most in the economic field, in so-called legal form. The gentile constitution was finished. It had been shattered by the division of labor and its result, the cleavage of society into classes. It was replaced by the state. . . .

The state is therefore by no means a power imposed on society from without; just as little is it "the reality of the moral idea," "the image and the reality of reason," as Hegel maintains. Rather, it is a product of society at a particular stage of development; it is the admission that this society has involved itself in insoluble self-contradiction and is cleft into irreconcilable antagonisms which it is powerless to exorcise. But in order that these antagonisms, classes with conflicting economic interests, shall not consume themselves and society in fruitless struggle, a power, apparently standing above society, has become necessary to moderate the conflict and keep it within the bounds of "order"; and this power, arisen out of society, but placing itself above it and increasingly alienating itself from it, is the state.

In contrast to the old gentile organization, the state is distinguished first by the grouping of its members on a territorial basis. The old gentile bodies, formed and held together by ties of blood, had, as we have seen, become inadequate largely because they presupposed that the gentile members were bound to one particular locality, whereas this had long ago ceased to be the case. The territory was still there, but the people had become mobile. The territorial division was therefore taken as the starting point and the system introduced by which citizens exercised their public rights and duties where they took up residence, without regard to gens or tribe. . . .

The second distinguishing characteristic is the institution of a public force which is no longer immediately identical with the people's own organization of themselves as an armed power. This special public force is needed because a self-acting armed organization of the people has become impossible since their cleavage into classes. . . . This public force exists in every state; it consists not merely of armed men, but also of material appendages, prisons and coercive institutions of all kinds, of which gentile society knew nothing. It may be very insignificant, practically negligible, in societies with still undeveloped class antagonisms and living in remote areas, as at times and in places in the United States of America. But it becomes stronger in proportion as the class antagonisms within the state become sharper and as adjoining states grow larger and more populous. It is enough to look at Europe today, where class struggle and rivalry in conquest have brought the public power to a pitch that it threatens to devour the whole of society and even the state itself.

In order to maintain this public power, contributions from the state citizens are necessary—taxes. These were completely unknown to gentile society. We know more than enough about them today. With advancing civilization, even taxes are not sufficient; the state draws drafts on the future, contracts loans, state debts. . . .

In possession of the public power and the right of taxation, the officials now present themselves as organs of society standing above society. The free, willing respect accorded to the organs of the gentile constitution is not enough for them, even if they could have it. Representatives of a power which estranges them from society, they have to be given prestige by means of special decrees, which invest them with a peculiar sanctity and inviolability. The lowest police officer of the civilized state has more "authority" than all the organs of gentile society put together; but the mightiest prince and the greatest statesman or general of civilization might envy the humblest of the gentile chiefs the unforced and unquestioned respect accorded to him. For the one stands in the midst of society; the other is forced to pose as something outside and above it.

As the state arose from the need to keep class antagonisms in check, but also arose in the thick of the fight between the classes, it is normally the state of the most powerful, economically ruling class, which by its means becomes also the politically ruling class, and so acquires new means of holding down and exploiting the oppressed class. The ancient state was, above all, the state of the slave-owners for holding down the slaves, just as the feudal state was the organ of the nobility for holding down the peasant serfs and bondsmen, and the modern representative state is the instrument for exploiting wage-labor by capital. Exceptional periods, however, occur when the warring classes are so nearly equal in forces that the state power, as apparent mediator, acquires for the moment a certain independence in relation to both. This applies to the absolute monarchy of the seventeenth and eighteenth centuries, which balances the nobility and the bourgeoisie against one another; and to the Bonapartism of the First and particularly of the Second French Empire, which played off the proletariat against the bourgeoisie and the bourgeoisie against the proletariat. The latest achievement in this line, in which ruler and ruled look equally comic, is the new German Empire of the Bismarckian nation; here the capitalists and the workers are balanced against one another and both of them fleeced for the benefit of the decayed Prussian cabbage Junkers.

Further, in most historical states the rights conceded to citizens are graded on a property basis, whereby it is directly admitted that the state is an organization for the protection of the possessing class against the non-possessing class. This is already the case in the Athenian and Roman property classes. Similarly in the medieval feudal state, in which the extent of political power was determined by the extent of landownership. Similarly, also, in the electoral qualifications in modern parliamentary states. This political recognition of property differences is, however, by no means essential. On the contrary, it marks a low stage in the development of the state. The highest form of the state, the democratic republic, which in our modern social conditions becomes more and more an unavoidable necessity and is the form of state in which alone the last decisive battle between proletariat and bourgeoisie can be fought out—the democratic republic no longer officially recognizes differences of property. Wealth here employs its power indirectly, but all the more surely. It does this in two ways: by plain corruption of officials, of which America is the classic example, and by an alliance between the government and the stock exchange, which is effected all the more easily the higher the state debt mounts and the more the joint-stock companies concentrate in their hands not only transport but also production itself, and themselves have their own center in the stock exchange. . . . And lastly the possessing class rules directly by means of universal suffrage. As long as the oppressed class—in our case, therefore, the proletariat—is not yet ripe for its self-liberation, so long will it, in its majority, recognize the existing order of society as the only possible one and remain politically the [tail] of the capitalist class, its extreme left wing. But in the measure in which it matures towards its self-emancipation, in the same measure it constitutes itself as its own party and votes for its own representatives, not those of the capitalists. Universal suffrage is thus the gauge of the maturity of the working class. It cannot and never will be anything more in the modern state; but that is enough. On the day when the thermometer of universal suffrage shows boiling-point among the workers, they as well as the capitalists will know where they stand. . . .

The state, therefore, has not existed from all eternity. There have been societies which have managed without it, which had no notion of the state or state power. At a definite stage of economic development, which necessarily involved the cleavage of society into classes, the state became a necessity because of this cleavage. We are now rapidly approaching a stage in the development of production at which the existence of these classes has not only ceased to be a necessity, but becomes a positive hindrance to production. They will fall as inevitably as they once arose. The state inevitably falls with them.

The society which organizes production anew on the basis of free and equal association of the producers will put the whole state machinery where it will then belong—into the museum of antiquities, next to the spinning wheel and the bronze ax.

Discussion Questions

1. According to Marx's materialist conception of history, what is the relationship between property or the division of labor and consciousness? How might property relations and ideas prevent or promote social change?

2. Do you think that truly communist societies have existed? Can they exist? What are some of the features that such a society must have in order for it to work?

3. What role does private property play in Marx's analysis of the inevitable communist revolution? In his emphasis on class, what factors might Marx have overlooked when accounting for revolutionary change or its absence?

4. Has the proletariat, or working class, sunk deeper and deeper with the advance of industry as Marx suggested? Why or why not? How prevalent is alienation in contemporary capitalist societies? Don't some people like their jobs? If so, have they been "fooled" somehow? Why or why not?

5. Discuss the prevalence of the fetishism of commodities in contemporary capitalist societies. What examples of commodity fetishism do you see in your own life and the lives of your family and friends?

6. To what degree does Engels's description of marriage and the position of women resonate with today's society? What has and has not changed? What forces have promoted or prevented change? What are the similarities and differences between the current discussion of and demands for sexual equality and Engels's analysis?

3 ÉMILE DURKHEIM (1858–1917)

Key Concepts

- ▥ Anomie
- ▥ Social facts
- ▥ Social solidarity
  - ❑ Mechanical solidarity
  - ❑ Organic solidarity
- ▥ Collective conscience
- ▥ Ritual
- ▥ Symbol
- ▥ Sacred and profane
- ▥ Collective representations

There can be no society which does not feel the need of upholding and reaffirming at regular intervals the collective sentiments and the collective ideas which makes its unity and its personality. Now this moral remaking cannot be achieved except by the means of reunions, assemblies and meetings where the individuals, being closely united to one another, reaffirm in common their common sentiments.

(Durkheim 1912/1995:474–75)

Have you ever been to a professional sports event in a stadium full of fans? Or to a religious service and taken communion, or to a concert and danced in the aisles (or maybe in a mosh pit)? How did these experiences make you feel? What do they have in common? Is it possible to have this same type of experience if or when you are alone? How so or why not?

These are the sorts of issues that intrigued Émile Durkheim. Above all, he sought to explain *what* held societies and social groups together—and *how*. In addressing these twin questions, Durkheim studied a wide variety of phenomena—from suicide and crime, to aboriginal religious totems and symbols. He was especially concerned about how modern, industrial societies can be held together when people don't even know each other and when their experiences and social positions are so varied. In other words, how can social ties, the very basis for society, be maintained in such an increasingly individualistic world?

Yet Durkheim is an important figure in the history of sociology not only because of his provocative theories about social cohesion, but also because he helped found the discipline of sociology. In contrast to some of the other figures whose works you will read in this book, Durkheim sought to delineate, both theoretically and methodologically, how sociology was different from existing schools of philosophy and history, which also examined social issues. Before we discuss his ideas and work, however, let's look at his biography because, like Marx, Durkheim's personal experiences and historical situation deeply influenced his perception and description of the social world.

A BIOGRAPHICAL SKETCH ▪▪

Émile Durkheim was born in a small town in northeastern France in 1858. In his youth, he followed family tradition, studying Hebrew and the Talmud in order to become a rabbi. However, in his adolescence, Durkheim apparently rejected Judaism. Though he did not disdain traditional religion, as a child of the Enlightenment (see Chapter 1) he came to consider both Christianity and Judaism outmoded in the modern world.

In 1879, Durkheim entered France's most prestigious college, the École Normale Supérieure in Paris, to study philosophy. However, by his third year, Durkheim had become disenchanted with the high-minded, literary, humanities curriculum at the Normale. He decided to pursue sociology, which he viewed as eminently more scientific, democratic, and practical. Durkheim still maintained his interest in complex philosophical questions, but he wanted to examine them through a "rational," "scientific" lens. His practical and scientific approach to central social issues would shape his ambition to use sociological methods as a means for reconstituting the moral order of French society, which he saw decaying in the aftermath of the French Revolution (Bellah 1973:xiii–xvi). Durkheim was especially concerned about the abuse of power by political and military leaders, increasing rates of divorce and suicide, and rising anti-Semitism. It seemed to Durkheim that social bonds and a sense of community had broken down and social disorder had come to prevail.[1]

Upon graduation from the École Normale, Durkheim began teaching in small lycées (secondary schools) near Paris. In 1887, he married Louise Dreyfus, from the Alsace region of France. In the same year, Durkheim began his career as a professor at the University of

---

[1]As indicated in Chapter 1, France had gone through numerous violent changes in government since the French Revolution in 1789. Between 1789 and 1870, there had been three monarchies, two empires, and two republics, culminating in the notorious reign of Napoleon III who overthrew the democratic government and ruled France for 20 years. Though the French Revolution had brought a brief period of democracy, it also sparked a terrifying persecution of all those who disagreed with the revolutionary leaders. Some 17,000 revolutionaries were executed in the infamous Reign of Terror, led by Maximilien Robespierre. Consequently, political and social divisions in France intensified. French conservatives called for a return to monarchy and a more prominent role for the Catholic Church. In direct contrast, a growing but still relatively small class of urban workers demanded political rights and a secular rather than religious education. At the same time, capitalists called for individual rights and free markets, while radical socialists advocated abolishing private property altogether.

Bordeaux, where he quickly gained the reputation for being a committed and exciting teacher. Émile and Louise soon had two children, Marie and André.

Durkheim was a serious and productive scholar. His first book, *The Division of Labor in Society*, which was based on his doctoral dissertation, came out in 1893; his second, *The Rules of Sociological Method*, appeared just two years later. In 1897, *Suicide: A Study in Sociology*, perhaps his most well known work, was published. The next year, Durkheim founded the journal *L'Année Sociologique*, which was one of the first sociology journals not only in France, but also in the world. *L'Année Sociologique* was produced annually until the outbreak of World War I in 1914.

In 1902, with his reputation as a leading social philosopher and scientist established, Durkheim was offered a position at the prestigious Sorbonne University in Paris. As he had done previously at Bordeaux, Durkheim quickly gained a large following at the Sorbonne. His education courses were compulsory for all students seeking teaching degrees in philosophy, history, literature, and languages. Durkheim also became an important administrator at the Sorbonne, serving on numerous councils and committees (Lukes 1985:372).

Yet not everyone was enamored with either Durkheim's substantial power or his ideas. Durkheim's notion that *any* social "thing"—including religion—could be studied sociologically (i.e., scientifically) was particularly controversial, as was his adamant insistence on providing students a moral, but secular, education. (These two issues will be discussed further below.) As Steven Lukes (1985:373), noted sociologist and Durkheim scholar, remarked, "To friends he was a prophet and an apostle, but to enemies he was a secular pope."

Moreover, Durkheim identified with some of the goals of socialism, but was unwilling to commit himself politically. He believed that sociologists should be committed to education, not political activism. His passion was for dispassionate, scientific research.

This apparent apoliticism, coupled with his focus on the moral constitution of societies (rather than conflict and revolution), has led some analysts to deem Durkheim politically conservative. However, as the eminent sociologist Robert Bellah (1973: xviii) points out, "to try to force Durkheim into the conservative side of some conservative/liberal dichotomy" is inappropriate. It ignores Durkheim's "lifelong preoccupation with orderly, continuous social change toward greater social justice" (ibid.:xvii). In addition, to consider Durkheim politically conservative is erroneous in light of how he was evaluated in his day. Durkheim was viewed as a radical modernist and liberal, who, though respectful of religion, was most committed to rationality, science, and humanism. Durkheim infuriated religious conservatives, who desired to replace democracy with a monarchy, and to strengthen the military. He also came under fire because he opposed instituting Catholic education as the basic curriculum.

Moreover, to label Durkheim "conservative" ignores his role in the "Dreyfus affair." Alfred Dreyfus was a Jewish army colonel who was charged and convicted on false charges of spying for Germany. The charges against Dreyfus were rooted in anti-Semitism, which was growing in the 1890s, alongside France's military losses and economic dissatisfaction. Durkheim was very active in the *Ligue des droits de l'homme* (League of the Rights of Men), which devoted itself to clearing Dreyfus of all charges.

Interestingly, Durkheim's assessment of the Dreyfus affair reflects his lifelong concern for the moral order of society. He saw the Dreyfus affair as symptomatic of a collective moral sickness, rather than merely anti-Semitism at the level of the individual. As Durkheim (1899, as cited by Lukes 1985:345) states,

[w]hen society undergoes suffering, it feels the need to find someone whom it can hold responsible for its sickness, on whom it can avenge its misfortunes; and those against whom public opinion already discriminates are naturally designated for this role. These are the pariahs who serve as expiatory victims. What confirms me in this interpretation

is the way in which the result of Dreyfus's trial was greeted in 1894. There was a surge of joy in the boulevards. People celebrated as a triumph what should have been a cause of public mourning. At least they knew whom to blame for the economic troubles and moral distress in which they lived. The trouble came from the Jews. The charge had been officially proved. By this very fact alone, things already seemed to be getting better and people felt consoled.

In 1912, Durkheim's culminating work, *The Elementary Forms of Religious Life*, was published. Shortly after that, World War I broke out, and Durkheim's life was thrown into turmoil. His son, André, was killed in battle, spiraling Durkheim into a grief from which he never fully recovered. On October 7, 1916, as he was leaving a committee meeting at the Sorbonne, Durkheim suffered a stroke. He spent the next year resting and seemed to have made much progress toward recovering. But on November 15, 1917, while in Fontainebleau where he had gone for peace and fresh air, Durkheim died. He was 59 years old (Lukes 1985:559).

INTELLECTUAL INFLUENCES AND CORE IDEAS ▪▪

As indicated previously, Durkheim wrote a number of books and articles on a wide variety of topics. Nevertheless, there are two major themes that transcend all of Durkheim's work. First, Durkheim sought to articulate the nature of society and, hence, his view of sociology as an academic discipline. Durkheim argued that society was a supraindividual force existing independently of the actors who compose it. The task of sociology, then, is to analyze **social facts**—conditions and circumstances external to the individual that, nevertheless, determine the individual's course of action. Durkheim argued that social facts can be ascertained by using collective data, such as suicide and divorce rates. In other words, through systematic collection of data, the patterns behind and within individual behavior can be uncovered. This emphasis on formal methods and objective data is what distinguished sociology from philosophy and put sociology "on the map" as a viable scientific discipline. The significance of Durkheim's position for the development of sociology as a distinct pursuit of knowledge cannot be overstated. As one of the first academics to hold a position in sociology, Durkheim was on the cutting edge of the birth of the discipline. Nevertheless, his conviction that society is sui generis (an objective reality that is irreducible to the individuals that compose it) and amenable to scientific investigation owes much to the work of **Auguste Comte** (1798–1857). Not only had Comte coined the term *sociology* in 1839, but he also contended that the social world could be studied in as rational and scientific a way as physical scientists (chemists, physicists, biologists, etc.) study their respective domains. Moreover, Durkheim's comparative and historical methodology was in large measure a continuation of the approach advocated earlier by Comte.

Significant Others

Auguste Comte (1798–1857): The Father of "Social Physics"

Born in southern France during a most turbulent period in French history, Auguste Comte was himself a turbulent figure. Though he excelled as a student, he had little patience for authority. Indeed, his obstinate temperament prevented him from completing his studies at the newly established École Polytechnique, Paris's elite university. Nevertheless, Comte was able to make a name for himself in the intellectual

(Continued)

(Continued)

circles of Paris. In 1817, he began working as a secretary and collaborator to Henri Saint-Simon. Their productive though fractious relationship came to an end seven years later in a dispute over assigning authorship to one of Comte's essays. Comte next set about developing his system of positivist philosophy while working in minor academic positions for meager wages. Beginning in 1926, Comte offered a series of private lectures in an effort to disseminate his views. Though attended by eminent thinkers, the grandiosity of his theoretical system led some to dismiss his ideas. Nevertheless, Comte continued undeterred: from 1830 to 1842, he worked single-mindedly on his magnum opus, the six-volume *The Positive Philosophy* (1830–42/1974). In the series, Comte not only outlines his "Law of Three Stages" (which posits that science develops through three mentally conceived stages: (1) the theological stage, (2) the metaphysical stage, and (3) the positive stage) but also delineates the proper methods for his new science of "social physics" as well as its fundamental task—the study of social statics (order) and dynamics (progress). The work was well received in some scientific quarters, and Comte seemed poised to establish himself as a first-rate scholar. Unfortunately, his temperament again proved to be a hindrance to his success, both personal and professional. His troubled marriage ended soon after *Positive Philosophy* was completed, and his petulance further alienated him from friends and colleagues while costing him a position at the École Polytechnique. Comte's life took a turn for the better, however, when in 1844 he met and fell in love with Clotilde de Vaux. Their affair did not last long; Clotilde developed tuberculosis and died within a year of their first meeting. Comte dedicated the rest of his life to "his angel." In her memory, he founded the Religion of Humanity for which he proclaimed himself the high priest. The new church was founded on the principle of universal love as Comte abandoned his earlier commitment to science and positivism. Until his death in 1857, Comte sought not supporters for his system of science, but converts to his Positive Church.

NOTE: This account of Comte's biography is based largely on Lewis Coser's (1977) discussion in *Masters of Sociological Thought.*

Significant Others

Herbert Spencer (1820–1903): Survival of the Fittest

Born in the English Midlands, Herbert Spencer's early years were shaped largely by his father and uncle. It was from these two men that Spencer received his education, an education that centered on math, physics, and chemistry. Moreover, it was from them that Spencer was exposed to the radical religious and social doctrines that would inform his staunch individualism. With little formal instruction in history, literature, and languages, Spencer conceded to the limits of his education, and at the age of sixteen declined to attend university, opting instead to pursue a "practical" career as an engineer for the London and Birmingham Railway. Nevertheless, he would prove to be an avid student of and a prolific writer on a range of social and philosophical topics.

With the completion of the railway in 1841, Spencer earned his living by writing essays for a number of radical journals and newspapers. Of particular note is a series of 12 letters he published through a dissenting newspaper, *The Nonconformist.* Titled "The Proper Sphere of Government," the letters are an early expression of Spencer's decidedly laissez-faire perspective. In them, Spencer argued that the role of government should be restricted solely to policing, while all other matters, including education, social welfare, and economic activities, should be left to the private sector. According to Spencer, government regulations interfere with the laws of human evolution that, if left unhampered, ensure the "survival of the fittest." It is not hard to see that Spencer's view of government still resonates with many American politicians and voters. Less sanguine, however, is the racism and sexism that was interjected into Spencer's argument. Following the logic of his view, those who don't survive—that is, succeed—are merely fulfilling their evolutionary destiny. To the extent that women and people of color are less "successful" than white males, their "success" and "failure" hinge not only on individual aptitude and effort, but also on institutional and cultural dynamics that sustain a less-than-level playing field.

A second major theme found in Durkheim's work is the issue of **social solidarity**, or the cohesion of social groups. As you will see, all of the selections in this chapter—from *The Division of Labor in Society, The Rules of Sociological Method, Suicide,* and *The Elementary Forms of Religious Life*—explore the nature of the bonds that hold individuals and social groups together. Durkheim was especially concerned about modern societies where people often don't know their neighbors (let alone everyone in the larger community) or worship together, and where people often hold jobs in impersonal companies and organizations. Durkheim wondered *how* individuals could feel tied to one another in such an increasingly individualistic world. This issue was of utmost importance, for he maintained that, without some semblance of solidarity and moral cohesion, society could not exist.

In his emphasis on the nature of solidarity in "traditional" and "modern" societies, Durkheim again drew on Comte's work as well as that of the British sociologist **Herbert Spencer** (1820–1903).[2] Both Comte and Spencer formulated an organic view of society to explain the developmental paths along which societies allegedly evolve. Such a view depicted society as a system of interrelated parts (religious institutions, the economy, government, the family) that work together to form a unitary, stable whole, analogous to how the parts of the human body (lungs, kidneys, brain) function interdependently to sustain its general well-being. Moreover, as the organism (society and the body) grows in size, it becomes increasingly complex, due to the differentiation of its parts.

However, Durkheim was only partially sympathetic to the organic, evolutionary models developed by Comte and Spencer. On the one hand, Durkheim's insistence that social solidarity is rooted in shared moral sentiments, and the sense of obligation they evoke, stems from Comte (as well as from Jean-Jacques Rousseau; see Chapter 1). Likewise, his notion that the specialized division of labor characteristic of modern societies leads to greater interdependency and integration owes much to Comte (as well as to Saint-Simon; see Chapter 2).

---

[2]Durkheim was influenced by a number of scholars, and not only by Comte and Spencer. Some of the more important figures in developing his views were the French Enlightenment intellectuals Charles Montesquieu (1689–75) and Jean-Jacques Rousseau (1712–78), Henri de Saint-Simon (1760–1825), Charles Renouvier (1815–1903), and the German experimental psychologist Wilhem Wundt (1832–1920).

Nevertheless, Durkheim did not embrace Comte's assertion that all societies progress through a series of identifiable evolutionary stages. In particular, he dismissed Comte's "Law of Three Stages," wherein all societies—as well as individual intellectual development—are said to pass from a theological stage characterized by "militaristic" communities led by priests, to a metaphysical stage organized according to "legalistic" principles and controlled by lawyers and clergy, and finally to a positivist or scientific stage in which "industrial" societies are governed by technocrats and, of course, sociologists.

In terms of Spencer, Durkheim was most influenced by Spencer's theory on the evolution of societies. According to Spencer, just as biological organisms become more differentiated as they grow and mature, so do small-scale, homogeneous communities become increasingly complex and diverse as a result of population growth. The individuals living in simple societies are minimally dependent on one another for meeting their survival and that of the community as they each carry out similar tasks. As the size of the population increases, however, similarity or likeness is replaced by heterogeneity and a specialized division of labor. Individuals become interdependent on one another as essential tasks are divided among the society's inhabitants. As a result, an individual's well-being becomes tied more and more to the general welfare of the larger society. Ensuring the functional integration of individuals now becomes the central issue for the survival of the society.

In this regard, Durkheim's perspective is compatible with that of Spencer. As further discussed below, Durkheim hypothesized that a different *kind* of solidarity was prevalent in modern—as opposed to smaller, more traditional—societies. Durkheim's equation of traditional societies with "mechanical" solidarity and modern societies with "organic" solidarity (discussed on pp. 103–105) shares an affinity with Spencer's classification of societies as either "simple" or "compound."

However, the two theorists diverge on the crucial point of integration. Spencer saw society as composed of atomistic individuals, each pursuing lines of self-interested conduct. In a classic expression of utilitarian philosophy, Spencer maintained that a stable, well-functioning social whole is the outgrowth of individuals freely seeking to maximize their advantages.

By contrast, Durkheim (and Comte) took a far less utilitarian approach than Spencer. Durkheim emphasized that society is not a result or aftereffect of individual conduct; rather, it exists prior to, and thus shapes, individual action. In other words, individual lines of conduct are the outgrowth of social arrangements, particularly those connected to the developmental stage of the division of labor. Social integration, then, cannot be an unintended consequence of an aggregate of individuals pursuing their self-interest. Instead, it is rooted in a shared moral code, for only it can sustain a harmonious social order. And it is this moral code, along with the feelings of solidarity it generates, that forms the basis of all societies. Without the restraints imposed by a sense of moral obligation to others, the selfish pursuit of interests would destroy the social fabric.

▪▪ DURKHEIM'S THEORETICAL ORIENTATION

As discussed previously, Durkheim was most concerned with analyzing "social facts": he sought to uncover the preexisting social conditions that shape the parameters for individual behavior. Consequently, Durkheim can be said to take a predominantly collectivist approach to order (see Figure 3.1).

This approach is most readily apparent in *Suicide*. In this study, Durkheim begins with one of the most seemingly individualistic, psychologically motivated acts there is—suicide—in order to illuminate the social and moral parameters behind and within this allegedly "individual" behavior. So too, Durkheim's emphasis on **collective conscience**

Figure 3.1   Durkheim's Basic Theoretical Orientation

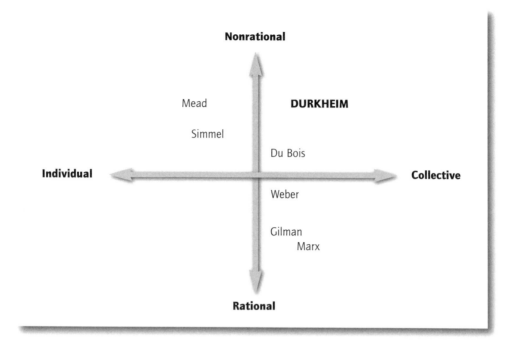

and **collective representations** indicates an interest in the collective level of society (see Figure 3.2). By *collective conscience,* Durkheim means the "totality of beliefs and sentiments common to average citizens of the same society" that "forms a determinate system which has it own life" (Durkheim 1893/1984:38–39). In later work, Durkheim used the term *collective representations* to refer to much the same thing. In any case, the point is that Durkheim's main concern is not with the conscious or psychological state of specific individuals, but rather with the collective beliefs and sentiments that exist "independent of the particular conditions in which individuals are placed; they pass on and it remains" (ibid.:80).

This leads us to one of the most common criticisms of Durkheim. Because of his pre-occupation with social facts and the collective conscience, it is often claimed that he overlooks the role of the individual in producing and reproducing the social order. Durkheim's emphasis on the power of the group makes it seem like we're just vessels for society's will. Yet this criticism ignores two essential points: First, Durkheim not only acknowledged individual autonomy, but also took it for granted as an inevitable condition of modern societies. Durkheim sought to show how, in modern societies, increasing individuation could produce detrimental effects because individuals are often torn between competing normative prescriptions and rules. For instance, in *Suicide,* Durkheim maintains that, rather than rest comfortably on all-pervasive norms and values, "a thirst arises for novelties, unfamiliar pleasures, nameless sensations, all of which lose their savor once known . . . [but that] all these new sensations in their infinite quantity cannot form a solid foundation for happiness to support one in days of trial" (Durkheim 1897/1951:256). To be sure, the criticism could still be made that Durkheim ignores individual agency in "traditional" societies based on mechanical solidarity. In these societies, Durkheim did in fact posit a lack of individual autonomy, perhaps reflecting the Enlightenment-driven, Eurocentric thinking of his day. (We discuss this issue more fully below.)

Figure 3.2   Durkheim's Core Concepts

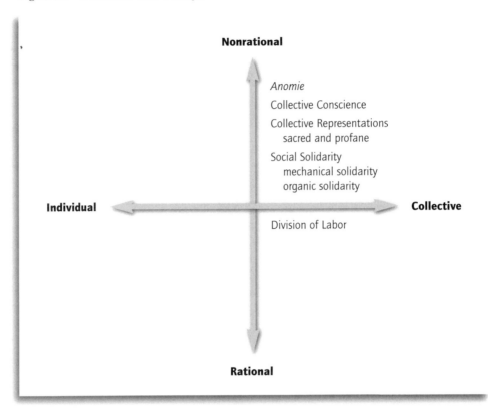

Relatedly, to assert that his orientation was singularly collectivist overlooks Durkheim's assumption that collective life emerges *in* social interaction. For instance, a major part of his analysis of the elementary forms of religious life involved showing how mundane objects, such as lizards and plants, take on the sacredness of the totem (the symbol of the tribe) by virtue of individuals coming together to participate in ritual practices. Similarly, in his study of suicide, Durkheim examined marriage and divorce rates not simply because he was fascinated by abstract, collective dimensions of social life, but also because he wanted to uncover objective factors that measure the extent to which individuals are bound together in an increasingly individualistic world.

This leads us to the issue of action. In our view, Durkheim is primarily nonrationalist in his orientation (see Figures 3.1 and 3.2). He focused on how collective representations and moral sentiments are a motivating force, much more so than "rational" or strategic interests connected to economic or political institutions. Yet it is important to point out that in emphasizing the external nature of social facts Durkheim also recognized that such facts are not confined to the realm of ideas or feelings, but often possess a concrete reality as well. For instance, educational institutions and penal systems are also decisive for shaping the social order and individuals' actions within it. Thus, social facts are capable of exerting both a moral and an institutional force. In the end, however, Durkheim stressed the nonrational aspect of social facts as suggested in his supposition that the penal system (courts, legal codes and their enforcement, etc.) ultimately rests on collective notions of morality, a complex symbolic system as to what is "right" and what is "wrong." This issue will be discussed further in the next section in relationship to the specific selections you will read.

Readings

In this section, you will read selections from the four major books that Durkheim published during his lifetime: *The Division of Labor in Society* (1893), *The Rules of Sociological Method* (1895), *Suicide* (1897), and *The Elementary Forms of Religious Life* (1912). We begin with *The Division of Labor in Society*, in which Durkheim set out the key concepts of mechanical and organic solidarity, and collective conscience. We then shift to excerpts from *The Rules of Sociological Method*. It is here, as you will see, that Durkheim first laid out his basic conceptualization of sociology as a discipline and delineated his concept of social facts. This is followed by excerpts from *Suicide: A Study in Sociology*, which is notable, first, in that it exemplifies Durkheim's distinctive approach to the study of the social world, and second, because it further delineates Durkheim's core concept of *anomie*. We conclude this chapter with excerpts from *The Elementary Forms of Religious Life*, which many theorists consider Durkheim's most theoretically significant work. In it, Durkheim takes an explicitly cultural turn, emphasizing the concepts of ritual and symbol, and the sacred and profane, and collective representations.

Introduction to *The Division of Labor in Society*

In Durkheim's first major work, *The Division of Labor in Society* (1893), which was based on his doctoral dissertation, Durkheim explains how the division of labor (or economic specialization) characteristic of modern societies affects individuals as well as society as a whole. As you may recall, this issue had been of utmost concern to Marx as well. Marx contended that modern, competitive capitalism, and the specialized division of labor that sustained it, resulted in alienation. In contrast, Durkheim argued that economic specialization was not necessarily "bad" for either the individual or the society as a whole. Instead, he argued that an extensive division of labor could exist without necessarily jeopardizing the moral cohesion of a society or the opportunity for individuals to realize their interests.

How is this possible? Durkheim argued that there were two basic types of solidarity: mechanical and organic.[1] **Mechanical solidarity** is typified by feelings of *likeness*. Mechanical solidarity is rooted in everyone doing/feeling the same thing. Durkheim maintained that this type of solidarity is characteristic of small, traditional societies. In these "simple" societies, circumstances compel individuals to be generalists involved in the production and distribution of a variety of goods. Indeed, in small, traditional societies, specialization in one task to the exclusion of others is not possible because the society depends on each individual providing a host of contributions to the group. For instance, men, women, and children are often all needed to pick crops at harvest time, and all partake in the harvest-time celebrations as well.

Durkheim argued that a significant social consequence of the shared work experience characteristic of traditional societies is a shared collective conscience. People in traditional

---

[1]Durkheim's distinction between mechanical and organic solidarity was developed, in part, as a critical response to the work of the German sociologist Ferdinand Tönnies (see Significant Others box, Chapter 6, pp. 274–275). In his book, *Gemeinschaft und Gesellschaft* (*Community and Society*), Tönnies argued that simpler, traditional societies (*Gemeinschaft*) were more "organic" and beneficial to the formation of social bonds. In contrast to Tönnies's conservative orientation, Durkheim contended that complex, modern societies were, in fact, more "organic" and thus more desirable because they promote individual liberties within a context of morally binding, shared social obligations.

Photo 3.1a   Durkheim maintained that different types of society exhibit different types of solidarity. Mechanical solidarity, based on likeness, is characteristic of small, traditional societies, such as this village in Namibia (Africa).

Photo 3.1b   Organic solidarity, based on specialization, is characteristic of large, modern industrial societies, such as Brasília (Brazil).

societies tend to feel "one and the same," and it is this feeling of "oneness" that is integral in the maintenance of social order.

Yet, Durkheim saw that in large, complex societies, this type of solidarity was waning. In large, modern societies, labor is specialized; people do not necessarily all engage in the same work or share the same ideas and beliefs. For Durkheim, **organic solidarity** refers to a type of solidarity in which each person is interdependent with others, forming a complex web of cooperative associations. In such situations, solidarity (or a feeling of "oneness") comes not from each person believing/doing the same thing, but from each person cultivating individual differences and knowing that each is doing her part for the good of the whole. Thus, Durkheim argued that the increasing specialization and individuation so readily apparent in modern industrial societies does not necessarily result in a decline in social stability or cohesion. Rather, the growth in a society's density (the number of people living in a community) and consequent increasingly specialized division of labor can result in simply a different *type* of social cohesion.

Significantly, however, Durkheim maintained that organic solidarity does not automatically emerge in modern societies. Rather, it arises only when the division of labor is "spontaneous" or voluntary. States Durkheim, "For the division of labor to produce solidarity, it is not sufficient, then, that each have his task; it is still necessary that this task be fitting to him" (Durkheim 1893/1984:375). Moreover, a "normal" division of labor exists only when the specialization of tasks is not exaggerated. If the division of labor is pushed too far, there is a danger for the individual to become "isolated in his special activity." In such cases, the division of labor becomes "a source of disintegration" for both the individual and society (ibid.). The individual "no longer feels the idea of common work being done by those who work side by side with him" (ibid.). Meanwhile, a rigid division of labor can lead to "the institution of classes and castes . . . [which] is often a source of dissension" (ibid.:374). Durkheim used the term **anomie** (a lack of moral regulation) to describe the "pathological" consequences of an overly specialized division of labor. This is a pivotal concept to which we will shortly return.

Most interestingly, then, the important point is not that Durkheim ignored the potentially harmful aspects of the division of labor in modern societies; on the contrary, Durkheim acknowledged that the division of labor is problematic when it is "forced" or pushed to an extreme. This position offers an important similarity as well as difference to that offered by Marx. As we noted previously, Marx saw both alienation and class conflict as inevitable (or "normal") in capitalist societies. By contrast, rather than seeing social conflict as a "normal" condition of capitalism, Durkheim maintained that anomie results only in "abnormal" conditions of overspecialization, when the rules of capitalism become too rigid and individuals are "forced" into a particular position in the division of labor.

——————— From *The Division of Labor in Society* (1893) ———————

Émile Durkheim

INTRODUCTION: THE PROBLEM

The division of labor is not of recent origin, but it was only at the end of the eighteenth century that social cognizance was taken of the principle, though, until then, unwitting submission had been rendered to it. To be sure, several thinkers from earliest times saw its importance;[i] but Adam Smith was the first to attempt a theory of it. Moreover, he

[i]Aristotle, *Nichomachean Ethics,* E, 1133a, 16.

adopted this phrase that social science later lent to biology.

Nowadays, the phenomenon has developed so generally it is obvious to all. We need have no further illusions about the tendencies of modern industry; it advances steadily towards powerful machines, towards great concentrations of forces and capital, and consequently to the extreme division of labor. Occupations are infinitely separated and specialized, not only inside the factories, but each product is itself a specialty dependent upon others. Adam Smith and John Stuart Mill still hoped that agriculture, at least, would be an exception to the rule, and they saw it as the last resort of small-scale industry. Although one must be careful not to generalize unduly in such matters, nevertheless it is hard to deny today that the principal branches of the agricultural industry are steadily being drawn into the general movement. Finally, business itself is ingeniously following and reflecting in all its shadings the infinite diversity of industrial enterprises; and, while this evolution is realizing itself with unpremeditated spontaneity, the economists, examining its causes and appreciating its results, far from condemning or opposing it, uphold it as necessary. They see in it the supreme law of human societies and the condition of their progress. But the division of labor is not peculiar to the economic world; we can observe its growing influence in the most varied fields of society. The political, administrative, and judicial functions are growing more and more specialized. It is the same with the aesthetic and scientific functions. It is long since philosophy reigned as the science unique; it has been broken into a multitude of special disciplines each of which has its object, method, and though. "Men working in the sciences have become increasingly more specialized."[ii]

Mechanical Solidarity

We are now in a position to come to a conclusion.

The totality of beliefs and sentiments common to average citizens of the same society forms a determinate system which has its own life; one may call it the *collective* or *common conscience*. No doubt, it has not a specific organ as a substratum; it is, by definition, diffuse in every reach of society. Nevertheless, it has specific characteristics which make it a distinct reality. It is, in effect, independent of the particular conditions in which individuals are placed; they pass on and it remains. It is the same in the North and in the South, in great cities and in small, in different professions. Moreover, it does not change with each generation, but, on the contrary, it connects successive generations with one another. It is, thus, an entirely different thing from particular consciences, although it can be realized only through them. It is the psychical type of society, a type which has its properties, its conditions of existence, its mode of development, just as individual types, although in a different way. Thus understood, it has the right to be denoted by a special word. The one which we have just employed is not, it is true, without ambiguity. As the terms, collective and social, are often considered synonymous, one is inclined to believe that the collective conscience is the total social conscience, that is, extend it to include more than the psychic life of society, although, particularly in advanced societies, it is only a very restricted part. Judicial, governmental, scientific, industrial, in short, all special functions are of a psychic nature, since they consist in systems of representations and actions. They, however, are surely outside the common conscience. To avoid the confusion[iii] into which some have fallen, the best way would be to create a technical expression especially to designate the totality of social similitudes. However, since the use of a new word, when not absolutely necessary, is not without inconvenience, we shall employ the well-worn expression, collective or common conscience, but we shall always mean the strict sense in which we have taken it.

We can, then, to resume the preceding analysis, say that an act is criminal when it

[ii]De Candolle, *Histoire des Sciences et des Savants,* 2nd ed., p. 263.

[iii]The confusion is not without its dangers. Thus, we sometimes ask if the individual conscience varies as the collective conscience. It all depends upon the sense in which the word is taken. If it represents social likenesses, the variation is inverse, as we shall see. If it signifies the total psychic life of society, the relation is direct. It is thus necessary to distinguish them.

offends strong and defined states of the collective conscience.[iv]

The statement of this proposition is not generally called into question, but it is ordinarily given a sense very different from that which it ought to convey. We take it as if it expressed, not the essential property of crime, but one of its repercussions. We well know that crime violates very pervasive and intense sentiments, but we believe that this pervasiveness and this intensity derive from the criminal character of the act, which consequently remains to be defined. We do not deny that every delict is universally reproved, but we take as agreed that the reprobation to which it is subjected results from its delictness. But we are hard put to say what this delictness consists of. In immorality which is particularly serious? I wish such were the case, but that is to reply to the question by putting one word in place of another, for it is precisely the problem to understand what this immorality is, and especially this particular immorality which society reproves by means of organized punishment and which constitutes criminality. It can evidently come only from one or several characteristics common to all criminological types. The only one which would satisfy this condition is that opposition between a crime, whatever it is, and certain collective sentiments. It is, accordingly, this opposition which makes crime rather than being a derivative of crime. In other words, we must not say that an action shocks the common conscience because it is criminal, but rather that it is criminal because it shocks the common conscience. We do not reprove it because it is a crime, but it is a crime because we reprove it. As for the intrinsic nature of these sentiments, it is impossible to specify them. They have the most diverse objects and cannot be encompassed in a single formula. We can say that they relate neither to vital interests of society nor to a minimum of justice. All these definitions are inadequate. By this alone can we recognize it: a sentiment, whatever its origin and end, is found in all consciences with a certain degree of force and precision, and every action which violates it is a crime. Contemporary psychology is more and more reverting to the idea of Spinoza, according to which things are good

because we like them, as against our liking them because they are good. What is primary is the tendency, the inclination; the pleasure and pain are only derivative facts. It is just so in social life. An act is socially bad because society disproves of it. But, it will be asked, are there not some collective sentiments which result from pleasure and pain which society feels from contact with their ends? No doubt, but they do not all have this origin. A great many, if not the larger part, come from other causes. Everything that leads activity to assume a definite form can give rise to habits, whence result tendencies which must be satisfied. Moreover, it is these latter tendencies which alone are truly fundamental. The others are only special forms and more determinate. Thus, to find charm in such and such an object, collective sensibility must already be constituted so as to be able to enjoy it. If the corresponding sentiments are abolished, the most harmful act to society will not only be tolerated, but even honored and proposed as an example. Pleasure is incapable of creating an impulse out of whole cloth; it can only link those sentiments which exist to such and such a particular end, provided that the end be in accord with their original nature. . . .

ORGANIC SOLIDARITY

Since negative solidarity does not produce any integration by itself, and since, moreover, there is nothing specific about it, we shall recognize only two kinds of positive solidarity which are distinguishable by the following qualities:

1. The first binds the individual directly to society without any intermediary. In the second, he depends upon society, because he depends upon the parts of which it is composed.

2. Society is not seen in the same aspect in the two cases. In the first, what we call society is a more or less organized totality of beliefs and sentiments common to all the members of the group: this is the collective type. On the other hand, the society in which we are solitary in the

---

[iv]We shall not consider the question whether the collective conscience is a conscience as is that of the individual. By this term, we simply signify the totality of social likenesses, without prejudging the category by which this system of phenomena ought to be defined.

second instance is a system of different, special functions which definite relations unite. These two societies really make up only one. They are two aspects of one and the same reality, but none the less they must be distinguished.

3. From this second difference there arises another which helps us to characterize and name the two kinds of solidarity.

The first can be strong only if the ideas and tendencies common to all the members of the society are greater in number and intensity than those which pertain personally to each member. It is as much stronger as the excess is more considerable. But what makes our personality is how much of our own individual qualities we have, what distinguishes us from others. This solidarity can grow only in inverse ratio to personality. There are in each of us, as we have said, two consciences: one which is common to our group in its entirety, which, consequently, is not ourself, but society living and acting within us; the other, on the contrary, represents that in us which is personal and distinct, that which makes us an individual.[v] Solidarity which comes from likenesses is at its maximum when the collective conscience completely envelops our whole conscience and coincides in all points with it. But, at that moment, our individuality is nil. It can be born only if the community takes smaller toll of us. There are, here, two contrary forces, one centripetal, the other centrifugal, which cannot flourish at the same time. We cannot, at one and the same time, develop ourselves in two opposite senses. If we have a lively desire to think and act for ourselves, we cannot be strongly inclined to think and act as others do. If our ideal is to present a singular and personal appearance, we do not want to resemble everybody else. Moreover, at the moment when this solidarity exercises its force, our personality vanishes, as our definition permits us to say, for we are no longer ourselves, but the collective life.

The social molecules which can be coherent in this way can act together only in the measure that they have no actions of their own, as the molecules of inorganic bodies. That is why we propose to call this type of solidarity mechanical. The term does not signify that it is produced by mechanical and artificial means. We call it that only by analogy to the cohesion which unites the elements of an inanimate body, as opposed to that which makes a unity out of the elements of a living body. What justifies this term is that the link which thus unites the individual to society is wholly analogous to that which attaches a thing to a person. The individual conscience, considered in this light, is a simple dependent upon the collective type and follows all of its movements, as the possessed object follows those of its owner. In societies where this type of solidarity is highly developed, the individual does not appear, as we shall see later. Individuality is something which the society possesses. Thus, in these social types, personal rights are not yet distinguished from real rights.

It is quite otherwise with the solidarity which the division of labor produces. Whereas the previous type implies that individuals resemble each other, this type presumes their difference. The first is possible only in so far as the individual personality is absorbed into the collective personality; the second is possible only if each one has a sphere of action which is peculiar to him; that is, a personality. It is necessary, then, that the collective conscience leave open a part of the individual conscience in order that special functions may be established there, functions which it cannot regulate. The more this region is extended, the stronger is the cohesion which results from this solidarity. In effect, on the one hand, each one depends as much more strictly on society as labor is more divided; and, on the other, the activity of each is as much more personal as it is more specialized. Doubtless, as circumscribed as it is, it is never completely original. Even in the exercise of our occupation, we conform to usages, to practices which are common to our whole professional brotherhood. But, even in this instance, the yoke that we submit to is much less heavy than when society completely controls us, and it leaves much more place open for the free play of our initiative. Here, then, the individuality of all

---

[v]However, these two consciences are not in regions geographically distinct from us, but penetrate from all sides.

grows at the same time as that of its parts. Society becomes more capable of collective movement, at the same time that each of its elements has more freedom of movement. This solidarity resembles that which we observe among the higher animals. Each organ, in effect, has its special physiognomy, its autonomy. And, moreover, the unity of the organism is as great as the individuation of the parts is more marked. Because of this analogy, we propose to call the solidarity which is due to the division of labor, organic. . . .

THE CAUSES

We can then formulate the following proposition: The division of labor varies in direct ratio with the volume and density of societies, and, if it progresses in a continuous manner in the course of social development, it is because societies become regularly denser and generally more voluminous.

At all times, it is true, it has been well understood that there was a relation between these two orders of fact, for, in order that functions be more specialized, there must be more co-operators, and they must be related to co-operate. But, ordinarily, this state of societies is seen only as the means by which the division of labor develops, and not as the cause of its development. The latter is made to depend upon individual aspirations toward well-being and happiness, which can be satisfied so much better as societies are more extensive and more condensed. The law we have just established is quite otherwise. We say, not that the growth and condensation of societies *permit,* but that they *necessitate* a greater division of labor. It is not

an instrument by which the latter is realized; it is its determining cause.[vi]

THE FORCED DIVISION OF LABOR

It is not sufficient that there be rules, however, for sometimes the rules themselves are the cause of evil. This is what occurs in class-wars. The institution of classes and of castes constitutes an organization of the division of labor, and it is a strictly regulated organization, although it often is a source of dissension. The lower classes not being, or no longer being, satisfied with the role which has devolved upon them from custom or by law aspire to functions which are closed to them and seek to dispossess those who are exercising these functions. Thus civil wars arise which are due to the manner in which labor is distributed.

There is nothing similar to this in the organism. No doubt, during periods of crises, the different tissues war against one another and nourish themselves at the expense of others. But never does one cell or organ seek to usurp a role different from the one which it is filling. The reason for this is that each anatomic element automatically executes its purpose. Its constitution, its place in the organism, determines its vocation; its task is a consequence of its nature. It can badly acquit itself, but it cannot assume another's task unless the latter abandons it, as happens in the rare cases of substitution that we have spoken of. It is not so in societies. Here the possibility is greater. There is a greater distance between the hereditary dispositions of the individual and the social function he will fill. The first do not imply the second with such immediate necessity. This space, open to striving and deliberation, is also at the mercy of a

---

[vi]On this point, we can still rely on Comte as authority. "I must," he said "now indicate the progressive condensation of our species as a last general concurrent element in regulating the effective speed of the social movement. We can first easily recognize that this influence contributes a great deal, especially in origin, in determining a more special division of human labor, necessarily incompatible with a small number of co-operators. *Besides, by a most intimate and little known property, although still most important, such a condensation stimulates directly, in a very powerful manner, the most rapid development of social evolution,* either in driving individuals to new efforts to assure themselves by more refined means of an existence which otherwise would become more difficult, or by obliging society with more stubborn and better concentrated energy to fight more stiffly against the more powerful effort of particular divergences. With one and the other, we see that it is not a question here of the absolute increase of the number of individuals, but especially of their more intense concourse in a given space." *Cours,* IV, p. 455.

multitude of causes which can make individual nature deviate from its normal direction and create a pathological state. Because this organization is more supple, it is also more delicate and more accessible to change. Doubtless, we are not, from birth, predestined to some special position; but we do have tastes and aptitudes which limit our choice. If no care is taken of them, if they are ceaselessly disturbed by our daily occupations, we shall suffer and seek a way of putting an end to our suffering. But there is no other way out than to change the established order and to set up a new one. For the division of labor to produce solidarity, it is not sufficient, then, that each have his task; it is still necessary that this task be fitting to him. Now, it is this condition which is not realized in the case we are examining. In effect, if the institution of classes or castes sometimes gives rise to anxiety and pain instead of producing solidarity, this is because the distribution of social functions on which it rests does not respond, or rather no longer responds, to the distribution of natural talents. . . .

CONCLUSION

But not only does the division of labor present the character by which we have defined morality; it more and more tends to become the essential condition of social solidarity. As we advance in the evolutionary scale, the ties which bind the individual to his family, to his native soil, to traditions which the past has given to him, to collective group usages, become loose. More mobile, he changes his environment more easily, leaves his people to go elsewhere to live a more autonomous existence, to a greater extent forms his own ideas and sentiments. Of course, the whole common conscience does not, on this account, pass out of existence. At least there will always remain this cult of personality, of individual dignity of which we have just been speaking, and which, today, is the rallying-point of so many people. But how little a thing it is when one contemplates the ever increasing extent of social life, and, consequently, of individual consciences! For, as they become more voluminous, as intelligence becomes richer, activity more varied, in order for morality to remain constant, that is to say, in order for the individual to remain attached to the group with a force equal to that of yesterday, the ties which bind him to it must become stronger and more numerous. If, then, he formed no others than those which come from resemblances, the effacement of the segmental type would be accompanied by a systematic debasement of morality. Man would no longer be sufficiently obligated; he would no longer feel about and above him this salutary pressure of society which moderates his egoism and makes him a moral being. This is what gives moral value to the division of labor. Through it, the individual becomes cognizant of his dependence upon society; from it come the forces which keep him in check and restrain him. In short, since the division of labor becomes the chief source of social solidarity, it becomes, at the same time, the foundation of the moral order.

We can then say that, in higher societies, our duty is not to spread our activity over a large surface, but to concentrate and specialize it. We must contract our horizon, choose a definite task and immerse ourselves in it completely, instead of trying to make ourselves a sort of creative masterpiece, quite complete, which contains its worth in itself and not in the services that it renders. Finally, this specialization ought to be pushed as far as the elevation of the social type, without assigning any other limit to it.[vii] No doubt, we ought so to work as to realize in ourselves the collective type as it exists. There are common sentiments, common ideas, without which, as has

---

[vii]There is, however, probably another limit which we do not have to speak of since it concerns individual hygiene. It may be held that, in the light of our organico-psychic constitution, the division of labor cannot go beyond a certain limit without disorders resulting. Without entering upon the question, let us straightaway say that the extreme specialization at which biological functions have arrived does not seem favorable to this hypothesis. Moreover, in the very order of psychic and social functions, has not the division of labor, in its historical development, been carried to the last stage in the relations of men and women? Have not there been faculties completely lost by both? Why cannot the same phenomenon occur between individuals of the same sex? Of course, it takes time for the organism to adapt itself to these changes, but we do not see why a day should come when this adaptation would become impossible.

been said, one is not a man. The rule which orders us to specialize remains limited by the contrary rule. Our conclusion is not that it is good to press specialization as far as possible, but as far as necessary. As for the part that is to be played by these two opposing necessities, that is determined by experience and cannot be calculated *a priori*. It is enough for us to have shown that the second is not of a different nature from the first, but that it also is moral, and that, moreover, this duty becomes ever more important and pressing, because the general qualities which are in question suffice less and less to socialize the individual. . . .

Let us first of all remark that it is difficult to see why it would be more in keeping with the logic of human nature to develop superficially rather than profoundly. Why would a more extensive activity, but more dispersed, be superior to a more concentrated, but circumscribed, activity? Why would there be more dignity in being complete and mediocre, rather than in living a more specialized, but more intense life, particularly if it is thus possible for us to find what we have lost in this specialization, through our association with other beings who have what we lack and who complete us? We take off from the principle that man ought to realize his nature as man, to accomplish his ὄικεῖον ἔργον, as Aristotle said. But this nature does not remain constant throughout history; it is modified with societies. Among lower peoples, the proper duty of man is to resemble his companions, to realize in himself all the traits of the collective type which are then confounded, much more than today, with the human type. But, in more advanced societies, his nature is, in large part, to be an organ of society, and his proper duty, consequently, is to play his role as an organ.

Moreover, far from being trammelled by the progress of specialization, individual personality develops with the division of labor.

To be a person is to be an autonomous source of action. Man acquires this quality only in so far as there is something in him which is his alone and which individualizes him, as he is something more than a simple incarnation of the generic type of his race and his group. It will be said that he is endowed with free will and that is enough to establish his personality. But although there may be some of this liberty in him, an object of so many discussions, it is not this metaphysical, impersonal, invariable attribute which can serve as the unique basis for concrete personality, which is empirical and variable with individuals. That could not be constituted by the wholly abstract power of choice between two opposites, but it is still necessary for this faculty to be exercised towards ends and aims which are proper to the agent. In other words, the very materials of conscience must have a personal character. But we have seen in the second book of this work that this result is progressively produced as the division of labor progresses. The effacement of the segmental type, at the same time that it necessitates a very great specialization, partially lifts the individual conscience from the organic environment which supports it, as from the social environment which envelops it, and, accordingly, because of this double emancipation, the individual becomes more of an independent factor in his own conduct. The division of labor itself contributes to this enfranchisement, for individual natures, while specializing, become more complex, and by that are in part freed from collective action and hereditary influences which can only enforce themselves upon simple, general things. . . .

▪▪

Introduction to *The Rules of Sociological Method*

In *The Rules of Sociological Method* (Durkheim 1895/1966:xiii), Durkheim makes at least three essential points. Durkheim insists, (1) sociology is a distinct field of study, and (2) although the social sciences are distinct from the natural sciences, the methods of the latter can be applied to the former. In addition, Durkheim maintains, (3) the social field is also distinct from the psychological realm. Thus, sociology is the study of social phenomena or "social facts," a very different enterprise from the study of an individual's own ideas or will.

Specifically, Durkheim maintains there are two different ways that social facts can be identified. First, social facts are "general throughout the extent of a given society" at a given stage in the evolution of that society (Durkheim 1895/1966:xv,13). Second, albeit related, a social fact is marked by "any manner of action . . . capable of exercising over the individual exterior constraint" (ibid.). In other words, a "social fact" is recognized by the "coercive power which it exercises or is capable of exercising over individuals" (ibid.:10). This does not mean that there are no "exceptions" to a social fact, but that it is potentially universal in the sense that, given specific conditions, it will be likely to emerge (ibid.:xv).

The "coercive power" of social facts brings us to a critical issue raised in *The Rules of Sociological Method:* crime. Durkheim argues that crime is inevitable or "normal" in all societies because crime defines the moral boundaries of a society and, in doing so, communicates to its inhabitants the range of acceptable behaviors. For Durkheim, crime is "normal"—*not* because there will always be "bad" or "wicked" individuals in society (i.e., not for idiosyncraticic, psychological reasons, though those may well exist too), but because crime is "indispensable to the normal evolution of morality and law" (Durkheim 1895/1966:69). As he maintains, "A society exempt from [crime] is utterly impossible" because crime affirms and reaffirms the collective sentiments on which it is founded and which are necessary for its existence (ibid.:67). The formation and reformation of the collective conscience is never complete. Indeed, Durkheim maintains that even in a hypothetical "society of saints," a "perfect cloister of exemplary individuals," "faults" will appear, which will cause the same "scandal that the ordinary offense does in ordinary consciousnesses" (ibid.:68,69). It is impossible for all to be alike . . . there cannot be a society in which the individuals do not differ more or less from the collective type" (ibid.:69,70). Simply put, you cannot have a society without "crime" for the same reason that you cannot have a game without rules (i.e., you can do A, but not B) and consequences to rule violations (if you do B, this will happen). Thus, when children make up a new game, they make up not only rules, but also *consequences* for rule infractions (e.g., you have to kick the ball between the tree and the mailbox; if the ball touches your hands, you're out). So too, one could argue, society is like a game. There are rules (norms and laws), and there are consequences or punishments if you break those norms/rules/laws (whether social ostracism or jail). Most importantly, it is the consequences of the action (crime and punishment) themselves that help clarify and reaffirm *what* the rules of the game *are* and thus the basis of society itself.

From *The Rules of Sociological Method* (1895)

Émile Durkheim

WHAT IS A SOCIAL FACT?

Before inquiring into the method suited to the study of social facts, it is important to know which facts are commonly called "social." This information is all the more necessary since the designation "social" is used with little precision. It is currently employed for practically all phenomena generally diffused within society, however small their social interest. But on that basis, there are, as it were, no human events that may not be called social. Each individual drinks, sleeps, eats, reasons; and it is to society's interest that these functions be exercised in an orderly manner. If, then, all these facts are counted as "social" facts, sociology would have no subject

SOURCE: Reprinted with permission of The Free Press, a Division of Simon & Schuster Adult Publishing Group, from *The Rules of Sociological Method* by Émile Durkheim, translated by Sarah A. Soloway and John H. Mueller, edited by George E. G. Catlin. Copyright © 1938 by George E. G. Catlin. Copyright renewed © 1966 by Sarah A. Soloway, John H. Mueller, George E. G. Catlin. All rights reserved.

matter exclusively its own, and its domain would be confused with that of biology and psychology.

But in reality there is in every society a certain group of phenomena which may be differentiated from those studied by the other natural sciences. When I fulfill my obligations as brother, husband, or citizen, when I execute my contracts, I perform duties which are defined, externally to myself and my acts, in law and in custom. Even if they conform to my own sentiments and I feel their reality subjectively, such reality is still objective, for I did not create them; I merely inherited them through my education. How many times it happens, moreover, that we are ignorant of the details of the obligations incumbent upon us, and that in order to acquaint ourselves with them we must consult the law and its authorized interpreters! Similarly, the church-member finds the beliefs and practices of his religious life ready-made at birth; their existence prior to his own implies their existence outside of himself. The system of signs I use to express my thought, the system of currency I employ to pay my debts, the instruments of credit I utilize in my commercial relations, the practices followed in my profession, etc., function independently of my own use of them. And these statements can be repeated for each member of society. Here, then, are ways of acting, thinking, and feeling that present the noteworthy property of existing outside the individual consciousness.

These types of conduct or thought are not only external to the individual but are, moreover, endowed with coercive power, by virtue of which they impose themselves upon him, independent of his individual will. Of course, when I fully consent and conform to them, this constraint is felt only slightly, if at all, and is therefore unnecessary. But it is, nonetheless, an intrinsic characteristic of these facts, the proof thereof being that it asserts itself as soon as I attempt to resist it. If I attempt to violate the law, it reacts against me so as to prevent my act before its accomplishment, or to nullify my violation by restoring the damage, if it is accomplished and reparable, or to make me expiate it if it cannot be compensated for otherwise.

In the case of purely moral maxims; the public conscience exercises a check on every act which offends it by means of the surveillance it exercises over the conduct of citizens, and the appropriate penalties at its disposal. In many cases the constraint is less violent, but nevertheless it always exists. If I do not submit to the conventions of society, if in my dress I do not conform to the customs observed in my country and in my class, the ridicule I provoke, the social isolation in which I am kept, produce, although in an attenuated form, the same effects as a punishment in the strict sense of the word. The constraint is nonetheless efficacious for being indirect. I am not obliged to speak French with my fellow-countrymen nor to use the legal currency, but I cannot possibly do otherwise. If I tried to escape this necessity, my attempt would fail miserably. As an industrialist, I am free to apply the technical methods of former centuries; but by doing so, I should invite certain ruin. Even when I free myself from these rules and violate them successfully, I am always compelled to struggle with them. When finally overcome, they make their constraining power sufficiently felt by the resistance they offer. The enterprises of all innovators, including successful ones, come up against resistance of this kind.

Here, then, is a category of facts with very distinctive characteristics: it consists of ways of acting, thinking, and feeling, external to the individual, and endowed with a power of coercion, by reason of which they control him. These ways of thinking could not be confused with biological phenomena, since they consist of representations and of actions; nor with psychological phenomena, which exist only in the individual consciousness and through it. They constitute, thus, a new variety of phenomena; and it is to them exclusively that the term "social" ought to be applied. And this term fits them quite well, for it is clear that, since their source is not in the individual, their substratum can be no other than society, either the political society as a whole or some one of the partial groups it includes, such as religious denominations, political, literary, and occupational associations, etc. On the other hand, this term "social" applies to them exclusively, for it has a distinct meaning only if it designates exclusively the phenomena which are not included in any of the categories of facts that have already been established and classified. These ways of thinking and acting therefore constitute the proper domain of sociology. It is true that, when we define them with this word "constraint," we risk shocking the zealous partisans of absolute individualism. For those who profess the complete autonomy of the individual, man's dignity is diminished whenever he is made to feel that he is not completely self-determinant. It is generally accepted today, however, that most of our ideas and our tendencies are not developed by ourselves but come to us from without. How can they become a part of us except by imposing themselves upon us? This is the whole meaning of our definition. And it is generally accepted, moreover, that social constraint is not necessarily incompatible with the individual personality.[i]

Since the examples that we have just cited (legal and moral regulations, religious faiths, financial systems, etc.) all consist of established beliefs and practices, one might be led to believe that social facts exist only where there is some social organization. But there are other facts without such crystallized form which have the same objectivity and the same ascendancy over the individual. These are called "social currents." Thus the great movements of enthusiasm, indignation, and pity in a crowd do not originate in any one of the particular individual consciousnesses. They come to each one of us from without and can carry us away in spite of ourselves. Of course, it may happen that, in abandoning myself to them unreservedly, I do not feel the pressure they exert upon me. But it is revealed as soon as I try to resist them. Let an individual attempt to oppose one of these collective manifestations, and the emotions that he denies will turn against him. Now, if this power of external coercion asserts itself so clearly in cases of resistance, it must exist also in the first-mentioned cases, although we are unconscious of it. We are then victims of the illusion of having ourselves created that which actually forced itself from without. If the complacency with which we permit ourselves to be carried along conceals the pressure undergone, nevertheless it does not abolish it. Thus, air is no less heavy because we do not detect its weight. So, even if we ourselves have spontaneously contributed to the production of the common emotion, the impression we have received differs markedly from that which we would have experienced if we had been alone. Also, once the crowd has dispersed, that is, once

these social influences have ceased to act upon us and we are alone again, the emotions which have passed through the mind appear strange to us, and we no longer recognize them as ours. We realize that these feelings have been impressed upon us to a much greater extent than they were created by us. It may even happen that they horrify us, so much were they contrary to our nature. Thus, a group of individuals, most of whom are perfectly inoffensive, may, when gathered in a crowd, be drawn into acts of atrocity. And what we say of these transitory outbursts applies similarly to those more permanent currents of opinion on religious, political, literary, or artistic matters which are constantly being formed around us, whether in society as a whole or in more limited circles.

To confirm this definition of the social fact by a characteristic illustration from common experience, one need only observe the manner in which children are brought up. Considering the facts as they are and as they have always been, it becomes immediately evident that all education is a continuous effort to impose on the child ways of seeing, feeling, and acting which he could not have arrived at spontaneously. From the very first hours of his life, we compel him to eat, drink, and sleep at regular hours; we constrain him to cleanliness, calmness, and obedience; later we exert pressure upon him in order that he may learn proper consideration for others, respect for customs and conventions, the need for work, etc. If, in time, this constraint ceases to be felt, it is because it gradually gives rise to habits and to internal tendencies that render constraint unnecessary; but nevertheless it is not abolished, for it is still the source from which these habits were derived. It is true that, according to Spencer, a rational education ought to reject such methods, allowing the child to act in complete liberty; but as this pedagogic theory has never been applied by any known people, it must be accepted only as an expression of personal opinion, not as a fact which can contradict the aforementioned observations. What makes these facts particularly instructive is that the aim of education is, precisely, the socialization of the human being; the process of education, therefore, gives us in a nutshell the historical fashion in which the social being is constituted. This unremitting pressure to which the child is subjected is the very pressure of the

---

[i]We do not intend to imply, however, that all constraints are normal. We shall return to this point later.

social milieu which tends to fashion him in its own image, and of which parents and teachers are merely the representatives and intermediaries.

It follows that sociological phenomena cannot be defined by their universality. A thought which we find in every individual consciousness, a movement repeated by all individuals, is not thereby a social fact. If sociologists have been satisfied with defining them by this characteristic, it is because they confused them with what one might call their reincarnation in the individual. It is, however, the collective aspects of the beliefs, tendencies, and practices of a group that characterize truly social phenomena. As for the forms that the collective states assume when refracted in the individual, these are things of another sort. This duality is clearly demonstrated by the fact that these two orders of phenomena are frequently found dissociated from one another. Indeed, certain of these social manners of acting and thinking acquire, by reason of their repetition, a certain rigidity which on its own account crystallizes them, so to speak, and isolates them from the particular events which reflect them. They thus acquire a body, a tangible form, and constitute a reality in their own right, quite distinct from the individual facts which produce it. Collective habits are inherent not only in the successive acts which they determine but, by a privilege of which we find no example in the biological realm, they are given permanent expression in a formula which is repeated from mouth to mouth, transmitted by education, and fixed even in writing. Such is the origin and nature of legal and moral rules, popular aphorisms and proverbs, articles of faith wherein religious or political groups condense their beliefs, standards of taste established by literary schools, etc. None of these can be found entirely reproduced in the applications made of them by individuals, since they can exist even without being actually applied.

No doubt, this dissociation does not always manifest itself with equal distinctness, but its obvious existence in the important and numerous cases just cited is sufficient to prove that the social fact is a thing distinct from its individual manifestations. Moreover, even when this dissociation is not immediately apparent, it may often be disclosed by certain devices of method. Such dissociation is indispensable if one wishes to separate social facts from their alloys in order to observe them in a state of purity. Currents of opinion, with an intensity varying according to the time and place, impel certain groups either to more marriages, for example, or to more suicides, or to a higher or lower birthrate, etc. These currents are plainly social facts. At first sight they seem inseparable from the forms they take in individual cases. But statistics furnish us with the means of isolating them. They are, in fact, represented with considerable exactness by the rates of births, marriages, and suicides, that is, by the number obtained by dividing the average annual total of marriages, births, suicides, by the number of persons whose ages lie within the range in which marriages, births, and suicides occur.[ii] Since each of these figures contains all the individual cases indiscriminately, the individual circumstances which may have had a share in the production of the phenomenon are neutralized and, consequently, do not contribute to its determination. The average, then, expresses a certain state of the group mind (*l'âme collective*).

Such are social phenomena, when disentangled from all foreign matter. As for their individual manifestations, these are indeed, to a certain extent, social, since they partly reproduce a social model. Each of them also depends, and to a large extent, on the organopsychological constitution of the individual and on the particular circumstances in which he is placed. Thus they are not sociological phenomena in the strict sense of the word. They belong to two realms at once; one could call them sociopsychological. They interest the sociologist without constituting the immediate subject matter of sociology. There exist in the interior of organisms similar phenomena, compound in their nature, which form in their turn the subject matter of the "hybrid sciences," such as physiological chemistry, for example.

The objection may be raised that a phenomenon is collective only if it is common to all members of society, or at least to most of them—in other words, if it is truly general. This may be true; but it is general because it is collective (that is, more or less obligatory), and certainly not collective because general. It is a group condition repeated in the individual because imposed on him. It is to be found in each part because it exists in the whole, rather than in the whole because it exists in the parts. This becomes conspicuously evident in those beliefs and practices which are transmitted to us

---

[ii]Suicides do not occur at every age, and they take place with varying intensity at the different ages in which they occur.

ready-made by previous generations; we receive and adopt them because, being both collective and ancient, they are invested with a particular authority that education has taught us to recognize and respect. It is, of course, true that a vast portion of our social culture is transmitted to us in this way; but even when the social fact is due in part to our direct collaboration, its nature is not different. A collective emotion which bursts forth suddenly and violently in a crowd does not express merely what all the individual sentiments had in common; it is something entirely different, as we have shown. It results from their being together, a product of the actions and reactions which take place between individual consciousnesses; and if each individual consciousness echoes the collective sentiment, it is by virtue of the special energy resident in its collective origin. If all hearts beat in unison, this is not the result of a spontaneous and pre-established harmony but rather because an identical force propels them in the same direction. Each is carried along by all.

We thus arrive at the point where we can formulate and delimit in a precise way the domain of sociology. It comprises only a limited group of phenomena. A social fact is to be recognized by the power of external coercion which it exercises or is capable of exercising over individuals, and the presence of this power may be recognized in its turn either by the existence of some specific sanction or by the resistance offered against every individual effort that tends to violate it. One can, however, define it also by its diffusion within the group, provided that, in conformity with our previous remarks, one takes care to add as a second and essential characteristic that its own existence is independent of the individual forms it assumes in its diffusion. This last criterion is perhaps, in certain cases, easier to apply than the preceding one. In fact, the constraint is easy to ascertain when it expresses itself externally by some direct reaction of society, as is the case in law, morals, beliefs, customs, and even fashions. But when it is only indirect, like the constraint which an economic organization exercises, it cannot always be so easily detected. Generality combined with externality may, then, be easier to establish. Moreover, this second definition is but another form of the first; for if a mode of behavior whose existence is external to individual consciousnesses becomes general, this can only be brought about by its being imposed upon them.

But these several phenomena present the same characteristic by which we defined the others. These "ways of existing" are imposed on the individual precisely in the same fashion as the "ways of acting" of which we have spoken. Indeed, when we wish to know how a society is divided politically, of what these divisions themselves are composed, and how complete is the fusion existing between them, we shall not achieve our purpose by physical inspection and by geographical observations; for these phenomena are social, even when they have some basis in physical nature. It is only by a study of public law that a comprehension of this organization is possible, for it is this law that determines the organization, as it equally determines our domestic and civil relations. This political organization is, then, no less obligatory than the social facts mentioned above. If the population crowds into our cities instead of scattering into the country, this is due to a trend of public opinion, a collective drive that imposes this concentration upon the individuals. We can no more choose the style of our houses than of our clothing—at least, both are equally obligatory. The channels of communication prescribe the direction of internal migrations and commerce, etc., and even their extent. Consequently, at the very most, it should be necessary to add to the list of phenomena which we have enumerated as presenting the distinctive criterion of a social fact only one additional category, "ways of existing"; and, as this enumeration was not meant to be rigorously exhaustive, the addition would not be absolutely necessary.

Such an addition is perhaps not necessary, for these "ways of existing" are only crystallized "ways of acting." The political structure of a society is merely the way in which its component segments have become accustomed to live with one another. If their relations are traditionally intimate, the segments tend to fuse with one another, or, in the contrary case, to retain their identity. The type of habitation imposed upon us is merely the way in which our contemporaries and our ancestors have been accustomed to construct their houses. The methods of communication are merely the channels which the regular currents of commerce and migrations have dug, by flowing in the same direction. To be sure, if the phenomena of a structural character alone presented this performance, one might believe that they constituted a distinct species. A legal regulation is an arrangement no less permanent than a type of architecture, and yet the regulation is a "physiological" fact. A simple moral maxim is assuredly somewhat more malleable, but it is much more rigid than a simple professional custom or a fashion. There is thus a whole series of degrees without a break in continuity between the facts of the most articulated structure and those free currents of social life which are not yet definitely molded. The differences between them are, therefore, only differences in the degree of consolidation they present. Both are simply life, more or less crystallized. No doubt, it may be of some advantage to reserve the term "morphological" for those social facts which concern the social substratum, but only on condition of not overlooking the fact that they are of the same nature as the others. Our definition will then include the whole relevant range of facts if we say: *A social fact is every way of acting, fixed or not, capable of exercising on the individual an external constraint; or again, every way of acting which is general throughout a given society, while at the same time existing in its own right independent of its individual manifestations. . . .*[iii]

THE NORMAL AND THE PATHOLOGICAL

If there is any fact whose pathological character appears incontestable, that fact is crime. All criminologists are agreed on this point. Although they explain this pathology differently, they are unanimous in recognizing it. But let us see if this problem does not demand a more extended consideration. . . .

Crime is present not only in the majority of societies of one particular species but in all societies of all types. There is no society that is not confronted with the problem of criminality. Its form changes; the acts thus characterized are not the same everywhere; but, everywhere and always, there have been men who have behaved in such a way as to draw upon themselves penal repression.

If, in proportion as societies pass from the lower to the higher types, the rate of criminality, i.e., the relation between the yearly number of crimes and the population, tended to decline, it might be believed that crime, while still normal, is tending to lose this character of normality. But we have no reason to believe that such a regression is substantiated. Many facts would seem rather to indicate a movement in the opposite direction. From the beginning of the [nineteenth] century, statistics enable us to follow the course of criminality. It has everywhere increased. In France the increase is nearly 300 per cent. There is, then, no phenomenon that presents more indisputably all the symptoms of normality, since it appears closely connected with the conditions of all collective life. To make of crime a form of social morbidity would be to admit that morbidity is not something accidental, but, on the contrary, that in certain cases it grows out of the fundamental constitution of the living organism; it would result in wiping out all distinction between the physiological and the pathological. No doubt it is possible that crime itself will have abnormal forms, as, for example, when its rate is unusually high. This excess is, indeed, undoubtedly morbid in nature. What is normal, simply, is the existence of criminality, provided that it attains and does not exceed, for each social type, a certain level, which it is perhaps not impossible to fix in conformity with the preceding rules.[iv]

Here we are, then, in the presence of a conclusion in appearance quite paradoxical. Let us make no mistake. To classify crime among the phenomena of normal sociology is not to say merely that it is an inevitable, although regrettable phenomenon, due to the incorrigible wickedness of men; it is to affirm that it is a factor in public health, an integral part of all healthy societies. This result is, at first glance, surprising enough to have puzzled even ourselves for a long time. Once this first surprise has been overcome, however, it is not difficult to find reasons explaining this normality and at the same time confirming it.

[iii]This close connection between life and structure, organ and function, may be easily proved in sociology because between these two extreme terms there exists a whole series of immediately observable intermediate stages which show the bond between them. Biology is not in the same favorable position. But we may well believe that the inductions on this subject made by sociology are applicable to biology and that, in organisms as well as in societies, only differences in degree exist between these two orders of facts.

[iv]From the fact that crime is a phenomenon of normal sociology, it does not follow that the criminal is an individual normally constituted from the biological and psychological points of view. The two questions are independent of each other. This independence will be better understood when we have shown, later on, the difference between psychological and sociological facts.

In the first place crime is normal because a society exempt from it is utterly impossible. Crime, we have shown elsewhere, consists of an act that offends certain very strong collective sentiments. In a society in which criminal acts are no longer committed, the sentiments they offend would have to be found without exception in all individual consciousnesses, and they must be found to exist with the same degree as sentiments contrary to them. Assuming that this condition could actually be realized, crime would not thereby disappear; it would only change its form, for the very cause which would thus dry up the sources of criminality would immediately open up new ones.

Indeed, for the collective sentiments which are protected by the penal law of a people at a specified moment of its history to take possession of the public conscience or for them to acquire a stronger hold where they have an insufficient grip, they must acquire an intensity greater than that which they had hitherto had. The community as a whole must experience them more vividly, for it can acquire from no other source the greater force necessary to control these individuals who formerly were the most refractory. . . .

Imagine a society of saints, a perfect cloister of exemplary individuals. Crimes, properly so called, will there be unknown; but faults which appear venial to the layman will create there the same scandal that the ordinary offense does in ordinary consciousnesses. If, then, this society has the power to judge and punish, it will define these acts as criminal and will treat them as such. For the same reason, the perfect and upright man judges his smallest failings with a severity that the majority reserve for acts more truly in the nature of an offense. Formerly, acts of violence against persons were more frequent than they are today, because respect for individual dignity was less strong. As this has increased, these crimes have become more rare; and also, many acts violating this sentiment have been introduced into the penal law which were not included there in primitive times.[v]

In order to exhaust all the hypotheses logically possible, it will perhaps be asked why this unanimity does not extend to all collective sentiments without exception. Why should not even the most feeble sentiment gather enough energy to prevent all dissent? The moral consciousness of the society would be present in its entirety in all the individuals, which a vitality sufficient to prevent all acts offending it—the purely conventional faults as well as the crimes. But a uniformity so universal and absolute is utterly impossible; for the immediate physical milieu in which each one of us is placed, the hereditary antecedents, and the social influences vary from one individual to the next, and consequently diversify consciousnesses. It is impossible for all to be alike, if only because each one has his own organism and that these organisms occupy different areas in space. That is why, even among the lower peoples, where individual originality is very little developed, it nevertheless does exist.

Thus, since there cannot be a society in which the individuals do not differ more or less from the collective type, it is also inevitable that, among these divergences, there are some with a criminal character. What confers this character upon them is not the intrinsic quality of a given act but that definition which the collective conscience lends them. If the collective conscience is stronger, if it has enough authority practically to suppress these divergences, it will also be more sensitive, more exacting; and, reacting against the slightest deviations with the energy it otherwise displays only against more considerable infractions, it will attribute to them the same gravity as formerly to crimes. In other words, it will designate them as criminal.

Crime is, then, necessary; it is bound up with the fundamental conditions of all social life, and by that very fact it is useful, because these conditions of which it is a part are themselves indispensable to the normal evolution of morality and law.

Indeed, it is no longer possible today to dispute the fact that law and morality vary from one social type to the next, nor that they change within the same type if the conditions of life are modified. But, in order that these transformations may be possible, the collective sentiments at the basis of morality must not be hostile to change, and consequently must have but moderate energy. If they were too strong, they would no longer be plastic. Every pattern is an obstacle to new patterns, to the extent that the first pattern is inflexible. The better a structure is articulated, the more it offers a healthy resistance to all modification; and this is equally true of

---

[v]Calumny, insults, slander, fraud, etc.

functional, as of anatomical, organization. If there were no crimes, this condition could not have been fulfilled; for such a hypothesis presupposes that collective sentiments have arrived at a degree of intensity unexampled in history. Nothing is good indefinitely and to an unlimited extent. The authority which the moral conscience enjoys must not be excessive; otherwise no one would dare criticize it, and it would too easily congeal into an immutable form. To make progress, individual originality must be able to express itself. In order that the originality of the idealist whose dreams transcend his century may find expression, it is necessary that the originality of the criminal, who is below the level of his time, shall also be possible. One does not occur without the other.

Nor is this all. Aside from this indirect utility, it happens that crime itself plays a useful role in this evolution. Crime implies not only that the way remains open to necessary changes but that in certain cases it directly prepares these changes. Where crime exists, collective sentiments are sufficiently flexible to take on a new form, and crime sometimes helps to determine the form they will take. How many times, indeed, it is only an anticipation of future morality—a step toward what will be! According to Athenian law, Socrates was a criminal, and his condemnation was no more than just. However, his crime, namely, the independence of his thought, rendered a service not only to humanity but to his country. It served to prepare a new morality and faith which the Athenians needed, since the traditions by which they had lived until then were no longer in harmony with the current conditions of life. Nor is the case of Socrates unique; it is reproduced periodically in history. It would never have been possible to establish the freedom of thought we now enjoy if the regulations prohibiting it had not been violated before being solemnly abrogated. At that time, however, the violation was a crime, since it was an offense against sentiments still very keen in the average conscience. And yet this crime was useful as a prelude to reforms which daily became more necessary. Liberal philosophy had as its precursors the heretics of all kinds who were justly punished by secular authorities during the entire course of the Middle Ages and until the eve of modern times.

From this point of view the fundamental facts of criminality present themselves to us in an entirely new light. Contrary to current ideas, the criminal no longer seems a totally unsociable being, a sort of parasitic element, a strange and inassimilable body, introduced into the midst of society.[vi] On the contrary, he plays a definite role in social life. . . .

Introduction to *Suicide: A Study in Sociology*

Suicide (1897) is both a theoretical and methodological exemplar. In this famous study, Durkheim examines a phenomenon that most people think of as an intensely individual act—suicide—and demonstrates its *social* (rather than psychological) roots. His method for doing this is to analyze *rates* of suicide between societies and historical periods and between different social groups within the same society. By linking the different suicide rates of particular societies and social groups to the specific characteristics of that society or social group, Durkheim not only demonstrates that individual pathologies are rooted in *social* conditions, but, in addition, shows how sociologists can scientifically study social behavior. His innovative examination of suicide rates lent credibility to his conviction that sociology should be considered a viable scientific discipline.

Most importantly, Durkheim argues that the places with the highest rates of alcoholism and mental illness are not the areas with the highest suicide rates (thereby undermining the notion that it is pathological psychological states that are solely determinative of the individual act of suicide). Rather, Durkheim maintains that suicide rates are highest in moments when and in places where individuals lack social and moral regulation or integration. In addition, as in his first book, *The Division of Labor in Society*, in *Suicide* Durkheim was particularly interested in delineating the fundamental differences

---

[vi]We have ourselves committed the error of speaking thus of the criminal, because of a failure to apply our rule (*Division du travail social*, pp. 395–96).

between traditional and modern societies. Durkheim sought to explain why suicide is rare in small, simple societies while much more frequent in modern, industrial ones. Parallel to his argument in *The Division of Labor*, Durkheim argues that traditional and modern societies differ not only in their rates of suicide, but in the *types* of suicide that are prevalent as well.

Specifically, Durkheim saw two main characteristics of modern, industrial society: (1) a lack of integration of the individual in the social group and (2) a lack of moral regulation. Durkheim used the term *egoism* to refer to the lack of integration of the individual in the social group. He used the term *anomie* to refer to a lack of moral regulation. Durkheim argued that both of these conditions—egoism and anomie—are "chronic" in modern, industrial society; and that in extreme, pathological form, both egoism and anomie can result in suicide. Let's look at these two different, albeit intimately interrelated, conditions in turn.

For Durkheim, egoistic suicide results from a pathological weakening of the bonds between the individual and the social group. This lack of integration is evident statistically, in that there are higher rates of suicide among single, divorced, and widowed persons than among married persons, and that there are higher rates of suicide among married persons without children than there are among married persons with children. Additionally, Durkheim argued that egoism helps explain why suicide rates are higher among Protestants than among Catholics or Jews: Protestantism emphasizes an *individual* relationship with God, which means that the individual is less bound to the religious clergy and members of the congregation. Interestingly, then, Durkheim maintains that it is not Catholic doctrine that inhibits the act of suicide; rather, it is Catholics' *social bonds*, their association with the priests, nuns, and other lay members of the congregation, that deters them from this act. Protestant rates of suicide are higher because Protestants are more socially and spiritually isolated than the more communally oriented Jews and Catholics.

Durkheim saw an increase in egoistic suicide as a "natural" outgrowth of the individuation of modern, industrial societies. For instance, today it is quite common—especially in big cities—for people to live alone. By contrast, in many traditional societies it is virtually unheard of for anyone to live alone. Children live with parents until they get married; parents move in with children (or vice versa) if a spouse dies; unmarried siblings live with either parents or other siblings. As we noted above, Durkheim argued that in its extreme form the type of social isolation found in modern societies can be—literally—fatal.

Intertwined with an increasing lack of social integration in modern, industrial societies is a lack of moral integration. Durkheim used the term *anomie* to refer to this lack of moral regulation. Anomic suicide is the pathological result of a lack of moral direction, when one feels morally adrift. Durkheim viewed modern societies as "chronically" anomic, or characterized by a lack of regulation of the individual by the collective.

Thus, for instance, modern industrial societies are religiously pluralistic, whereby people are more able to freely choose among a variety of religious faiths—or to choose not to "believe" at all. Similarly, today many people choose to "identify"—or not—with a specific part of their ethnic heritage. That we spend much time and energy searching for "identity"—"I'm a punk!" "I'm Irish!"—reflects a lack of moral regulation. To be sure, there are many wonderful benefits from this increasing individuation, which contrasts significantly from small, traditional, homogeneous societies in which "who" you are is taken for granted. In small, closed, indigenous societies without so many (or any) options, where there is one religion and one ethnic group, your place in that society is a cultural given—a "place" that may be quite oppressive. Not surprisingly, then, Durkheim asserts that suppressing individuation also can produce pathological consequences. (This point will be discussed further below.)

The lack of moral regulation in modern societies is especially prevalent in times of intense social and personal change. During such periods, the authority of the family, the church, and the community may be challenged or questioned; without moral guidance and authority, individuals may feel like they have no moral anchor. The pursuit of individual desires and goals can overtake moral concerns. However, Durkheim maintains that anomie can result not only from "bad" social change, such as losing one's job or a political crisis, but from "positive" social change as well. Consider, for instance, what happens when someone wins the lottery. Most people think that if they were to win the lottery, they would experience only happiness. Indeed, some people buy lottery tickets thinking, "If I win the 'big one,' all my problems will be solved!" However, Durkheim contends that sudden life-changing events can bring on a battery of social and personal issues that one might not expect.

First, after winning the lottery, you might suddenly find yourself confronted with weighty existential issues. Before the lottery, you may have simply worked—and worked hard— because you needed to earn a living. But now that you've won the lottery, you don't know what to do. By not having to work, you might start thinking about things, such as the meaning of life, that you had never thought about before. This feeling that you don't know "what to do" and "how to act" is a state of anomie.

In addition, you might start to wonder how much friends and family should get from your winnings. You might begin to feel like everyone just wants your money and that it is hard to tell who likes *you* and who just likes your newfound fame and fortune. You might feel like you can't talk to your friends about your dilemma, that no one in your previous social circle really "understands" you anymore. You may begin to find that you can't relate to the people from your old socioeconomic class, but that you can't relate to anyone in your new class either. Thus, the sudden change brought about by winning the lottery can lead not only to feeling morally "anchorless" (anomie), but also to feeling socially alone (egoism). A most extreme outcome of feeling this moral and social isolation would be suicide.

As we noted previously, Durkheim argued that traditional and modern societies are rooted in different social conditions. Compared to modern societies, social regulation is intensive in traditional societies, thus limiting the development of individuality. In extreme form, such restrictions can lead to altruistic suicide, where an individual gives his life for

Photo 3.2 In a modern-day incident of altruistic suicide, a number of South Vietnamese Buddhist monks used self-immolation to protest the persecution of the country's majority Buddhist population at the hands of the Catholic president, Ngô Đình Diệm. Here, Thích Quảng Đức burns himself to death on a Saigon street, June 11, 1963.

the social group. According to Durkheim, this is the primary type of suicide that occurs in small, traditional societies where individuation is minimal. The classic type of altruistic suicide was the Aztecs' practice of human sacrifice, in which a person was sacrificed for the moral or spiritual benefit of the group.[1]

Today many sociologists find fault with Durkheim's distinction between "modern" and "traditional" societies. This binary opposition seems to be a function of the Eurocentrism of his day: social scientists tended to imagine that their societies were extremely "complex," while "traditional" societies were just "simple." Indeed, "traditional" and "modern" societies may have more in common than Durkheim let on. The degree of integration of the individual into the collective social group is a complex process rather than a permanent state. For instance, even though Durkheim saw altruistic suicide as more prevalent in "primitive" societies, sadly, it is far from absent in "modern" societies as well. Not unlike the altruistic suicides in primitive societies, modern-day wars and suicide bombings are carried out on the premise that sacrificing one's life is necessary for the fight to preserve or attain a sacred way of life for the group as a whole. Nowhere are the similarities between these expressions of altruistic suicide (soldiers, suicide bombers, and "primitive" human sacrifice) more readily apparent than in the tragic case of the Japanese kamikaze (suicide) pilots of World War II. Shockingly, kamikaze flights were a principal tactic of Japan in the last year of the war.[2]

From *Suicide: A Study in Sociology* (1897)

Émile Durkheim

ANOMIC SUICIDE

I

But society is not only something attracting the sentiments and activities of individuals with unequal force. It is also a power controlling them. There is a relation between the way this regulative action is performed and the social suicide-rate.

It is a well-known fact that economic crises have an aggravating effect on the suicidal tendency.

In Vienna, in 1873 a financial crisis occurred which reached its height in 1874; the number of suicides immediately rose. From 141 in 1872, they rose to 153 in 1873 and 216 in 1874. The increase

---

[1]Durkheim briefly mentioned another type of suicide prevalent in "primitive" societies—"fatalistic suicide." For Durkheim, fatalistic suicide was rooted in *hopelessness*—the hopelessness of oppressed people, such as slaves, who had not the slightest chance of changing their personal situation.

[2]In October 1944, some 1,200 kamikaze (which translates from Japanese as "god wind") plunged to their deaths in an attack on a U.S. naval fleet in the Leyte Gulf in the Philippines. Six months later, some 1,900 kamikaze dove to their deaths in the battle of Okinawa, resulting in the death of more than 5,000 American sailors. Most of the kamikaze pilots involved were men in their teens or early 20s. They were said to have gone to their deaths "joyfully," having followed specific rituals of cleanliness, and equipped with books with uplifting thoughts to "transcend life and death" and "[b]e always pure-hearted and cheerful" (Daniel Ford, Review of *Kamikaze: Japan's Suicide Gods*, by Albert Axell and Hideaki Kase, *Wall Street Journal*, September 10, 2002).

in 1874 is 53 per cent[i] above 1872 and 41 per cent above 1873. What proves this catastrophe to have been the sole cause of the increase is the special prominence of the increase when the crisis was acute, or during the first four months of 1874. From January 1 to April 30 there had been 48 suicides in 1871, 44 in 1872, 43 in 1873; there were 73 in 1874. The increase is 70 per cent.[ii] The same crisis occurring at the same time in Frankfurt-on-Main produced the same effects there. In the years before 1874, 22 suicides were committed annually on the average; in 1874 there were 32, or 45 per cent more.

The famous crash is unforgotten which took place on the Paris Bourse during the winter of 1882. Its consequences were felt not only in Paris but throughout France. From 1874 to 1886 the average annual increase was only 2 per cent; in 1882 it was 7 per cent. Moreover, it was unequally distributed among the different times of year, occurring principally during the first three months or at the very time of the crash. Within these three months alone 59 per cent of the total rise occurred. So distinctly is the rise the result of unusual circumstances that it not only is not encountered in 1881 but has disappeared in 1883, although on the whole the latter year had a few more suicides than the preceding one:

This relation is found not only in some exceptional cases, but is the rule. The number of bankruptcies is a barometer of adequate sensitivity, reflecting the variations of economic life. When they increase abruptly from year to year, some serious disturbance has certainly occurred. From 1845 to 1869 there were sudden rises, symptomatic of crises, on three occasions. While the annual increase in the number of bankruptcies during this period is 3.2 per cent, it is 26 per cent in 1847, 37 per cent in 1854 and 20 per cent in 1861. At these three moments, there is also to be observed an unusually rapid rise in the number of suicides. While the average annual increase

during these 24 years was only 2 per cent, it was 17 per cent in 1847, 8 per cent in 1854 and 9 per cent in 1861.

But to what do these crises owe their influence? Is it because they increase poverty by causing public wealth to fluctuate? Is life more readily renounced as it becomes more difficult? The explanation is seductively simple; and it agrees with the popular idea of suicide. But it is contradicted by facts.

Actually, if voluntary deaths increased because life was becoming more difficult, they should diminish perceptibly as comfort increases. Now, although when the price of the most necessary foods rises excessively, suicides generally do the same, they are not found to fall below the average in the opposite case. In Prussia, in 1850 wheat was quoted at the lowest point it reached during the entire period of 1848–81; it was at 6.91 marks per 50 kilograms; yet at this very time suicides rose from 1,527 where they were in 1849 to 1,736, or an increase of 13 per cent, and continued to increase during the years 1851, 1852 and 1853 although the cheap market held. In 1858–59 a new fall took place; yet suicides rose from 2,038 in 1857 to 2,126 in 1858, and to 2,146 in 1859. From 1863 to 1866 prices which had reached 11.04 marks in 1861 fell progressively to 7.95 marks in 1864 and remained very reasonable for the whole period; suicides during the same time increased 17 per cent (2,112 in 1862, 2,485 in 1866).[iii] Similar facts are observed in Bavaria. According to a curve

	1881	**1882**	**1883**
Annual total	6,741	7,213 (plus 7%)	7,267
First three months	1,589	1,770 (plus 11%)	1,604

constructed by Mayr[iv] for the period 1835–61, the price of rye was lowest during the years 1857–58 and 1858–59; now suicides, which in 1857 numbered only 286, rose to 329 in 1858, to 387 in 1859. The same phenomenon had already occurred during the years 1848–50; at that time wheat had been very cheap in Bavaria as well as throughout Europe. Yet, in spite of a slight

---

[i]Durkheim incorrectly gives this figure as 51 percent.—Ed.

[ii]In 1874 over 1873.—Ed.

[iii]See Starck, *Verbrechen und Vergehen in Preussen,* Berlin, 1884, p. 55.

[iv]*Die Gesetzmässigkeit im Gesellschaftsleben,* p. 345.

temporary drop due to political events, which we have mentioned, suicides remained at the same level. There were 217 in 1847, there were still 215 in 1848, and if they dropped for a moment to 189 in 1849, they rose again in 1850 and reached 250.

So far is the increase in poverty from causing the increase in suicide that even fortunate crises, the effect of which is abruptly to enhance a country's prosperity, affect suicide like economic disasters.

The conquest of Rome by Victor-Emmanuel in 1870, by definitely forming the basis of Italian unity, was the starting point for the country of a process of growth which is making it one of the great powers of Europe. Trade and industry received a sharp stimulus from it and surprisingly rapid changes took place. Whereas in 1876, 4,459 steam boilers with a total of 54,000 horse-power were enough for industrial needs, the number of machines in 1887 was 9,983 and their horse-power of 167,000 was threefold more. Of course the amount of production rose proportionately during the same time.[v] Trade followed the same rising course; not only did the merchant marine, communications and transportation develop, but the number of persons and things transported doubled.[vi] As this generally heightened activity caused an increase in salaries (an increase of 35 per cent is estimated to have taken place from 1873 to 1889), the material comfort of workers rose, especially since the price of bread was falling at the same time.[vii] Finally, according to calculations by Bodio, private wealth rose from 45 and a half billions on the average during the period 1875–80 to 51 billions during the years 1880–85 and 54 billions and a half in 1885–90.[viii]

Now, an unusual increase in the number of suicides is observed parallel with this collective renaissance. From 1866 to 1870 they were

roughly stable; from 1871 to 1877 they increased 36 per cent. There were in

1864–70	29 suicides per million	1874	37 suicides per million
1871	31 suicides per million	1875	34 suicides per million
1872	33 suicides per million	1876	36.5 suicides per million
1873	36 suicides per million	1877	40.6 suicides per million

And since then the movement has continued. The total figure, 1,139 in 1877, was 1,463 in 1889, a new increase of 28 per cent.

In Prussia the same phenomenon occurred on two occasions. In 1866 the kingdom received a first enlargement. It annexed several important provinces, while becoming the head of the Confederation of the North. Immediately this growth in glory and power was accompanied by a sudden rise in the number of suicides. There had been 123 suicides per million during the period 1856–60 per average year and only 122 during the years 1861–65. In the five years, 1866–70, in spite of the drop in 1870, the average rose to 133. The year 1867, which immediately followed victory, was that in which suicide achieved the highest point it had reached since 1816 (1 suicide per 5,432 inhabitants, while in 1864 there was only one case per 8,739).

On the morrow of the war of 1870 a new accession of good fortune took place. Germany was unified and placed entirely under Prussian hegemony. An enormous war indemnity addedto the public wealth; commerce and industry made great strides. The development of suicide was never so rapid. From 1875 to 1886 it increased 90 per cent, from 3,278 cases to 6,212.

World expositions, when successful, are considered favorable events in the existence of a society. They stimulate business, bring more money into the country and are thought to increase public prosperity, especially in the city where they take place. Yet, quite possibly, they ultimately take their toll in a considerably higher number of suicides. Especially does this seem

---

[v]See Fornasari di Verce, *La criminalita e le vicende economiche d'Italia,* Turin 1894, pp. 7783.

[vi]Ibid., pp. 108–117.

[vii]Ibid., pp. 86–104.

[viii]The increase is less during the period 1885–90 because of a financial crisis.

to have been true of the Exposition of 1878. The rise that year was the highest occurring between 1874 and 1886. It was

	1888	1889	1890
The seven months of the Exposition	517	567	540
The five other months	319	311	356

8 per cent, that is, higher than the one caused by the crash of 1882. And what almost proves the Exposition to have been the cause of this increase is that 86 per cent of it took place precisely during the six months of the Exposition.

In 1889 things were not identical all over France. But quite possibly the Boulanger crisis neutralized the contrary effects of the Exposition by its depressive influence on the growth of suicides. Certainly at Paris, although the political feeling aroused must have had the same effect as in the rest of the country, things happened as in 1878. For the 7 months of the Exposition, suicides increased almost 10 per cent, 9.66 to be exact, while through the remainder of the year they were below what they had been in 1888 and what they afterwards were in 1890.

It may well be that but for the Boulanger influence the rise would have been greater.

What proves still more conclusively that economic distress does not have the aggravating influence often attributed to it, is that it tends rather to produce the opposite effect. There is very little suicide in Ireland, where the peasantry leads so wretched a life. Poverty-stricken Calabria has almost no suicides; Spain has a tenth as many as France. Poverty may even be considered a protection. In the various French departments the more people there are who have independent means, the more numerous are suicides. . . .

If therefore industrial or financial crises increase suicides, this is not because they cause poverty, since crises of prosperity have the same result; it is because they are crises, that is, disturbances of the collective order.[ix]

Departments Where Suicides Were Committed (1878–1887; per 100,000 Inhabitants)		Average Number of Persons of Independent Means per 1,000 Inhabitants in Each Group of Department (1886)
Suicides	Number of Departments	
From 48 to 43	5	127
From 38 to 31	6	73
From 30 to 24	6	69
From 23 to 18	15	59
From 17 to 13	18	49
From 12 to 8	26	49
From 7 to 3	10	42

Every disturbance of equilibrium, even though it achieves greater comfort and a heightening of general vitality, is an impulse to voluntary death. Whenever serious readjustments take place in the social order, whether or not due to a sudden growth or to an unexpected catastrophe, men are more inclined to self-destruction. How is this possible? How can something considered generally to improve existence serve to detach men from it?

For the answer, some preliminary considerations are required.

[ix]To prove that an increase in prosperity diminishes suicides, the attempt has been made to show that they become less when emigration, the escape-valve of poverty, is widely practiced (See Legoyt, pp. 257–59). But cases are numerous where parallelism instead of inverse proportions exist between the two. In Italy from 1876 to 1890 the number of emigrants rose from 76 per 100,000 inhabitants to 335, a figure itself exceeded between 1887 and 1889. At the same time suicides did not cease to grow in nnumbers.

II

No living being can be happy or even exist unless his needs are sufficiently proportioned to his means. In other words, if his needs require more than can be granted, or even merely something of a different sort, they will be under continual friction and can only function painfully. Movements incapable of production without pain tend not to be reproduced. Unsatisfied tendencies atrophy, and as the impulse to live is merely the result of all the rest, it is bound to weaken as the others relax.

In the animal, at least in a normal condition, this equilibrium is established with automatic spontaneity because the animal depends on purely material conditions. All the organism needs is that the supplies of substance and energy constantly employed in the vital process should be periodically renewed by equivalent quantities; that replacement be equivalent to use. When the void created by existence in its own resources is filled, the animal, satisfied, asks nothing further. Its power of reflection is not sufficiently developed to imagine other ends than those implicit in its physical nature. On the other hand, as the work demanded of each organ itself depends on the general state of vital energy and the needs of organic equilibrium, use is regulated in turn by replacement and the balance is automatic. The limits of one are those of the other; both are fundamental to the constitution of the existence in question, which cannot exceed them.

This is not the case with man, because most of his needs are not dependent on his body or not to the same degree. Strictly speaking, we may consider that the quantity of material supplies necessary to the physical maintenance of a human life is subject to computation, though this be less exact than in the preceding case and a wider margin left for the free combinations of the will; for beyond the indispensable minimum which satisfies nature when instinctive, a more awakened reflection suggests better conditions, seemingly desirable ends craving fulfillment. Such appetites, however, admittedly sooner or later reach a limit which they cannot pass. But how determine the quantity of well-being, comfort or luxury legitimately to be craved by a human being? Nothing appears in man's organic nor in his psychological constitution which sets a limit to such tendencies. The functioning of individual life does not require them to cease at one point rather than at another; the proof being that they have constantly increased since the beginnings of history, receiving more and more complete satisfaction, yet with no weakening of average health. Above all, how establish their proper variation with different conditions of life, occupations, relative importance of services, etc.? In no society are they equally satisfied in the different stages of the social hierarchy. Yet human nature is substantially the same among all men, in its essential qualities. It is not human nature which can assign the variable limits necessary to our needs. They are thus unlimited so far as they depend on the individual alone. Irrespective of any external regulatory force, our capacity for feeling is in itself an insatiable and bottomless abyss.

But if nothing external can restrain this capacity, it can only be a source of torment to itself. Unlimited desires are insatiable by definition and insatiability is rightly considered a sign of morbidity. Being unlimited, they constantly and infinitely surpass the means at their command; they cannot be quenched. Inextinguishable thirst is constantly renewed torture. It has been claimed, indeed, that human activity naturally aspires beyond assignable limits and sets itself unattainable goals. But how can such an undetermined state be any more reconciled with the conditions of mental life than with the demands of physical life? All man's pleasure in acting, moving and exerting himself implies the sense that his efforts are not in vain and that by walking he has advanced. However, one does not advance when one walks toward no goal, or—which is the same thing—when his goal is infinity. Since the distance between us and it is always the same, whatever road we take, we might as well have made the motions without progress from the spot. Even our glances behind and our feeling of pride at the distance covered can cause only deceptive satisfaction, since the remaining distance is not proportionately reduced. To pursue a goal which is by definition unattainable is to condemn oneself to a state of perpetual unhappiness. Of course, man may hope contrary to all reason, and hope has its pleasures even when unreasonable. It may sustain him for a time; but it cannot survive the repeated disappointments of experience indefinitely. What more can the future offer him than the past, since he can never reach a tenable condition nor even approach the glimpsed ideal? Thus, the more one has, the more one wants, since satisfactions received only stimulate instead of filling needs. Shall action as such be considered agreeable? First, only on condition of blindness to its uselessness. Secondly, for this pleasure to be felt and to temper and half veil the accompanying painful unrest, such unending motion must at least always be easy and unhampered. If it is interfered with only restlessness is left, with the lack of ease which it, itself, entails. But it would be a miracle if no insurmountable obstacle were never encountered. Our thread of life on these conditions is pretty thin, breakable at any instant.

To achieve any other result, the passions first must be limited. Only then can they be harmonized with the faculties and satisfied. But since the individual has no way of limiting them, this must be done by some force exterior to him. A regulative force must play the same role for moral needs which the organism plays for physical needs. This means that the force can only be moral. The awakening of conscience interrupted the state of equilibrium of the animal's dormant existence; only conscience, therefore, can furnish the means to re-establish it. Physical restraint would be ineffective; hearts cannot be touched by physio-chemical forces. So far as the appetites are not automatically restrained by physiological mechanisms, they can be halted only by a limit that they recognize as just. Men would never consent to restrict their desires if they felt justified in passing the assigned limit. But, for reasons given above, they cannot assign themselves this law of justice. So they must receive it from an authority which they respect, to which they yield spontaneously. Either directly and as a whole or through the agency of one of its organs, society alone can play this moderating role; for it is the only moral power superior to the individual, the authority of which he accepts. It alone has the

power necessary to stipulate law and to set the point beyond which the passions must not go. Finally, it alone can estimate the reward to be prospectively offered to every class of human functionary, in the name of the common interest.

As a matter of fact, at every moment of history there is a dim perception, in the moral consciousness of societies, of the respective value of different social services, the relative reward due to each, and the consequent degree of comfort appropriate on the average to workers in each occupation. The different functions are graded in public opinion and a certain coefficient of well-being assigned to each, according to its place in the hierarchy. According to accepted ideas, for example, a certain way of living is considered the upper limit to which a workman may aspire in his efforts to improve his existence, and there is another limit below which he is not willingly permitted to fall unless he has seriously bemeaned himself. Both differ for city and country workers, for the domestic servant and the day-laborer, for the business clerk and the official, etc. Likewise the man of wealth is reproved if he lives the life of a poor man, but also if he seeks the refinements of luxury overmuch. Economists may protest in vain; public feeling will always be scandalized if an individual spends too much wealth for wholly superfluous use, and it even seems that this severity relaxes only in times of moral disturbance.[x] A genuine regimen exists, therefore, although not always legally formulated, which fixes with relative precision the maximum degree of ease of living to which each social class may legitimately aspire. However, there is nothing immutable about such a scale. It changes with the increase or decrease of collective revenue and the changes occurring in the moral ideas of society. Thus what appears luxury to one period no longer does so to another; and the well-being which for long periods was granted to a class only by exception and supererogation, finally appears strictly necessary and equitable.

Under this pressure, each in his sphere vaguely realizes the extreme limit set to his ambitions and aspires to nothing beyond. At least if he respects regulations and is docile to

---

[x]Actually, this is a purely moral reprobation and can hardly be judicially implemented. We do not consider any reestablishment of sumptuary laws desirable or even possible.

collective authority, that is, has a wholesome moral constitution, he feels that it is not well to ask more. Thus, an end and goal are set to the passions. Truly, there is nothing rigid nor absolute about such determination. The economic ideal assigned each class of citizens is itself confined to certain limits, within which the desires have free range. But it is not infinite. This relative limitation and the moderation it involves, make men contented with their lot while stimulating them moderately to improve it; and this average contentment causes the feeling of calm, active happiness, the pleasure in existing and living which characterizes health for societies as well as for individuals. Each person is then at least, generally speaking, in harmony with his condition, and desires only what he may legitimately hope for as the normal reward of his activity. Besides, this does not condemn man to a sort of immobility. He may seek to give beauty to his life; but his attempts in this direction may fail without causing him to despair. For, loving what he has and not fixing his desire solely on what he lacks, his wishes and hopes may fail of what he has happened to aspire to, without his being wholly destitute. He has the essentials. The equilibrium of his happiness is secure because it is defined, and a few mishaps cannot disconcert him.

But it would be of little use for everyone to recognize the justice of the hierarchy of functions established by public opinion, if he did not also consider the distribution of these functions just. The workman is not in harmony with his social position if he is not convinced that he has his deserts. If he feels justified in occupying another, what he has would not satisfy him. So it is not enough for the average level of needs for each social condition to be regulated by public opinion, but another, more precise rule, must fix the way in which these conditions are open to individuals. There is no society in which such regulation does not exist. It varies with times and places. Once it regarded birth as the almost exclusive principle of social classification; today it recognizes no other inherent inequality than hereditary fortune and merit. But in all these various forms its object is unchanged. It is also only possible, everywhere, as a restriction upon individuals imposed by superior authority, that is, by collective authority. For it can be established only by requiring of one or another group of men, usually of all, sacrifices and concessions in the name of the public interest.

Some, to be sure, have thought that this moral pressure would become unnecessary if men's economic circumstances were only no longer determined by heredity. If inheritance were abolished, the argument runs, if everyone began life with equal resources and if the competitive struggle were fought out on a basis of perfect equality, no one could think its results unjust. Each would instinctively feel that things are as they should be.

Truly, the nearer this ideal equality were approached, the less social restraint will be necessary. But it is only a matter of degree. One sort of heredity will always exist, that of natural talent. Intelligence, taste, scientific, artistic, literary or industrial ability, courage and manual dexterity are gifts received by each of us at birth, as the heir to wealth receives his capital or as the nobleman formerly received his title and function. A moral discipline will therefore still be required to make those less favored by nature accept the lesser advantages which they owe to the chance of birth. Shall it be demanded that all have an equal share and that no advantage be given those more useful and deserving? But then there would have to be a discipline far stronger to make these accept a treatment merely equal to that of the mediocre and incapable.

But like the one first mentioned, this discipline can be useful only if considered just by the peoples subject to it. When it is maintained only by custom and force, peace and harmony are illusory; the spirit of unrest and discontent are latent; appetites superficially restrained are ready to revolt. This happened in Rome and Greece when the faiths underlying the old organization of the patricians and plebeians were shaken, and in our modern societies when aristocratic prejudices began to lose their old ascendancy. But this state of upheaval is exceptional; it occurs only when society is passing through some abnormal crisis. In normal conditions the collective order is regarded as just by the great majority of persons. Therefore, when we say that an authority is necessary to impose this order on individuals, we certainly do not mean that violence is the only means of establishing it. Since this regulation is meant to restrain individual passions, it must come from a power which dominates individuals; but this power must also be obeyed through respect, not fear.

It is not true, that human activity can be released from all restraint. Nothing in the world can enjoy such a privilege. All existence being a part of the universe is relative to the remainder; its nature and method of manifestation accordingly depend not only on itself but on other beings, who consequently restrain and regulate it. Here there are only differences of degree and form between the mineral realm and the thinking person. Man's characteristic privilege is that the bond he accepts is not physical but moral; that is, social. He is governed not by a material environment brutally imposed on him, but by a conscience superior to his own, the superiority of which he feels. Because the greater, better part of his existence transcends the body, he escapes the body's yoke, but is subject to that of society.

But when society is disturbed by some painful crisis or by beneficent but abrupt transitions, it is momentarily incapable of exercising this influence; thence come the sudden rises in the curve of suicides which we have pointed out above.

In the case of economic disasters, indeed, something like a declassification occurs which suddenly casts certain individuals into a lower state than their previous one. Then they must reduce their requirements, restrain their needs, learn greater self-control. All the advantages of social influence are lost so far as they are concerned; their moral education has to be recommenced. But society cannot adjust them instantaneously to this new life and teach them to practice the increased self-repression to which they are unaccustomed. So they are not adjusted to the condition forced on them, and its very prospect is intolerable; hence the suffering which detaches them from a reduced existence even before they have made trial of it.

It is the same if the source of the crisis is an abrupt growth of power and wealth. Then, truly, as the conditions of life are changed, the standard according to which needs were regulated can no longer remain the same; for it varies with social resources, since it largely determines the share of each class of producers. The scale is upset; but a new scale cannot be immediately improvised. Time is required for the public conscience to reclassify men and things. So long as the social forces thus freed have not regained equilibrium, their respective values are unknown and so all regulation is lacking for a time. The limits are unknown between the possible and the impossible, what is just and what is unjust, legitimate claims and hopes and those which are immoderate. Consequently, there is no restraint upon aspirations. If the disturbance is profound, it affects even the principles controlling the distribution of men among various occupations. Since the relations between various parts of society are necessarily modified, the ideas expressing these relations must change. Some particular class especially favored by the crisis is no longer resigned to its former lot, and, on the other hand, the example of its greater good fortune arouses all sorts of jealousy below and about it. Appetites, not being controlled by a public opinion become disoriented, no longer recognize the limits proper to them. Besides, they are at the same time seized by a sort of natural erethism simply by the greater intensity of public life. With increased prosperity desires increase. At the very moment when traditional rules have lost their authority, the richer prize offered these appetites stimulates them and makes them more exigent and impatient of control. The state of de-regulation or anomy is thus further heightened by passions being less disciplined, precisely when they need more disciplining.

But then their very demands make fulfillment impossible. Overweening ambition always exceeds the results obtained, great as they may be, since there is no warning to pause here. Nothing gives satisfaction and all this agitation is uninterruptedly maintained without appeasement. Above all, since this race for an unattainable goal can give no other pleasure but that of the race itself, if it is one, once it is interrupted the participants are left empty-handed. At the same time the struggle grows more violent and painful, both from being less controlled and because competition is greater. All classes contend among themselves because no established classification any longer exists. Effort grows, just when it becomes less productive. How could the desire to live not be weakened under such conditions?

This explanation is confirmed by the remarkable immunity of poor countries. Poverty protects against suicide because it is a restraint in itself. No matter how one acts, desires have to depend upon resources to some extent; actual possessions are partly the criterion of those aspired to. So the less one has the less he is tempted to extend the range of his needs indefinitely. Lack of power, compelling moderation, accustoms men to it, while nothing excites envy if no one has superfluity. Wealth, on the other hand, by the power it bestows, deceives us into believing that we depend on ourselves only. Reducing the resistance we encounter from objects, it suggests the possibility of unlimited success against them. The less limited one feels, the more intolerable all limitation appears. Not without reason, therefore, have so many religions dwelt on the advantages and moral value of poverty. It is actually the best school for teaching self-restraint. Forcing us to constant self-discipline, it prepares us to accept collective discipline with equanimity, while wealth, exalting the individual, may always arouse the spirit of rebellion which is the very source of immorality. This, of course, is no reason why humanity should not improve its material condition. But though the moral danger involved in every growth of prosperity is not irremediable, it should not be forgotten.

III

If anomy never appeared except, as in the above instances, in intermittent spurts and acute crisis, it might cause the social suicide-rate to vary from time to time, but it would not be a regular, constant factor. In one sphere of social life, however—the sphere of trade and industry—it is actually in a chronic state.

For a whole century, economic progress has mainly consisted in freeing industrial relations from all regulation. Until very recently, it was the function of a whole system of moral forces to exert this discipline. First, the influence of religion was felt alike by workers and masters, the poor and the rich. It consoled the former and taught them contentment with their lot by informing them of the providential nature of the social order, that the share of each class was assigned by God himself, and by holding out the hope for just compensation in a world to come in return for the inequalities of this world. It governed the latter, recalling that worldly interests are not man's entire lot, that they must be subordinate to other and higher interests, and that they should therefore not be pursued without rule or measure. Temporal power, in turn, restrained the scope of economic functions by its supremacy over them and by the relatively subordinate role it assigned them. Finally, within the business world proper, the occupational groups by regulating salaries, the price of products and production itself, indirectly fixed the average level of income on which needs are partially based by the very force of circumstances. However, we do not mean to propose this organization as a model. Clearly it would be inadequate to existing societies without great changes. What we stress is its existence, the fact of its useful influence, and that nothing today has come to take its place.

Actually, religion has lost most of its power. And government, instead of regulating economic life, has become its tool and servant. The most opposite schools, orthodox economists and extreme socialists, unite to reduce government to the role of a more or less passive intermediary among the various social functions. The former wish to make it simply the guardian of individual contracts; the latter leave it the task of doing the collective bookkeeping, that is, of recording the demands of consumers, transmitting them to producers, inventorying the total revenue and distributing it according to a fixed formula. But both refuse it any power to subordinate other social organs to itself and to make them converge toward one dominant aim. On both sides nations are declared to have the single or chief purpose of achieving industrial prosperity; such is the implication of the dogma of economic materialism, the basis of both apparently opposed systems. And as these theories merely express the state of opinion, industry, instead of being still regarded as a means to an end transcending itself, has become the supreme end of individuals and societies alike. Thereupon the appetites thus excited have become freed of any limiting authority. By sanctifying them, so to speak, this apotheosis of well-being has placed them above all human law. Their restraint seems like a sort of sacrilege. For this reason, even the purely utilitarian regulation of them exercised by the industrial world itself through the medium of occupational groups has been unable to persist. Ultimately, this liberation of desires has been made worse by the very development of industry and the almost infinite extension of the market. So long as the producer could gain his profits only in his immediate neighborhood, the restricted amount of possible gain could not much overexcite ambition. Now that he may assume to have almost the entire world as his customer, how could passions accept their former confinement in the face of such limitless prospects?

Such is the source of the excitement predominating in this part of society, and which has thence extended to the other parts. There, the state of crisis and anomy is constant and, so to speak, normal. From top to bottom of the ladder, greed is aroused without knowing where to find ultimate foothold. Nothing can calm it, since its goal is far beyond all it can attain. Reality seems valueless by comparison with the dreams of fevered imaginations; reality is therefore abandoned, but so too is possibility abandoned when it in turn becomes reality. A thirst arises for novelties, unfamiliar pleasures, nameless sensations, all of which lose their savor once known. Henceforth one has no strength to endure the least reverse. The whole fever subsides and the sterility of all the tumult is apparent, and it is seen that all these new sensations in their infinite quantity cannot form a solid foundation of happiness to support one during days of trial. The wise man, knowing how to enjoy achieved results without having constantly to replace them with others, finds in them an attachment to life in the hour of difficulty. But the man who has always pinned all his hopes on the future and lived with his eyes fixed upon it, has nothing in the past as a comfort against the present's afflictions, for the past was nothing to him but a series of hastily experienced stages. What blinded him to himself was his expectation always to find further on the happiness he had so far missed. Now he is stopped in his tracks; from now on nothing remains behind or ahead of him to fix his gaze upon. Weariness alone, moreover, is enough to bring disillusionment, for he cannot in the end escape the futility of an endless pursuit.

We may even wonder if this moral state is not principally what makes economic catastrophes of our day so fertile in suicides. In societies where a man is subjected to a healthy discipline, he submits more readily to the blows of chance. The necessary effort for sustaining a little more discomfort costs him relatively little, since he is used to discomfort and constraint. But when every constraint is hateful in itself, how can closer constraint not seem intolerable? There is no tendency to resignation in the feverish impatience of men's lives. When there is no other aim but to outstrip constantly the point arrived at, how painful to be thrown back! Now this very lack of organization characterizing our economic condition throws the door wide to every sort of adventure. Since imagination is hungry for novelty, and ungoverned, it gropes at random. Setbacks necessarily increase with risks and thus crises multiply, just when they are becoming more destructive.

Yet these dispositions are so inbred that society has grown to accept them and is accustomed to think them normal. It is everlastingly repeated that it is man's nature to be eternally dissatisfied, constantly to advance, without relief or rest, toward an indefinite goal. The longing for infinity is daily represented as a mark of moral distinction, whereas it can only appear within unregulated consciences which elevate to a rule the lack of rule from which they suffer. The doctrine of the most ruthless and swift progress has become an article of faith. But other theories appear parallel with those praising the advantages of instability, which, generalizing the situation that gives them birth, declare life evil, claim that it is richer in grief than in pleasure and that it attracts men only by false claims. Since this disorder is greatest in the economic world, it has most victims there.

Industrial and commercial functions are really among the occupations which furnish the greatest number of suicides (see Table XXIV).

Table XXIV   Suicides per Million Persons of Different Occupations

	Trade	Transportation	Industry	Agriculture	Liberal* Professions
France (1878–87)[†]	440	–	340	240	300
Switzerland (1876)	664	1,514	577	304	558
Italy (1866–76)	277	152.6	80.4	26.7	618[‡]
Prussia (1883–90)	754	–	456	315	832
Bavaria (1884–91)	465	–	369	153	454
Belgium (1886–90)	421	–	160	160	100
Wurttemberg (1873–78)	273	–	190	206	–
Saxony (1878)		341.59[§]		71.17	–

* When statistics distinguish several different sorts of liberal occupation, we show as a specimen the one in which the suicide-rate is highest.
† From 1826 to 1880 economic functions seem less affected (see Compte-rendu of 1880); but were occupational statistics very accurate?
‡ This figure is reached only by men of letters.
§ Figure represents Trade, Transportation, and Industry combined for Saxony. Ed.

Almost on a level with the liberal professions, they sometimes surpass them; they are especially more afflicted than agriculture, where the old regulative forces still make their appearance felt most and where the fever of business has least penetrated. Here is best called what was once the general constitution of the economic order. And the divergence would be yet greater if, among the suicides of industry, employers were distinguished from workmen, for the former are probably most stricken by the state of anomy. The enormous rate of those with independent means (720 per million) sufficiently shows that the possessors of most comfort suffer most. Everything that enforces subordination attenuates the effects of this state. At least the horizon of the lower classes is limited by those above them, and for this same reason their desires are more modest. Those who have only empty space above them are almost inevitably lost in it, if no force restrains them.

Anomy, therefore, is a regular and specific factor in suicide in our modern societies; one of the springs from which the annual contingent feeds. So we have here a new type to distinguish from the others. It differs from them in its dependence, not on the way in which individuals are attached to society, but on how it regulates them. Egoistic suicide results from man's no longer finding a basis for existence in life; altruistic suicide, because this basis for existence appears to man situated beyond life itself. The third sort of suicide, the existence of which has just been shown, results from man's activity's lacking regulation and his consequent sufferings. By virtue of its origin we shall assign this last variety the name of *anomic suicide.*

Certainly, this and egoistic suicide have kindred ties. Both spring from society's insufficient presence in individuals. But the sphere of its absence is not the same in both cases. In egoistic suicide it is deficient in truly collective activity, thus depriving the latter of object and meaning. In anomic suicide, society's influence is lacking in the basically individual passions, thus leaving them without a check-rein. In spite of their relationship, therefore, the two types are independent of each other. We may offer society everything social in us, and still be unable to control our desires;

one may live in an anomic state without being egoistic, and vice versa. These two sorts of suicide therefore do not draw their chief recruits from the same social environments; one has its principal field among intellectual careers, the world of thought—the other, the industrial or commercial world.

IV

But economic anomy is not the only anomy which may give rise to suicide.

The suicides occurring at the crisis of widowhood, of which we have already spoken[xi] are really due to domestic anomy resulting from the death of husband or wife. A family catastrophe occurs which affects the survivor. He is not adapted to the new situation in which he finds himself and accordingly offers less resistance to suicide.

But another variety of anomic suicide should draw greater attention, both because it is more chronic and because it will serve to illustrate the nature and functions of marriage.

In the *Annales de demographie internationale* (September 1882), Bertillon published a remarkable study of divorce, in which he proved the following proposition: throughout Europe the number of suicides varies with that of divorces and separations [Table XXV illustrates such variations]. . . .

INDIVIDUAL FORMS OF THE DIFFERENT TYPES OF SUICIDE

One result now stands out prominently from our investigation: namely, that there are not one but various forms of suicide. Of course, suicide is always the act of a man who prefers death to life. But the causes determining him are not of the same sort in all cases: they are even sometimes mutually opposed. Now, such difference in causes must reappear in their effects. We may therefore be sure that there are several sorts of suicide which are distinct in quality from one another. But the certainty that these differences exist is not enough; we need to observe them directly and know of what they consist. We need to see the

characteristics of special suicides grouped in distinct classes corresponding to the types just distinguished. Thus we would follow the various currents which generate suicide from their social origins to their individual manifestations.

This morphological classification, which was hardly possible at the commencement of this study, may be undertaken now that an aetiological classification forms its basis. Indeed, we only need to start with the three kinds of factors which we have just assigned to suicide and discover whether the distinctive properties it assumes in manifesting itself among individual persons may be derived from them, and if so, how. Of course, not all the peculiarities which suicide may present can be deduced in this fashion; for some may exist which depend solely on the person's own nature. Each victim of suicide gives his act a personal stamp which expresses his temperament, the special conditions in which he is involved, and which, consequently, cannot be explained by the social and general causes of the phenomenon. But these causes in turn must stamp the suicides they determine with a shade all their own, a special mark expressive of them. This collective mark we must find.

To be sure, this can be done only approximately. We are not in a position to describe methodically all the suicides daily committed by men or committed in the course of history. We can only emphasize the most general and striking characteristics without even having an objective criterion for making the selection. Moreover, we can only proceed deductively in relating them to the respective causes from which they seem to spring. All that we can do is to show their logical implication, though the reasoning may not always be able to receive experimental confirmation. We do not forget that a deduction uncontrolled by experiment is always questionable. Yet this research is far from being useless, even with these reservations. Even though it may be considered only a method of illustrating the preceding results by examples, it would still have the worth of giving them a more concrete character by connecting them more closely with the data of sense-perception and with the details of daily experience. It will also introduce some little distinctiveness into this mass of facts usually lumped together as though varying only by

---

[xi]See above, Book II, Ch. 3.

Table XXV   Comparison of European States from the Point of View of Both Divorce and Suicide

	Annual Divorces per 1,000 Marriages		Suicides per Million Inhabitants
I. Countries Where Divorce and Separation Are Rare			
Norway	0.54	(1875–80)	73
Russia	1.6	(1871–77)	30
England and Wales	1.3	(1871–79)	68
Scotland	2.1	(1871–81)	–
Italy	3.05	(1871–73)	31
Finland	3.9	(1875–79)	30.8
Averages	2.07		46.5
II. Countries Where Divorce and Separation Are of Average Frequency			
Bavaria	5.0	(1881)	90.5
Belgium	5.1	(1871–80)	68.5
Holland	6.0	(1871–80)	35.5
Sweden	6.4	(1871–80)	81
Baden	6.5	(1874–79)	156.6
France	7.5	(1871–79)	150
Wurttemberg	8.4	(1876–78)	162.4
Prussia	–		133
Averages	6.4		109.6
III. Countries Where Divorce and Separation Are Frequent			
Kingdom of Saxony	26.9	(1876–80)	299
Denmark	38	(1871–80)	258
Switzerland	47	(1876–80)	216
Averages	37.3		257

shades, though there are striking differences among them. Suicide is like mental alienation. For the popular mind the latter consists in a single state, always identical, capable only of superficial differentiation according to circumstances. For the alienist, on the contrary, the word denotes many nosological types. Every suicide is, likewise, ordinarily considered a victim of melancholy whose life has become a burden to him. Actually, the acts by which a man renounces life belong to different species, of wholly different moral and social significance.

▪▪
▪▪

Introduction to *The Elementary Forms of Religious Life*

In his final and most theoretically acclaimed book, *The Elementary Forms of Religious Life* (1912), Durkheim sought to explain the way the moral realm worked by focusing on religion. Durkheim saw religious ceremonies not merely as a celebration of supernatural deities, but as a worshipping of social life itself, such that as long as there are societies, there will be religion (Robertson 1970:13).

In other words, for Durkheim, social life—whether in traditional or modern society—is inherently religious, for "religious force is nothing other than the collective and anonymous force" of society (Durkheim 1912/1995:210). The worship of transcendent gods or spirits and the respect and awe accorded to their power is in actuality the worship of the social group and the force it exerts over the individual. No matter how "simple" or "complex" the society, religion is thus a "system of ideas with which the individuals represent to themselves the society of which they are members, and the obscure but intimate relations which they have with it . . . for it is an eternal truth that outside of us there exists something greater than us, with which we enter into communion" (ibid.:257). For Durkheim, this outside power, this "something greater" is society.

In saying that social life is inherently religious, Durkheim defined religion in a very broad way. For Durkheim, "religion" does not mean solely "churchly" or institutional things; rather, religion is a system of symbols and rituals about the sacred that is practiced by a community of believers. This definition of religion is often called "functionalist" rather than "substantive" because it emphasizes not the substantive content of religion, such as particular rituals or doctrines (e.g., baptisms or bar mitzvahs, or belief in an afterlife, higher beings, etc.), but the social *function* of religion.

For Durkheim, the primary function of religion is to encode the system of relations of the group (Eliade and Couliano 1991:2). It focuses and reaffirms the collective sentiments and ideas that hold the group together. Religious practices, accordingly, serve to bind participants together in celebration of the society (Robertson 1970:15). As Durkheim (1912/1995:429) states,

> [t]here can be no society which does not feel the need of upholding and reaffirming at regular intervals the collective sentiments and the collective ideas which makes its unity and its personality. Now this moral remaking cannot be achieved except by the means of reunions, assemblies and meetings where the individuals, being closely united to one another, reaffirm in common their common sentiments.

This communal function of religion is carried out through the dual processes of ritualization and symbolization. A **ritual** is a highly routinized act such as taking communion. As the name reveals, the Christian ritual of communion not only commemorates a historical event in the life of Jesus, but also represents participation in the unity ("communion") of believers (McGuire 1997:187). Most interestingly, because they are practices (not beliefs or values), rituals can unite a social group regardless of individual differences in beliefs or strength of convictions. It is the common *experience* and *focus* that binds the participants together (see Photos 3.3a and 3.3b).

Because the central issue for Durkheim is communal practice and experience (rather than symbolic content), Durkheim sees no essential difference between "religious" and "secular" ritual acts. "Let us pray" (an opening moment in a religious service) and "Let us stand for the national anthem" (an opening moment of a baseball game) are both ritual acts that bond the individual to a community. In exactly the same way, Durkheim suggested that there is no essential difference between religious holidays, such as Passover or Christmas, and secular holidays, such as Independence Day or Thanksgiving. Both are collective celebrations of identity and community (see Edles 2002:27–30). Individuals know they are moved; they just don't understand the real causes for their feelings. Religious ritual moments are ones in which the moral authority of the group is perceived as (or chalked up to) a *spiritual* force.

As noted above, in addition to ritual practices, there is another important means through which the communal function of religion is achieved: symbolization. A **symbol** is something that stands for something else. It is a representation that calls up collective ideas and meanings. Thus, for instance, a "cross" is a marker that symbolizes Christian spirituality or tradition. Wearing a cross on a necklace often *means* that one is a Christian. It identifies the wearer as a member of a specific religious community or specific shared ideas (e.g., a religious tradition in which Jesus Christ is understood as the son of God). Most importantly, symbols such as the cross are capable of calling up and reaffirming shared meaning and the

Photo 3.3a   Congregation Taking Communion at a Catholic Church

Photo 3.3b   Fans at Sporting Event Doing "the Wave"

Both church goers and sports fans engage in communal ritual acts. As Durkheim (1912/1995:262) states, "It is by uttering the same cry, pronouncing the same word, or performing the same gesture in regard to some object that they become and feel themselves to be in unison."

feeling of community in between periodic ritual acts (such as religious celebrations and weekly church services). As Durkheim (1912/1995:232) states, "Without symbols, social sentiments could have only a precarious existence."

In *The Elementary Forms of Religious Life* (1912), Durkheim explains that symbols are classified as fundamentally sacred or profane. The **sacred** refers to the extraordinary, that which is set apart from and "above and beyond" the everyday world. In direct contrast to the sacred realm, is the realm of the everyday world of the mundane or routine, or the **profane**. Most importantly, objects are intrinsically neither sacred nor profane; rather, their meaning or classification is continually produced and reproduced (or altered) in collective processes of ritualization and symbolization. Thus, for instance, lighting a candle can either be a relatively mundane task to provide light or it can be a sacred act, as in the case of the Jewish ritual of lighting a candle to commemorate the Sabbath (McGuire 1997:17). In the latter context, this act denotes a sacred *moment* as well as a celebration. This points to the central function of the distinction between the sacred and the profane. It imposes an orderly system on the inherently untidy experience of living (Gamson 1998:141). Thus, for instance, ritual practices (e.g., standing for the national anthem or lighting a candle to commemorate the Sabbath) transform a profane moment into a sacred moment, while sacred sites (churches, mosques, synagogues) differentiate "routine" places from those that compel attitudes of awe and inspiration. The symbolic plasticity of time and space is especially apparent in the way devout Muslims (who often must pray in everyday, mundane settings in order to fulfill their religious duties) carry out the frequent prayers required by their religion. They lay down a (sacred) prayer carpet in their office or living room, thereby enabling them to convert a profane time and space into a sacred time and space. This temporal and spatial reordering transforms the profane realm of work or home into a spiritual, sacred domain. Such acts, and countless others, help order and organize our experience of the world by carving it into that which is extraordinary or sacred and that which is unremarkable or profane.

—— From *The Elementary Forms of Religious Life* (1912) ——

Émile Durkheim

PRELIMINARY QUESTIONS

Religious phenomena are naturally arranged in two fundamental categories: beliefs and rites. The first are states of opinion, and consist in representations; the second are determined modes of action. Between these two classes of facts, there is all the difference that separates thought from action.

The rites can be defined and distinguished from other human practices, moral practices, for example, only by the special nature of their object. A moral rule prescribes certain manners of acting to us, just as a rite does, but which are addressed to a different class of objects. So it is the object of the rite that must be characterized, if we are to characterize the rite itself. Now it is in the beliefs that the special nature of this object is expressed. It is possible to define the rite only after we have defined the belief.

All known religious beliefs, whether simple or complex, present one common characteristic: they presuppose a classification of all the things, real and ideal, of which men think, into two classes or opposed groups, generally designated by two distinct terms which are translated well enough by the words *profane* and *sacred* (*profane, sacré*). This division of the world into two domains, the one containing all that is sacred, the other all that is profane, is the distinctive trait of religious thought; the beliefs, myths, dogmas and legends are either representations or systems of representations which express the nature of sacred things, the virtues and powers which are attributed to them, or their relations with each other and with profane things, But by sacred things one must not understand simply those personal beings which are called gods or spirits; a rock, a tree, a spring, a pebble, a piece of wood, a house, in a word, anything can be sacred. A rite can have this character; in fact, the rite does not exist which does not have it to a certain degree. There are words, expressions and formulae which can be pronounced only by the mouths of

SOURCE: Reprinted from *The Elementary Forms of Religious Life* by Émile Durkheim, translated from the French by Joseph Ward Swain. (2008). Mineola, NY: Dover Publications, Inc.

consecrated persons; there are gestures and movements which everybody cannot perform. If the Vedic sacrifice has had such an efficacy that, according to mythology, it was the creator of the gods, and not merely a means of winning their favour, it is because it possessed a virtue comparable to that of the most sacred beings. The circle of sacred objects cannot be determined, then, once for all. Its extent varies infinitely, according to the different religions. That is how Buddhism is a religion: in default of gods, it admits the existence of sacred things, namely, the four noble truths and the practices derived from them.[i]

Up to the present we have confined ourselves to enumerating a certain number of sacred things as examples: we must now show by what general characteristics they are to be distinguished from profane things.

One might be tempted, first of all, to define them by the place they are generally assigned in the hierarchy of things. They are naturally considered superior in dignity and power to profane things, and particularly to man, when he is only a man and has nothing sacred about him. One thinks of himself as occupying an inferior and dependent position in relation to them; and surely this conception is not without some truth. Only there is nothing in it which is really characteristic of the sacred. It is not enough that one thing be subordinated to another for the second to be sacred in regard to the first. Slaves are inferior to their masters, subjects to their king, soldiers to their leaders, the miser to his gold, the man ambitious for power to the hands which keep it from him; but if it is sometimes said of a man that he makes a religion of those beings or things whose eminent value and superiority to himself he thus recognizes, it is clear that in any case the word is taken in a metaphorical sense, and that there is nothing in these relations which is really religious.[ii]

On the other hand, it must not be lost to view that there are sacred things of every degree, and that there are some in relation to which a man feels himself relatively at his ease. An amulet has a sacred character, yet the respect which it inspires is nothing exceptional. Even before his gods, a man is not always in

such a marked state of inferiority; for it very frequently happens that he exercises a veritable physical constraint upon them to obtain what he desires. He beats the fetish with which he is not contented, but only to reconcile himself with it again, if in the end it shows itself more docile to the wishes of its adorer. . . . To have rain, he throws stones into the spring or sacred lake where the god of rain is thought to reside; he believes that by this means he forces him to come out and show himself. . . . Moreover, if it is true that man depends upon his gods, this dependence is reciprocal. The gods also have need of man; without offerings and sacrifices they would die. We shall even have occasion to show that this dependence of the gods upon their worshippers is maintained even in the most idealistic religions.

But if a purely hierarchic distinction is a criterium at once too general and too imprecise, there is nothing left with which to characterize the sacred in its relation to the profane except their heterogeneity. However, this heterogeneity is sufficient to characterize this classification of things and to distinguish it from all others, because it is very particular: *it is absolute.* In all the history of human thought there exists no other example of two categories of things so profoundly differentiated or so radically opposed to one another. The traditional opposition of good and bad is nothing beside this; for the good and the bad are only two opposed species of the same class, namely morals, just as sickness and health are two different aspects of the same order of facts, life, while the sacred and the profane have always and everywhere been conceived by the human mind as two distinct classes, as two worlds between which there is nothing in common. The forces which play in one are not simply those which are met with in the other, but a little stronger; they are of a different sort. In different religions, this opposition has been conceived in different ways. Here, to separate these two sorts of things, it has seemed sufficient to localize them in different parts of the physical universe; there, the first have been put into an ideal and transcendental world, while the material world is left in full possession of

---

[i] Not to mention the sage and the saint who practice these truths and who for that reason are sacred.

[ii] This is not saying that these relations cannot take a religious character. But they do not do so necessarily.

the others. But howsoever much the forms of the contrast may vary,[iii] the fact of the contrast is universal.

This is not equivalent to saying that a being can never pass from one of these worlds into the other: but the manner in which this passage is effected when it does take place, puts into relief the essential duality of the two kingdoms. In fact, it implies a veritable metamorphosis. This is notably demonstrated by the initiation rites, such as they are practiced by a multitude of peoples. This initiation is a long series of ceremonies with the object of introducing the young man into the religious life: for the first time, he leaves the purely profane world where he passed his first infancy, and enters into the world of sacred things. Now this change of state is thought of, not as a simple and regular development of pre-existent germs, but as a transformation *totius substantiae*—of the whole being. It is said that at this moment the young man dies, that the person that he was ceases to exist, and that another is instantly substituted for it. He is re-born under a new form. Appropriate ceremonies are felt to bring about this death and re-birth, which are not understood in a merely symbolic sense, but are taken literally. . . . Does this not prove that between the profane being which he was and the religious being which he becomes, there is a break of continuity?

This heterogeneity is even so complete that it frequently degenerates into a veritable antagonism. The two worlds are not only conceived of as separate, but as even hostile and jealous rivals of each other. Since men cannot fully belong to one except on condition of leaving the other completely, they are exhorted to withdraw themselves completely from the profane world, in order to lead an exclusively religious life. Hence comes the monasticism which is artificially organized outside of and apart from the natural environment in which the ordinary man leads the life of this world, in a different one, closed to the first, and nearly its contrary. Hence comes the mystic asceticism whose object is to root out from man all the attachment for the profane world that remains in him. From that come all the forms of religious suicide, the logical working-out of this asceticism; for the only manner of fully escaping the profane life is, after all, to forsake all life.

The opposition of these two classes manifests itself outwardly with a visible sign by which we can easily recognize this very special classification, wherever it exists. Since the idea of the sacred is always and everywhere separated from the idea of the profane in the thought of men, and since we picture a sort of logical chasm between the two, the mind irresistibly refuses to allow the two corresponding things to be confounded, or even to be merely put in contact with each other; for such a promiscuity, or even too direct a contiguity, would contradict too violently the dissociation of these ideas in the mind. The sacred thing is *par excellence* that which the profane should not touch, and cannot touch with impunity. . . .

Thus we arrive at the first criterium of religious beliefs. Undoubtedly there are secondary species within these two fundamental classes which, in their turn, are more or less incomparable with each other. . . . But the real characteristic of religious phenomena is that they always suppose a bipartite division of the whole universe, known and knowable, into two classes which embrace all that exists, but which radically exclude each other. Sacred things are those which the interdictions protect and isolate; profane things, those to which these interdictions are applied and which must remain at a distance from the first. Religious beliefs are the representations which express the nature of sacred things and the relations which they sustain, either with each other or with profane things. Finally, rites are the rules of conduct which prescribe how a man should comport himself in the presence of these sacred objects.

When a certain number of sacred things sustain relations of co-ordination or subordination with each other in such a way as to form a system having a certain unity, but which is not comprised within any other system of the same

---

[iii]The conception according to which the profane is opposed to the sacred, just as the irrational is to the rational, or the intelligible is to the mysterious, is only one of the forms under which this opposition is expressed. Science being once constituted, it has taken a profane character, especially in the eyes of the Christian religions; from that it appears as though it could not be applied to sacred things.

sort, the totality of these beliefs and their corresponding rites constitutes a religion. From this definition it is seen that a religion is not necessarily contained within one sole and single idea, and does not proceed from one unique principle which, though varying according to the circumstances under which it is applied, is nevertheless at bottom always the same: it is rather a whole made up of distinct and relatively individualized parts. Each homogeneous group of sacred things, or even each sacred thing of some importance, constitutes a centre of organization about which gravitate a group of beliefs and rites, or a particular cult; there is no religion, howsoever unified it may be, which does not recognize a plurality of sacred things. . . .

However, this definition is not yet complete, for it is equally applicable to two sorts of facts which, while being related to each other, must be distinguished nevertheless: these are magic and religion.

Magic, too, is made up of beliefs and rites. Like religion, it has its myths and its dogmas; only they are more elementary, undoubtedly because, seeking technical and utilitarian ends, it does not waste its time in pure speculation. It has its ceremonies, sacrifices, lustrations, prayers, chants and dances as well. The beings which the magician invokes and the forces which he throws in play are not merely of the same nature as the forces and beings to which religion addresses itself; very frequently, they are identically the same. Thus, even with the most inferior societies, the souls of the dead are essentially sacred things, and the object of religious rites. But at the same time, they play a considerable role in magic. . . .

Then will it be necessary to say that magic is hardly distinguishable from religion; that magic is full of religion just as religion is full of magic, and consequently that it is impossible to separate them and to define the one without the other? It is difficult to sustain this thesis, because of the marked repugnance of religion for magic, and in return, the hostility of the second towards the first. Magic takes a sort of professional pleasure in profaning holy things. . . . On its side, religion, when it has not condemned and prohibited magic rites, has always looked upon them with disfavor. . . . Whatever relations there may be

between these two sorts of institutions, it is difficult to imagine their not being opposed somewhere; and it is still more necessary for us to find where they are differentiated, as we plan to limit our researches to religion, and to stop at the point where magic commences.

Here is how a line of demarcation can be traced between these two domains.

The really religious beliefs are always common to a determined group, which makes profession of adhering to them and of practicing the rites connected with them. They are not merely received individually by all the members of this group; they are something belonging to the group, and they make its unity. The individuals which compose it feel themselves united to each other by the simple fact that they have a common faith. A society whose members are united by the fact that they think in the same way in regard to the sacred world and its relations with the profane world, and by the fact that they translate these common ideas into common practices, is what is called a Church. In all history, we do not find a single religion without a Church. Sometimes the Church is strictly national, sometimes it passes the frontiers; sometimes it embraces an entire people (Rome, Athens, the Hebrews), sometimes it embraces only a part of them (the Christian societies since the advent of Protestantism); sometimes it is directed by a corps of priests, sometimes it is almost completely devoid of any official directing body. But wherever we observe the religious life, we find that it has a definite group as its foundation. . . .

It is quite another matter with magic. To be sure, the belief in magic is always more or less general; it is very frequently diffused in large masses of the population, and there are even peoples where it has as many adherents as the real religion. But it does not result in binding together those who adhere to it, nor in uniting them into a group leading a common life. *There is no Church of magic.* Between the magician and the individuals who consult him, as between these individuals themselves, there are no lasting bonds which make them members of the same moral community, comparable to that formed by the believers in the same god or the observers of the same cult. The magician has a clientele and not a Church, and it is very possible that his clients have no other relations between each other, or even do not know each other; even the relations which they have with him are generally accidental and transient; they are just like those of a sick man with his physician. The official and public character with which he is sometimes invested changes nothing in this situation; the fact that he works openly does not unite him more regularly or more durably to those who have recourse to his services.

It is true that in certain cases, magicians form societies among themselves: it happens that they assemble more or less periodically to celebrate certain rites in common; it is well known what a place these assemblies of witches hold in European folk-lore. But it is to be remarked that these associations are in no way indispensable to the working of the magic; they are even rare and rather exceptional. The magician has no need of uniting himself to his fellows to practice his art. More frequently, he is a recluse; in general, far from seeking society, he flees it. . . .

Religion, on the other hand, is inseparable from the idea of a Church. From this point of view, there is an essential difference between magic and religion. But what is especially important is that when these societies of magic are formed, they do not include all the adherents to magic, but only the magicians; the laymen, if they may be so called, that is to say, those for whose profit the rites are celebrated, in fine those who represent the worshippers in the regular cults, are excluded. Now the magician is for magic what the priest is for religion, but a college of priests is not a Church, any more than a religious congregation which should devote itself to some particular saint in the shadow of a cloister, would be a particular cult. A Church is not a fraternity of priests; it is a moral community formed by all the believers in a single faith, laymen as well as priests. But magic lacks any such community. . . .

But if the idea of a Church is made to enter into the definition of religion, does that not exclude the private religions which the individual establishes for himself and celebrates by himself? There is scarcely a society where these are not found. Every Ojibway . . . has his own personal *manitou*, which he chooses himself and to which he renders special religious services;

the Melanesian of the Banks Islands has his *tamaniu* . . . the Christian, his patron saint and guardian angel, etc. By definition all these cults seem to be independent of all idea of the group. Not only are these individual religions very frequent in history, but nowadays many are asking if they are not destined to be the pre-eminent form of the religious life, and if the day will not come when there will be no other cult than that which each man will freely perform within himself. . . .

But if we leave these speculations in regard to the future aside for the moment, and confine ourselves to religions such as they are at present or have been in the past, it becomes clearly evident that these individual cults are not distinct and autonomous religious systems, but merely aspects of the common religion of the whole Church, of which the individuals are members. The patron saint of the Christian is chosen from the official list of saints recognized by the Catholic Church; there are even canonical rules prescribing how each Catholic should perform this private cult. In the same way, the idea that each man necessarily has a protecting genius is found, under different forms, at the basis of a great number of American religions, as well as of the Roman religion (to cite only these two examples); for, as will be seen later, it is very closely connected with the idea of the soul, and this idea of the soul is not one of those which can be left entirely to individual choice. In a word, it is the Church of which he is a member which teaches the individual what these personal gods are, what their function is, how he should enter into relations with them and how he should honour them. When a methodical analysis is made of the doctrines of any Church whatsoever, sooner or later we come upon those concerning private cults. So these are not two religions of different types, and turned in opposite directions; both are made up of the same ideas and the same principles, here applied to circumstances which are of interest to the group as a whole, there to the life of the individual. . . .

There still remain those contemporary aspirations towards a religion which would consist entirely in internal and subjective states, and which would be constructed freely by each of us. But howsoever real these aspirations may be, they cannot affect our definition, for this is to be applied only to facts already realized, and not to uncertain possibilities. One can define religions such as they are, or such as they have been, but not such as they more or less vaguely tend to become. It is possible that this religious individualism is destined to be realized in facts; but before we can say just how far this may be the case, we must first know what religion is, of what elements it is made up, from what causes it results, and what function it fulfils—all questions whose solution cannot be foreseen before the threshold of our study has been passed. It is only at the close of this study that we can attempt to anticipate the future.

Thus we arrive at the following definition: *A religion is a unified system of beliefs and practices relative to sacred things, that is to say, things set apart and forbidden—beliefs and practices which unite into one single moral community called a Church, all those who adhere to them.* The second element which thus finds a place in our definition is no less essential than the first; for by showing that the idea of religion is inseparable from that of the Church, it makes it clear that religion should be an eminently collective thing. . . .

Origins of These Beliefs

It is obviously not out of the sensations which the things serving as totems are able to arouse in the mind; we have shown that these things are frequently insignificant. The lizard, the caterpillar, the rat, the ant, the frog, the turkey, the bream-fish, the plum-tree, the cockatoo, etc., to cite only those names which appear frequently in the lists of Australian totems, are not of a nature to produce upon men these great and strong impressions which in a way resemble religious emotions and which impress a sacred character upon the objects they create. It is true that this is not the case with the stars and the great atmospheric phenomena, which have, on the contrary, all that is necessary to strike the imagination forcibly; but as a matter of fact, these serve only very exceptionally as totems. It is even probable that they were very slow in taking this office. So it is not the intrinsic nature of the thing whose name the clan bears that marked it out to become the object of a cult. Also, if the sentiments which

it inspired were really the determining cause of the totemic rites and beliefs, it would be the pre-eminently sacred thing; the animals or plants employed as totems would play an eminent part in the religious life. But we know that the centre of the cult is actually elsewhere. It is the figurative representations of this plant or animal and the totemic emblems and symbols of every sort, which have the greatest sanctity; so it is in them that is found the source of that religious nature, of which the real objects represented by these emblems receive only a reflection.

Thus the totem is before all a symbol, a material expression of something else. But of what?

From the analysis to which we have been giving our attention, it is evident that it expresses and symbolizes two different sorts of things. In the first place, it is the outward and visible form of what we have called the totemic principle or god. But it is also the symbol of the determined society called the clan. It is its flag; it is the sign by which each clan distinguishes itself from the others, the visible mark of its personality, a mark borne by everything which is a part of the clan under any title whatsoever, men, beasts or things. So if it is at once the symbol of the god and of the society, is that not because the god and the society are only one? How could the emblem of the group have been able to become the figure of this quasi-divinity, if the group and the divinity were two distinct realities? The god of the clan, the totemic principle, can therefore be nothing else than the clan itself, personified and represented to the imagination under the visible form of the animal or vegetable which serves as totem.

But how has this apotheosis been possible, and how did it happen to take place in this fashion?

II

In a general way, it is unquestionable that a society has all that is necessary to arouse the sensation of the divine in minds, merely by the power that it has over them; for to its members it is what a god is to his worshippers. In fact, a god is, first of all, a being whom men think of as superior to themselves, and upon whom they feel that they depend. Whether it be a conscious personality, such as Zeus or Jahveh, or merely abstract forces such as those in play in totemism, the worshipper, in the one case as in the other, believes himself held to certain manners of acting which are imposed upon him by the nature of the sacred principle with which he feels that he is in communion. Now society also gives us the sensation of a perpetual dependence. Since it has a nature which is peculiar to itself and different from our individual nature, it pursues ends which are likewise special to it; but, as it cannot attain them except through our intermediacy, it imperiously demands our aid. It requires that, forgetful of our own interest, we make ourselves its servitors, and it submits us to every sort of inconvenience, privation and sacrifice, without which social life would be impossible. It is because of this that at every instant we are obliged to submit ourselves to rules of conduct and of thought which we have neither made nor desired, and which are sometimes even contrary to our most fundamental inclinations and instincts.

Even if society were unable to obtain these concessions and sacrifices from us except by a material constraint, it might awaken in us only the idea of a physical force to which we must give way of necessity, instead of that of a moral power such as religious adore. But as a matter of fact, the empire which it holds over consciences is due much less to the physical supremacy of which it has the privilege than to the moral authority with which it is invested. If we yield to its orders, it is not merely because it is strong enough to triumph over our resistance; it is primarily because it is the object of a venerable respect.

We say that an object, whether individual or collective, inspires respect when the representation expressing it in the mind is gifted with such a force that it automatically causes or inhibits actions, *without regard for any consideration relative to their useful or injurious effects.* When we obey somebody because of the moral authority which we recognize in him, we follow out his opinions, not because they seem wise, but because a certain sort of physical

energy is imminent in the idea that we form of this person, which conquers our will and inclines it in the indicated direction. Respect is the emotion which we experience when we feel this interior and wholly spiritual pressure operating upon us. Then we are not determined by the advantages or inconveniences of the attitude which is prescribed or recommended to us; it is by the way in which we represent to ourselves the person recommending or prescribing it. This is why commands generally take a short, peremptory form leaving no place for hesitation; it is because, in so far as it is a command and goes by its own force, it excludes all idea of deliberation or calculation; it gets its efficacy from the intensity of the mental state in which it is placed. It is this intensity which creates what is called a moral ascendancy.

Now the ways of action to which society is strongly enough attached to impose them upon its members, are, by that very fact, marked with a distinctive sign provocative of respect. Since they are elaborated in common, the vigour with which they have been thought of by each particular mind is retained in all the other minds, and reciprocally. The representations which express them within each of us have an intensity which no purely private states of consciousness could ever attain; for they have the strength of the innumerable individual representations which have served to form each of them. It is society who speaks through the mouths of those who affirm them in our presence; it is society whom we hear in hearing them; and the voice of all has an accent which that of one alone could never have.[iv] The very violence with which society reacts, by way of blame or material suppression, against every attempted dissidence, contributes to strengthening its empire by manifesting the common conviction through this burst of ardour[v] In a word, when something is the object of such a state of opinion, the representation which each individual has of it gains a power of action from its origins and the conditions in which it was born, which even those feel who do not submit themselves to it. It tends to repel the representations which contradict it, and it keeps them at a distance; on

---

[iv]See our *Division du travail social,* 3rd ed., pp. 64 ff.
[v]Ibid., p. 76.

the other hand, it commands those acts which will realize it, and it does so, not by a material coercion or by the perspective of something of this sort, but by the simple radiation of the mental energy which it contains. It has an efficacy coming solely from its psychical properties, and it is by just this sign that moral authority is recognized. So opinion, primarily a social thing, is a source of authority, and it might even be asked whether all authority is not the daughter of opinion.[vi] It may be objected that science is often the antagonist of opinion, whose errors it combats and rectifies. But it cannot succeed in this task if it does not have sufficient authority, and it can obtain this authority only from opinion itself. If a people did not have faith in science, all the scientific demonstrations in the world would be without any influence whatsoever over their minds. Even to-day, if science happened to resist a very strong current of public opinion, it would risk losing its credit there.[vii]

Since it is in spiritual ways that social pressure exercises itself, it could not fail to give men the idea that outside themselves there exist one or several powers, both moral and, at the same time, efficacious, upon which they depend. They must think of these powers, at least in part, as outside themselves, for these address them in a tone of command and sometimes even order them to do violence to their most natural inclinations. It is undoubtedly true that if they were able to see that these influences which they feel emanate from society, then the mythological system of inter-pretations would never be born. But social action follows ways that are too circuitous and obscure, and employs psychical mechanisms that are too complex to allow the ordinary observer to see when it comes. As long as scientific analysis does not come to teach it to them, men know well that they are acted upon, but they do not know by whom. So they must invent by themselves the idea of these powers with which they feel themselves in connection, and from that, we are able to catch a glimpse of the way by which they were led to represent them under forms that are really foreign to their nature and to transfigure them by thought.

But a god is not merely an authority upon whom we depend; it is a force upon which our strength relies. The man who has obeyed his god and who for this reason, believes the god is with him, approaches the world with confidence and with the feeling of an increased energy. Likewise, social action does not confine itself to demanding sacrifices, privations and efforts from us. For the collective force is not entirely outside of us; it does not act upon us wholly from without; but rather, since society cannot exist except in and through individual con-sciousness,[viii] this force must also penetrate us and organize itself within us; it thus becomes an integral part of our being and by that very fact this is elevated and magnified.

There are occasions when this strengthening and vivifying action of society is especially apparent. In the midst of an assembly animated by a common passion, we become susceptible of acts and sentiments of which we are incapable when reduced to our own forces; and when the

---

[vi]This is the case at least with all moral authority recognized as such by the group as a whole.

[vii]We hope that this analysis and those which follow will put an end to an inexact interpretation of our thought, from which more than one misunderstanding has resulted. Since we have made constraint the *outward sign* by which social facts can be the most easily recognized and distinguished from the facts of individual psychology, it has been assumed that according to our opinion, physical constraint is the essential thing for social life. As a matter of fact, we have never considered it more than the material and apparent expression of an interior and profound fact which is wholly ideal: this is *moral authority*. The problem of sociology—if we can speak of a sociological problem—consists in seeking, among the different forms of external constraint, the different sorts of moral authority corresponding to them and in discovering the causes which have determined these latter. The particular question which we are treating in this present work has as its principal object, the discovery of the form under which that particular variety of moral authority which is inherent in all that is religious has been born, and out of what elements it is made. It will be seen presently that even if we do make social pressure one of the distinctive characteristics of sociological phenomena, we do not mean to say that it is the only one. We shall show another aspect of the collective life, nearly opposite to the preceding one, but none the less real.

[viii]Of course this does not mean to say that the collective consciousness does not have distinctive characteristics of its own (on this point, see *Représentations individuelles et représentations collectives*, in *Revue de Métaphysique et de Morale*, 1898, pp. 273 ff.).

assembly is dissolved and when, finding ourselves alone again, we fall back to our ordinary level, we are then able to measure the height to which we have been raised above ourselves. History abounds in examples of this sort. It is enough to think of the night of the Fourth of August, 1789, when an assembly was suddenly led to an act of sacrifice and abnegation which each of its members had refused the day before, and at which they were all surprised the day after.[ix] This is why all parties political, economic or confessional, are careful to have periodical reunions where their members may revivify their common faith by manifesting it in common. To strengthen those sentiments which, if left to themselves, would soon weaken, it is sufficient to bring those who hold them together and to put them into closer and more active relations with one another. This is the explanation of the particular attitude of a man speaking to a crowd, at least if he has succeeded in entering into communion with it. His language has a grandiloquence that would be ridiculous in ordinary circumstances; his gestures show a certain domination; his very thought is impatient of all rules, and easily falls into all sorts of excesses. It is because he feels within him an abnormal over-supply of force which overflows and tries to burst out from him; sometimes he even has the feeling that he is dominated by a moral force which is greater than he and of which he is only the interpreter. It is by this trait that we are able to recognize what has often been called the demon of oratorical inspiration. Now this exceptional increase of force is something very real; it comes to him from the very group which he addresses. The sentiments provoked by his words come back to him, but enlarged and amplified, and to this degree they strengthen his own sentiment. The passionate energies he arouses re-echo within him and quicken his vital tone. It is no longer a simple individual who speaks; it is a group incarnate and personified.

Besides these passing and intermittent states, there are other more durable ones, where this strengthening influence of society makes itself felt with greater consequences and frequently even with greater brilliancy. There are periods in history when, under the influence of some great collective shock, social interactions have become much more frequent and active. Men look for each other and assemble together more than ever. That general effervescence results which is characteristic of revolutionary or creative epochs. Now this greater activity results in a general stimulation of individual forces. Men see more and differently now than in normal times. Changes are not merely of shades and degrees; men become different. The passions moving them are of such an intensity that they cannot be satisfied except by violent and unrestrained actions, actions of superhuman heroism or of bloody barbarism. This is what explains the Crusades,[x] for example, or many of the scenes, either sublime or savage, of the French Revolution.[xi] Under the influence of the general exaltation, we see the most mediocre and inoffensive bourgeois become either a hero or a butcher.[xii] And so clearly are all these mental processes the ones that are also at the root of religion that the individuals themselves have often pictured the pressure before which they thus gave way in a distinctly religious form. The Crusaders believed that they felt God present in the midst of them, enjoining them to go to the conquest of the Holy Land; Joan of Arc believed that she obeyed celestial voices.[xiii]

But it is not only in exceptional circumstances that this stimulating action of society makes itself felt; there is not, so to speak, a moment in our lives when some current of

---

[ix]This is proved by the length and passionate character of the debates where a legal form was given to the resolutions made in a moment of collective enthusiasm. In the clergy as in the nobility, more than one person called this celebrated night the dupe's night, or, with Rivarol, the St. Bartholomew of the estates (see Stoll, *Suggestion und Hypnotismus in de Völkerpsychologie,* 2nd ed., p. 618, n. 2).

[x]See Stoll, *op. cit.,* pp. 353 ff.

[xi]Ibid., pp. 619, 635.

[xii]Ibid., pp. 622 ff.

[xiii]The emotions of fear and sorrow are able to develop similarly and to become intensified under these same conditions. As we shall see, they correspond to quite another aspect of the religious life (Bk. III, ch. v).

energy does not come to us from without. The man who has done his duty finds, in the manifestations of every sort expressing the sympathy, esteem or affection which his fellows have for him, a feeling of comfort, of which he does not ordinarily take account, but which sustains him, none the less. The sentiments which society has for him raise the sentiments which he has for himself. Because he is in moral harmony with his comrades, he has more confidence, courage and boldness in action, just like the believer who thinks that he feels the regard of his god turned graciously towards him. It thus produces, as it were, a perpetual sustenance of our moral nature. Since this varies with a multitude of external circumstances, as our relations with the groups about us are more or less active and as these groups themselves vary, we cannot fail to feel that this moral support depends upon an external cause; but we do not perceive where this cause is nor what it is. So we ordinarily think of it under the form of a moral power which, though immanent in us, represents within us something not ourselves: this is the moral conscience, of which, by the way, men have never made even a slightly distinct representation except by the aid of religious symbols.

In addition to these free forces which are constantly coming to renew our own, there are others which are fixed in the methods and traditions which we employ. We speak a language that we did not make; we use instruments that we did not invent; we invoke rights that we did not found; a treasury of knowledge is transmitted to each generation that it did not gather itself, etc. It is to society that we owe these varied benefits of civilization, and if we do not ordinarily see the source from which we get them, we at least know that they are not our own work. Now it is these things that give man his own place among things; a man is a man only because he is civilized. So he could not escape the feeling that outside of him there are active causes from which he gets the characteristic attributes of his nature and which,

as benevolent powers, assist him, protect him and assure him of a privileged fate. And of course he must attribute to these powers a dignity corresponding to the great value of the good things he attributes to them.[xiv]

Thus the environment in which we live seems to us to be peopled with forces that are at once imperious and helpful, august and gracious, and with which we have relations. Since they exercise over us a pressure of which we are conscious, we are forced to localize them outside ourselves, just as we do for the objective causes of our sensations. But the sentiments which they inspire in us differ in nature from those which we have for simple visible objects. As long as these latter are reduced to their empirical characteristics as shown in ordinary experience, and as long as the religious imagination has not metamorphosed them, we entertain for them no feeling which resembles respect, and they contain within them nothing that is able to raise us outside ourselves. Therefore, the representations which express them appear to us to be very different from those aroused in us by collective influences. The two form two distinct and separate mental states in our consciousness, just as do the two forms of life to which they correspond. Consequently, we get the impression that we are in relations with two distinct sorts of reality and that a sharply drawn line of demarcation separates them from each other: on the one hand is the world of profane things, on the other, that of sacred things.

Also, in the present day just as much as in the past, we see society constantly creating sacred things out of ordinary ones. If it happens to fall in love with a man and if it thinks it has found in him the principal aspirations that move it, as well as the means of satisfying them, this man will be raised above the others and, as it were, deified. Opinion will invest him with a majesty exactly analogous to that protecting the gods. This is what has happened to so many sovereigns in whom their age had faith: if they were not made gods, they were at least regarded

---

[xiv]This is the other aspect of society which, while being imperative, appears at the same time to be good and gracious. It dominates us and assists us. If we have defined the social fact by the first of these characteristics rather than the second, it is because it is more readily observable, for it is translated into outward and visible signs; but we have never thought of denying the second (see our *Règles de la Méthode Sociologique*, preface to the second edition, p. xx, n. 1).

as direct representatives of the deity. And the fact that it is society alone which is the author of these varieties of apotheosis, is evident since it frequently chances to consecrate men thus who have no right to it from their own merit. The simple deference inspired by men invested with high social functions is not different in nature from religious respect. It is expressed by the same movements: a man keeps at a distance from a high personage; he approaches him only with precautions; in conversing with him, he uses other gestures and language than those used with ordinary mortals. The sentiment felt on these occasions is so closely related to the religious sentiment that many peoples have confounded the two. In order to explain the consideration accorded to princes, nobles and political chiefs, a sacred character has been attributed to them. In Melanesia and Polynesia, for example, it is said that an influential man has *mana,* and that his influence is due to this *mana.*[xv] However, it is evident that his situation is due solely to the importance attributed to him by public opinion. Thus the moral power conferred by opinion and that with which sacred beings are invested are at bottom of a single origin and made up of the same elements. That is why a single word is able to designate the two.

In addition to men, society also consecrates things, especially ideas. If a belief is unanimously shared by a people, then, for the reason which we pointed out above, it is forbidden to touch it, that is to say, to deny it or to contest it. Now the prohibition of criticism is an interdiction like the others and proves the presence of something sacred. Even to-day, howsoever great may be the liberty which we accord to others, a man who should totally deny progress or ridicule the human ideal to which modern societies are attached, would produce the effect of a sacrilege.

There is at least one principle which those the most devoted to the free examination of everything tend to place above discussion and to regard as untouchable, that is to say, as sacred: this is the very principle of free examination.

This aptitude of society for setting itself up as a god or for creating gods was never more apparent than during the first years of the French Revolution. At this time, in fact, under the influence of the general enthusiasm, things purely laïcal by nature were transformed by public opinion into sacred things: these were the Fatherland, Liberty, Reason.[xvi] A religion tended to become established which had its dogmas,[xvii] symbols,[xviii] altars[xix] and feasts.[xx] It was to these spontaneous aspirations that the cult of Reason and the Supreme Being attempted to give a sort of official satisfaction. It is true that this religious renovation had only an ephemeral duration. But that was because the patriotic enthusiasm which at first transported the masses soon relaxed.[xxi] The cause being gone, the effect could not remain. But this experiment, though short-lived, keeps all its sociological interest. It remains true that in one determined case we have seen society and its essential ideas become, directly and with no transfiguration of any sort, the object of a veritable cult.

All these facts allow us to catch glimpses of how the clan was able to awaken within its members the idea that outside of them there exist forces which dominate them and at the same time sustain them, that is to say in fine, religious forces: it is because there is no society with which the primitive is more directly and closely connected. The bonds uniting him to the tribe are much more lax and more feebly felt. Although this is not at all strange or foreign to him, it is with the people of his own clan that he has the greatest number of things in common; it is the action of this group that he feels the most directly; so it is

---

[xv]Codrington, *The Melanesians,* pp. 50, 103, 120. It is also generally thought that in the Polynesian languages, the word *mana* primitively had the sense of authority (see Tregear, *Maori Comparative Dictionary, s.v.*).

[xvi]See Albert Mathiez, *Les origines des cultes révolutionnaires* (1789–1792).

[xvii]Ibid., p. 24.

[xviii]Ibid., pp. 29, 32.

[xix]Ibid., p. 30.

[xx]Ibid., p. 46.

[xxi]See Mathiez, *La Théophilanthropie et la Culte décadaire,* p. 36.

this also which, in preference to all others, should express itself in religious symbols. . . .

III

One can readily conceive how, when arrived at this state of exaltation, a man does not recognize himself any longer. Feeling himself dominated and carried away by some sort of an external power which makes him think and act differently than in normal times, he naturally has the impression of being himself no longer. It seems to him that he has become a new being: the decorations he puts on and the masks that cover his face and figure materially in this interior transformation, and to a still greater extent, they aid in determining its nature. And as at the same time all his companions feel themselves transformed in the same way and express this sentiment by their cries, their gestures and their general attitude, everything is just as though he really were transported into a special world, entirely different from the one where he ordinarily lives, and into an environment filled with exceptionally intense forces that take hold of him and metamorphose him. How could such experiences as these, especially when they are repeated every day for weeks, fail to leave in him the conviction that there really exist two heterogeneous and mutually incomparable worlds? One is that where his daily life drags wearily along; but he cannot penetrate into the other without at once entering into relations with extraordinary powers that excite him to the point of frenzy. The first is the profane world, the second, that of sacred things.

So it is in the midst of these effervescent social environments and out of this effervescence itself that the religious idea seems to be born. The theory that this is really its origin is confirmed by the fact that in Australia the really religious activity is almost entirely confined to the moments when these assemblies are held. To be sure, there is no people among whom the great solemnities of the cult are not more or less periodic; but in the more advanced societies, there is not, so to speak, a day when some prayer or offering is not addressed to the gods and some ritual act is not performed. But in Australia, on the contrary, apart from the celebrations of the clan and tribe, the time is nearly all filled with lay and profane occupations. Of course there are prohibitions that should be and are preserved even during these periods of temporal activity; it is never permissible to kill or eat freely of the totemic animal, at least in those parts where the interdiction has retained its original vigour; but almost no positive rites are then celebrated, and there are no ceremonies of any importance. These take place only in the midst of assembled groups. The religious life of the Australian passes through successive phases of complete lull and of superexcitation, and social life oscillates in the same rhythm. This puts clearly into evidence the bond uniting them to one another, but among the peoples called civilized, the relative continuity of the two blurs their relations. It might even be asked whether the violence of this contrast was not necessary to disengage the feeling of sacredness in its first form. By concentrating itself almost entirely in certain determined moments, the collective life has been able to attain its greatest intensity and efficacy, and consequently to give men a more active sentiment of the double existence they lead and of the double nature in which they participate. . . .

Now the totem is the flag of the clan. It is therefore natural that the impressions aroused by the clan in individual minds—impressions of dependence and of increased vitality—should fix themselves to the idea of the totem rather than that of the clan: for the clan is too complex a reality to be represented clearly in all its complex unity by such rudimentary intelligences. More than that, the primitive does not even see that these impressions come to him from the group. He does not know that the coming together of a number of men associated in the same life results in disengaging new energies, which transform each of them. All that he knows is that he is raised above himself and that he sees a different life from the one he ordinarily leads. However, he must connect these sensations to some external object as their cause. Now what does he see about him? On every side those things which appeal to his senses and strike his imagination are the numerous images of the totem. They are the waninga and the nurtunja, which are symbols of the sacred being. They are churinga and bull-roarers, upon which are generally carved combinations of lines having the same significance.

They are the decorations covering the different parts of his body, which are totemic marks. How could this image, repeated everywhere and in all sorts of forms, fail to stand out with exceptional relief in his mind? Placed thus in the centre of the scene, it becomes representative. The sentiments experienced fix themselves upon it, for it is the only concrete object upon which they can fix themselves. It continues to bring them to mind and to evoke them even after the assembly has dissolved, for it survives the assembly, being carved upon the instruments of the cult, upon the sides of rocks, upon bucklers, etc. By it, the emotions experienced are perpetually sustained and revived. Everything happens just as if they inspired them directly. It is still more natural to attribute them to it for, since they are common to the group, they can be associated only with something that is equally common to all. Now the totemic emblem is the only thing satisfying this condition. By definition, it is common to all. During the ceremony, it is the centre of all regards. While generations change, it remains the same; it is the permanent element of the social life. So it is from it that those mysterious forces seem to emanate with which men feel that they are related, and thus they have been led to represent these forces under the form of the animate or inanimate being whose name the clan bears.

When this point is once established, we are in a position to understand all that is essential in the totemic beliefs.

Since religious force is nothing other than the collective and anonymous force of the clan, and since this can be represented in the mind only in the form of the totem, the totemic emblem is like the visible body of the god. Therefore, it is from it that those kindly and dreadful actions seem to emanate, which the cult seeks to provoke or prevent; consequently, it is to it that the cult is addressed. This is the explanation of why it holds the first place in the series of sacred things.

But the clan, like every other sort of society, can live only in and through the individual consciousnesses that compose it. So if religious force, in so far as it is conceived as incorporated in the totemic emblem, appears to be outside of the individuals and to be endowed with a sort of transcendence over them, it, like the clan of which it is the symbol, can be realized only in and through them; in this sense, it is imminent in them and they necessarily represent it as such.

They feel it present and active within them, for it is this which raises them to a superior life. This is why men have believed that they contain within them a principle comparable to the one residing in the totem, and consequently, why they have attributed a sacred character to themselves, but one less marked than that of the emblem. It is because the emblem is the pre-eminent source of the religious life; the man participates in it only indirectly, as he is well aware; he takes into account the fact that the force that transports him into the world of sacred things is not inherent in him, but comes to him from the outside. . . .

But if this theory of totemism has enabled us to explain the most characteristic beliefs of this religion, it rests upon a fact not yet explained. When the idea of the totem, the emblem of the clan, is given, all the rest follows; but we must still investigate how this idea has been formed. This is a double question and may be subdivided as follows: What has led the clan to choose an emblem? and why have these emblems been borrowed from the animal and vegetable worlds, and particularly from the former?

That an emblem is useful as a rallying-centre for any sort of a group it is superfluous to point out. By expressing the social unity in a material form, it makes this more obvious to all, and for that very reason the use of emblematic symbols must have spread quickly when once thought of. But more than that, this idea should spontaneously arise out of the conditions of common life; for the emblem is not merely a convenient process for clarifying the sentiment society has of itself: it also serves to create this sentiment; it is one of its constituent elements.

In fact, if left to themselves, individual consciousnesses are closed to each other; they can communicate only by means of signs which express their internal states. If the communication established between them is to become a real communion, that is to say, a fusion of all particular sentiments into one common sentiment, the signs expressing them must themselves be fused into one single and unique resultant. It is the appearance of this that informs individuals that they are in harmony and makes them conscious of their moral unity. It is by uttering the same cry, pronouncing the same word, or performing the same gesture in regard to some object that they become and feel themselves to be in unison. It is true that individual representations also cause reactions in the organism that are not without importance; however, they can be thought of apart from these physical reactions which accompany them or follow them, but which do not constitute them. But it is quite another matter with collective representations. They presuppose that minds act and react upon one another; they are the product of these actions and reactions which are themselves possible only through material intermediaries. These latter do not confine themselves to revealing the mental state with which they are associated; they aid in creating it. Individual minds cannot come in contact and communicate with each other except by coming out of themselves; but they cannot do this except by movements. So it is the homogeneity of these movements that gives the group consciousness of itself and consequently makes it exist. When this homogeneity is once established and these movements have once taken a stereotyped form, they serve to symbolize the corresponding representations. But they symbolize them only because they have aided in forming them.

Moreover, without symbols, social sentiments could have only a precarious existence. Though very strong as long as men are together and influence each other reciprocally, they exist only in the form of recollections after the assembly has ended, and when left to themselves, these become feebler and feebler; for since the group is now no longer present and active, individual temperaments easily regain the upper hand. The violent passions which may have been released in the heart of a crowd fall away and are extinguished when this is dissolved, and men ask themselves with astonishment how they could ever have been so carried away from their normal character. But if the movements by which these sentiments are expressed are connected with something that endures, the sentiments themselves become more durable. These other things are constantly bringing them to mind and arousing them; it is as though the cause which excited them in the first place continued to act. Thus these systems

of emblems, which are necessary if society is to become conscious of itself, are no less indispensable for assuring the continuation of this consciousness.

So we must refrain from regarding these symbols as simple artifices, as sorts of labels attached to representations already made, in order to make them more manageable: they are an integral part of them. Even the fact that collective sentiments are thus attached to things completely foreign to them is not purely conventional: it illustrates under a conventional form a real characteristic of social facts, that is, their transcendence over individual minds. In fact, it is known that social phenomena are born, not in individuals, but in the group. Whatever part we may take in their origin, each of us receives them from without.[xxii] So when we represent them to ourselves as emanating from a material object, we do not completely misunderstand their nature. Of course they do not come from the specific thing to which we connect them, but nevertheless, it is true that their origin is outside of us. If the moral force sustaining the believer does not come from the idol he adores or the emblem he venerates, still it is from outside of him, as he is well aware. The objectivity of its symbol only translates its eternalness.

Thus social life, in all its aspects and in every period of its history, is made possible only by a vast symbolism. The material emblems and figurative representations with which we are more especially concerned in our present study, are one form of this; but there are many others. Collective sentiments can just as well become incarnate in persons or formulæ: some formulæ are flags, while there are persons, either real or mythical, who are symbols. . . .

CONCLUSION

As we have progressed, we have established the fact that the fundamental categories of thought, and consequently of science, are of religious origin. We have seen that the same is true for magic and consequently for the different processes which have issued from it. On the other hand, it

---

[xxii]On this point see *Régles de la méthode sociologique*, pp. 5 ff.

has long been known that up until a relatively advanced moment of evolution, moral and legal rules have been indistinguishable from ritual pre-scriptions. In summing up, then, it may be said that nearly all the great social institutions have been born in religion.[xxiii] Now in order that these principal aspects of the collective life may have commenced by being only varied aspects of the religious life, it is obviously necessary that the religious life be the eminent form and, as it were, the concentrated expression of the whole collective life. If religion has given birth to all that is essential in society, it is because the idea of society is the soul of religion.

Religious forces are therefore human forces, moral forces. It is true that since collective sentiments can become conscious of themselves only by fixing themselves upon external objects, they have not been able to take form without adopting some of their characteristics from other things: they have thus acquired a sort of physical nature; in this way they have come to mix themselves with the life of the material world, and then have considered themselves capable of explaining what passes there. But when they are considered only from this point of view and in this role, only their most superficial aspect is seen. In reality, the essential elements of which these collective sentiments are made have been borrowed by the understanding. It ordinarily seems that they should have a human character only when they are conceived under human forms;[xxiv] but even the most impersonal and the most anonymous are nothing else than objectified sentiments.

It is only by regarding religion from this angle that it is possible to see its real significance. If we stick closely to appearances, rites often give the effect of purely manual operations: they are anointings, washings, meals. To consecrate something, it is put in contact with a source of religious energy, just as to-day a body is put in contact with a source of heat or electricity to warm

or electrize it; the two processes employed are not essentially different. Thus understood, religious technique seems to be a sort of mystic mechanics. But these material manoeuvres are only the external envelope under which the mental operations are hidden. Finally, there is no question of exercising a physical constraint upon blind and, incidentally, imaginary forces, but rather of reaching individual consciousnesses of giving them a direction and of disciplining them. It is sometimes said that inferior religions are materialistic. Such an expression is inexact. All religions, even the crudest, are in a sense spiritualistic: for the powers they put in play are before all spiritual, and also their principal object is to act upon the moral life. Thus it is seen that whatever has been done in the name of religion cannot have been done in vain: for it is necessarily the society that did it, and it is humanity that has reaped the fruits. . . .

II

Thus there is something eternal in religion which is destined to survive all the particular symbols in which religious thought has successively enveloped itself. There can be no society which does not feel the need of upholding and reaffirming at regular intervals the collective sentiments and the collective ideas which make its unity and its personality. Now this moral remaking cannot be achieved except by the means of reunions, assemblies and meetings where the individuals, being closely united to one another, reaffirm in common their common sentiments; hence come ceremonies which do not differ from regular religious ceremonies, either in their object, the results which they produce, or the processes employed to attain these results. What essential difference is there between an assembly of Christians celebrating the principal dates of the life of Christ, or of Jews remembering the exodus from Egypt or the

---

[xxiii]Only one form of social activity has not yet been expressly attached to religion: that is economic activity. Sometimes processes that are derived from magic have, by that fact alone, an origin that is indirectly religious. Also, economic value is a sort of power or efficacy, and we know the religious origins of the idea of power. Also, richness can confer *mana;* therefore it has it. Hence it is seen that the ideas of economic value and of religious value are not without connection. But the question of the nature of these connections has not yet been studied.

[xxiv]It is for this reason that Frazer and even Preuss set impersonal religious forces outside of, or at least on the threshold of religion, to attach them to magic.

promulgation of the decalogue, and a reunion of citizens commemorating the promulgation of a new moral or legal system or some great event in the national life?. . .

In summing up, then, we must say that society is not at all the illogical or a-logical, incoherent and fantastic being which it has too often been considered. Quite on the contrary, the collective consciousness is the highest form of the psychic life, since it is the consciousness of the consciousnesses. Being placed outside of and above individual and local contingencies, it sees things only in their permanent and essential aspects, which it crystallizes into communicable ideas. At the same time that it sees from above, it sees farther; at every moment of time, it embraces all known reality; that is why it alone can furnish the mind with the moulds which are applicable to the totality of things and which make it possible to think of them. It does not create these moulds artificially; it finds them within itself; it does nothing but become conscious of them. . . .

Discussion Questions

1. Outline the two forms of solidarity discussed by Durkheim. What are the distinguishing features of each type of solidarity? What is the relationship between the two forms of solidarity and the division of labor? Do these concepts help explain the division of labor in your family of origin? In a current or previous place of employment? How so or why not? Be specific.

2. Discuss the various types of suicide that Durkheim delineates using specific examples. To what extent do you agree or disagree with the notion that different *types* of suicide prevail in "modern" as opposed to "traditional" societies? Give concrete examples.

3. Define, compare, and contrast Marx's concept of alienation and Durkheim's concept of anomie. How exactly do these concepts overlap? How are they different?

4. Discuss Durkheim's notion of collective conscience. What does Durkheim mean by saying that the collective conscience is *not* just a "sum" of individual consciousnesses? How does collective conscience compare to such notions as "group think" or "mob mentality"? Use concrete examples to explain.

5. Discuss specific moments of collective effervescence that you have experienced (e.g., concerts, church, etc.). What particular symbols and rituals were called up and used to arouse this social state?

6. Discuss Durkheim's definition of religion as well as the sacred and profane, using concrete examples. What are the advantages and disadvantages of Durkheim's definition of religion, both for understanding the essence of religion and for doing research on religion? How does Durkheim distinguish "religion" and "magic"? Do you agree or disagree with this distinction?

4 MAX WEBER (1864–1920)

Key Concepts

- *Verstehen*
- Ideal types
- Protestant ethic
- Calling
- Iron cage
- Rationalization
- Bureaucracy
- Authority
- Charisma
- Class, status, and party

No one knows who will live in this cage in the future, or whether at the end of this tremendous development entirely new prophets will arise, or there will be a great rebirth of old ideas and ideals, or, if neither, mechanized petrification embellished with a sort of convulsive self-importance. For of the last stage of this cultural development it might well be truly said: "Specialists without spirit, sensualists without heart; this nullity imagines that it has attained a level of civilization never before achieved."

(Weber 1904–05/1958:182)

From the course requirements necessary to earn your degree, to the paperwork and tests you must complete in order to receive your driver's license, to the record keeping and mass of files that organize most every business enterprise, our everyday life is channeled in large measure through formalized, codified procedures. Indeed, in Western cultures few aspects of life have been untouched by the general tendency toward rationalization and the adoption of methodical practices. So, whether it's developing a long-term financial plan for one's business, following the advice written in sex manuals, or even planning for one's own death, little in modern life is left to chance. It was toward an examination of the causes and consequences of this "disenchantment" of everyday life that Max Weber's wide-ranging work crystallized. In this chapter, we explore Weber's study of this general trend in modern society as well as other aspects of his writings. But while Weber did not self-consciously set out to develop a unified theoretical model, making his intellectual path unlike that followed by both Marx and Durkheim, it is this characteristic of his work that has made it a continual wellspring of inspiration for other scholars. Perhaps the magnitude of Weber's impact on the development of sociology is captured best by the prominent social theorist, Raymond Aron, who described Weber as "the greatest of the sociologists" (Aron 1965/1970:294).

⠿ A BIOGRAPHICAL SKETCH

Max Weber, Jr., was born in Erfurt, Germany, in 1864. He was the eldest of eight children born to Max Weber, Sr., and Helene Fallenstein Weber, although only six survived to adulthood. Max Jr. was a sickly child. When he was four years old, he became seriously ill with meningitis. Though he eventually recovered, throughout the rest of his life he suffered the physical and emotional aftereffects of the disease, most apparently anxiety and nervous tension. From an early age, books were central in Weber's life. He read whatever he could get his hands on, including Kant, Machiavelli, Spinoza, Goethe, and Schopenhauer, and he wrote two historical essays before his 14th birthday. But Weber paid little attention in class and did almost no work for school. According to his widow Marianne, although "he was not uncivil to his teachers, he did not respect them. . . . If there was a gap in his knowledge, he went to the root of the matter and then gladly shared what he knew" (Marianne Weber 1926/1975:48).

In 1882, at 18 years old, Weber took his final high school examinations. His teachers acknowledged his outstanding intellectual accomplishments and thirst for knowledge, but expressed doubts about his "moral maturity." Weber went to the University of Heidelberg for three semesters and then completed one year of military service in Strasbourg. When his service ended, he enrolled at the University of Berlin and, for the next eight years, lived at his parents' home. Upon passing his first examination in law in 1886, Weber began work as a full-time legal apprentice. While working as a junior barrister, he earned a Ph.D. in economic and legal history in 1889. He then took a position as lecturer at the University of Berlin.

Throughout his life, Weber was torn by the personal struggles between his mother and his father. Weber admired his mother's extraordinary religious piety and devotion to her family, and loathed his father's abusive treatment of her. At the same time, Weber admired his father's intellectual prowess and achievements and reviled his mother's passivity. Weber followed in his father's footsteps by becoming a lawyer and joining the same organizations as his father had at the University of Heidelberg. Like his father, he was active in government affairs as well. As a member of the National Liberal Party, Max Sr. was elected to the Reichstag (national legislature) and later appointed by Chancellor Bismarck to the Prussian House of Deputies. For his part, Max Jr. was a committed nationalist and served the government in numerous capacities, including as a delegate to the German Armistice Commission in Versailles following Germany's defeat in World War I. But he was also imbued with a sense of moral duty quite similar to that of his mother. Weber's feverish work ethic—he drove himself mercilessly, denying himself all leisure—can be understood as an inimitable combination of his father's intellectual accomplishments and his mother's moral resolve.

In 1893, at the age of 29, Weber married Marianne Schnitger, a distant cousin, and finally left his childhood home. Today, Marianne Weber is recognized as an important feminist, intellectual, and sociologist in her own right. She was a popular public speaker on social and sexual ethics and wrote many books and articles. Her most influential works, *Marriage and Motherhood in the Development of Law* (1907) and *Women and Love* (1935), examined feminist issues and the reform of marriage. However, Marianne is known best as the intellectual partner of her husband. She and Max made a conscious effort to establish an egalitarian relationship, and worked together on intellectual projects. Interestingly, Marianne referred to Max as her "companion" and implied that theirs was an unconsummated marriage. (It is rumored that Max had a long-lasting affair with a woman of Swiss nobility who was a member of the Tobleron family.) Despite her own intellectual accomplishments, Marianne's 700-page treatise, *Max Weber: A Biography*, first published in 1926, has received the most attention, serving as the central source of biographical information on her husband (and vital to this introduction as well).

In 1894, Max Weber joined the faculty at Freiburg University as a full professor of economics. Shortly thereafter, in 1896, Weber accepted a position as chair of economics at the University of Heidelberg, where he first began his academic career. But in 1897, he suffered a serious nervous breakdown. According to Marianne, the breakdown was triggered by the inexorable guilt Weber experienced after his father's sudden death. Just seven weeks before he died, Weber had rebuked his father over his tyrannical treatment of his mother. The senior Weber had prohibited his wife Helene from visiting Max and Marianne at their home in Heidelberg without him. When he and Helene showed up together for the visit, his son forced him to leave. Unfortunately, that was the last time father and son ever spoke.

Weber experienced debilitating anxiety and insomnia throughout the rest of his life. He often resorted to taking opium in order to sleep. Despite resigning his academic posts, traveling, and resting, the anxiety could not be dispelled. Nevertheless, he had spurts of manic intellectual activity and continued to write as an independent scholar. In 1904, Weber traveled to the United States and began to formulate the argument of what would be his most celebrated work, *The Protestant Ethic and the Spirit of Capitalism* (Weber 1904–05/1958).

After returning to Europe, Weber resumed his intellectual activity. He met with the brilliant thinkers of his day, including Werner Sombart, Paul Hensel, Ferdinand Tönnies, Ernst Troeltsch, and Georg Simmel (see Chapter 6). He helped establish the Heidelberg Academy of the Sciences in 1909 and the Sociological Society in 1910 (Marianne Weber 1926/1975:425). However, Weber was still plagued by compulsive anxiety. In 1918, he helped draft the constitution of the Weimar Republic while giving his first university lectures in 19 years at the University of Vienna. He suffered tremendously, however, and turned down an offer for a permanent post (Weber 1958:23). In 1920, at the age of 56, Max Weber died of pneumonia. Marianne lived for another 34 years and completed several important manuscripts left unfinished at her husband's death.

INTELLECTUAL INFLUENCES AND CORE IDEAS

Weber's work encompasses a wide scope of substantive interests. Most, if not all, of his writing has had a profound impact on sociology. As such, an attempt to fully capture the breadth and significance of his scholarship exceeds the limitations of a single chapter. Nevertheless, we can isolate several aspects of his work that, taken together, serve as a foundation for understanding the impetus behind much of his writing. To this end, we divide our discussion in this section into two major parts: (1) Weber's view of the science of sociology and (2) his engagement with the work of Friedrich Nietzsche and Karl Marx.

Sociology

Weber defined sociology as "a science which attempts the interpretive understanding of social action in order thereby to arrive at a causal explanation of its course and effects" (Weber 1947:88). In casting "interpretive understanding," or **Verstehen**, as the principal objective, Weber's vision of sociology offers a distinctive counter to those who sought to base the young discipline on the effort to uncover universal laws applicable to all societies. Thus, unlike Durkheim, who analyzed objective, sui generis "social facts" that operated independently of the individuals making up a society, Weber turned his attention to the subjective dimension of social life, seeking to understand the states of mind or motivations that guide individuals' behavior.

In delimiting the subject matter of sociology, Weber further specified "social action" to mean that which, "by virtue of the subjective meaning attached to it by the acting individual (or individuals), it takes account of the behaviour of others and is thereby oriented in its course" (Weber 1947:88). Such action can be either observable or internal to the actor's imagination, and it can involve a deliberate intervening in a given situation, an abstaining of involvement, or acquiescence. The task for the sociologist is to understand the meanings individuals assign to the contexts in which they are acting and the consequences that such meanings have for their conduct.

To systematize interpretive analyses of meaning, Weber distinguished four types of social action. In doing so, he clearly demonstrates his multidimensional approach to the problem of action (see Figure 4.1). First is *instrumental-rational action*. Such action is geared toward the efficient pursuit of goals through calculating the advantages and disadvantages associated with the possible means for realizing them. Under this category would fall the decision of a labor union to strike in order to bargain for greater employment benefits. Rehearsing one's performance for an upcoming job interview is another example of instrumental-rational action.

Figure 4.1   Weber's Four Types of Social Action

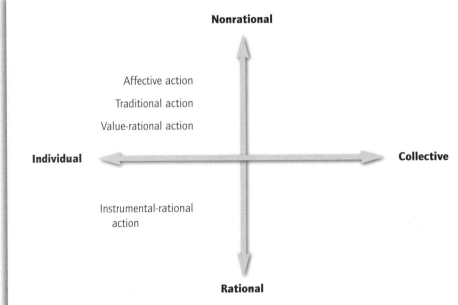

Like instrumental-rational action, *value-rational action* involves the strategic selection of means capable of effectively achieving one's goals. However, value-rational action is pursued as an end in itself, not because it serves as a means for achieving an ulterior goal. As such, it "always involves 'commands' or 'demands'" that compel the individual to follow a line of conduct for its own sake—because it is the "right" thing to do (Weber 1947:116). Examples of this type of action include risking arrest to further an environmental cause, or refraining from cheating on exams.

The third type of social action outlined by Weber is *traditional action*, where behaviors are determined by habit or longstanding custom. Here, an individual's conduct is shaped not by a concern with maximizing efficiency or commitment to an ethical principle, but rather by an unreflective adherence to established routines. This category includes religious rites of passage such as confirmations and bar mitzvahs, singing the national anthem at the start of sporting events, and eating turkey at Thanksgiving with one's family.

The fourth type is *affective action*, which is marked by impulsiveness or a display of unchecked emotions. Absent from this behavior is the calculated weighing of means for a given end. Examples of affective action are a baseball player arguing an umpire's called strike or parents crying at their child's wedding ceremony.

It is important to point out that in everyday life a given behavior or course of conduct is likely to exhibit characteristics of more than one type of social action. Thus, a person may pursue a career in social work not only because it is a means for earning a salary, but also because he is committed to the goal of helping others as a value in its own right. Weber's categories of social action, then, serve as **ideal types** or analytical constructs against which real-life cases can be compared. Such "pure" categories are not realized in concrete cases, but instead are a conceptual yardstick for examining differences and similarities, as well as causal connections, between the social processes under investigation. Thus, "ideal" refers to an emphasis on particular aspects of social life specified by the researcher, not to a value judgment as to whether something is "good" or "bad." As you will read in the selections that follow, Weber's work is guided in large measure by constructing ideal types. For instance, his essay on bureaucracy consists in the main of a discussion of the ideal characteristics of such an organization. Similarly, his essay on the three forms of domination involves isolating the features specific to each ideal type, none of which actually exists in pure form.

Weber's notion of sociology as an interpretive science based on *Verstehen* (understanding) and his focus on constructing ideal types marks his ties to important intellectual debates that were taking shape in German universities (Bendix 1977). At the heart of the debates was the distinction drawn between the natural and social sciences, and the methodologies appropriate to each. The boundary separating biology, chemistry, and physics from history, economics, psychology, and sociology was an outgrowth of German Idealism and the philosophy of **Immanuel Kant** (1724–1804). Kant argued that the realm of mind and "spirit" was radically different from the external, physical world of objects. According to Kant, because individuals create meaning and ultimately are free to choose their course of action, it is not possible to construct universal laws regarding human behavior. As a result, social life is not amenable to scientific investigation. On the other hand, absent of consciousness, objects and processes occurring in the natural world are open to scientific analysis and the development of general laws regarding their actions.

Among the scholars grappling with the implications of the Kantian division were the historical economists **Wilhelm Dilthey** (1833–1911) and **Heinrich Rickert** (1863–1936), whose work would have a profound impact on Weber. It was Dilthey who articulated the view that historical studies, and the social sciences more generally, should seek to *understand* particular events and their relationship to the specific contexts in which they occur. The task of history, then, is to interpret the subjective meanings actors assign to their conduct, not to search for causal explanations couched in terms of universal laws. According to Dilthey, any attempt to produce general causal laws regarding human behavior would not capture the unique historical conditions that shaped the events in question or a society's development. Moreover, such efforts would fail to study the very things that separate social life from the physical world of objects—human intent and motivation. Unlike the natural sciences and their analyses of the regularities governing observable objects and events, the social sciences aim to understand the internal states of actors and their relationship to behaviors.

In Weber's own definition of sociology, quoted above, we clearly see his indebtedness to Dilthey's work. Following Dilthey, Weber cast the social sciences as a branch of knowledge dedicated to developing an interpretive understanding of the subjective meanings actors attach to their conduct. However, Weber maintained a view not shared by Dilthey—that the social sciences, like the natural sciences, are conducted by making use of abstract and generalizing concepts. Here lies the impetus behind Weber's development of ideal types as a method for producing generalizable findings based on the study of historically specific events. For Weber, scientific knowledge is distinguished from nonscientific analyses not on the basis of the subject matter under consideration, but rather on how such studies are carried out. Thus, in constructing ideal types of action, Weber argued that analyses of the social world were not inherently less scientific or generalizable than investigations of the physical world. Nevertheless, Weber's *Verstehen* approach led him to contend that the search for universal laws of human action would lose sight of what is human—the production of meaningful behavior as it is grounded within a specific historical context.

It is in his notion of ideal types that we find Weber's links to the work of Heinrich Rickert. As a neo-Kantian thinker, Rickert accepted the distinction between the natural and social sciences as self-evident. However, he saw the differences between the two branches of knowledge as tied to the method of inquiry appropriate to each, not to any inherent differences in subject matter, as did Dilthey. According to Rickert, regardless of whether an investigator is trying to understand the meanings that motivate actors or attempting to uncover universal laws that govern the world of physical objects, they would study both subjects by way of concepts. Moreover, it is through the use of concepts that the investigator is able to select the aspects of the social or natural world most relevant to the purpose of her inquiry. The difference between the sciences lies, then, in *how* concepts are used to generate knowledge.

While the natural sciences used concepts as a way to generate abstract principles that explain the uniformities that shape the physical world, Rickert maintained that concepts used in the social sciences are best directed toward detailing the particular features that account for the uniqueness of an event or a society's development. In short, for Rickert the natural sciences were driven by the deductive search for universal laws. On the other hand, the social sciences were committed to producing inductive descriptions of historically specific phenomena.

For example, in subjecting molecules to changes in temperature and pressure, a physicist is interested in explaining the molecules' reactions in terms of causal laws whose validity is not restricted to any specific time period or setting. Conversely, social scientists studying episodes of protests, for instance, should seek to understand why individuals chose to act and how the cultural and institutional contexts shaped their behaviors. But because the contexts in which, for instance, the French Revolution, the Boston Tea Party, and the women's suffrage movement occurred were historically unique, it is not possible to formulate generalized explanations of protests on the basis of such specific, unreplicable events. Attempts to do so would require a level of conceptual abstraction that would necessarily lose sight of the particulars that made the events historically meaningful.

Weber's use of ideal types as a method for framing his analyses stems in important respects from Rickert's discussions on the role of concepts in the sciences. However, he did not share Rickert's view that the social sciences are unable to construct general causal explanations of historical events or societal development. Here, Weber sought to forge a middle ground between the generating of abstract laws characteristic of the natural sciences and the accumulation of historically specific facts that some contended must guide the social sciences. To this end, he cast the determination of causality as an attempt to establish the *probability* that a series of actions or events are related or have an *elective affinity.* Hence, Weber's notion of causality is fundamentally different from the conventional scientific usage, which sees it as the positing of invariant and necessary relationships between variables. According to Weber, the complexities of social life make it unamenable to formulating strict causal arguments such as those found in the natural sciences. While it can be stated that temperatures above 32 degrees Fahrenheit (x) will cause ice to melt (y), such straightforward, universal relationships between variables cannot be isolated when analyzing social processes; individual conduct and societal developments are not carried out with the constancy and singular causal "elegance" that characterizes the physical world. Thus, a sociologist cannot say with the same degree of certainty that an increase in educational attainment (x) will cause a rise in income (y), because while this relationship between the two variables may be probable, it is not inevitable. One need only keep in mind that a university professor with a Ph.D. typically makes far less money than a corporate executive with a bachelor's degree. As a result, sociologists should set out to determine the set of factors that, when taken together, have an elective affinity with a particular outcome. Armed with ideal types, the sociologist can then develop general arguments that establish the probable relationship between a combination of causes and a particular consequence.

Of Nietzsche and Marx

> The honesty of a contemporary scholar . . . can be measured by the position he takes vis-à-vis Nietzsche and Marx. Whoever fails to acknowledge that he could not carry out the most important part of his own work without the work done by both . . . deceives himself and others. The intellectual world in which we live is a world which to a large extent bears the imprint of Marx and Nietzsche.[1]

Such were the words spoken by Max Weber to his students shortly before his death. While his vision of sociology as a discipline was shaped in large measure by his links to German Idealism and the controversies surrounding historical studies, his substantive interests bear important connections to the work of Friedrich Nietzsche (1844–1900) and Karl Marx (1818–1883).

Evidencing his connection to Nietzsche, a major theme running throughout the whole of Weber's work is **rationalization**. By rationalization, Weber was referring to an ongoing process in which social interaction and institutions become increasingly governed by methodical procedures and calculable rules. Thus, in steering the course of societal development, values, traditions, and emotions were being displaced in favor of formal and impersonal bureaucratic practices. While such practices may breed greater efficiency in obtaining designated ends, they also lead to the "disenchantment of the world" where "there are no mysterious incalculable forces that come into play, but rather that one can, in principle, master all things by calculation" (Weber 1919/1958:139).

---

[1]Quoted in Robert J. Antonio (1995:3).

Few domains within modern Western societies have escaped from the trend toward rationalization. For instance, music became thoroughly codified by the 1500s with the development of scales derived from mathematical formulas and tonal and rhythmic notation. While musical improvisation by no means disappeared, it henceforth was based on underlying systematized principles of melody and harmony. The visual arts likewise became codified according to principles of perspective, composition, and color against which the avant-garde purposively rebels (and is thus no less subject to). Sex as an "irrational" bodily pleasure or as a rite tied to orgiastic rituals has been replaced by sex as a rational practice necessary for procreation. And procreation has itself come under increasing scientific control as advances in birth control and in vitro fertilization make it possible to plan when a birth will occur, to circumvent a person's natural infertility, and even to prenatally select specific traits. The transformation of sex was itself part of the broader displacement of magical belief systems by doctrinal religions, which were themselves later marginalized by an instrumental, scientific worldview. With each step, the work of fortune and fate, and mysterious and unknown powers were further removed from everyday life. The pantheon of gods and spirits that once ruled the universe would be distilled and simplified into the one all-knowing, omnipotent God, who would eventually lose his throne to the all-seeing telescope. And finally, Weber places special emphasis on the changes to social life brought on by the rationalization of capitalistic economic activity, as you will read below.

The ambivalence with which Weber viewed the process of rationalization stems from the loss of ultimate meaning that accompanied the growing dominance of an instrumental and scientific orientation to life. While science can provide technological advances that enable us to address more efficiently *how* to do things, it cannot provide us with a set of meanings and values that answer the more fundamental question: *Why?* Unlike those who saw in the Enlightenment's debunking of magical superstitions and religious beliefs the road to progress, Weber maintained that rationalization—and the scientific, calculative outlook in which it is rooted—does not generate "an increased and general knowledge of the conditions under which one lives" (Weber 1919/1958:139). They offer, instead, techniques empty of ultimate meaning.

Weber's reluctance to champion the progress brought by science and technological advances was influenced by Nietzsche's own nihilistic view of modernity expressed most boldly in his assertion, "God is dead" (Nietzsche 1866/1966). Nietzsche's claim reflected his conviction that the eclipse of religious and philosophical absolutes brought on by the rise of science and instrumental reasoning had created an era of nihilism or meaninglessness. Without religious or philosophical doctrines to provide a foundation for moral direction, life itself would cease to have an ultimate purpose. No longer could ethical distinctions be made between what one ought to do and what one can do (Nietzsche 1866/1966).

Weber was unwilling to assign a determinative end to history, however. Whether or not the spiritual void created by the disenchantment of the modern world would continue was, for him, an open question. The search for meaning—which Weber saw as the essence of the human condition—carried out in a meaningless world sparked the rise of charismatic leaders who were capable of offering their followers purpose and direction in their lives. (See "The Types of Legitimate Domination" below.) Ruling over others by virtue of their professed "state of grace," such figures were capable of radically transforming the existing social order. Weber's depiction of the power of charismatic leaders, with their ability to transcend the conventions and expectations imposed by the social order, bears important similarities to Nietzsche's notion of the *Übermensch*, or "superman." For Nietzsche, the fate of humanity and what is truly human lay in the hands of the *Übermenschen*, who alone are capable of overcoming the moral and spiritual bankruptcy that he believed corrupted the modern age (Nietzsche 1883/1978).

Significant Others

Friedrich Nietzsche (1844–1900): Is God Dead?

It is difficult to overstate the influence that the work of German philosopher and social critic Friedrich Nietzsche has had on twentieth-century thought. From theologians and psychologists, to philosophers and sociologists, to poets and playwrights, Nietzsche's ideas have penetrated virtually every domain of modern intellectual culture. It was not until after his death, however, that he would earn such acclaim, for during his life his writings attracted but the smallest of audiences.

Beset with a host of physical ailments, and stricken by a complete mental breakdown at the age of 45, Nietzsche, nevertheless, managed to develop a number of themes that would usher in a thoroughgoing critique of seemingly unassailable truths. Rejecting the Enlightenment notion that reason offers the pathway to human emancipation, Nietzsche believed that the essence of humanity lies in emotional and physical experiences. Moreover, he repudiated Christianity's ascetic ethic as a renunciation or avoidance of life, and championed, instead, the embracing of all that life offers, even the most tragic of sufferings, as the ultimate expression of greatness.

The man who declared, "God is dead" and who argued that truth, values, and morals are not based on some intrinsic, ahistorical criteria, but, instead, are established by the victors in the unending struggle for power, did not enter the canon of liberal academia without controversy. Owing to the intentional distortions and forgeries of some of his writing by his sister, Elisabeth, Nietzsche was often interpreted as an anti-Semitic fascist. Though he abhorred such hatred as "slavish" and "herd-like," Hitler's Third Reich reinvented Nietzsche's notion of the "will to power" and the *Übermensch* or "superman" as a justification for its military aggression and genocidal practices. Fortunately, contemporary scholars of Nietzsche's work have corrected many of Elisabeth's falsities, allowing the true intention of his piercing, original insights into modern culture to be realized.

In addition to drawing inspiration from Nietzsche's work, much of Weber's writing reflects a critical engagement with and extension of Marx's theory of historical materialism.[2] As we noted in Chapter 2, Marx saw class struggles as the decisive force in the evolution of history. Class struggles were, in turn, the inevitable outcome of the inherent contradictions found in all precommunist economic systems. While finding much convincing in Marx's argument, Weber nevertheless did not embrace it in its entirety. In constructing his own theoretical framework, Weber departed from Marx in a number of respects, three of which we outline here.

First, Weber maintained that social life did not evolve according to some immanent or necessary law. Thus, unlike Marx, Weber did not foresee a definitive "end of prehistory" toward which social evolution progressed. Instead, he saw the future of modern society as an open question, the answer to which it is impossible to foretell. This position, coupled with his view that rationalizing processes had transformed modern society into an "iron cage" (see below), accounts for Weber's unwillingness to accept a utopian vision of humanity's future.

---

[2]It is important to point out that Weber's critique of Marx was based more on secondary interpretations of Marx's work than on a thorough, firsthand encounter with his writings, since much of it was unavailable. In Weber's time, and continuing today, Marx was (is) often miscast by his followers and critics alike as an economic determinist. Perhaps more accurately, then, Weber was responding to a "crude," reductionist version of Marxism.

Second, he contended that the development of societies could not be adequately explained on the basis of a single or primary causal mechanism. The analysis of economic conditions and class dynamics alone could not capture the complex social and cultural processes responsible for shaping a society's trajectory. In particular, Weber maintained that Marx, in emphasizing economic factors and class-based interests, underestimated the role that *ideas* play in determining a society's course of development. On this point, Weber sought to incorporate Marx's argument into his own work while offering what he saw as a necessary corrective, remarking, "Not ideas, but material and ideal interests, directly govern men's conduct. Yet very frequently the 'world images' that have been created by 'ideas' have, like switchmen, determined the tracks along which action has been pushed by the dynamic of interest" (Weber 1915/1958:280).

Acknowledging the powerful sway that "interests" hold over individuals as they chart their course of action, Weber nevertheless argued that ideas play a central role in shaping the paths along which interests are realized. He saw ideas as an independent cultural force and not as a reflection of material conditions or the existing mode of production. As the source for constructing meaning and purposeful lines of action, ideas are not simply one element among others confined to the "superstructure." Instead, they serve as the bases on which individuals carve out possible avenues of action, and, more dramatically, when advanced by a charismatic leader ideas can inspire revolution.

A third difference lies in where the two theorists located the fundamental problems facing modern industrial society. As you read previously, Marx identified capitalism as the primary source of humanity's inhumanity. The logic of capitalism necessarily led to the exploitation of the working class as well as to the alienation of the individual from his work, himself, and others. For Weber, however, it was not capitalism but the process of rationalization and the increasing dominance of bureaucracies that threatened to destroy creativity and individuality. By design, bureaucratic organizations—and the rational procedures that govern them—routinize and standardize people and products. Though making for greater efficiency and predictability in the spheres of life they have touched, the impersonality of bureaucracies, their indifference to difference, has created a "cold" and empty world. (See Weber's essay "Bureaucracy," excerpted below.)

Not surprisingly, then, Weber, unlike many of his contemporaries, did not see in socialism the cure for society's ills. In taking control of a society's productive forces, socialist forms of government would only further bureaucratize the social order, offering a poor alternative to capitalism. Indeed, Weber believed capitalism was a "better" economic system to the extent that its competitiveness allowed more opportunities to express one's individuality and creative impulses. Clearly, Weber did not embrace Marx's or his followers' calls for a communist revolution, because such a movement, to the extent that it led to an expansion of the scope of bureaucracies, would accelerate the hollowing out of human life.

> ### Significant Others
>
> #### Robert Michels (1876–1936): The Iron Law of Oligarchy
>
> Political activist and sociologist, Robert Michels is best known for his studies on the organization of political parties. Influenced by the ideas of his teacher and mentor, Max Weber, Michels argued that all large-scale organizations have a tendency to evolve into hierarchical bureaucracies regardless of their original formation and ultimate goals. Even organizations that adopt an avowedly democratic agenda are inevitably subject to this "iron law of oligarchy" because leadership is necessarily transferred to an elite decision-making body.

Michels developed his argument in *Political Parties* (1911) in which he examined the organizational structures of western European socialist trade unions and political parties. During the late 1800s, the democratic ethos was particularly strong within these revolutionary socialist parties whose principal aim was to overthrow aristocratic or oligarchic regimes and replace them with governing bodies controlled directly by the people. Despite their intent on destroying elite rule—the rule of the many by the few—these parties were themselves unable to escape the tendency toward oligarchy.

Michels advanced his ideas in part as a response to his disillusionment with the German Social Democratic Party. An active member of the party, he witnessed first-hand its growing political conservatism. (Michels was censured by German government authorities for his political radicalism, compelling him to take positions at universities in Italy and Switzerland.) Established in the 1870s as an advocate for the working class, the Marxist-inspired party abandoned its revolutionary program soon after its formation, as its ambitions to wrest control of the means of production into the hands of the people was replaced by the conservative goals of increasing its membership, amassing funds for its war chest, and winning electoral seats in the German legislature through which piece-meal reform might be gained. Considered a vanguard of the proletariat revolution, this dramatic shift in party tactics signaled a rejection of Marxist principles and the abandonment of the struggle for realizing an ideal democracy where workers controlled their labor and freedom from want existed throughout society.

What led to the cooptation of this and similarly driven parties' ideals? The answer lies in the working classes' lack of economic and political power. In order to effect democratic change, the otherwise powerless working-class individuals must first organize; their strength as a movement is directly related to their strength in numbers. Numbers, however, require representation through individual delegates who are entrusted by the mass to act on its behalf. Despite Marx's utopian promise, the growth in numbers necessary to achieve power makes it impossible for the people to exercise direct control over their destinies. Instead, the success of working-class parties hinges on creating an organization committed to representing its interests: "Organization is . . . the source from which the conservative currents flow over the plain of democracy, occasioning there disastrous floods and rendering the plain unrecognizable" (Michels 1911/1958:26). The inevitable rise of an organization brings with it the equally inevitable need for technical expertise, centralized authority, and a professional staff to ensure its efficient functioning. Bureaucratization transforms the party from a means to an end, to an end in itself. The preservation of the organization itself becomes the essential aim, and its original democratic ambitions are preserved only in talk, because aggressive action against the state would surely threaten its continued existence.

Thus, while "[d]emocracy is inconceivable without organization" (Michels 1911/1958:25), the inherently oligarchic and bureaucratic nature of party organizations saps its revolutionary zeal and replaces it with the pursuit of disciplined, cautious policies intended to defend its own long-term interests, which do not necessarily coincide with the interests of the class it represents. As Michels notes, it would seem

> society cannot exist without a "dominant" or "political" class [that] . . . constitutes the only factor of sufficiently durable efficacy in the history of human development. . . . [T]he state, cannot be anything other than the organization of a minority. It is the aim of this minority to impose upon the rest of society a

(Continued)

(Continued)

"legal order," which is the outcome of the exigencies of dominion and of the exploitation of the mass of helots effected by the ruling minority, and can never be truly representative of the majority. The majority is thus permanently incapable of self-government. Even when the discontent of the masses culminates in a successful attempt to deprive the bourgeoisie of power, this is . . . effected only in appearance; always and necessarily there springs from the masses a new organized minority which raises itself to the rank of a governing class. Thus the majority of human beings, in a condition of eternal tutelage, are predestined by tragic necessity to submit to the dominion of a small minority, and must be content to constitute the pedestal of an oligarchy. (Michels 1911/1958:406, 407)

Figure 4.2 Weber's Basic Theoretical Orientation

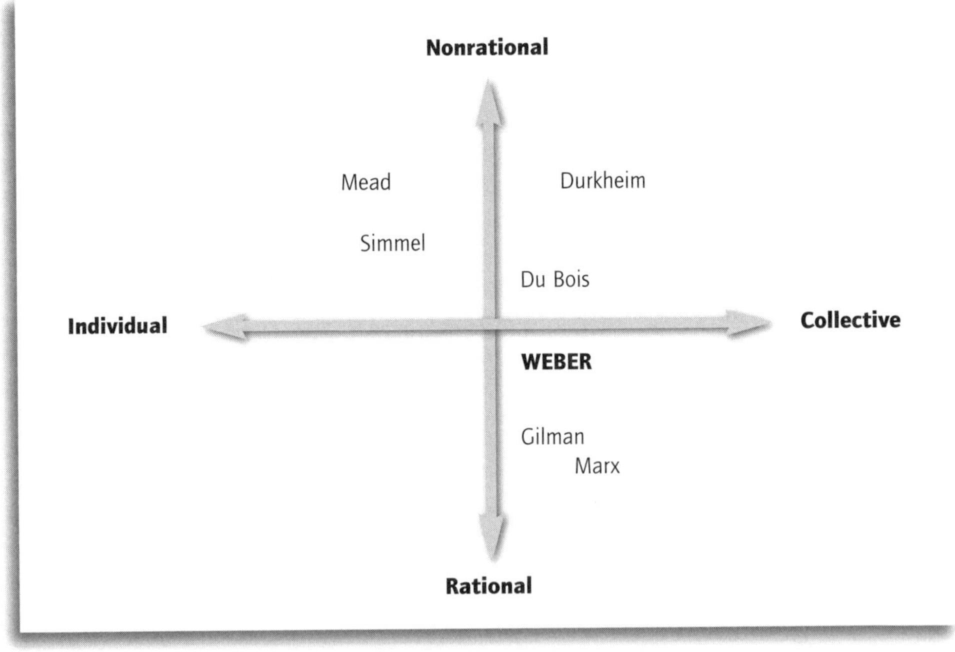

WEBER'S THEORETICAL ORIENTATION

Weber's work is avowedly multidimensional. This is depicted in Figure 4.2 by his positioning relative to the other theorists discussed in this text. He explicitly recognized that individual action is channeled through a variety of motivations that encompass both rationalist and nonrationalist dimensions. Moreover, his definition of sociology as a science aimed at the interpretive understanding of social action squarely places the individual and her conduct at the center of analysis. Complementing this position are Weber's substantive interests that led him to study religious idea systems, institutional arrangements, class and status structures, forms of domination, and broad historical trends; in short, elements aligned with the collective dimension of social life.

Of course, not every essay incorporates elements from each of the four dimensions. For instance, Weber's discussion of bureaucracy (excerpted below) focuses on the administrative functions and rules that account for the efficiency and impersonality that mark this organizational form. As a result, he emphasizes the structural or collectivist aspects of bureaucracies and how they work down to shape a given individual's behaviors and attitudes within them. Thus, you will find Weber remarking, "The individual bureaucrat cannot squirm out of the apparatus into which he has been harnessed. . . . [H]e is only a small cog in a ceaselessly moving mechanism which prescribes to him an essentially fixed route of march" (Weber 1925d/1978:988). Weber's interest, then, lies here in describing the bureaucratic apparatus replete with its institutionalized demands for technical expertise and leveling of social differences.[3]

In Figure 4.3, we have highlighted a number of key concepts found in our preceding remarks or in the primary selections that follow. From the chart, it is readily apparent that

Figure 4.3 Weber's Core Concepts

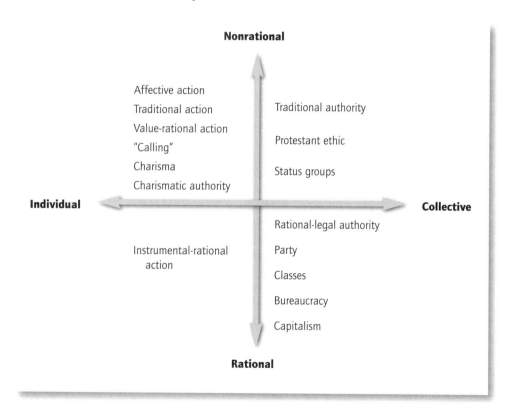

---

[3]While Weber's approach is clearly multidimensional, it is due to arguments like the one expressed in his essay on bureaucracy that we position the body of his work "off-center," ultimately in the collectivist/rationalist quadrant of our diagram. In the end, his emphasis lies in examining the rationalizing (i.e., rationalist) processes that have shaped the development of modern Western institutions (i.e., collectivist).

Weber's theoretical orientation spans each of the four dimensions. Because some of these concepts were discussed previously (for instance, those regarding the types of action) and others will be addressed later in our introductions to the selections, we will restrict our comments in this section to a single example that underscores Weber's multidimensional approach.

In *The Protestant Ethic and the Spirit of Capitalism* (1904–05), Weber discusses the importance of the **calling** in motivating individuals to pursue worldly success. A doctrine first espoused by the Protestant reformer **Martin Luther** (1483–1546), the idea that each individual has a calling or "life-task" has its roots in a religious quest for salvation. In terms of our theoretical map, then, the calling reflects a nonrationalist orientation to action. The actions of the religiously faithful were motivated by the *moral* obligation to perform the duties of his labor to the best of his abilities. Here, the individual's actions are inspired by his desire to glorify God and thus gain confidence in the certainty of His grace, not by a desire to accumulate wealth as a means for purchasing material goods. Moreover, the calling is an individualist concept. It serves as the basis on which individuals make sense of their life circumstances as they chart their chances for attaining worldly success and eternal salvation.

Weber's analysis of the calling, however, was not tied solely to an examination of how religious ideas motivate individual conduct. For Weber, the significance of the calling also lies in its fueling a dramatic social transformation: the growth and eventual dominance of capitalism and the accompanying rationalization of much of social life. While oversimplifying his argument, Weber contended that the development of modern forms of capitalism was tied to the ascetic lifestyle demanded by the pursuit of one's calling. Originally a religious injunction to lead a life freed from the "temptations of the flesh," the secularization of the calling was a major force contributing to the explosive growth of capitalism in the West as businesses were increasingly organized on the basis of impersonal, methodical practices aimed at the efficient production of goods and services. Profit was now sought not to ensure one's state of grace, but because it was in one's self-interest to do so. Stripped of its religious impulses and spiritual moorings, the calling was further transformed into an overarching rationalist orientation to action that, as we remarked earlier, introduced methodical and calculative procedures into not only economic practices, but also into numerous spheres of life including politics, art, and sex, to name only a few.

Last, Weber's argument reveals a decidedly collectivist element as well. The ascetic ideals lying at the heart of the **Protestant ethic** were carried into the practical affairs of economic activity and social life more generally. This unleashed the process of rationalization, disenchanting Western society and creating an **iron cage** from which the individual is left with little power to escape. The dominance of capitalism and impersonal, bureaucratic forms of organization was a collective force that determined the life-chances of the individual. This dynamic is illustrated in the following passage taken from *The Protestant Ethic* and with which we end this section:

> The Puritan wanted to work in a calling; we are forced to do so. For when asceticism was carried out of monastic cells into everyday life, and began to dominate worldly morality, it did its part in building the tremendous cosmos of the modern economic order. This order is now bound to the technical and economic conditions of machine production which to-day [sic] determine the lives of all the individuals who are born into this mechanism, not only those directly concerned with economic acquisition, with irresistible force. (Weber 1904–05/1958:181)

Readings

In the selections that follow you will be introduced to five of Weber's most influential writings. In the first reading, excerpts from *The Protestant Ethic and the Spirit of Capitalism* (Weber 1904–05) offer Weber's analysis of the relationship between Protestantism and the economic and cultural life of modern Western society. In the second reading, from "The Social Psychology of the World Religions," Weber expands this theme in an examination of the psychological motivations underlying the "world religions." In the third reading, Weber investigates the crosscutting sources of power: class, status, and party. A parallel theme is addressed in the fourth selection, "The Types of Legitimate Domination," in which Weber outlines three distinct types of domination or authority. Finally, in "Bureaucracy," we end with Weber's description of bureaucracy, the predominant form of modern social organizations.

Introduction to *The Protestant Ethic and the Spirit of Capitalism*

Beyond doubt, one of the most influential sociology books ever written, *The Protestant Ethic* masterfully captures the two subjects that preoccupied Weber's intellectual activities: (1) the rationalizing tendencies so prevalent in Western society and (2) the role of ideas in shaping them. In addressing these twin issues, Weber argues that a religious belief system, intended to explain the path to a transcendent eternal salvation, paradoxically fueled the creation of a secular world in which "material goods have gained an increasing and finally an inexorable power over the lives of men as at no previous period in history" (Weber 1904–05/1958:181).

Unlike Marx, who viewed religion as "the opiate of masses," or as an ideology that served the economic interests of the ruling class, and unlike Durkheim, who saw in religion humanity's worship of itself, Weber saw in religious beliefs a system of meaning aimed at explaining the existence of suffering and evil in the world. For Weber, such explanations have a profound impact on individuals' actions, and consequently on the broader social order. Of particular import is whether in addressing these ultimate issues, a belief system orients its adherents toward a "mastery" of the world or a mystical or contemplative escape from it. Thus, Protestantism, and Calvinism in particular, demanded that its followers serve as the "instruments" of God in order to fashion the world in His image. Conversely, Eastern religions such as Buddhism and Hinduism required their faithful to become "vessels" for the divine spirit in order to commune with otherworldly cosmic powers. The active engagement with the external, secular world called for by the Protestant belief system functioned as a potent impetus for social change, while the inward search for spiritual awakening characteristic of the major Eastern religions proved to be a socially conservative force.

In developing a scientifically based account of the independent role religious ideas can play in shaping the social order and, in particular, economic systems, Weber offered a powerful critique of Marxist theories of capitalism. As we discussed previously, he saw in historical materialism a one-sided causal interpretation, and, in several passages of *The Protestant Ethic*, you will read Weber clearly setting his sights on piercing this doctrine. As a counter to Marx's emphasis on property relations and class struggle, Weber maintained that the extraordinarily methodical *attitude* that characterized Protestant asceticism was integral to the rise and eventual dominance of Western capitalism.[1] Thus, Weber sought to demonstrate that not only "material" factors, but also "ideal" factors can be instrumental in producing social change. In doing so, he sparked one of the most important and enduring debates in the history of sociology.

Having already highlighted several key elements of *The Protestant Ethic* when we outlined Weber's theoretical orientation, we briefly call attention to the book's main ideas. Weber traced the rise of individualism to the late sixteenth century and the Protestant Reformation, which, among other things, redefined the nature of the relationship between man and God. Led by **Martin Luther** (1483–1546), the Protestant Reformers insisted that each individual must methodically strive to realize a moral and righteous life each and every day in all their practical activities, as a constant expression of their devotion to the glorification of God. This methodical individualism challenged the previously dominant religious practice in which a handful of religious professionals (clergy) performed rituals in order to appease the gods either on behalf of the whole society or on behalf of those who paid them for their services. But Luther maintained that these token, periodic rituals (for instance, the Catholic confessional) or occasional "good works" could never placate or gain the favor of a great and all-powerful God. Instead, it was the duty of each to *submit* to the will of God through faithful dedication to his calling. It was demanded of rich and poor alike to be content with their lot, for it was God's unfathomable will that had assigned to each his station in life.

With its emphasis on submission and faith, Luther's view of the calling, like the Catholicism it rebelled against, promoted a traditional economic ethic that discouraged both laboring and profit seeking beyond what had long been established through custom. Workers and merchants sought simply to maintain the level of productivity and standard of living associated with the vocation in which they were engaged. However, in the hands of later Puritans leaders, the meaning of the calling was transformed. Under **John Calvin** (1509–64) and **Richard Baxter** (1615–91), the calling was interpreted as God's commandment to *work* for His divine glory. With submission and faith no longer sufficient for gaining confidence in one's salvation, how could the believer know that he was fulfilling his calling and thus might be one of His elect? Existing beyond the influence of mortals, only God knows who will be saved; there could be no certainty of proof of one's state of grace. The best one could hope for was a divinely granted sign. And that sign?: success and profit in worldly affairs, the pursuit of which was now religiously enjoined. Baxter stated the injunction thusly, "If God show you a way in which you may lawfully get more than in another way . . . if you refuse this, and choose the less gainful way, you cross one of the ends of your calling, and you refuse to be God's steward, and to accept His gifts and use them for Him when He requireth it" (Weber 1904–05/1958:162). Profit was now understood

---

[1]Significantly, Weber's central point was not that the Protestant ethic caused the emergence and growth of Western capitalism. Protestantism alone was not sufficient for creating this profound economic change. Rather, he argued that Protestant asceticism combined with a number of other important structural and social factors to produce the dominance of Western capitalism. In particular, Weber pointed to the separation of business pursuits from the home; the development of rational bookkeeping methods; technological advances in methods of production, distribution, and communication; the development of a rational legal system based on impersonal, formal rules; and most importantly, the rational organization of free labor.

to be a visible blessing from God that allowed the faithful to answer the most burning of all questions: Am I saved? Possessed by this "new spirit," one's predestined, eternal fate was now tied to the success of his conduct in work, a sphere of activity that was catapulted to the center of the believer's existence.

It was not success itself that offered proof, however. Rather, it was *how* success was achieved that marked a person as one of God's elect. Baxter cautioned his followers that, "You may labour to be rich for God, though not for the flesh and sin" (Weber 1904–05/1958:162). In this proscription lay the seeds for the subjective disposition that would ignite the growth of capitalism. Wealth served as confirmation of one's salvation only if it did not lead to idleness or the enjoyment of luxuries. Profitableness, moreover, was best guaranteed when economic pursuits were carried out on the basis of methodical and rational planning. Thus, ascetic restrictions on consumption were combined with the religiously derived compulsion to increase one's wealth. The ethical imperative to save and invest one's wealth would become the spiritual foundation for the spread of capitalism.

It would not be long, however, before the rational pursuit of wealth and bureaucratic structures necessary to modern capitalism would render obsolete the religious ethic that first had imbued work with a sense of meaning and purpose.[2] Chained by unquenchable consumption, modern humanity is now left to live in a disenchanted world where "material goods have gained an increasing and finally inexorable power over the lives of men" (Weber 1904–05/1958:180). "In Baxter's view the care for external goods should only lie on the shoulders of the 'saint like a light cloak, which can be thrown aside at any moment.' But fate decreed that the cloak should become an iron cage" (ibid.:181).

And what of the iron cage today? Consider some statistics from the U.S. Commerce Department and the Federal Reserve Board: The average household is saddled with a credit card debt of $8,000, while the nation's credit card debt currently stands at $880 million. Not including home mortgages, in 2003, the average household was faced with more than $18,000 in total debt. As a nation, consumer debt soared to nearly 2 trillion dollars, an increase of 40 percent from 1998's total. Not surprisingly, personal savings rates have declined. After essential expenditures, Americans saved 9 percent of their disposable income during the 1980s. This rate fell to 5 percent during the 1990s, and in 2006 Americans registered a negative savings rate (–1 percent) for the first time since the Great Depression. Currently, 40 percent of Americans spend more than they earn. Far from being a "light cloak," our "care for external goods" has become central to our personal identity and sense of self. We define ourselves through the cars we drive, the clothes we wear, the places we vacation, and the neighborhoods we live in rather than through a sense of ultimate purpose or meaning to life. Whether it's trying to keep up with the Joneses or to distinguish ourselves from the herd, we are in continual "need" of new and better products, the purchasing of which requires ever-longer working hours in order to earn more money, so we can spend more money. To keep pace with the growing accumulation of products, over the last 50 years the average home size has doubled. Still, we can't seem to fit everything in so we hire companies to organize our closets and garages, or, when that fails, we pay to pack our "unessential" belongings into one of the thousands of self-storage spaces that dot the landscape. Like Marx's views on the fetishism of commodities and Veblen's notion of conspicuous consumption, the iron cage has imprisoned us in the pursuit of the "lifestyles of the rich and famous" whether or not we can afford to live like the affluent.

---

[2]One need merely note the spread of capitalism to countries and regions of the world that have not been exposed in any significant degree to Protestantism.

From *The Protestant Ethic and the Spirit of Capitalism* (1904)

Max Weber

THE SPIRIT OF CAPITALISM

In the title of this study is used the somewhat pretentious phrase, the *spirit* of capitalism. What is to be understood by it? The attempt to give anything like a definition of it brings out certain difficulties which are in the very nature of this type of investigation. . . .

Thus, if we try to determine the object, the analysis and historical explanation of which we are attempting, it cannot be in the form of a conceptual definition, but at least in the beginning only a provisional description of what is here meant by the spirit of capitalism. Such a description is, however, indispensable in order clearly to understand the object of the investigation. For this purpose we turn to a document of that spirit which contains what we are looking for in almost classical purity, and at the same time has the advantage of being free from all direct relationship to religion, being thus, for our purposes, free of preconceptions.

"Remember, that *time* is money. He that can earn ten shillings a day by his labour, and goes abroad, or sits idle, one half of that day, though he spends but sixpence during his diversion or idleness, ought not to reckon *that* the only expense; he has really spent, or rather thrown away, five shillings besides.

"Remember, that *credit* is money. If a man lets his money lie in my hands after it is due, he gives me the interest, or so much as I can make of it during that time. This amounts to a considerable sum where a man has good and large credit, and makes good use of it.

"Remember, that money is of the prolific, generating nature. Money can beget money, and its offspring can beget more, and so on. Five shillings turned is six, turned again it is seven and threepence, and so on, till it becomes a hundred pounds. The more there is of it, the more it produces every turning, so that the profits rise quicker and quicker. He that kills a breeding-sow, destroys all her offspring to the thousandth generation. He that murders a crown, destroys all that it might have produced, even scores of pounds."

"Remember this saying, *The good paymaster is lord of another man's purse.* He that is known to pay punctually and exactly to the time he promises, may at any time, and on any occasion, raise all the money his friends can spare. This is sometimes of great use. After industry and frugality, nothing contributes more to the raising of a young man in the world than punctuality and justice in all his dealings; therefore never keep borrowed money an hour beyond the time you promised, lest a disappointment shut up your friend's purse for ever.

"The most trifling actions that affect a man's credit are to be regarded. The sound of your hammer at five in the morning, or eight at night, heard by a creditor, makes him easy six months longer; but if he sees you at a billiard-table, or hears your voice at a tavern, when you should be at work, he sends for his money the next day; demands it, before he can receive it, in a lump.

"It shows, besides, that you are mindful of what you owe; it makes you appear a careful as well as an honest man, and that still increases your credit.

"Beware of thinking all your own that you possess, and of living accordingly. It is a mistake that many people who have credit fall into. To prevent this, keep an exact account for some time both of your expenses and your income. If you take the pains at first to mention particulars, it will have this good effect: you will discover how wonderfully small, trifling expenses mount up to large sums, and will discern what might have been, and may for the future be saved, without occasioning any great inconvenience.

"For six pounds a year you may have the use of one hundred pounds, provided you are a man of known prudence and honesty.

SOURCE: *The Protestant Ethic and the Spirit of Capitalism,* 1st edition, by Max Weber. Copyright © 1958. Reprinted by permission of Pearson Education, Inc., Upper Saddle River, New Jersey.

"He that spends a groat a day idly, spends idly above six pounds a year, which is the price for the use of one hundred pounds.

"He that wastes idly a groat's worth of his time per day, one day with another, wastes the privilege of using one hundred pounds each day.

"He that idly loses five shillings' worth of time, loses five shillings, and might as prudently throw five shillings into the sea.

"He that loses five shillings, not only loses that sum, but all the advantage that might be made by turning it in dealing, which by the time that a young man becomes old, will amount to a considerable sum of money."

It is Benjamin Franklin who preaches to us in these sentences, the same which Ferdinand Kürnberger satirizes in his clever and malicious *Picture of American Culture* as the supposed confession of faith of the Yankee. That it is the spirit of capitalism which here speaks in characteristic fashion, no one will doubt, however little we may wish to claim that everything which could be understood as pertaining to that spirit is contained in it. Let us pause a moment to consider this passage, the philosophy of which Kürnberger sums up in the words, "They make tallow out of cattle and money out of men." The peculiarity of this philosophy of avarice appears to be the ideal of the honest man of recognized credit, and above all the idea of a duty of the individual toward the increase of his capital, which is assumed as an end in itself. Truly what is here preached is not simply a means of making one's way in the world, but a peculiar ethic. The infraction of its rules is treated not as foolishness but as forgetfulness of duty. That is the essence of the matter. It is not mere business astuteness, that sort of thing is common enough, it is an ethos. *This* is the quality which interests us. . . .

Now, all Franklin's moral attitudes are coloured with utilitarianism. Honesty is useful, because it assures credit; so are punctuality, industry, frugality, and that is the reason they are virtues. A logical deduction from this would be that where, for instance, the appearance of honesty serves the same purpose, that would suffice, and an unnecessary surplus of this virtue would evidently appear to Franklin's eyes as unproductive waste. And as a matter of fact, the story in his autobiography of his conversion to those virtues, or the discussion of the value of a strict maintenance of the appearance of modesty, the assiduous belittlement of one's own deserts in order to gain general recognition later, confirms this impression. According to Franklin, those virtues, like all others, are only in so far virtues as they are actually useful to the individual, and the surrogate of mere appearance is always sufficient when it accomplishes the end in view. It is a conclusion which is inevitable for strict utilitarianism. The impression of many Germans that the virtues professed by Americanism are pure hypocrisy seems to have been confirmed by this striking case. But in fact the matter is not by any means so simple. Benjamin Franklin's own character, as it appears in the really unusual candidness of his autobiography, belies that suspicion. The circumstance that he ascribes his recognition of the utility of virtue to a divine revelation which was intended to lead him in the path of righteousness, shows that something more than mere garnishing for purely egocentric motives is involved.

In fact, the *summum bonum* of this ethic, the earning of more and more money, combined with the strict avoidance of all spontaneous enjoyment of life, is above all completely devoid of any eudæmonistic, not to say hedonistic, admixture. It is thought of so purely as an end in itself, that from the point of view of the happiness of, or utility to, the single individual, it appears entirely transcendental and absolutely irrational. Man is dominated by the making of money, by acquisition as the ultimate purpose of his life. Economic acquisition is no longer subordinated to man as the means for the satisfaction of his material needs. This reversal of what we should call the natural relationship, so irrational from a naïve point of view, is evidently as definitely a leading principle of capitalism as it is foreign to all peoples not under capitalistic influence. At the same time it expresses a type of feeling which is closely connected with certain religious ideas. If we thus ask, *why* should "money be made out of men," Benjamin Franklin himself, although he was a colourless deist, answers in his autobiography with a quotation from the Bible, which his strict Calvinistic father drummed into him again and again in his youth: "Seest thou a man diligent in his business? He shall stand before kings" (Prov. xxii. 29). The earning of money within the modern economic order is, so long as it is done legally, the result and the expression of virtue and proficiency in a calling; and this virtue and proficiency are, as it is now not difficult to see, the real Alpha and Omega of Franklin's ethic, as expressed in the passages we have quoted, as well as in all his works without exception.

And in truth this peculiar idea, so familiar to us to-day, but in reality so little a matter of course, of one's duty in a calling, is what is most characteristic of the social ethic of capitalistic culture, and is in a sense the fundamental basis of it. It is an obligation which the individual is supposed to feel and does feel towards the content of his professional activity, no matter in what it consists, in particular no matter whether it appears on the surface as a utilization of his personal powers, or only of his material possessions (as capital).

Of course, this conception has not appeared only under capitalistic conditions. On the contrary, we shall later trace its origins back to a time previous to the advent of capitalism. Still less, naturally, do we maintain that a conscious acceptance of these ethical maxims on the part of the individuals, entrepreneurs or labourers, in modern capitalistic enterprises, is a condition of the further existence of present-day capitalism. The capitalistic economy of the present day is an immense cosmos into which the individual is born, and which presents itself to him, at least as an individual, as an unalterable order of things in which he must live. It forces the individual, in so far as he is involved in the system of market relationships, to conform to capitalistic rules of action. The manufacturer who in the long run acts counter to these norms, will just as inevitably be eliminated from the economic scene as the worker who cannot or will not adapt himself to them will be thrown into the streets without a job.

Thus the capitalism of to-day, which has come to dominate economic life, educates and selects the economic subjects which it needs through a process of economic survival of the fittest. But here one can easily see the limits of the concept of selection as a means of historical explanation. In order that a manner of life so well adapted to the peculiarities of capitalism could be selected at all, i.e. should come to dominate others, it had to originate somewhere, and not in isolated individuals alone, but as a way of life common to whole groups of men. This origin is what really needs explanation. Concerning the doctrine of the more naïve historical materialism, that such ideas originate as a reflection or superstructure of economic situations, we shall speak more in detail below. At this point it will suffice for our purpose to call attention to the fact that without doubt, in the country of Benjamin Franklin's birth (Massachusetts), the spirit of capitalism (in the sense we have attached to it) was present before the capitalistic order. . . . It is further undoubted that capitalism remained far less developed in some of the neighbouring colonies, the later Southern States of the United States of America, in spite of the fact that these latter were founded by large capitalists for business motives, while the New England colonies were founded by preachers and seminary graduates with the help of small bourgeois, craftsmen and yoemen, for religious reasons. In this case the causal relation is certainly the reverse of that suggested by the materialistic standpoint.

But the origin and history of such ideas is much more complex than the theorists of the superstructure suppose. The spirit of capitalism, in the sense in which we are using the term, had to fight its way to supremacy against a whole world of hostile forces. A state of mind such as that expressed in the passages we have quoted from Franklin, and which called forth the applause of a whole people, would both in ancient times and in the Middle Ages have been proscribed as the lowest sort of avarice and as an attitude entirely lacking in self-respect. It is, in fact, still regularly thus looked upon by all those social groups which are least involved in or adapted to modern capitalistic conditions. This is not wholly because the instinct of acquisition was in those times unknown or undeveloped, as has often been said. Nor because the *auri sacra fames*, the greed for gold, was then, or now, less powerful outside of bourgeois capitalism than within its peculiar sphere, as the illusions of modern romanticists are wont to believe. The difference between the capitalistic and pre-capitalistic spirits is not to be found at this point. The greed of the Chinese Mandarin, the old Roman aristocrat, or the modern peasant, can stand up to any comparison. And the *auri sacra fames* of a Neapolitan cab-driver or *barcaiuolo*, and certainly of Asiatic representatives of similar trades, as well as of the craftsmen of southern European or Asiatic countries, is, as anyone can find out for himself, very much more intense, and especially more unscrupulous than that of, say, an Englishman in similar circumstances. . . .

The most important opponent with which the spirit of capitalism, in the sense of a definite standard of life claiming ethical sanction, has had to struggle, was that type of attitude and reaction to new situations which we may designate as traditionalism. . . .

One of the technical means which the modern employer uses in order to secure the greatest possible amount of work from his men is the device of piece-rates. In agriculture, for instance, the gathering of the harvest is a case where the greatest possible intensity of labour is called for, since, the weather being uncertain, the difference between high profit and heavy loss may depend on the speed with which the harvesting can be done. Hence a system of piece-rates is almost universal in this case. And since the interest of the employer in a speeding-up of harvesting increases with the increase of the results and the intensity of the work, the attempt has again and again been made, by increasing the piece-rates of the workmen, thereby giving them an opportunity to earn what is for them a very high wage, to interest them in increasing their own efficiency. But a peculiar difficulty has been met with surprising frequency: raising the piece-rates has often had the result that not more but less has been accomplished in the same time, because the worker reacted to the increase not by increasing but by decreasing the amount of his work. A man, for instance, who at the rate of 1 mark per acre mowed 2½ acres per day and earned 2½ marks, when the rate was raised to 1.25 marks per acre mowed, not 3 acres, as he might easily have done, thus earning 3.75 marks, but only 2 acres, so that he could still earn the 2½ marks to which he was accustomed. The opportunity of earning more was less attractive than that of working less. He did not ask: how much can I earn in a day if I do as much work as possible? but: how much must I work in order to earn the wage, 2½ marks, which I earned before and

which takes care of my traditional needs? This is an example of what is here meant by traditionalism. A man does not "by nature" wish to earn more and more money, but simply to live as he is accustomed to live and to earn as much as is necessary for that purpose. Wherever modern capitalism has begun its work of increasing the productivity of human labour by increasing its intensity, it has encountered the immensely stubborn resistance of this leading trait of pre-capitalistic labour. And to-day it encounters it the more, the more backward (from a capitalistic point of view) the labouring forces are with which it has to deal.

Another obvious possibility, to return to our example, since the appeal to the acquisitive instinct through higher wage-rates failed, would have been to try the opposite policy, to force the worker by reduction of his wage-rates to work harder to earn the same amount than he did before. Low wages and high profits seem even to-day to a superficial observer to stand in correlation; everything which is paid out in wages seems to involve a corresponding reduction of profits. That road capitalism has taken again and again since its beginning. For centuries it was an article of faith, that low wages were productive, i.e. that they increased the material results of labour so that, as Pieter de la Cour, on this point, as we shall see, quite in the spirit of the old Calvinism, said long ago, the people only work because and so long as they are poor.

But the effectiveness of this apparently so efficient method has its limits. Of course the presence of a surplus population which it can hire cheaply in the labour market is a necessity for the development of capitalism. But though too large a reserve army may in certain cases favour its quantitative expansion, it checks its qualitative development, especially the transition to types of enterprise which make more intensive use of labour. Low wages are by no means identical with cheap labour. From a purely quantitative point of view the efficiency of labour decreases with a wage which is physiologically insufficient, which may in the long run even mean a survival of the unfit. . . . Low wages fail even from a purely business point of view wherever it is a question of producing goods which require any sort of skilled labour, or the use of expensive machinery which is easily damaged, or in general wherever any great amount of sharp attention or of initiative is required. Here low wages do not pay, and their effect is the opposite of what was intended. For not only is a developed sense of responsibility absolutely indispensable, but in general also an attitude which, at least during working hours, is freed from continual calculations of how the customary wage may be earned with a maximum of comfort and a minimum of exertion. Labour must, on the contrary, be performed as if it were an absolute end in itself, a calling. But such an attitude is by no means a product of nature. It cannot be evoked by low wages or high ones alone, but can only be the product of a long and arduous process of education. To-day, capitalism, once in the saddle, can recruit its labouring force in all industrial countries with comparative ease. In the past this was in every case an extremely difficult problem. And even to-day it could probably not get along without the support of a powerful ally along the way, which, as we shall see below, was at hand at the time of its development. . . .

Now, how could activity, which was at best ethically tolerated, turn into a calling in the sense of Benjamin Franklin? The fact to be explained historically is that in the most highly capitalistic centre of that time, in Florence of the fourteenth and fifteenth centuries, the money and capital market of all the great political Powers, this attitude was considered ethically unjustifiable, or at best to be tolerated. But in the backwoods small bourgeois circumstances of Pennsylvania in the eighteenth century, where business threatened for simple lack of money to fall back into barter, where there was hardly a sign of large enterprise, where only the earliest beginnings of banking were to be found, the same thing was considered the essence of moral conduct, even commanded in the name of duty. To speak here of a reflection of material conditions in the ideal superstructure would be patent nonsense. What was the background of ideas which could account for the sort of activity apparently directed toward profit alone as a calling toward which the individual feels himself to have an ethical obligation? For it was this idea which gave the way of life of the new entrepreneur its ethical foundation and justification. . . .

Asceticism and the Spirit of Capitalism

In order to understand the connection between the fundamental religious ideas of ascetic Protestantism and its maxims for everyday economic conduct, it is necessary to examine with especial care such writings as have evidently been derived from ministerial practice. For in a time in which the beyond meant everything, when the social position of the Christian depended upon his admission to the communion, the clergyman, through his ministry, Church discipline, and preaching, exercised and influence (as a glance at collections of *consilia, casus conscientiæ*, etc., shows) which we modern men are entirely unable to picture. In such a time the religious forces which express themselves through such channels are the decisive influences in the formation of national character.

For the purposes of this chapter, though by no means for all purposes, we can treat ascetic Protestantism as a single whole. But since that side of English Puritanism which was derived from Calvinism gives the most consistent religious basis for the idea of the calling, we shall, following our previous method, place one of its representatives at the centre of the discussion. Richard Baxter stands out above many other writers on Puritan ethics, both because of his eminently practical and realistic attitude, and, at the same time, because of the universal recognition accorded to his works, which have gone through many new editions and translations. He was a Presbyterian and an apologist of the Westminster Synod, but at the same time, like so many of the best spirits of his time, gradually grew away from the dogmas of pure Calvinism. . . . His *Christian Directory* is the most complete compendium of Puritan ethics, and is continually adjusted to the practical experiences of his own ministerial activity. In comparison we shall make use of Spener's *Theologische Bedenken*, as representative of German Pietism, Barclay's *Apology* for the Quakers, and some other representatives of ascetic ethics, which, however, in the interest of space, will be limited as far as possible.

Now, in glancing at Baxter's *Saints' Everlasting Rest*, or his *Christian Directory*, or similar works of others, one is struck at first glance by the emphasis placed, in the discussion of wealth and its acquisition, on the ebionitic elements of the New Testament. Wealth as such is a great danger; its temptations never end, and its pursuit is not only senseless as compared with the dominating importance of the Kingdom of God, but it is morally suspect. Here asceticism seems to have turned much more sharply against the acquisition of earthly goods than it did in Calvin, who saw no hindrance to the effectiveness of the clergy in their wealth, but rather a thoroughly desirable enhancement of their prestige. Hence he permitted them to employ their means profitably. Examples of the condemnation of the pursuit of money and goods may be gathered without end from Puritan writings, and may be contrasted with the late mediæval ethical literature, which was much more open-minded on this point.

Moreover, these doubts were meant with perfect seriousness; only it is necessary to examine them somewhat more closely in order to understand their true ethical significance and implications. The real moral objection is to relaxation in the security of possession, the enjoyment of wealth with the consequence of idleness and the temptations of the flesh, above all of distraction from the pursuit of a righteous life. In fact, it is only because possession involves this danger of relaxation that it is objectionable at all. For the saints' everlasting rest in the next world; on earth man must, to be certain of his state of grace, "do the works of him who sent him, as long as it is yet day." Not leisure and enjoyment, but only activity serves to increase the glory of God, according to the definite manifestations of His will.

Waste of time is thus the first and in principle the deadliest of sins. The span of human life is infinitely short and precious to make sure of one's own election. Loss of time through sociability, idle talk, luxury, even more sleep than is necessary for health, six to at most eight hours, is worthy of absolute moral condemnation. It does not yet hold, with Franklin, that time is money, but the proposition is true in a certain spiritual sense. It is infinitely valuable because every hour lost is lost to labour for the glory of God. Thus inactive contemplation is also valueless, or even directly reprehensible if it is at the expense of one's daily work. For it is less pleasing to God than the active performance of His will in a calling. Besides, Sunday is provided for

that, and, according to Baxter, it is always those who are not diligent in their callings who have no time for God when the occasion demands it.

Accordingly, Baxter's principal work is dominated by the continually repeated, often almost passionate preaching of hard, continuous bodily or mental labour. It is due to a combination of two different motives. Labour is, on the one hand, an approved ascetic technique, as it always has been in the Western Church, in sharp contrast not only to the Orient but to almost all monastic rules the world over. It is in particular the specific defence against all those temptations which Puritanism united under the name of the unclean life, whose rôle for it was by no means small. The sexual asceticism of Puritanism differs only in degree, not in fundamental principle, from that of monasticism; and on account of the Puritan conception of marriage; its practical influence is more far-reaching than that of the latter. For sexual intercourse is permitted, even within marriage, only as the means willed by God for the increase of His glory according to the commandment, "Be fruitful and multiply." Along with a moderate vegetable diet and cold baths, the same prescription is given for all sexual temptations as is used against religious doubts and a sense of moral unworthiness: "Work hard in your calling." But the most important thing was that even beyond that labour came to be considered in itself the end of life, ordained as such by God. St. Paul's "He who will not work shall not eat" holds unconditionally for everyone. Unwillingness to work is symptomatic of the lack of grace. . . .

[Not] only do these exceptions to the duty to labour naturally no longer hold for Baxter, but he holds most emphatically that wealth does not exempt anyone from the unconditional command. Even the wealthy shall not eat without working, for even though they do not need to labour to support their own needs, there is God's commandment which they, like the poor, must obey. For everyone without exception God's Providence has prepared a calling, which he should profess and in which he should labour. And this calling is not, as it was for the Lutheran, a fate to which he must submit and which he must make the best of, but God's commandment to the individual to work for the divine glory. This seemingly subtle difference had far-reaching psychological consequences, and became connected with a further development of the providential interpretation of the economic order which had begun in scholasticism.

The phenomenon of the division of labour and occupations in society had, among others, been interpreted by Thomas Aquinas, to whom we may most conveniently refer, as a direct consequence of the divine scheme of things. But the places assigned to each man in this cosmos follow *ex causis naturalibus* and are fortuitous (contingent in the Scholastic terminology). The differentiation of men into the classes and occupations established through historical development became for Luther, as we have seen, a direct result of the divine will. The perseverance of the individual in the place and within the limits which God had assigned to him was a religious duty. . . .

But in the Puritan view, the providential character of the play of private economic interests takes on a somewhat different emphasis. True to the Puritan tendency to pragmatic interpretations, the providential purpose of the division of labour is to be known by its fruits. . . .

But the characteristic Puritan element appears when Baxter sets at the head of his discussion the statement that "outside of a well-marked calling the accomplishments of a man are only casual and irregular, and he spends more time in idleness than at work," and when he concludes it as follows: "and he [the specialized worker] will carry out his work in order while another remains in constant confusion, and his business knows neither time nor place. . . therefore is a certain calling the best for everyone." Irregular work, which the ordinary labourer is often forced to accept, is often unavoidable, but always an unwelcome state of transition. A man without a calling thus lacks the systematic, methodical character which is, as we have seen, demanded by worldly asceticism.

The Quaker ethic also holds that a man's life in his calling is an exercise in ascetic virtue, a proof of his state of grace through his conscientiousness, which is expressed in the care and method with which he pursues his calling. What God demands is not labour in itself, but rational labour in a calling. In the Puritan concept of the calling the emphasis is always placed on this methodical character of worldly asceticism, not, as with Luther, on the acceptance of the lot which God has irretrievably assigned to man.

Hence the question whether anyone may combine several callings is answered in the affirmative, if it is useful for the common good or one's own, and not injurious to anyone, and if it does not lead to unfaithfulness in one of the callings. Even a change of calling is by no means regarded as objectionable, if it is not thoughtless and is made for the purpose of pursuing a calling more pleasing to God, which means, on general principles, one more useful.

It is true that the usefulness of a calling, and thus its favour in the sight of God, is measured primarily in moral terms, and thus in terms of the importance of the goods produced in it for the community. But a further, and, above all, in practice the most important, criterion is found in private profitableness. For if that God, whose hand the Puritan sees in all the occurrences of life, shows one of His elect a chance of profit, he must do it with a purpose. Hence the faithful Christian must follow the call by taking advantage of the opportunity. "If God show you a way in which you may lawfully get more than in another way (without wrong to your soul or to any other), if you refuse this, and choose the less gainful way, you cross one of the ends of your calling, and you refuse to be God's steward, and to accept His gifts and use them for Him when He requireth it: you may labour to be rich for God, though not for the flesh and sin."

Wealth is thus bad ethically only in so far as it is a temptation to idleness and sinful enjoyment of life, and its acquisition is bad only when it is with the purpose of later living merrily and without care. But as a performance of duty in a calling it is not only morally permissible, but actually enjoined. The parable of the servant who was rejected because he did not increase the talent which was entrusted to him seemed to say so directly. To wish to be poor was, it was often argued, the same as wishing to be unhealthy; it is objectionable as a glorification of works and derogatory to the glory of God. Especially begging, on the part of one able to work, is not only the sin of slothfulness, but a violation of the duty of brotherly love according to the Apostle's own word.

The emphasis on the ascetic importance of a fixed calling provided an ethical justification of the modern specialized division of labour. In a similar way the providential interpretation of profit-making justified the activities of the business man. The superior indulgence of the *seigneur* and the parvenu ostentation of the *nouveau riche* are equally detestable to asceticism. But, on the other hand, it has the highest ethical appreciation of the sober, middle-class, self-made man. "God blesseth His trade" is a stock remark about those good men who had successfully followed the divine hints. The whole power of the God of the Old Testament, who rewards His people for their obedience in this life, necessarily exercised a similar influence on the Puritan who, following Baxter's advice, compared his own state of grace with that of the heroes of the Bible, and in the process interpreted the statements of the Scriptures as the articles of a book of statutes. . . .

Let us now try to clarify the points in which the Puritan idea of the calling and the premium it placed upon ascetic conduct was bound directly to influence the development of a capitalistic way of life. As we have seen, this asceticism turned with all its force against one thing: the spontaneous enjoyment of life and all it had to offer. . . .

As against this the Puritans upheld their decisive characteristic, the principle of ascetic conduct. For otherwise the Puritan aversion to sport, even for the Quakers, was by no means simply one of principle. Sport was accepted if it served a rational purpose, that of recreation necessary for physical efficiency. But as a means for the spontaneous expression of undisciplined impulses, it was under suspicion; and in so far as it became purely a means of enjoyment, or awakened pride, raw instincts or the irrational gambling instinct, it was of course strictly condemned. Impulsive enjoyment of life, which leads away both from work in a calling and from religion, was as such the enemy of rational asceticism, whether in the form of seigneurial sports, or the enjoyment of the dance-hall or the public-house of the common man. . . .

The theatre was obnoxious to the Puritans, and with the strict exclusion of the erotic and of nudity from the realm of toleration, a radical view of either literature or art could not exist. The conceptions of idle talk, of superfluities, and of vain ostentation, all designations of an irrational attitude without objective purpose, thus not ascetic, and especially not serving the glory of God, but of man, were always at hand to serve in deciding in favour of sober utility as against any artistic tendencies. This was especially true in the case of decoration of the person, for instance clothing. That powerful tendency toward uniformity of life, which to-day so immensely aids the capitalistic interest in the standardization of production, had its ideal foundations in the repudiation of all idolatry of the flesh. . . .

Although we cannot here enter upon a discussion of the influence of Puritanism in all these directions, we should call attention to the fact that the toleration of pleasure in cultural goods, which contributed to purely aesthetic or athletic enjoyment, certainly always ran up against one characteristic limitation: they must not cost anything. Man is only a trustee of the goods which have come to him through God's grace. He must, like the servant in the parable, give an account of every penny entrusted to him, and it is at least hazardous to spend any of it for a purpose which does not serve the glory of God but only one's own enjoyment. What person, who keeps his eyes open, has not met representatives of this view-point even in the present? The idea of a man's duty to his possessions, to which he subordinates himself as an obedient steward, or even as an acquisitive machine, bears with chilling weight on his life. The greater the possessions the heavier, if the ascetic attitude toward life stands the test, the feeling of responsibility for them, for holding them undiminished for the glory of God and increasing them by restless effort. The origin of this type of life also extends in certain roots, like so many aspects of the spirit of capitalism, back into the Middle Ages. But it was in the ethic of ascetic Protestantism that it first found a consistent ethical foundation. Its significance for the development of capitalism is obvious.

This worldly Protestant asceticism, as we may recapitulate up to this point, acted powerfully against the spontaneous enjoyment of possessions; it restricted consumption, especially of luxuries. On the other hand, it had the psychological effect of freeing the acquisition of goods from the inhibitions of traditionalistic ethics. It broke the bonds of the impulse of acquisition in that it not only legalized it, but (in the sense discussed) looked upon it as directly willed by God. The campaign against the temptations of the flesh, and the dependence on external things, was, as besides the Puritans the great Quaker apologist Barclay expressly says, not a struggle against the rational acquisition, but against the irrational use of wealth.

But this irrational use was exemplified in the outward forms of luxury which their code condemned as idolatry of the flesh, however natural they had appeared to the feudal mind. On the other hand, they approved the rational and utilitarian uses of wealth which were willed by God for the needs of the individual and the community. They did not wish to impose mortification on the man of wealth, but the use of his means for necessary and practical things. The idea of comfort characteristically limits the extent of ethically permissible expenditures. It is naturally no accident that the development of a manner of living consistent with that idea may be observed earliest and most clearly among the most consistent representatives of this whole attitude toward life. Over against the glitter and ostentation of feudal magnificence which, resting on an unsound economic basis, prefers a sordid elegance to a sober simplicity, they set the clean and solid comfort of the middle-class home as an ideal.

On the side of the production of private wealth, asceticism condemned both dishonesty and impulsive avarice. What was condemned as covetousness, Mammonism, etc., was the pursuit of riches for their own sake. For wealth in itself was a temptation. But here asceticism was the power "which ever seeks the good but ever creates evil"; what was evil in its sense was possession and its temptations. For, in conformity with the Old Testament and in analogy to the ethical valuation of good works, asceticism looked upon the pursuit of wealth as an end in itself as highly reprehensible; but the attainment of it as a fruit of labour in a calling was a sign of God's blessing. And even more important: the religious valuation of restless, continuous, systematic work in a worldly calling, as the highest means to asceticism, and at the same time the surest and most evident proof of rebirth and genuine faith, must have been the most powerful conceivable lever for the expansion

of that attitude toward life which we have here called the spirit of capitalism.

When the limitation of consumption is combined with this release of acquisitive activity, the inevitable practical result is obvious: accumulation of capital through ascetic compulsion to save. The restraints which were imposed upon the consumption of wealth naturally served to increase it by making possible the productive investment of capital. . . .

As far as the influence of the Puritan outlook extended, under all circumstances—and this is, of course, much more important than the mere encouragement of capital accumulation—it favoured the development of a rational bourgeois economic life; it was the most important, and above all the only consistent influence in the development of that life. It stood at the cradle of the modern economic man.

To be sure, these Puritanical ideals tended to give way under excessive pressure from the temptations of wealth, as the Puritans themselves knew very well. With great regularity we find the most genuine adherents of Puritanism among the classes which were rising from a lowly status, the small bourgeois and farmers, while the *beati possidentes*, even among Quakers, are often found tending to repudiate the old ideals. It was the same fate which again and again befell the predecessor of this worldly asceticism, the monastic asceticism of the Middle Ages. In the latter case, when rational economic activity had worked out its full effects by strict regulation of conduct and limitation of consumption, the wealth accumulated either succumbed directly to the nobility, as in the time before the Reformation, or monastic discipline threatened to break down, and one of the numerous reformations became necessary.

In fact the whole history of monasticism is in a certain sense the history of a continual struggle with the problem of the secularizing influence of wealth. The same is true on a grand scale of the worldly asceticism of Puritanism. The great revival of Methodism, which preceded the expansion of English industry toward the end of the eighteenth century, may well be compared with such a monastic reform. We may hence quote here a passage from John Wesley himself which might well serve as a motto for everything which has been said above. For it shows that the leaders of these ascetic movements

understood the seemingly paradoxical relationships which we have here analysed perfectly
well, and in the same sense that we have given
them. He wrote:

"I fear, wherever riches have increased, the
essence of religion has decreased in the same
proportion. Therefore I do not see how it is
possible, in the nature of things, for any revival
of true religion to continue long. For religion
must necessarily produce both industry and
frugality, and these cannot but produce riches.
But as riches increase, so will pride, anger, and
love of the world in all its branches. How then
is it possible that Methodism, that is, a religion
of the heart, though it flourishes now as a
green bay tree, should continue in this state?
For the Methodists in every place grow diligent
and frugal; consequently they increase in
goods. Hence they proportionately increase in
pride, in anger, in the desire of the flesh, the
desire of the eyes, and the pride of life. So,
although the form of religion remains, the
spirit is swiftly vanishing away. Is there no way
to prevent this—this continual decay of pure
religion? We ought not to prevent people from
being diligent and frugal; *we must exhort all
Christians to gain all they can, and to save all
they can; that is, in effect, to grow rich.*"

There follows the advice that those who gain
all they can and save all they can should also
give all they can, so that they will grow in grace
and lay up a treasure in heaven. It is clear that
Wesley here expresses, even in detail, just what
we have been trying to point out.

As Wesley here says, the full economic effect
of those great religious movements, whose significance for economic development lay above
all in their ascetic educative influence, generally
came only after the peak of the purely religious
enthusiasm was past. Then the intensity of the
search for the Kingdom of God commenced
gradually to pass over into sober economic
virtue; the religious roots died out slowly, giving
way to utilitarian worldliness. . . .

A specifically bourgeois economic ethic had
grown up. With the consciousness of standing in
the fullness of God's grace and being visibly
blessed by Him, the bourgeois business man, as
long as he remained within the bounds of formal

correctness, as long as his moral conduct was
spotless and the use to which he put his wealth
was not objectionable, could follow his pecuniary interests as he would and feel that he was
fulfilling a duty in doing so. The power of religious asceticism provided him in addition with
sober, conscientious, and unusually industrious
workmen, who clung to their work as to a life
purpose willed by God.

Finally, it gave him the comforting assurance
that the unequal distribution of the goods of this
world was a special dispensation of Divine
Providence, which in these differences, as in
particular grace, pursued secret ends unknown
to men. Calvin himself had made the much-
quoted statement that only when the people, i.e.
the mass of labourers and craftsmen, were poor
did they remain obedient to God. In the
Netherlands (Pieter de la Court and others), that
had been secularized to the effect that the mass
of men only labour when necessity forces them
to do so. This formulation of a leading idea of
capitalistic economy later entered into the current theories of the productivity of low wages.
Here also, with the dying out of the religious
root, the utilitarian interpretation crept in unnoticed, in the line of development which we have
again and again observed. . . .

Now naturally the whole ascetic literature of
almost all denominations is saturated with the
idea that faithful labour, even at low wages, on
the part of those whom life offers no other opportunities, is highly pleasing to God. In this respect
Protestant Asceticism added in itself nothing
new. But it not only deepened this idea most powerfully, it also created the force which was alone
decisive for its effectiveness: the psychological
sanction of it through the conception of this
labour as a calling, as the best, often in the last
analysis the only means of attaining certainty of
grace. And on the other hand it legalized the
exploitation of this specific willingness to work,
in that it also interpreted the employer's business
activity as a calling. It is obvious how powerfully
the exclusive search for the Kingdom of God
only through the fulfilment of duty in the calling,
and the strict asceticism which Church discipline
naturally imposed, especially on the propertyless
classes, was bound to affect the productivity of
labour in the capitalistic sense of the word.
The treatment of labour as a calling became as

characteristic of the modern worker as the corresponding attitude toward acquisition of the business man. It was a perception of this situation, new at his time, which caused so able an observer as Sir William Petty to attribute the economic power of Holland in the seventeenth century to the fact that the very numerous dissenters in that country (Calvinists and Baptists) "are for the most part thinking, sober men, and such as believe that Labour and Industry is their duty towards God." . . .

One of the fundamental elements of the spirit of modern capitalism, and not only of that but of all modern culture: rational conduct on the basis of the idea of the calling, was born—that is what this discussion has sought to demonstrate—from the spirit of Christian asceticism. One has only to re-read the passage from Franklin, quoted at the beginning of this essay, in order to see that the essential elements of the attitude which was there called the spirit of capitalism are the same as what we have just shown to be the content of the Puritan worldly asceticism, only without the religious basis, which by Franklin's time had died away. The idea that modern labour has an ascetic character is of course not new. Limitation to specialized work, with a renunciation of the Faustian universality of man which it involves, is a condition of any valuable work in the modern world; hence deeds and renunciation inevitably condition each other today. This fundamentally ascetic trait of middle-class life, if it attempts to be a way of life at all, and not simply the absence of any, was what Goethe wanted to teach, at the height of his wisdom, in the *Wanderjahren*, and in the end which he gave to the life of his *Faust*. For him the realization meant a renunciation, a departure from an age of full and beautiful humanity, which can no more be repeated in the course of our cultural development than can the flower of the Athenian culture of antiquity.

The Puritan wanted to work in a calling; we are forced to do so. For when asceticism was carried out of monastic cells into everyday life, and began to dominate worldly morality, it did its part in building the tremendous cosmos of the modern economic order. This order is now bound to the technical and economic conditions of machine production which to-day determine the lives of all the individuals who are born into this mechanism, not only those directly concerned with economic acquisition, with irresistible force. Perhaps it will so determine them until the last ton of fossilized coal is burnt. In Baxter's view the care for external goods should only lie on the shoulders of the "saint like a light cloak, which can be thrown aside at any moment." But fate decreed that the cloak should become an iron cage.

Since asceticism undertook to remodel the world and to work out its ideals in the world, material goods have gained an increasing and finally an inexorable power over the lives of men as at no previous period in history. To-day the spirit of religious asceticism—whether finally, who knows?—has escaped from the cage. But victorious capitalism, since it rests on mechanical foundations, needs its support no longer. The rosy blush of its laughing heir, the Enlightenment, seems also to be irretrievably fading, and the idea of duty in one's calling prowls about in our lives like the ghost of dead religious beliefs. Where the fulfilment of the calling cannot directly be related to the highest spiritual and cultural values, or when, on the other hand, it need not be felt simply as economic compulsion, the individual generally abandons the attempt to justify it at all. In the field of its highest development, in the United States, the pursuit of wealth, stripped of its religious and ethical meaning, tends to become associated with purely mundane passions, which often actually give it the character of sport.

No one knows who will live in this cage in the future, or whether at the end of this tremendous development entirely new prophets will arise, or there will be a great rebirth of old ideas and ideals, or, if neither, mechanized petrification, embellished with a sort of convulsive self-importance. For of the last stage of this cultural development, it might well be truly said: "Specialists without spirit, sensualists without heart; this nullity imagines that it has attained a level of civilization never before achieved."

But this brings us to the world of judgments of value and of faith, with which this purely historical discussion need not be burdened. The next task would be rather to show the significance of ascetic rationalism, which has only been touched in the foregoing sketch, for the content of practical social ethics, thus for the types of organization and the functions of social groups from the conventicle to the State. Then its relations to humanistic rationalism, its ideals of life and cultural influence; further to the development of philosophical and scientific empiricism, to technical development and to spiritual ideals would have to be analysed. Then its historical development from the mediæval beginnings of worldly asceticism to its dissolution into pure utilitarianism would have to be traced out through all the areas of ascetic religion. Only then could the quantitative cultural significance of ascetic Protestantism in its relation to the other plastic elements of modern culture be estimated.

Here we have only attempted to trace the fact and the direction of its influence to their motives in one, though a very important point. But it would also further be necessary to investigate how Protestant Asceticism was in turn influenced in its development and its character by the totality of social conditions, especially economic. The modern man is in general, even with the best will, unable to give religious ideas a significance for culture and national character which they deserve. But it is, of course, not my aim to substitute for a one-sided materialistic an equally one-sided spiritualistic causal interpretation of culture and of history. Each is equally possible, but each, if it does not serve as the preparation, but as the conclusion of an investigation, accomplishes equally little in the interest of historical truth.

Introduction to "The Social Psychology of the World Religions"

In this essay, Weber extends his analysis developed in *The Protestant Ethic* by taking up five major world religions—Confucianism, Hinduism, Buddhism, Islam, and Christianity—to address more generally the relationship between religion and "economic ethics." (He was completing his studies on Judaism when he died.) In doing so, he again provides an account of religious experience that diverges from those offered by Marx and Durkheim. Drawing a contrast with Marxist views, Weber asserts that religion is not a "simple 'function' of the social situation of the stratum which appears as its characteristic bearer" nor does it represent "the stratum's 'ideology' [nor is it] a 'reflection' of a stratum's material or ideal interest-situation" (Weber 1958:269–70). Religion, instead, shapes economic, practical behavior just as much as such behavior shapes religious doctrines. Most importantly, religions address the psychological need of the fortunate to legitimate their good fortune, while for the less fortunate they offer the promise of a future salvation. While this "religious need" may be universal, the form in which it is met varies across different social strata (warriors, peasants, political officials, intellectuals, "civic") that exhibit an affinity for particular religious worldviews. Nevertheless, these worldviews have their own impact on behavior that cannot be understood simply as a reflection of its bearer's material position. This is particularly the case for religious virtuosos whose quest for salvation is guided by authentically spiritual motives. For the devout, actively proving oneself as an instrument or tool of God's will, communing contemplatively with the cosmic love of Nirvana, or striving for orgiastic ecstasy, represents genuine religious aims that cannot be reduced to some sort of underlying "distorted" class interest. Nor can the motives of the devout be understood as misguided intentions to deify society or as expressions of the collective conscience, as Durkheim would contend.

Weber also notes how religions have fostered the "rationalization of reality." Offering a promise of redemption, whether it be from social oppression, evil spirits, the cycle of rebirths, human imperfections, or any number of other forces, all religions counter a "senseless" world with the belief that "the world in its totality is, could, and should somehow be a meaningful 'cosmos'" (Weber 1958:281). The specific religious form of meaning is derived from a "systematic and rationalized 'image of the world" that determines "'[f]rom what' and 'for what' one wished to be redeemed and . . . 'could be' redeemed" (ibid.:280). Religion declares that the world is not a playground for chance; instead, it is ruled by reasons and fates that can be "known." Knowing how to redeem oneself and how to obtain salvation requires that one knows how the world "works." In devising answers for such concerns, religions have developed along two primary paths: "exemplary" prophecy and "emissary" prophecy.

Exemplary prophecy is rooted in the conception of a supreme, impersonal being accessible only through contemplation, while emissary prophecy conceives of a personal God who is vengeful and loving, forgiving and punishing, and who demands of the faithful active, ethical conduct in order to serve His commandments. Though the masses may be religiously "unmusical," the religiosity of the devout (monks, prophets, shamans, ascetics) nevertheless "has been of decisive importance for the development of the way of life of the masses," particularly with regard to regulating practical, economic activity (Weber 1958:289). Thus, religions grounded in an exemplary prophecy (e.g., Buddhism, Hinduism) lead adherents away from workaday life by seeking salvation through extraordinary psychic states attained through mystical, orgiastic, or ecstatic experiences. The virtuoso's hostility toward economic activity discourages this-worldly practical conduct by viewing it as "religiously inferior," a distraction from communing with the divine. Absent from the contemplative, mystical "flight from the world" is any psychological motivation to engage in worldly action as a path for redemption. As a result, a rationalized economic ethic remains underdeveloped.

Conversely, religions based on an emissary prophecy (e.g., Judaism, Christianity, Islam) require the devout to actively fashion the world according to the will of their god. Not contemplative "flight from," but, rather, ascetic "work in" this world is the path for redemption according this prophecy. Seeking mystical union with the cosmos is understood here as an irrational act of hedonism that devalues the God-created world. The virtuoso is instead compelled to "prove" himself as a worthy instrument of God through the ethical quality of his everyday activity. This psychological imperative leads to the development of rational, economic ethic that transforms work into a "holy," worldly calling. Everyday life is here the setting for the "methodical and rationalized routine-activities of workaday life in the service of the Lord" (Weber 1958:289). Yet, as Weber argued in *The Protestant Ethic*, this worldview, while faithful to God's commandments and devoted to creating His Kingdom on earth, leads to a thoroughgoing "disenchantment of the world."

From "The Social Psychology of the World Religions" (1915)

Max Weber

By "world religions," we understand the five religions or religiously determined systems of life-regulation which have known how to gather multitudes of confessors around them. The term is used here in a completely value-neutral sense. The Confucian, Hinduist, Buddhist, Christian, and Islamist religious ethics all belong to the category of world religion. A sixth religion, Judaism, will also be dealt with. It is included because it contains historical preconditions decisive for understanding Christianity and Islamism, and because of its historic and autonomous significance for the development of the modern economic ethic of the Occident—a significance, partly real and partly alleged, which has been discussed several times recently. . . .

What is meant by the "economic ethic" of a religion will become increasingly clear during the course of our presentation. . . . The term

SOURCE: Translation of the Introduction to *The Economic Ethic of the World Religions* by Max Weber, 1915.

"economic ethic" points to the practical impulses for action which are founded in the psychological and pragmatic contexts of religions. The following presentation may be sketchy, but it will make obvious how complicated the structures and how many-sided the conditions of a concrete economic ethic usually are. Furthermore, it will show that externally similar forms of economic organization may agree with very different economic ethics and, according to the unique character of their economic ethics, how such forms of economic organization may produce very different historical results. An economic ethic is not a simple "function" of a form of economic organization; and just as little does the reverse hold, namely, that economic ethics unambiguously stamp the form of the economic organization.

No economic ethic has ever been determined solely by religion. In the face of man's attitudes towards the world—as determined by religious or other (in our sense) "inner" factors—an economic ethic has, of course, a high measure of autonomy. Given factors of economic geography and history determine this measure of autonomy in the highest degree. The religious determination of life-conduct, however, is also one—note this—only one, of the determinants of the economic ethic. Of course, the religiously determined way of life is itself profoundly influenced by economic and political factors operating within given geographical, political, social, and national boundaries. We should lose ourselves in these discussions if we tried to demonstrate these dependencies in all their singularities. Here we can only attempt to peel off the directive elements in the life-conduct of those social *strata* which have most strongly influenced the practical ethic of their respective religions. These elements have stamped the most characteristic features upon practical ethics, the features that distinguish one ethic from others; *and*, at the same time, they have been important for the respective economic ethics. . . .

It is not our thesis that the specific nature of a religion is a simple "function" of the social situation of the stratum which appears as its characteristic bearer, or that it represents the stratum's "ideology," or that it is a "reflection" of a stratum's material or ideal interest-situation.

On the contrary, a more basic misunderstanding of the standpoint of these discussions would hardly be possible.

However incisive the social influences, economically and politically determined, may have been upon a religious ethic in a particular case, it receives its stamp primarily from religious sources, and, first of all, from the content of its annunciation and its promise. Frequently the very next generation reinterprets these annunciations and promises in a fundamental fashion. Such reinterpretations adjust the revelations to the needs of the religious community. If this occurs, then it is at least usual that religious doctrines are adjusted to *religious needs*. Other spheres of interest could have only a secondary influence; often, however, such influence is very obvious and sometimes it is decisive.

For every religion we shall find that a change in the socially decisive strata has usually been of profound importance. On the other hand, the type of a religion, once stamped, has usually exerted a rather far-reaching influence upon the life-conduct of very heterogeneous strata. In various ways people have sought to interpret the connection between religious ethics and interest-situations in such a way that the former appear as mere "functions" of the latter. Such interpretation occurs in so-called historical materialism—which we shall not here discuss—as well as in a purely psychological sense. . . .

In treating suffering as a symptom of odiousness in the eyes of the gods and as a sign of secret guilt, religion has psychologically met a very general need. The fortunate is seldom satisfied with the fact of being fortunate. Beyond this, he needs to know that he has a *right* to his good fortune. He wants to be convinced that he "deserves" it, and above all, that he deserves it in comparison with others. He wishes to be allowed the belief that the less fortunate also merely experience his due. Good fortune thus wants to be "legitimate" fortune.

If the general term "fortune" covers all the "good" of honor, power, possession, and pleasure, it is the most general formula for the service of legitimation, which religion has had to accomplish for the external and the inner interests of all ruling men, the propertied, the victorious, and the healthy. In short, religion provides the theodicy of good fortune for those who are fortunate. This theodicy is anchored in highly robust ("pharisaical") needs of man and is therefore easily understood, even if sufficient attention is often not paid to its effects. . . .

The annunciation and the promise of religion have naturally been addressed to the masses of those who were in need of salvation. They and their interests have moved into the center of the professional organization for the "cure of the soul," which, indeed, only therewith originated. The typical service of magicians and priests becomes the determination of the factors to be blamed for suffering, that is, the confession of "sins." At first, these sins were offenses against ritual commandments. The magician and priest also give counsel for behavior fit to remove the suffering. The material and ideal interests of magicians and priests could thereby actually and increasingly enter the service of specifically *plebeian* motives. A further step along this course was signified when, under the pressure of typical and ever-recurrent distress, the religiosity of a "redeemer" evolved. This religiosity presupposed the myth of a savior, hence (at least relatively) of a *rational* view of the world. Again, suffering became the most important topic. The primitive mythology of nature frequently offered a point of departure for this religiosity. The spirits who governed the coming and going of vegetation and the paths of celestial bodies important for the seasons of the year became the preferred carriers of the myths of the suffering, dying, and resurrecting god to needful men. The resurrected god guaranteed the return of good fortune in this world or the security of happiness in the world beyond. . . .

The need for an ethical interpretation of the "meaning" of the distribution of fortunes among men increased with the growing rationality of conceptions of the world. As the religious and ethical reflections upon the world were increasingly rationalized and primitive, and magical notions were eliminated, the theodicy of suffering encountered increasing difficulties. Individually "undeserved" woe was all too frequent; not "good" but "bad" men succeeded—even when "good" and "bad" were measured by the yardstick of the master stratum and not by that of a "slave morality."

One can explain suffering and injustice by referring to individual sin committed in a former life (the migration of souls), to the guilt of ancestors, which is avenged down to the third and fourth generation, or—the most principled—to the wickedness of all creatures *per se*. As compensatory promises, one can refer to hopes of the individual for a better life in the future in this world (transmigration of souls) or to hopes for the successors (Messianic realm), or to a better life in the hereafter (paradise). . . .

The distrust of wealth and power, which as a rule exists in genuine religions of salvation, has had its natural basis primarily in the experience of redeemers, prophets, and priests. They understood that those strata which were "satiated" and favored in this world had only a small urge to be saved, regardless of the kind of salvation offered. Hence, these master strata have been less. "devout" in the sense of salvation religions. The development of a rational religious ethic has had positive and primary roots in the inner conditions of those social strata which were less socially valued.

Strata in solid possession of social honor and power usually tend to fashion their status-legend in such a way as to claim a special and intrinsic quality of their own, usually a quality of blood; their sense of dignity feeds on their actual or alleged being. The sense of dignity of socially repressed strata or of strata whose status is negatively (or at least not positively) valued is nourished most easily on the belief that a special "mission" is entrusted to them; their worth is guaranteed or constituted by an *ethical imperative*, or by their own functional *achievement*. Their value is thus moved into something beyond themselves, into a "task" placed before them by God. One source of the ideal power of ethical prophecies among socially disadvantaged strata lies in this fact. . . .

Psychologically considered, man in quest of salvation has been primarily preoccupied by attitudes of the here and now. The puritan *certitudo salutis*, the permanent state of grace that rests in the feeling of "having proved oneself," was psychologically the only concrete object among the sacred values of this ascetic religion. The Buddhist monk, certain to enter Nirvana, seeks the sentiment of a cosmic love; the devout Hindu seeks either Bhakti (fervent love in the possession of God) or apathetic ecstasy. The Chlyst with his radjeny, as well as the dancing

Dervish, strives for orgiastic ecstasy. Others seek to be possessed by God and to possess God, to be a bridegroom of the Virgin Mary, or to be the bride of the Savior. The Jesuit's cult of the heart of Jesus, quietistic edification, the pietists' tender love for the child Jesus and its "running sore," the sexual and semi-sexual orgies at the wooing of Krishna, the sophisticated cultic dinners of the Vallabhacharis, the gnostic onanist cult activities, the various forms of the *unio mystica*, and the contemplative submersion in the All-one—these states undoubtedly have been sought, first of all, for the sake of such emotional value as they directly offered the devout. In this respect, they have in fact been absolutely equal to the religious and alcoholic intoxication of the Dionysian or the soma cult; to totemic meat-orgies, the cannibalistic feasts, the ancient and religiously consecrated use of hashish, opium, and nicotine; and, in general, to all sorts of magical intoxication. They have been considered specifically consecrated and divine because of their psychic extraordinariness and because of the intrinsic value of the respective states conditioned by them. . . .

The two highest conceptions of sublimated religious doctrines of salvation are "rebirth" and "redemption." Rebirth, a primeval magical value, has meant the acquisition of a new soul by means of an orgiastic act or through methodically planned asceticism. Man transitorily acquired a new soul in ecstasy; but by means of magical asceticism, he could seek to gain it permanently. The youth who wished to enter the community of warriors as a hero, or to participate in its magical dances or orgies, or who wished to commune with the divinities in cultic feasts, had to have a new soul The heroic and magical asceticism, the initiation rites of youths, and the sacramental customs of rebirth at important phases of private and collective life are thus quite ancient. The means used in these activities varied, as did their ends: that is, the answers to the question, "For what should I be reborn?" . . .

The kind of empirical state of bliss or experience of rebirth that is sought after as the supreme value by a religion has obviously and necessarily varied according to the character of the stratum which was foremost in adopting it. The chivalrous, warrior class, peasants, business classes, and intellectuals with literary education have naturally pursued different religious tendencies. As will become evident, these tendencies have not by themselves determined the psychological character of religion; they have, however, exerted a very lasting influence upon it. The contrast between warrior and peasant classes, and intellectual and business classes, is of special importance. Of these groups, the intellectuals have always been the exponents of a rationalism which in their case has been relatively theoretical. The business classes (merchants and artisans) have been at least possible exponents of rationalism of a more practical sort. Rationalism of either kind has borne very different stamps, but has always exerted a great influence upon the religious attitude.

Above all, the peculiarity of the intellectual strata in this matter has been in the past of the greatest importance for religion. At the present time, it matters little in the development of a religion whether or not modern intellectuals feel the need of enjoying a "religious" state as an "experience," in addition to all sorts of other sensations, in order to decorate their internal and stylish furnishings with paraphernalia guaranteed to be genuine and old. A religious revival has never sprung from such a source. In the past, it was the work of the intellectuals to sublimate the possession of sacred values into a belief in "redemption." The conception of the idea of redemption, as such, is very old, if one understands by it a liberation from distress, hunger, drought, sickness, and ultimately from suffering and death. Yet redemption attained a specific significance only where it expressed a systematic and rationalized "image of the world" and represented a stand in the face of the world. For the meaning as well as the intended and actual psychological quality of redemption has depended upon such a world image and such a stand. Not ideas, but material and ideal interests, directly govern men's conduct. Yet very frequently the "world images" that have been created by "ideas" have, like switchmen, determined the tracks along which action has been pushed by the dynamic of interest. "From what" and "for what" one wished to be redeemed and, let us not forget, "could be" redeemed, depended upon one's image of the world.

There have been very different possibilities in this connection: One could wish to be saved from political and social servitude and lifted into a Messianic realm in the future of this world; or one could wish to be saved from being defiled by ritual impurity and hope for the pure beauty of psychic and bodily existence. One could wish to escape being incarcerated in an impure body and hope for a purely spiritual existence. One could wish to be saved from the eternal and senseless play of human passions and desires and hope for the quietude of the pure beholding of the divine. One could wish to be saved from radical evil and the servitude of sin and hope for the eternal and free benevolence in the lap of a fatherly god. One could wish to be saved from peonage under the astrologically conceived determination of stellar constellations and long for the dignity of freedom and partaking of the substance of the hidden deity. One could wish to be redeemed from the barriers to the finite, which express themselves in suffering, misery and death, and the threatening punishment of hell, and hope for an eternal bliss in an earthly or paradisical future existence. One could wish to be saved from the cycle of rebirths with their inexorable compensations for the deeds of the times past and hope for eternal rest. One could wish to be saved from senseless brooding and events and long for the dreamless sleep. Many more varieties of belief have, of course, existed. Behind them always lies a stand towards something in the actual world which is experienced as specifically "senseless." Thus, the demand has been implied: that the world order in its totality is, could, and should somehow be a meaningful "cosmos." This quest, the core of genuine religious rationalism, has been borne precisely by strata of intellectuals. The avenues, the results, and the efficacy of this metaphysical need for a meaningful cosmos have varied widely. Nevertheless, some general comments may be made.

The general result of the modern form of thoroughly rationalizing the conception of the world and of the way of life, theoretically and practically, in a purposive manner, has been that religion has been shifted into the realm of the irrational. This has been the more the case the further the purposive type of rationalization has progressed, if one takes the standpoint of an intellectual articulation of an image of the world. This shift of religion into the irrational

realm has occurred for several reasons. On the one hand, the calculation of consistent rationalism has not easily come out even with nothing left over. In music, the Pythagorean "comma" resisted complete rationalization oriented to tonal physics. The various great systems of music of all peoples and ages have differed in the manner in which they have either covered up or bypassed this inescapable irrationality or, on the other hand, put irrationality into the service of the richness of tonalities. The same has seemed to happen to the theoretical conception of the world, only far more so; and above all, it has seemed to happen to the rationalization of practical life. The various great ways of leading a rational and methodical life have been characterized by irrational presuppositions, which have been accepted simply as "given" and which have been incorporated into such ways of life. What these presuppositions have been is historically and socially determined, at least to a very large extent, through the peculiarity of those strata that have been the carriers of the ways of life during its formative and decisive period. The *interest* situation of these strata, as determined socially and psychologically, has made for their peculiarity, as we here understand it.

Furthermore, the irrational elements in the rationalization of reality have been the *loci* to which the irrepressible quest of intellectualism for the possession of supernatural values has been compelled to retreat. That is the more so the more denuded of irrationality the world appears to be. The unity of the primitive image of the world, in which everything was concrete magic, has tended to split into rational cognition and mastery of nature, on the one hand, and into "mystic" experiences, on the other. The inexpressible contents of such experiences remain the only possible "beyond," added to the mechanism of a world robbed of gods. In fact, the beyond remains an incorporeal and metaphysical realm in which individuals intimately possess the holy. Where this conclusion has been drawn without any residue, the individual can pursue his quest for salvation only as an individual. This phenomenon appears in some form, with progressive intellectualist rationalism, wherever men have ventured to rationalize the image of the world as being a cosmos governed by impersonal rules. Naturally it has occurred most strongly among religions and religious ethics which have been quite strongly determined by genteel strata of intellectuals devoted to the purely cognitive comprehension of the world and of its "meaning." This was the case with Asiatic and, above all, Indian world religions. For all of them, contemplation became the supreme and ultimate religious value accessible to man. Contemplation offered them entrance into the profound and blissful tranquillity [sic] and immobility of the All-one. All other forms of religious states, however, have been at best considered a relatively valuable *Ersatz* for contemplation. This has had far-reaching consequences for the relation of religion to life, including economic life, as we shall repeatedly see. Such consequences flow from the general character of "mystic" experiences, in the contemplative sense, and from the psychological preconditions of the search for them.

The situation in which strata decisive for the development of a religion were active in practical life has been entirely different. Where they were chivalrous warrior heroes, political officials, economically acquisitive classes, or, finally, where an organized hierocracy dominated religion, the results were different than where genteel intellectuals were decisive.

The rationalism of hierocracy grew out of the professional preoccupation with cult and myth or—to a far higher degree—out of the cure of souls, that is, the confession of sin and counsel to sinners. Everywhere hierocracy has sought to monopolize the administration of religious values. They have also sought to bring and to temper the bestowal of religious goods into the form of "sacramental" or "corporate grace," which could be ritually bestowed only by the priesthood and could not be attained by the individual. The individual's quest for salvation or the quest of free communities by means of contemplation, orgies, or asceticism, has been considered highly suspect and has had to be regulated ritually and, above all, controlled hierocratically. From the standpoint of the interests of the priesthood in power, this is only natural.

Every body of *political* officials, on the other hand, has been suspicious of all sorts of individual pursuits of salvation and of the free formation of communities as sources of emancipation from domestication at the hands of the institution of the state. Political officials have distrusted the competing priestly corporation of grace and, above all, at bottom they have despised the very quest for these impractical values lying beyond utilitarian and worldly ends. For all political bureaucracies, religious duties have ultimately been simply official or social obligations of the citizenry and of status groups. . . .

It is also usual for a stratum of *chivalrous* warriors to pursue absolutely worldly interests and to be remote from all "mysticism." Such strata, however, have lacked—and this is characteristic of heroism in general—the desire as well as the capacity for a rational mastery of reality. The irrationality of "fate" and, under certain conditions, the idea of a vague and deterministically conceived "destiny" (the Homeric *Moira*) has stood above and behind the divinities and demons who were conceived of as passionate and strong heroes, measuring out assistance and hostility, glory and booty, or death to the human heroes.

Peasants have been inclined towards magic. Their whole economic existence has been specifically bound to nature and has made them dependent upon elemental forces. They readily believe in a compelling sorcery directed against spirits who rule over or through natural forces, or they believe in simply buying divine benevolence. Only tremendous transformations of life-orientation have succeeded in tearing them away from this universal and primeval form of religiosity. Such transformations have been derived either from other strata or from mighty prophets, who, through the power of miracles, legitimize themselves as sorcerers. Orgiastic and ecstatic states of "possession," produced by means of toxics or by the dance, are strange to the status honor of knights because they are considered undignified. Among the peasants, however, such states have taken the place that "mysticism" holds among the intellectuals.

Finally, we may consider the strata that in the western European sense are called "civic," as well as those which elsewhere correspond to them: artisans, traders, enterprisers engaged in cottage industry, and their derivatives existing only in the modern Occident. Apparently these strata have been the most ambiguous with regard to the religious stands open to them. And this is especially important to us. . . .

Of course, the religions of all strata are certainly far from being unambiguously dependent upon the character of the strata we have presented as having special affinities with them. Yet, at first sight, civic strata appear, in this respect and on the whole, to lend themselves to a more varied determination. Yet it is precisely among these strata that elective affinities for special types of religion stand out. The tendency towards a *practical* rationalism in conduct is common to all civic strata; it is conditioned by the nature of their way of life, which is greatly detached from economic bonds to nature. Their whole existence has been based upon technological or economic calculations and upon the mastery of nature and of man, however primitive the means at their disposal. The technique of living handed down among them may, of course, be frozen in traditionalism, as has occurred repeatedly and everywhere. But precisely for these, there has always existed the possibility—even though in greatly varying measure—of letting an *ethical* and rational regulation of life arise. This may occur by the linkage of such an ethic to the tendency of technological and economic rationalism. Such regulation has not always been able to make headway against traditions which, in the main, were magically stereotyped. But where prophecy has provided a religious basis, this basis could be one of two fundamental types of prophecy which we shall repeatedly discuss: "exemplary" prophecy, and "emissary" prophecy.

Exemplary prophecy points out the path to salvation by exemplary living, usually by a contemplative and apathetic-ecstatic life. The emissary type of prophecy addresses its *demands* to the world in the name of a god. Naturally these demands are ethical; and they are often of an active ascetic character.

It is quite understandable that the more weighty the civic strata as such have been, and the more they have been torn from bonds of taboo and from divisions into sibs and castes, the more favorable has been the soil for religions that call for action in this world. Under these conditions, the preferred religious attitude could become the attitude of active asceticism, of God-willed *action* nourished by the sentiment of being God's "tool," rather than the possession of the deity or the inward and contemplative surrender to God, which has appeared as the supreme value to religions influenced by strata of genteel intellectuals. In the Occident the attitude of active asceticism has repeatedly retained supremacy over contemplative mysticism and orgiastic or apathetic ecstasy, even though these latter types have been well known in the Occident. . . .

In the missionary prophecy the devout have not experienced themselves as vessels of the divine but rather as instruments of a god. This emissary prophecy has had a profound elective affinity to a special conception of God: the conception of a supra-mundane, personal, wrathful, forgiving, loving, demanding, punishing Lord of Creation. Such a conception stands in contrast to the supreme being of exemplary prophecy. As a rule, though by no means without exception, the supreme being of an exemplary prophecy is an impersonal being because, as a static state, he is accessible only by means of contemplation. The conception of an active God, held by emissary prophecy, has dominated the Iranian and Mid-Eastern religions and those Occidental religions which are derived from them. The conception of a supreme and static being, held by exemplary prophecy, has come to dominate Indian and Chinese religiosity.

These differences are not primitive in nature. On the contrary, they have come into existence only by means of a far-reaching sublimation of primitive conceptions of animist spirits and of heroic deities which are everywhere similar in nature. Certainly the connection of conceptions of God with religious states, which are evaluated and desired as sacred values, have also been strongly influential in this process of sublimation. These religious states have simply been interpreted in the direction of a different conception of God, according to whether the holy states, evaluated as supreme, were contemplative mystic experiences or apathetic ecstasy, or whether they were the orgiastic possession of god, or visionary inspirations and "commands." . . .

The rational elements of a religion, its "doctrine," also have an autonomy: for instance, the Indian doctrine of Kharma, the Calvinist belief in predestination, the Lutheran justification through faith, and the Catholic doctrine of sacrament. The rational religious pragmatism of salvation, flowing from the nature of the images of God and of the world, have under certain conditions had far-reaching results for the fashioning of a practical way of life.

These comments presuppose that the nature of the desired sacred values has been strongly influenced by the nature of the external interest-situation and the corresponding way of life of the ruling strata and thus by the social stratification itself. But the reverse also holds: wherever the direction of the whole way of life has been methodically rationalized, it has been profoundly determined by the ultimate values toward which this rationalization has been directed. These values and positions were thus *religiously* determined. Certainly they have not always, or exclusively, been decisive; however, they have been decisive in so far as an *ethical* rationalization held sway, at least so far as its influence reached. As a rule, these religious values have been also, and frequently absolutely, decisive. . . .

The empirical fact, important for us, that men are *differently qualified* in a religious way stands at the beginning of the history of religion. This fact had been dogmatized in the sharpest rationalist form in the "particularism of grace," embodied in the doctrine of predestination by the Calvinists. The sacred values that have been most cherished, the ecstatic and visionary capacities of shamans, sorcerers, ascetics, and pneumatics of all sorts, could not be attained by everyone. The possession of such faculties is a "charisma," which, to be sure, might be awakened in some but not in all. It follows from this that all intensive religiosity has a tendency toward a sort of *status stratification*, in accordance with differences in the charismatic qualifications. "Heroic" or "virtuoso" religiosity is opposed to mass religiosity. By "mass" we understand those who are religiously "unmusical"; we do not, of course, mean those who occupy an inferior position in the secular status order. . . .

Now, every hierocratic and official authority of a "church"—that is, a community organized by officials into an institution which bestows gifts of grace—fights principally against all virtuoso-religion and against its autonomous development. For the church, being the holder of institutionalized grace, seeks to organize the religiosity of the masses and to put its own officially monopolized and mediated sacred values in the place of the autonomous and religious status qualifications of the religious virtuosos. By its nature, that is, according to the interest-situation of its officeholders, the church must be "democratic" in the sense of making the sacred values generally accessible. This means that the church stands for a universalism of grace and for the ethical sufficiency of all those who are enrolled under its institutional authority. Sociologically, the process of leveling constitutes a complete parallel with the political struggles of the bureaucracy against the political privileges of the aristocratic estates. As with hierocracy, every full-grown political bureaucracy is necessarily and in a quite similar sense "democratic"—namely, in the sense of leveling and of fighting against status privileges that compete with its power. . . .

The religious virtuosos saw themselves compelled to adjust their demands to the possibilities of the religiosity of everyday life in order to gain and to maintain ideal and material mass-patronage. The nature of their concessions have naturally been of primary significance for the way in which they have religiously influenced everyday life. In almost all Oriental religions, the virtuosos allowed the masses to remain stuck in magical tradition. Thus, the influence of religious virtuosos has been infinitely smaller than was the case where religion has undertaken ethically and generally to rationalize everyday life. This has been the case even when religion has aimed precisely at the masses and has cancelled however many of its ideal demands. Besides the relations between the religiosity of the virtuosos and the religion of the masses, which finally resulted from this struggle, the peculiar nature of the concrete religiosity of the virtuosos has been of decisive importance for the development of the way of life of the masses. This virtuoso religiosity has therefore also been important for the economic ethic of the respective religion. The religion of the virtuoso has been the genuinely "exemplary" and practical religion. According to the way of life his religion prescribed to the virtuoso, there have been various possibilities of establishing a rational ethic of everyday life. The relation of virtuoso religion to *workaday life* in the locus of the economy has varied, especially according to the peculiarity of the sacred values desired by such religions.

Wherever the sacred values and the redemptory means of a virtuoso religion bore a contemplative or orgiastic-ecstatic character, there has been no bridge between religion and the practical action of the workaday world. In such cases, the economy and all other action in the world has been considered religiously inferior, and no psychological motives for worldly action could be derived from the attitude cherished as the supreme value. In their innermost beings, contemplative and ecstatic religions have been rather specifically hostile to economic life. Mystic, orgiastic, and ecstatic experiences are extraordinary psychic states; they lead away from everyday life and from all expedient conduct. Such experiences are, therefore, deemed to be "holy." With such religions, a deep abyss separates the way of life of the laymen from that of the community of virtuosos. The rule of the status groups of religious virtuosos over the religious community readily shifts into a magical anthropolatry; the virtuoso is directly worshipped as a Saint, or at least laymen buy his blessing and his magical powers as a means of promoting mundane success or religious salvation. As the peasant was to the landlord, so the layman was to the Buddhist and Jainist bhikshu: ultimately, mere sources of tribute. Such tribute allowed the virtuosos to live entirely for religious salvation without themselves performing profane work, which always would endanger their salvation. Yet the conduct of the layman could still undergo a certain ethical regulation, for the virtuoso was the layman's spiritual adviser, his father confessor and *directeur de l'âme*. Hence, the virtuoso frequently exercises a powerful influence over the religiously "unmusical" laymen; this influence might not be in the direction of his (the virtuoso's) own religious way of life; it might be an influence in merely ceremonious, ritualist, and conventional particulars. For action in this world remained in principle religiously insignificant; and compared with the desire for the religious end, action lay in the very opposite direction.

In the end, the charisma of the pure "mystic" serves only himself. The charisma of the genuine magician serves others.

Things have been quite different where the religiously qualified virtuosos have combined into an ascetic sect, striving to mould life in this world according to the will of a god. To be sure, two things were necessary before this could happen in a genuine way. First, the supreme and sacred value must not be of a contemplative nature; it must not consist of a union with a supra-mundane being who, in contrast to the world, lasts forever; nor in a *unia mystica* to be grasped orgiastically or apathetic-ecstatically. For these ways always lie apart from everyday life and beyond the real world and lead away from it. Second, such a religion must, so far as possible, have given up the purely magical or sacramental character of the *means* of grace. For these means always devalue action in this world as, at best, merely relative in their religious significance, and they link the decision about salvation to the success of processes which are *not* of a rational everyday nature.

When religious virtuosos have combined into an active asceticist sect, two aims are completely attained: the disenchantment of the world and the blockage of the path to salvation by a flight from the world. The path to salvation is turned away from a contemplative "flight from the world" and towards an active ascetic "work in this world." If one disregards the small rationalist sects, such as are found all over the world, this has been attained only in the great church and sect organizations of Occidental and asceticist Protestantism. The quite distinct and the purely historically determined destinies of Occidental religions have co-operated in this matter. Partly the social environment exerted an influence, above all, the environment of the stratum that was decisive for the development of such religion. Partly, however—and just as strongly—the intrinsic character of Christianity exerted an influence: the supra-mundane God and the specificity of the means and paths of salvation as determined historically, first by Israelite prophecy and the thora doctrine.

The religious virtuoso can be placed in the world as the instrument of a God and cut off from all magical means of salvation. At the same time, it is imperative for the virtuoso that he "prove" himself before God, as being called *solely* through the ethical quality of his conduct in this world. This actually means that he "prove" himself to himself as well. No matter how much the "world" as such is religiously devalued and rejected as being creatural and a vessel of sin, yet psychologically the world is all the more affirmed as the theatre of God-willed activity in one's worldly "calling." For this inner-worldly asceticism rejects the world in the sense that it despises and taboos the values of dignity and beauty, of the beautiful frenzy and the dream, purely secular power, and the purely worldly pride of the hero. Asceticism outlawed these values as competitors of the kingdom of God. Yet precisely because of this rejection, asceticism did not fly from the world, as did contemplation. Instead, asceticism has wished to rationalize the world ethically in accordance with God's commandments. It has therefore remained oriented towards the world in a more specific and thoroughgoing sense than did the naive "affirmation of the world" of unbroken humanity, for instance, in Antiquity and in lay-Catholicism. In inner-worldly asceticism, the grace and the chosen state of the religiously qualified man prove themselves in everyday life. To be sure, they do so not in the everyday life as it is given, but in methodical and rationalized routine-activities of workaday life in the service of the Lord. Rationally raised into a vocation, everyday conduct becomes the locus for proving one's state of grace. The Occidental sects of the religious virtuosos have fermented the methodical rationalization of conduct, including economic conduct. These sects have not constituted valves for the longing to escape from the senselessness of work in this world, as did the Asiatic communities of the ecstatics: contemplative, orgiastic, or apathetic. . . .

We have to remind ourselves in advance that "rationalism" may mean very different things. It means one thing if we think of the kind of rationalization the systematic thinker performs on the image of the world: an increasing theoretical mastery of reality by means of increasingly precise and abstract concepts. Rationalism means another thing if we think of the methodical attainment of a definitely given and practical end by means of an increasingly precise calculation of adequate means. These types of rationalism are very different, in spite of the fact that ultimately they belong inseparately together. . . .

"Rational" may also mean a "systematic arrangement." In this sense, the following methods are rational: methods of mortificatory or of magical asceticism, of contemplation in its most consistent forms—for instance, in *yoga*—or in the manipulations of the prayer machines of later Buddhism.

In general, all kinds of practical ethics that are systematically and unambiguously oriented to fixed goals of salvation are "rational," partly in the same sense as formal method is rational, and partly in the sense that they distinguish between "valid" norms and what is empirically given.

Introduction to "The Distribution of Power Within the Political Community: Class, Status, Party"

In "Class, Status, Party," we again find Weber engaged in an implicit debate with Marx. While Marx saw interests, and the power to realize them, tied solely to class position, Weber saw the two as flowing from several sources. In fact, he argued that distinct interests and forms of power were connected to economic **classes, status** groups, and political **parties.** (See Table 4.1.) The result is a discarding of Marx's model in favor of a more complex view of how interests shape individuals' actions and the organization of societies.

Weber begins this essay with a definition of power, a definition that to this day guides work in political sociology. He defines it as "the chance of a man or of a number of men to realize their own will in a social action even against the resistance of others" (Weber 1925b/1978:926). Such chances, however, are not derived from a single source, nor is power valued for any one particular reason. Power may be exercised for economic gain, to increase one's "social honor" (or **status),** or for its own sake. Moreover, power stemming from one source, for instance economic power, may not translate into other domains. Thus, a person who has achieved substantial economic wealth through criminal activity will not have a high degree of status in the general society. Conversely, academics have a relatively high degree of status, but little economic power. Whatever power intellectuals have stems from their social honor, not from their ability to "realize their own will" through financial influence.

This essay is significant not only for its picture of the crosscutting sources of interests and power. Weber also offers here a distinct definition of class as well as his conception of status groups and parties. Recall that for Marx classes are based on a group's more or less stable relationship to the means of production (owners of capital versus owners of labor power). For Weber, however, classes are not stable groups or "communities" produced by existing property relations. Instead, they are people who share "life chances" or possibilities that are determined by "economic interests in the possession of goods and opportunities for income" within the commodity and labor markets (Weber 1925b/1978:927). While recognizing with Marx that "property" and "lack of property" form the basic distinction between classes, Weber nevertheless argued that classes are themselves the product of a shared "class situation"—a situation that reflects the type and amount of exchanges one can pursue in the market.

Status groups, on the other hand, are communities. The fate of such communities is determined not by their chances on the commodity or labor markets, however, but by "a specific, positive or negative, social estimation of *honor*" (Weber 1925b/1978:932, emphasis in the original). Such "honor" is expressed through "styles of life" or "conventions" that identify individuals with specific social circles. Race, ethnicity, religion, taste in fashion and the arts, and occupation have often formed a basis for making status distinctions. More than anything, membership in status groups serves to restrict an individual's chances for social interaction. For instance, the selection of marriage partners has frequently depended on a potential mate's religion or ethnicity. Even in modern, "egalitarian" societies like the United States, interracial marriages are relatively uncommon.

Additionally, regardless of possessing significant economic power or material wealth, one's race or religion can either close or open a person to educational and professional opportunities, as well as to membership in various clubs or associations.[1] Indeed, once membership into a style of life or institution can be bought, its ability to function as an expression of social honor or sign of exclusivity is threatened. This dynamic can be seen in shifting fashions in clothes and tastes in music, as well as in the democratization of education whereby proper "breeding" is no longer a prerequisite for getting a college diploma.

The third domain from which distinct interests are generated and power is exercised is the "legal order." Here, "parties' reside in the sphere of power" (Weber 1925b/1978:938). They include not only explicitly political groups, but also rationally organized groups more generally. As such, parties are characterized by the strategic pursuit of goals and the maintenance of a staff capable of implementing their objectives. Moreover, they are not necessarily tied to either class or status group interests, but are aimed instead at "influencing a communal action no matter what its content may be" (ibid.). Examples of parties include labor unions, which, through bureaucratic channels and the election of officers, seek to win

---

[1]During the early years of unionizing in the United States, trade unions were segregated racially, and at times ethnically. Thus, while sharing a common "class situation," workers, nevertheless, were divided by status group memberships. Some sociologists and labor historians have argued that the overriding salience of racial (i.e., status group) divisions fractured the working class, preventing workers from achieving more fully their class-based interests. Similar arguments have been made with regard to the feminist movement. In this case, white, middle-class women are charged with forsaking the plight of non-white and lower-class women in favor of pursuing goals that derive from their unique class situation.

economic benefits on behalf of workers, and, of course, the Republican and Democratic parties, which pursue legislative action that alternates between serving the class interests of their constituents (e.g., tax policy, trade regulations) and the interests of varying status groups (e.g., affirmative action, abortion rights, and gun control).

Table 4.1   Weber's Notions of Class, Status, and Party

		ORDER	
		Individual	**Collective**
ACTION	**Nonrational**	Status	*Status Groups:* "A specific, positive or negative, social estimation of *honor.*"
	Rational	Interests	*Class:* People who share "life chances" or possibilities that are "determined by economic interests in the possession of goods and opportunities for income." *Party:* Aimed at "influencing a communal action no matter what its content may be."

"The Distribution of Power Within the Political Community: Class, Status, Party" (1925)

Max Weber

A. Economically determined power and the status order. The structure of every legal order directly influences the distribution of power, economic or otherwise, within its respective community. This is true of all legal orders and not only that of the state. In general, we understand by "power" the chance of a man or a number of men to realize their own will in a social action even against the resistance of others who are participating in the action.

"Economically conditioned" power is not, of course, identical with "power" as such. On the contrary, the emergence of economic power may be the consequence of power existing on other grounds. Man does not strive for power only in order to enrich economically. Power, including economic power, may be valued for its own sake. Very frequently the striving for power is also conditioned by the social honor it entails. Not all power, however, entails social honor: The typical

American Boss, as well as the typical big speculator, deliberately relinquishes social honor. Quite generally, "mere economic" power, and especially "naked" money power, is by no means a recognized basis of social honor. Nor is power the only basis of social honor. Indeed, social honor, or prestige, may even be the basis of economic power, and very frequently has been. Power, as well as honor, may be guaranteed by the legal order, but, at least normally, it is not their primary source. The legal order is rather an additional factor that enhances the chance to hold power or honor; but it can not always secure them.

The way in which social honor is distributed in a community between typical groups participating in this distribution we call the "status order." The social order and the economic order are related in a similar manner to the legal order. However, the economic order merely defines the way in which economic goods and services are distributed and

SOURCE: Excerpts from Max Weber's *Economy and Society,* 2 vols. Translated and edited by Guenther Roth and Claus Wittich. Copyright © 1978 the Regents of the University of California. Published by the University of California Press.

used. Of course, the status order is strongly influenced by it, and in turn reacts upon it.

Now: "classes," "status groups," and "parties" are phenomena of the distribution of power within a community.

B. Determination of class situation by market situation. In our terminology, "classes" are not communities; they merely represent possible, and frequent, bases for social action. We may speak of a "class" when (1) a number of people have in common a specific causal component of their life chances, insofar as (2) this component is represented exclusively by economic interests in the possession of goods and opportunities for income, and (3) is represented under the conditions of the commodity or labor markets. This is "class situation."

It is the most elemental economic fact that the way in which the disposition over material property is distributed among a plurality of people, meeting competitively in the market for the purpose of exchange, in itself creates specific life chances. The mode of distribution, in accord with the law of marginal utility, excludes the non-wealthy from competing for highly valued goods; it favors the owners and, in fact, gives to them a monopoly to acquire such goods. Other things being equal, the mode of distribution monopolizes the opportunities for profitable deals for all those who, provided with goods, do not necessarily have to exchange them. It increases, at least generally, their power in the price struggle with those who, being propertyless, have nothing to offer but their labor or the resulting products, and who are compelled to get rid of these products in order to subsist at all. The mode of distribution gives to the propertied a monopoly on the possibility of transferring property from the sphere of use as "wealth" to the sphere of "capital," that is, it gives them the entrepreneurial function and all chances to share directly or indirectly in returns on capital. All this holds true within the area in which pure market conditions prevail. "Property" and "lack of property" are, therefore, the basic categories of all class situations. It does not matter whether these two categories become effective in the competitive struggles of the consumers or of the producers.

Within these categories, however, class situations are further differentiated: on the one hand, according to the kind of property that is usable for returns; and, on the other hand, according to the kind of services that can be offered in the market. Ownership of dwellings; workshops; warehouses; stores; agriculturally usable land in large or small holdings—a quantitative difference with possibly qualitative consequences; ownership of mines; cattle; men (slaves); disposition over mobile instruments of production, or capital goods of all sorts, especially money or objects that can easily be exchanged for money; disposition over products of one's own labor or of others' labor differing according to their various distances from consumability; disposition over transferable monopolies of any kind—all these distinctions differentiate the class situations of the propertied just as does the "meaning" which they can give to the use of property, especially to property which has money equivalence. Accordingly, the propertied, for instance, may belong to the class of rentiers or to the class of entrepreneurs.

Those who have no property but who offer services are differentiated just as much according to their kinds of services as according to the way in which they make use of these services, in a continuous or discontinuous relation to a recipient. But always this is the generic connotation of the concept of class: that the kind of chance in the *market* is the decisive moment which presents a common condition for the individual's fate. Class situation is, in this sense, ultimately market situation. The effect of naked possession *per se*, which among cattle breeders gives the non-owning slave or serf into the power of the cattle owner, is only a fore-runner of real "class" formation. However, in the cattle loan and in the naked severity of the law of debts in such communities for the first time mere "possession" as such emerges as decisive for the fate of the individual; this is much in contrast to crop-raising communities, which are based on labor. The creditor-debtor relation becomes the basis of "class situations" first in the cities, where a "credit market," however primitive, with rates of interest increasing according to the extent of dearth and factual monopolization of lending in the hands of a plutocracy could develop. Therewith "class struggles" begin.

Those men whose fate is not determined by the chance of using goods or services for themselves on the market, e.g., slaves, are not, however, a class in the technical sense of the term. They are, rather, a status group.

C. Social action flowing from class interest. According to our terminology, the factor that creates "class" is unambiguously economic interest, and indeed, only those interests involved in the existence of the market. Nevertheless, the concept of class-interest is an ambiguous one: even as an empirical concept it is ambiguous as soon as one understands by it something other than the factual direction of interests following with a certain probability from the class situation for a certain average of those people subjected to the class situation. The class situation and other circumstances remaining the same, the direction in which the individual worker, for instance, is likely to pursue his interests may vary widely, according to whether he is constitutionally qualified for the task at hand to a high, to an average, or to a low degree. In the same way, the direction of interests may vary according to whether or not social action of a larger or smaller portion of those commonly affected by the class situation, or even an association among them, e.g., a trade union, has grown out of the class situation, from which the individual may expect promising results for himself. The emergence of an association or even of mere social action from a common class situation is by no means a universal phenomenon.

The class situation may be restricted in its efforts to the generation of essentially *similar* reactions, that is to say, within our terminology, of "mass behavior." However, it may not even have this result. Furthermore, often merely amorphous social action emerges. For example, the "grumbling" of workers known in ancient Oriental ethics: The moral disapproval of the work-master's conduct, which in its practical significance was probably equivalent to an increasingly typical phenomenon of precisely the latest industrial development, namely, the slowdown of laborers by virtue of tacit agreement. The degree in which "social action" and possibly associations emerge from the mass behavior of the members of a class is linked to general cultural conditions, especially to those of an intellectual sort. It is also linked to the extent of the contrasts that have already evolved, and is especially linked to the transparency of the connections between the causes and the consequences of the class situation. For however different life chances may be, this fact in itself, according to all experience, by no means gives

birth to "class action" (social action by the members of a class). For that, the real conditions and the results of the class situation must be distinctly recognizable. For only then the contrast of life chances can be felt not as an absolutely given fact to be accepted, but as a resultant from either (1) the given distribution of property, or (2) the structure of the concrete economic order. It is only then that people may react against the class structure not only through acts of intermittent and irrational protest, but in the form of rational association. There have been "class situations" of the first category (1), of a specifically naked and transparent sort, in the urban centers of Antiquity and during the Middle Ages; especially then when great fortunes were accumulated by factually monopolized trading in local industrial products or in foodstuffs; furthermore, under certain conditions, in the rural economy of the most diverse periods, when agriculture was increasingly exploited in a profit-making manner. The most important historical example of the second category (2) is the class situation of the modern proletariat.

D. Types of class struggle. Thus every class may be the carrier of any one of the innumerable possible forms of class action, but this is not necessarily so. In any case, a class does not in itself constitute a group (*Gemeinschaft*). To treat "class" conceptually as being equivalent to "group" leads to distortion. That men in the same class situation regularly react in mass actions to such tangible situations as economic ones in the direction of those interests that are most adequate to their average number is an important and after all simple fact for the understanding of historical events. However, this fact must not lead to that kind of pseudo-scientific operation with the concepts of class and class interests which is so frequent these days and which has found its most classic expression in the statement of a talented author, that the individual may be in error concerning his interests but that the class is infallible about its interests.

If classes as such are not groups, nevertheless class situations emerge only on the basis of social action. However, social action that brings forth class situations is not basically action among members of the identical class; it is an action among members of different classes. Social actions that directly determine the class situation of the worker and the entrepreneur are: the labor market, the commodities market, and the capitalistic enterprise. But, in its turn, the existence of a capitalistic enterprise presupposes that a very specific kind of social action exists to protect the possession of goods *per se*, and especially the power of individuals to dispose, in principle freely, over the means of production: a certain kind of legal order. Each kind of class situation, and above all when it rests upon the power of property *per se*, will become most clearly efficacious when all other determinants of reciprocal relations are, as far as possible, eliminated in their significance. It is in this way that the use of the power of property in the market obtains its most sovereign importance.

Now status groups hinder the strict carrying through of the sheer market principle. In the present context they are of interest only from this one point of view. Before we briefly consider them, note that not much of a general nature can be said about the more specific kinds of antagonism between classes (in our meaning of the term). The great shift, which has been going on continuously in the past, and up to our times, may be summarized, although at a cost of some precision: the struggle in which class situations are effective has progressively shifted from consumption credit toward, first, competitive struggles in the commodity market and then toward wage disputes on the labor market. The class struggles of Antiquity—to the extent that they were genuine class struggles and not struggles between status groups—were initially carried on by peasants and perhaps also artisans threatened by debt bondage and struggling against urban creditors. . . .

The propertyless of Antiquity and of the Middle Ages protested against monopolies, preemption, forestalling, and the withholding of goods from the market in order to raise prices. Today the central issue is the determination of the price of labor. The transition is represented by the fight for access to the market and for the determination of the price of products. Such fights went on between merchants and workers in the putting-out system of domestic handicraft during the transition to modern times. Since it is quite a general phenomenon we must mention here that the class antagonisms that are conditioned through the market situations are usually most bitter between those who actually and directly participate as opponents in price wars. It

is not the rentier, the share-holder, and the banker who suffer the ill will of the worker, but almost exclusively the manufacturer and the business executives who are the direct opponents of workers in wage conflicts. This is so in spite of the fact that it is precisely the cash boxes of the rentier, the shareholder, and the banker into which the more or less unearned gains flow, rather than into the pockets of the manufacturers or of the business executives. This simple state of affairs has very frequently been decisive for the role the class situation has played in the formation of political parties. For example, it has made possible the varieties of patriarchal socialism and the frequent attempts—formerly, at least—of threatened status groups to form alliances with the proletariat against the bourgeoisie.

E. Status honor. In contrast to classes, *Stände* (*status groups*) are normally groups. They are, however, often of an amorphous kind. In contrast to the purely economically determined "class situation," we wish to designate as *status situation* every typical component of the life of men that is determined by a specific, positive or negative, social estimation of *honor*. This honor may be connected with any quality shared by a plurality, and, of course, it can be knit to a class situation: class distinctions are linked in the most varied ways with status distinctions. Property as such is not always recognized as a status qualification, but in the long run it is, and with extraordinary regularity. In the subsistence economy of neighborhood associations, it is often simply the richest who is the "chieftain." However, this often is only an honorific preference. For example, in the so-called pure modern democracy, that is, one devoid of any expressly ordered status privileges for individuals, it may be that only the families coming under approximately the same tax class dance with one another. This example is reported of certain smaller Swiss cities. But status honor need not necessarily be linked with a class situation. On the contrary, it normally stands in sharp opposition to the pretensions of sheer property.

Both propertied and propertyless people can belong to the same status group, and frequently they do with very tangible consequences. This equality of social esteem may, however, in the long run become quite precarious. The equality of status among American gentlemen, for instance, is expressed by the fact that outside the subordination determined by the different functions of business, it would be considered strictly repugnant—wherever the old tradition still prevails—if even the richest boss, while playing billiards or cards in his club would not treat his clerk as in every sense fully his equal in birthright, but would bestow upon him the condescending status-conscious "benevolence" which the German boss can never dissever from his attitude. This is one of the most important reasons why in America the German clubs have never been able to attain the attraction that the American clubs have.

In content, status honor is normally expressed by the fact that above all else a specific *style of life* is expected from all those who wish to belong to the circle. Linked with this expectation are restrictions on social intercourse (that is, intercourse which is not subservient to economic or any other purposes). These restrictions may confine normal marriages to within the status circle and may lead to complete endogamous closure. Whenever this is not a mere individual and socially irrelevant imitation of another style of life, but consensual action of this closing character, the status development is under way.

In its characteristic form, stratification by status groups on the basis of conventional styles of life evolves at the present time in the United States out of the traditional democracy. For example, only the resident of a certain street ("the Street") is considered as belonging to "society," is qualified for social intercourse, and is visited and invited. Above all, this differentiation evolves in such a way as to make for strict submission to the fashion that is dominant at a given time in society. This submission to fashion also exists among men in America to a degree unknown in Germany; it appears as an indication of the fact that a given man puts forward a *claim* to qualify as a gentleman. This submission decides, at least *prima facie*, that he will be treated as such. And this recognition becomes just as important for his employment chances in swank establishments, and above all, for social intercourse and marriage with "esteemed" families, as the qualification for dueling among Germans. As for the rest, status honor is usurped by certain families resident for a long time, and, of course, correspondingly wealthy (e.g., F.F.V., the First Families of Virginia), or by

the actual or alleged descendants of the "Indian Princess" Pocahontas, of the Pilgrim fathers, or of the Knickerbockers, the members of almost inaccessible sects and all sorts of circles setting themselves apart by means of any other characteristics and badges. In this case stratification is purely conventional and rests largely on usurpation (as does almost all status honor in its beginning). But the road to legal privilege, positive or negative, is easily traveled as soon as a certain stratification of the social order has in fact been "lived in" and has achieved stability by virtue of a stable distribution of economic power.

F. Ethnic segregation and caste. Where the consequences have been realized to their full extent, the status group evolves into a closed caste. Status distinctions are then guaranteed not merely by conventions and laws, but also by religious sanctions. This occurs in such a way that every physical contact with a member of any caste that is considered to be lower by the members of a higher caste is considered as making for a ritualistic impurity and a stigma which must be expiated by a religious act. In addition, individual castes develop quite distinct cults and gods.

In general, however, the status structure reaches such extreme consequences only where there are underlying differences which are held to be "ethnic." The caste is, indeed, the normal form in which ethnic communities that believe in blood relationship and exclude exogamous marriage and social intercourse usually associate with one another. Such a caste situation is part of the phenomenon of pariah peoples and is found all over the world. These people form communities, acquire specific occupational traditions of handicrafts or of other arts, and cultivate a belief in their ethnic community. They live in a diaspora strictly segregated from all personal intercourse, except that of an unavoidable sort, and their situation is legally precarious. Yet, by virtue of their economic indispensability, they are tolerated, indeed frequently privileged, and they live interspersed in the political communities. The Jews are the most impressive historical example.

A status segregation grown into a caste differs in its structure from a mere ethnic segregation: the caste structure transforms the horizontal and unconnected coexistences of ethnically segregated groups into a vertical social system of super- and subordination. Correctly formulated: a comprehensive association integrates the ethnically divided communities into one political unit. They differ precisely in this way: ethnic coexistence, based on mutual repulsion and disdain, allows each ethnic community to consider its own honor as the highest one; the caste structure brings about a social subordination and an acknowledgement of "more honor" in favor of the privileged caste and status groups. This is due to the fact that in the caste structure ethnic distinctions as such have become "functional" distinctions within the political association (warriors, priests, artisans that are politically important for war and for building, and so on). But even pariah peoples who are most despised (for example, the Jews) are usually apt to continue cultivating the belief in their own specific "honor," a belief that is equally peculiar to ethnic and to status groups.

However, with the negatively privileged status groups the sense of dignity takes a specific deviation. A sense of dignity is the precipitation in individuals of social honor and of conventional demands which a positively privileged status group raises for the deportment of its members. The sense of dignity that characterizes positively privileged status groups is naturally related to their "being" which does not transcend itself, that is, it is related to their "beauty and excellence" (ὄικεῖον ἔργον). Their kingdom is "of this world." They live for the present and by exploiting their great past. The sense of dignity of the negatively privileged strata naturally refers to a future lying beyond the present, whether it is of this life or of another. In other words, it must be nurtured by the belief in a providential mission and by a belief in a specific honor before God. The chosen people's dignity is nurtured by a belief either that in the beyond "the last will be the first," or that in this life a Messiah will appear to bring forth into the light of the world which has cast them out the hidden honor of the pariah people. This simple state of affairs, and not the resentment which is so strongly emphasized in Nietzsche's much-admired construction in the *Genealogy of Morals*, is the source of the religiosity cultivated by pariah status groups. . . .

For the rest, the development of status groups from ethnic segregations is by no means the normal phenomenon. On the contrary. Since objective "racial differences" are by no means behind every subjective sentiment of an ethnic community, the question of an ultimately racial foundation of status structure is rightly a question of the concrete individual case. Very frequently a status group is instrumental in the production of a thoroughbred anthropological type. Certainly status groups are to a high degree effective in producing extreme types, for they select personally qualified individuals (e.g., the knighthood selects those who are fit for warfare, physically and psychically). But individual selection is far from being the only, or the predominant, way in which status groups are formed: political membership or class situation has at all times been at least as frequently decisive. And today the class situation is by far the predominant factor. After all, the possibility of a style of life expected for members of a status group is usually conditioned economically.

G. Status privileges. For all practical purposes, stratification by status goes hand in hand with a monopolization of ideal and material goods or opportunities, in a manner we have come to know as typical. Besides the specific status honor, which always rests upon distance and exclusiveness, honorific preferences may consist of the privilege of wearing special costumes, of eating special dishes taboo to others, of carrying arms—which is most obvious in its consequences—, the right to be a dilettante, for example, to play certain musical instruments. However, material monopolies provide the most effective motives for the exclusiveness of a status group; although, in themselves, they are rarely sufficient, almost always they come into play to some extent. Within a status circle there is the question of intermarriage: the interest of the families in the monopolization of potential bridegrooms is at least of equal importance and is parallel to the interest in the monopolization of daughters. The daughters of the members must be provided for. With an increased closure of the status group, the conventional preferential opportunities for special employment grow into a legal monopoly of special offices for the members. Certain goods become objects for monopolization by status groups, typically, entailed estates, and frequently also the possession of serfs or bondsmen and, finally, special trades. This monopolization occurs positively when the status group is exclusively entitled to own and to manage them; and negatively when, in order to maintain its specific way of life, the status group must not own and manage them. For the decisive role of a style of life in status honor means that status groups are the specific bearers of all conventions. In whatever way it may be manifest, all stylization of life either originates in status groups or is at least conserved by them. Even if the principles of status conventions differ greatly, they reveal certain typical traits, especially among the most privileged strata. Quite generally, among privileged status groups there is a status disqualification that operates against the performance of common physical labor. This disqualification is now "setting in" in America against the old tradition of esteem for labor. Very frequently every rational economic pursuit, and especially entrepreneurial activity, is looked upon as a disqualification of status. Artistic and literary activity is also considered degrading work as soon as it is exploited for income, or at least when it is connected with hard physical exertion. An example is the sculptor working like a mason in his dusty smock as over against the painter in his salon-like studio and those forms of musical practice that are acceptable to the status group.

H. Economic conditions and effects of status stratification. The frequent disqualification of the gainfully employed as such is a direct result of the principle of status stratification, and of course, of this principle's opposition to a distribution of power which is regulated exclusively through the market. These two factors operate along with various individual ones, which will be touched upon below.

We have seen above that the market and its processes knows no personal distinctions: "functional" interests dominate it. It knows nothing of honor. The status order means precisely the reverse: stratification in terms of honor and styles of life peculiar to status groups as such. The status order would be threatened at its very root if mere economic acquisition and naked economic power still bearing the stigma of its extra-status origin could bestow upon anyone who has won them the same or even greater honor as the vested interests claim for themselves. After all, given equality of status honor, property *per se* represents an addition even if it is not overtly acknowledged to be such. Therefore all groups having

interest in the status order react with special sharpness precisely against the pretensions of purely economic acquisition. In most cases they react the more vigorously the more they feel themselves threatened. . . . Precisely because of the rigorous reactions against the claims of property *per se*, the "parvenu" is never accepted, personally and without reservation, by the privileged status groups, no matter how completely his style of life has been adjusted to theirs. They will only accept his descendants who have been educated in the conventions of their status group and who have never besmirched its honor by their own economic labor.

As to the general *effect* of the status order, only one consequence can be stated, but it is a very important one: the hindrance of the free development of the market. This occurs first for those goods that status groups directly withhold from free exchange by monopolization, which may be effected either legally or conventionally. For example, in many Hellenic cities during the "status era" and also originally in Rome, the inherited estate (as shown by the old formula for placing spendthrifts under a guardian) was monopolized, as were the estates of knights, peasants, priests, and especially the clientele of the craft and merchant guilds. The market is restricted, and the power of naked property *per se*, which gives its stamp to class formation, is pushed into the background. The results of this process can be most varied. Of course, they do not necessarily weaken the contrasts in the economic situation. Frequently they strengthen these contrasts, and in any case, where stratification by status permeates a community as strongly as was the case in all political communities of Antiquity and of the Middle Ages, one can never speak of a genuinely free market competition as we understand it today. There are wider effects than this direct exclusion of special goods from the market. From the conflict between the status order and the purely economic order mentioned above, it follows that in most instances the notion of honor peculiar to status absolutely abhors that which is essential to the market: hard bargaining. Honor abhors hard bargaining among peers and occasionally it taboos it for the members of a status group in general. Therefore, everywhere some status groups, and usually the most influential, consider almost any kind of overt participation in economic acquisition as absolutely stigmatizing.

With some over-simplification, one might thus say that classes are stratified according to their relations to the production and acquisition of goods; whereas status groups are stratified according to the principles of their *consumption* of goods as represented by special styles of life.

An "occupational status group," too, is a status group proper. For normally, it successfully claims social honor only by virtue of the special style of life which may be determined by it. The differences between classes and status groups frequently overlap. It is precisely those status communities most strictly segregated in terms of honor (viz., the Indian castes) who today show, although within very rigid limits, a relatively high degree of indifference to pecuniary income. However, the Brahmins seek such income in many different ways.

As to the general economic conditions making for the predominance of stratification by status, only the following can be said. When the bases of the acquisition and distribution of goods are relatively stable, stratification by status is favored. Every technological repercussion and economic transformation threatens stratification by status and pushes the class situation into the foreground. Epochs and countries in which the naked class situation is of predominant significance are regularly the periods of technical and economic transformations. And every slowing down of the change in economic stratification leads, in due course, to the growth of status structures and makes for a resuscitation of the important role of social honor.

I. Parties. Whereas the genuine place of classes is within the economic order, the place of status groups is within the social order, that is, within the sphere of the distribution of honor. From within these spheres, classes and status groups influence one another and the legal order and are in turn influenced by it. *"Parties"* reside in the sphere of power. Their action is oriented toward the acquisition of social power, that is to say, toward influencing social action no matter what its content may be. In principle, parties may exist in a social club as well as in a state. As over against the actions of classes and status groups, for which this is not necessarily the case, party-oriented social action always involves association. For it is always directed toward a goal which is striven for in a planned manner. This goal may be a cause (the party may aim at realizing a program for ideal or material purposes), or the goal may be personal (sinecures, power, and from these, honor for the leader and the followers of the party). Usually the party aims at all these simultaneously. Parties are, therefore, only possible within groups that have an associational character, that is, some rational order and a staff of persons available who are ready to enforce it. For parties aim precisely at influencing this staff, and if possible, to recruit from it party members.

In any individual case, parties may represent interests determined through class situation or status situation, and they may recruit their following respectively from one or the other. But they need be neither purely class nor purely status parties; in fact, they are more likely to be mixed types, and sometimes they are neither. They may represent ephemeral or enduring structures. Their means of attaining power may be quite varied, ranging from naked violence of any sort to canvassing for votes with coarse or subtle means: money, social influence, the force of speech, suggestion, clumsy hoax, and so on to the rougher or more artful tactics of obstruction in parliamentary bodies.

The sociological structure of parties differs in a basic way according to the kind of social action which they struggle to influence; that means, they differ according to whether or not the community is stratified by status or by classes. Above all else, they vary according to the structure of domination. For their leaders normally deal with its conquest. In our general terminology, parties are not only products of modern forms of domination. We shall also designate as parties the ancient and medieval ones, despite the fact that they differ basically from modern parties. Since a party always struggles for political control (*Herrschaft*), its organization too is frequently strict and "authoritarian." Because of these variations between the forms of domination, it is impossible to say anything about the structure of parties without discussing them first. Therefore, we shall now turn to this central phenomenon of all social organization.

Before we do this, we should add one more general observation about classes, status groups and parties: The fact that they presuppose a larger association, especially the framework of a polity, does not mean that they are confined to it. On the contrary, at all times it has been the order of the day that such association (even when it aims at the use of military force in common) reaches beyond the state boundaries. This can be seen in the [interlocal] solidarity of interests of oligarchs and democrats in Hellas, of Guelphs and Ghibellines in the Middle Ages, and within the Calvinist party during the age of religious struggles; and all the way up to the solidarity of landlords (International Congresses of Agriculture), princes (Holy Alliance, Karlsbad Decrees [of 1819]), socialist workers, conservatives (the longing of Prussian conservatives for Russian intervention in 1850). But their aim is not necessarily the establishment of a new territorial dominion. In the main they aim to influence the existing polity.

Introduction to "The Types of Legitimate Domination"

In this selection, Weber defines three "ideal types" of legitimate domination: **rational** or **legal authority, traditional authority,** and **charismatic authority.** (See Table 4.2.) As abstract constructs, none of the ideal types actually exists in pure form. Instead, public authority is based on some mixture of the three types. Nevertheless, social systems generally exhibit a predominance of one form or another of domination.

Before briefly describing the forms of legitimate authority, we first need to clarify Weber's definition of legitimacy. By "legitimacy," Weber was referring to the belief systems on which valid commands issuing from authority figures are based. Such belief systems supply the justifications and motives for demanding obedience and allow those in authority to rightfully exercise domination over others. It is to these justifications that authority figures turn when seeking to legitimate their actions and the actions of those subjected to their commands.

Modern states are ruled through rational-legal authority. This form of domination is based on the rule of rationally established laws. Legitimacy thus rests "on a belief in the legality of enacted rules and the right of those elevated to authority under such rules to issue commands" (Weber 1925c/1978:215). Obedience is owed not to the person who exercises authority, but to the office or position in which authority is vested. It is the impersonal, legal order that vests the superior with the authority to demand compliance, a right that is ceded on vacating the office. Once retired, a police officer or judge is but another civilian and as such no longer has the power to enforce the law.

Traditional authority is the authority of "eternal yesterday." It rests on an "established belief in the sanctity of immemorial traditions" (ibid.:215). This is the rule of kings and tribal chieftains. Leadership is attained not on the basis of impersonally measured merit, but on lines of heredity or rites of passage. Subjects owe their allegiance not to bureaucratically imposed rules and laws that are open to change, but to their personal "master" whose demands for compliance and loyalty are legitimated by sacred, inviolable traditions.

Weber's third type of authority derives from the **charisma** possessed by the leader. Demands for obedience are legitimated by the leader's "gift of grace," which is demonstrated through extraordinary feats, acts of heroism, or revelations—in short, the miracles of heroes and prophets. Like traditional authority, loyalty is owed to the person and not to an office defined through impersonal rules. But unlike traditional authority, legitimacy is not based on appeals to sacred traditions or on the exalting of "what has always been." Instead, compliance from "disciples" is demanded on the basis of the "conception that it is the duty of those subject to charismatic authority to recognize its genuineness and to act accordingly" (ibid.:242).

History is replete with charismatic leaders who have inspired intense personal devotion to themselves and their cause. From Jesus and Muhammad, Joan of Arc and Gandhi, to Napoleon and Hitler, such leaders have proved to be a powerful force for social change, both good and bad. Indeed, in its rejection of both tradition and rational, formal rules, charismatic authority, by its very nature, poses a challenge to existing political order. In breaking from history as well as objective laws, charisma is a creative force that carries the commandment: "It is written, but I say unto you."

Photo 4.1a   William Jefferson Clinton, the 42nd President of the United States

Embodiments of legitimate domination: President Clinton exercised rational-legal authority; Queen Victoria ruled on the basis of traditional authority; Mahatma Gandhi possessed charismatic authority.

Photo 4.1b   England's Queen Victoria (1819–1901)

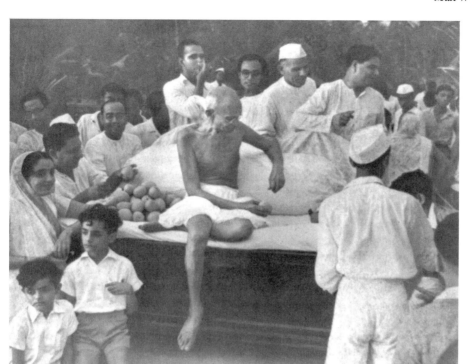

Photo 4.1c   Mahatma Gandhi (1869–1948), India's Past Spiritual and Political Leader

However, the revolutionary potential of charismatic authority makes it inherently unstable. Charisma lasts only as long as its possessor is able to provide benefits to his followers. If the leader's prophecies are proved wrong, if enemies are not defeated, if miraculous deeds begin to "dry up," then his legitimacy will be called into question. On the other hand, even if such deeds or benefits provide a continued source of legitimacy, the leader at some point will die. With authority resting solely in the charismatic individual, the movement he inspired will collapse along with his rule, unless designs for a successor are developed. Often, the transferring of authority eventually leads to the "routinization of charisma" and the transformation of legitimacy into either a rational-legal or traditional type—witness the Catholic Church.

Table 4.2   Weber's Types of Legitimate Domination

		ORDER	
		Individual	**Collective**
ACTION	**Nonrational**	*Charismatic:* "Gift of grace" of leader	*Traditional:* "Established belief in the sanctity of immemorial traditions"
	Rational		*Rational-Legal:* "Belief in the legality of enacted rules and the right of those elevated to authority under such rules to issue commands"

── From "The Types of Legitimate Domination" (1925) ──

Max Weber

DOMINATION AND LEGITIMACY

Domination was defined as the probability that certain specific commands (or all commands) will be obeyed by a given group of persons. It thus does not include every mode of exercising "power" or "influence" over other persons. Domination ("authority") in this sense may be based on the most diverse motives of compliance: all the way from simple habituation to the most purely rational calculation of advantage. Hence every genuine form of domination implies a minimum of voluntary compliance, that is, an *interest* (based on ulterior motives or genuine acceptance) in obedience.

Not every case of domination makes use of economic means; still less does it always have economic objectives. However, normally the rule over a considerable number of persons requires a staff, that is, a *special* group which can normally be trusted to execute the general policy as well as the specific commands. The members of the administrative staff may be bound to obedience to their superior (or superiors) by custom, by affectual ties, by a purely material complex of interests, or by ideal (*wertrationale*) motives. The quality of these motives largely determines the type of domination. *Purely* material interests and calculations of advantages as the basis of solidarity between the chief and his administrative staff result, in this as in other connexions, in a relatively unstable situation. Normally other elements, affectual and ideal, supplement such interests. In certain exceptional cases the former alone may be decisive. In everyday life these relationships, like others, are governed by custom and material calculation of advantage. But custom, personal advantage, purely affectual or ideal motives of solidarity, do not form a sufficiently reliable basis for a given domination. In addition there is normally a further element, the belief in *legitimacy*.

Experience shows that in no instance does domination voluntarily limit itself to the appeal to material or affectual or ideal motives as a basis for its continuance. In addition every such system attempts to establish and to cultivate the belief in its legitimacy. But according to the kind of legitimacy which is claimed, the type of obedience, the kind of administrative staff developed to guarantee it, and the mode of exercising authority, will all differ fundamentally. Equally fundamental is the variation in effect. Hence, it is useful to classify the types of domination according to the kind of claim to legitimacy typically made by each. In doing this, it is best to start from modern and therefore more familiar examples. . . .

The Three Pure Types of Authority

There are three pure types of legitimate domination. The validity of the claims to legitimacy may be based on:

1. Rational grounds—resting on a belief in the legality of enacted rules and the right of those elevated to authority under such rules to issue commands (legal authority).

2. Traditional grounds—resting on an established belief in the sanctity of immemorial traditions and the legitimacy of those exercising authority under them (traditional authority); or finally,

3. Charismatic grounds—resting on devotion to the exceptional sanctity, heroism or exemplary character of an individual person, and of the normative patterns or order revealed or ordained by him (charismatic authority).

In the case of legal authority, obedience is owed to the legally established impersonal order.

SOURCE: Excerpts from Max Weber's *Economy and Society,* 2 vols. Translated and edited by Guenther Roth and Claus Wittich. Copyright © 1978 the Regents of the University of California. Published by the University of California Press.

It extends to the persons exercising the authority of office under it by virtue of the formal legality of their commands and only within the scope of authority of the office. In the case of traditional authority, obedience is owed to the *person* of the chief who occupies the traditionally sanctioned position of authority and who is (within its sphere) bound by tradition. But here the obligation of obedience is a matter of personal loyalty within the area of accustomed obligations. In the case of charismatic authority, it is the charismatically qualified leader as such who is obeyed by virtue of personal trust in his revelation, his heroism or his exemplary qualities so far as they fall within the scope of the individual's belief in his charisma. . . .

LEGAL AUTHORITY WITH A BUREAUCRATIC STAFF

Legal Authority: The Pure Type

Legal authority rests on the acceptance of the validity of the following mutually interdependent ideas.

1. That any given legal norm may be established by agreement or by imposition, on grounds of expediency or value-rationality or both, with a claim to obedience at least on the part of the members of the organization. This is, however, usually extended to include all persons within the sphere of power in question—which in the case of territorial bodies is the territorial area—who stand in certain social relationships or carry out forms of social action which in the order governing the organization have been declared to be relevant.

2. That every body of law consists essentially in a consistent system of abstract rules which have normally been intentionally established. Furthermore, administration of law is held to consist in the application of these rules to particular cases; the administrative process in the rational pursuit of the interests which are specified in the order governing the organization within the limits laid down by legal precepts and following principles which are capable of generalized formulation and are approved in the order governing the group, or at least not disapproved in it.

3. That thus the typical person in authority, the "superior," is himself subject to an impersonal order by orienting his actions to it in his own dispositions and commands. (This is true not only for persons exercising legal authority who are in the usual sense "officials," but, for instance, for the elected president of a state.)

4. That the person who obeys authority does so, as it is usually stated, only in his capacity as a "member" of the organization and what he obeys is only "the law." (He may in this connection be the member of an association, of a community, of a church, or a citizen of a state.)

5. In conformity with point 3, it is held that the members of the organization, insofar as they obey a person in authority, do not owe this obedience to him as an individual, but to the impersonal order. Hence, it follows that there is an obligation to obedience only within the sphere of the rationally delimited jurisdiction which, in terms of the order, has been given to him. . . .

The purest type of exercise of legal authority is that which employs a bureaucratic administrative staff. Only the supreme chief of the organization occupies his position of dominance (*Herrenstellung*) by virtue of appropriation, of election, or of having been designated for the succession. But even *his* authority consists in a sphere of legal "competence." The whole administrative staff under the supreme authority then consist, in the purest type, of individual officials (constituting a "monocracy" as opposed to the "collegial" type, which will be discussed below) who are appointed and function according to the following criteria:

(1) They are personally free and subject to authority only with respect to their impersonal official obligations.

(2) They are organized in a clearly defined hierarchy of offices.

(3) Each office has a clearly defined sphere of competence in the legal sense.

(4) The office is filled by a free contractual relationship. Thus, in principle, there is free selection.

(5) Candidates are selected on the basis of technical qualifications. In the most rational case, this is tested by examination

or guaranteed by diplomas certifying technical training, or both. They are *appointed*, not elected.

(6) They are remunerated by fixed salaries in money, for the most part with a right to pensions. Only under certain circumstances does the employing authority, especially in private organizations, have a right to terminate the appointment, but the official is always free to resign. The salary scale is graded according to rank in the hierarchy; but in addition to this criterion, the responsibility of the position and the requirements of the incumbent's social status may be taken into account.

(7) The office is treated as the sole, or at least the primary, occupation of the incumbent.

(8) It constitutes a career. There is a system of "promotion" according to seniority or to achievement, or both. Promotion is dependent on the judgment of superiors.

(9) The official works entirely separated from ownership of the means of administration and without appropriation of his position.

(10) He is subject to strict and systematic discipline and control in the conduct of the office.

This type of organization is in principle applicable with equal facility to a wide variety of different fields. It may be applied in profit-making business or in charitable organizations, or in any number of other types of private enterprises serving ideal or material ends. It is equally applicable to political and to hierocratic organizations. With the varying degrees of approximation to a pure type, its historical existence can be demonstrated in all these fields. . . .

TRADITIONAL AUTHORITY: THE PURE TYPE

Authority will be called traditional if legitimacy is claimed for it and believed in by virtue of the sanctity of age-old rules and powers. The masters are designated according to traditional rules and are obeyed because of their traditional status (*Eigenwürde*). This type of organized rule is, in the simplest case, primarily based on personal loyalty which results from common upbringing. The person exercising authority is not a "superior," but a personal master, his administrative staff does not consist mainly of officials but of personal retainers, and the ruled are not "members" of an association but are either his traditional "comrades" or his "subjects." Personal loyalty, not the official's impersonal duty, determines the relations of the administrative staff to the master.

Obedience is owed not to enacted rules but to the person who occupies a position of authority by tradition or who has been chosen for it by the traditional master. The commands of such a person are legitimized in one of two ways:

a) partly in terms of traditions which themselves directly determine the content of the command and are believed to be valid within certain limits that cannot be overstepped without endangering the master's traditional status;

b) partly in terms of the master's discretion in that sphere which tradition leaves open to him; this traditional prerogative rests primarily on the fact that the obligations of personal obedience tend to be essentially unlimited.

Thus there is a double sphere:

a) that of action which is bound to specific traditions;

b) that of action which is free of specific rules.

In the latter sphere, the master is free to do good turns on the basis of his personal pleasure and likes, particularly in return for gifts—the historical sources of dues (*Gebühren*). So far as his action follows principles at all, these are governed by considerations of ethical common sense, of equity or of utilitarian expediency. They are not formal principles, as in the case of legal authority. The exercise of power is oriented toward the consideration of how far master and staff can go in view of the subjects' traditional compliance without arousing their resistance. When resistance occurs, it is directed against the master or his servant personally, the accusation being that he failed to observe the traditional limits of his power. Opposition is not directed against the system as such—it is a case of "traditionalist revolution."

In the pure type of traditional authority it is impossible for law or administrative rule to be deliberately created by legislation. Rules which in fact are innovations can be legitimized only by the claim that they have been "valid of yore," but have only now been recognized by means of "Wisdom" [the *Weistum* of ancient Germanic law]. Legal decisions as "finding of the law" (*Rechtsfindung*) can refer only to documents of tradition, namely to precedents and earlier decisions. . . .

In the pure type of traditional rule, the following features of a bureaucratic administrative staff are absent:

a) a clearly defined sphere of competence subject to impersonal rules,

b) a rationally established hierarchy,

c) a regular system of appointment on the basis of free contract, and orderly promotion,

d) technical training as a regular requirement,

e) (frequently) fixed salaries, in the type case paid in money. . . .

CHARISMATIC AUTHORITY

The term "charisma" will be applied to a certain quality of an individual personality by virtue of which he is considered extraordinary and treated as endowed with supernatural, superhuman, or at least specifically exceptional powers or qualities. These are such as are not accessible to the ordinary person, but are regarded as of divine origin or as exemplary, and on the basis of them the individual concerned is treated as a "leader." In primitive circumstances this peculiar kind of quality is thought of as resting on magical powers, whether of prophets, persons with a reputation for therapeutic or legal wisdom, leaders in the hunt, or heroes in war. How the quality in question would be ultimately judged from any ethical, aesthetic, or other such point of view is naturally entirely indifferent for purposes of

definition. What is alone important is how the individual is actually regarded by those subject to charismatic authority, by his "followers" or "disciples.". . .

I. It is recognition on the part of those subject to authority which is decisive for the validity of charisma. This recognition is freely given and guaranteed by what is held to be a proof, originally always a miracle, and consists in devotion to the corresponding revelation, hero worship, or absolute trust in the leader. But where charisma is genuine, it is not this which is the basis of the claim to legitimacy. This basis lies rather in the conception that it is the duty of those subject to charismatic authority to recognize its genuineness and to act accordingly. Psychologically this recognition is a matter of complete personal devotion to the possessor of the quality, arising out of enthusiasm, or of despair and hope. . . .

II. If proof and success elude the leader for long, if he appears deserted by his god or his magical or heroic powers, above all, if his leadership fails to benefit his followers, it is likely that his charismatic authority will disappear. This is the genuine meaning of the divine right of kings (*Gottesgnadentum*). . . .

III. An organized group subject to charismatic authority will be called a charismatic community (*Gemeinde*). It is based on an emotional form of communal relationship (*Vergemeinschaftung*). The administrative staff of a charismatic leader does not consist of "officials"; least of all are its members technically trained. It is not chosen on the basis of social privilege nor from the point of view of domestic or personal dependency. It is rather chosen in terms of the charismatic qualities of its members. The prophet has his disciples; the warlord his bodyguard; the leader, generally, his agents (*Vertrauensmänner*). There is no such thing as appointment or dismissal, no career, no promotion. There is only a call at the instance of the leader on the basis of the charismatic qualification of those he summons. There is no hierarchy; the leader merely intervenes in general or in individual cases when he considers the members of his staff lacking in charismatic qualification for a given task. There is no such thing as a bailiwick or definite sphere of competence, and no appropriation of official powers on the basis of social privileges. There may, however, be territorial or functional limits to charismatic powers and to the individual's mission. There is no such thing as a salary or a benefice.

Disciples or followers tend to live primarily in a communistic relationship with their leader on means which have been provided by voluntary gift. There are no established administrative organs. In their place are agents who have been provided with charismatic authority by their chief or who possess charisma of their own. There is no system of formal rules, of abstract legal principles, and hence no process of rational judicial decision oriented to them. But equally there is no legal wisdom oriented to judicial precedent. Formally concrete judgments are newly created from case to case and are originally regarded as divine judgments and revelations. From a substantive point of view, every charismatic authority would have to subscribe to the proposition, "It is written . . . but I say unto you . . ." The genuine prophet, like the genuine military leader and every true leader in this sense, preaches, creates, or demands *new* obligations—most typically, by virtue of revelation, oracle, inspiration, or of his own will, which are recognized by the members of the religious, military, or party group because they come from such a source. Recognition is a duty. When such an authority comes into conflict with the competing authority of another who also claims charismatic sanction, the only recourse is to some kind of a contest, by magical means or an actual physical battle of the leaders. In principle, only one side can be right in such a conflict; the other must be guilty of a wrong which has to be expiated.

Since it is "extra-ordinary," charismatic authority is sharply opposed to rational, and particularly bureaucratic, authority, and to traditional authority, whether in its patriarchal, patrimonial, or estate variants, all of which are everyday forms of domination; while the charismatic type is the direct antithesis of this. Bureaucratic authority is specifically rational in the sense of being bound to intellectually analysable rules; while charismatic authority is specifically irrational in the sense of being foreign to all rules. Traditional authority is bound to the precedents handed down from the past and to this extent is also oriented to rules. Within the sphere of its claims, charismatic authority repudiates the past, and is in this sense a specifically revolutionary force. It recognizes no appropriation of positions of power by virtue of the possession of property, either on the part of a chief or of socially privileged groups. The only basis of legitimacy for it is personal charisma so long as it is proved; that is, as long as it receives recognition and as long as the followers and disciples prove their usefulness charismatically. . . .

IV. Pure charisma is specifically foreign to economic considerations. Wherever it appears, it constitutes a "call" in the most emphatic sense of the word, a "mission" or a "spiritual duty." In the pure type, it disdains and repudiates economic exploitation of the gifts of grace as a source of income, though, to be sure, this often remains more an ideal than a fact. It is not that charisma always demands a renunciation of property or even of acquisition, as under certain circumstances prophets and their disciples do. The heroic warrior and his followers actively seek booty; the elective ruler or the charismatic party leader requires the material means of power. The former in addition requires a brilliant display of his authority to bolster his prestige. What is despised, so long as the genuinely charismatic type is adhered to, is traditional or rational everyday economizing, the attainment of a regular income by continuous economic activity devoted to this end. Support by gifts, either on a grand scale involving donation, endowment, bribery and honoraria, or by begging, constitute the voluntary type of support. On the other hand, "booty" and extortion, whether by force or by other means, is the typical form of charismatic provision for needs. From the point of view of rational economic activity, charismatic want satisfaction is a typical anti-economic force. It repudiates any sort of involvement in the everyday routine world. It can only tolerate, with an attitude of complete emotional indifference, irregular, unsystematic acquisitive acts. In that it relieves the recipient of economic concerns, dependence on property income can be the economic basis of a charismatic mode of life for some groups; but that is unusual for the normal charismatic "revolutionary." . . .

V. In traditionalist periods, charisma is *the* great revolutionary force. The likewise revolutionary force of "reason" works from *without:* by altering the situations of life and hence its problems, finally in this way changing men's attitudes toward them; or it intellectualizes the individual. Charisma, on the other hand, *may* effect a subjective or *internal* reorientation born out of suffering, conflicts, or enthusiasm. It may then result in a radical alteration of the central attitudes and directions of action with a completely new orientation of all attitudes toward the different problems of the "world." In prerationalistic periods, tradition and charisma between them have almost exhausted the whole of the orientation of action.

The Routinization of Charisma

In its pure form charismatic authority has a character specifically foreign to everyday routine structures. The social relationships directly involved are strictly personal, based on the validity and practice of charismatic personal qualities. If this is not to remain a purely transitory phenomenon, but to take on the character of a permanent relationship, a "community" of disciples or followers or a party organization or any sort of political or hierocratic organization, it is necessary for the character of charismatic authority to become radically changed. Indeed, in its pure form charismatic authority may be said to exist only *in statu nascendi.* It cannot remain stable, but becomes either traditionalized or rationalized, or a combination of both.

The following are the principal motives underlying this transformation: (a) The ideal and also the material interests of the followers in the continuation and the continual reactivation of the community, (b) the still stronger ideal and also stronger material interests of the members of the administrative staff, the disciples, the party workers, or others in continuing their relationship. Not only this, but they have an interest in continuing it in such a way that both from an ideal and a material point of view, their own position is put on a stable everyday basis. This means, above all, making it possible to participate in normal family relationships or at least to enjoy a secure social position in place of the kind of discipleship which is cut off from ordinary worldly connections, notably in the family and in economic relationships.

These interests generally become conspicuously evident with the disappearance of the personal charismatic leader and with the problem of *succession.* The way in which this problem is met—if it is met at all and the charismatic community continues to exist or now begins to emerge—is of crucial importance for the character of the subsequent social relationships. . . .

Concomitant with the routinization of charisma with a view to insuring adequate succession, go the interests in its routinization on the part of the administrative staff. It is only in the initial stages and so long as the charismatic leader acts in a way which is completely outside everyday social organization, that it is possible for his followers to live communistically in a community of faith and enthusiasm, on gifts, booty, or sporadic acquisition. Only the members of the small group of enthusiastic disciples and followers are prepared to devote their lives purely idealistically to their call. The great majority of disciples and followers will in the long run "make their living" out of their "calling" in a material sense as well. Indeed, this must be the case if the movement is not to disintegrate.

Hence, the routinization of charisma also takes the form of the appropriation of powers and of economic advantages by the followers or disciples, and of regulating recruitment. This process of traditionalization or of legalization, according to whether rational legislation is involved or not, may take any one of a number of typical forms. . . .

For charisma to be transformed into an everyday phenomenon, it is necessary that its antieconomic character should be altered. It must be adapted to some form of fiscal organization to provide for the needs of the group and hence to the economic conditions necessary for raising taxes and contributions. When a charismatic movement develops in the direction of prebendal provision, the "laity" becomes differentiated from the "clergy"—derived from κλῆρος, meaning a "share"—, that is, the participating members of the charismatic administrative staff which has now become routinized. These are the priests of the developing "church." Correspondingly, in a developing political body—the "state" in the rational case—vassals, benefice-holders, officials or appointed party officials (instead of voluntary party workers and functionaries) are differentiated from the "tax payers.". . .

It follows that, in the course of routinization, the charismatically ruled organization is largely transformed into one of the everyday authorities, the patrimonial form, especially in its estate-type or bureaucratic variant. Its original peculiarities are apt to be retained in the charismatic status honor acquired by heredity or office-holding. This applies to all who participate in the appropriation, the chief himself and the members of his staff. It is thus a matter of the type of prestige enjoyed by ruling groups. A hereditary monarch by "divine right" is not a simple patrimonial chief, patriarch, or sheik; a vassal is not a mere household retainer or official. Further details must be deferred to the analysis of status groups.

As a rule, routinization is not free of conflict. In the early stages personal claims on the charisma of the chief are not easily forgotten and the conflict between the charisma of the office or of hereditary status with personal charisma is a typical process in many historical situations.

Introduction to "Bureaucracy"

In this essay, Weber defines the "ideal type" of bureaucracy, outlining its unique and most significant features. The salience of Weber's description lies in the fact that bureaucracies have become the dominant form of social organization in modern society. Indeed, bureaucracies are indispensable to modern life. Without them, a multitude of necessary tasks could not be performed with the degree of efficiency required for serving large numbers of individuals. For instance, strong and effective armies could not be maintained, the mass production of goods and their sale would slow to a trickle, the thousands of miles of public roadways could not be paved, hospitals could not treat the millions of patients in need of care, and establishing a university capable of educating 20,000 students would be impossible. Of course, all of these tasks and countless others are themselves dependent on a bureaucratic organization capable of collecting tax dollars from millions of people.

Despite whatever failings particular bureaucracies may exhibit, the form of organization is as essential to modern life as the air we breathe. In accounting for the ascendancy of bureaucracies, Weber is clear:

> The decisive reason for the advance of bureaucratic organization has also been its purely *technical* superiority over any other form of organization. . . . Precision, speed, unambiguity, knowledge of the files, continuity, discretion, unity, strict subordination, reduction of friction and of material and personal costs—these are raised to the optimum point in the strictly bureaucratic administration. . . . As compared with all [other] forms of administration, trained bureaucracy is superior on all these points. (Weber 1925d/1978:973, emphasis in the original)

A number of features ensure the technical superiority of bureaucracies. First, authority is hierarchically structured, making for a clear chain of command. Second, selection of personnel is competitive and based on demonstrated merit. This reduces the likelihood of incompetence that can result from appointing officials through nepotism or by virtue of tradition. Third, a specialized division of labor allows for the more efficient completion of assigned tasks. Fourth, bureaucracies are governed by formal, impersonal rules that regulate all facets of the organization. As a result, predictability of action and the strategic planning that it makes possible are better guaranteed.

Photo 4.2   Look Familiar? Waiting in Line at a University Student Services Building

As the epitome of the process of rationalization, however, Weber by no means embraced unequivocally the administrative benefits provided by bureaucracies. While in important respects, bureaucracies are dependent on the development of mass democracy for their fullest expression, they nevertheless create new elite groups of experts and technocrats. Moreover, he contended that their formal rules and procedures led to the loss of individual freedom.[1] For those working in bureaucracies (and countless do), Weber saw the individual "chained to his activity in his entire economic and ideological existence" (Weber 1925d/1978:988). The bureaucrat adopts as his own the detached, objective attitudes on which the efficiency and predictability of bureaucracies depend. Operating "'[w]ithout regard for persons . . . [b]ureaucracy develops the more perfectly, the more it is 'dehumanized,' the more completely it succeeds in eliminating from official business love, hatred, and all purely personal, irrational, and emotional elements which escape calculation" (ibid.:975). Whether as an employee or as a client, who among us has not been confronted with the faceless impersonality of a bureaucracy immune to the "special circumstances" that, after all, make up the very essence of our individuality?

---

[1]As we noted earlier, Weber's analysis of bureaucratic organizations offers an important critique of Marx's perspective. While Marx argued that capitalism is the source of alienation in modern society, Weber saw the source lying in bureaucracies and the rational procedures they embody. Additionally, in recognizing that bureaucracies create elite groups of technocrats who pursue their own professional interests, Weber also suggested that such organizational leaders (i.e., state officials) do not necessarily advance the interests of a ruling capitalist class. A related theme can likewise be found in "Class, Status, Party."

From "Bureaucracy" (1925)

Max Weber

CHARACTERISTICS OF MODERN BUREAUCRACY

Modern officialdom functions in the following manner:

I. There is the principle of official *jurisdictional areas*, which are generally ordered by rules, that is, by laws or administrative regulations. This means:

(1) The regular activities required for the purposes of the bureaucratically governed structure are assigned as official duties.

(2) The authority to give the commands required for the discharge of these duties is distributed in a stable way and is strictly delimited by rules concerning the coercive means, physical, sacerdotal, or otherwise, which may be placed at the disposal of officials.

(3) Methodical provision is made for the regular and continuous fulfillment of these duties and for the exercise of the corresponding rights; only persons who qualify under general rules are employed.

In the sphere of the state these three elements constitute a bureaucratic *agency*, in the sphere of the private economy they constitute a bureaucratic *enterprise*. Bureaucracy, thus understood, is fully developed in political and ecclesiastical communities only in the modern state, and in the private economy only in the most advanced institutions of capitalism. Permanent agencies, with fixed jurisdiction, are not the historical rule but rather the exception. This is even true of large political structures such as those of the ancient Orient, the Germanic and Mongolian empires of conquest, and of many feudal states. In all these cases, the ruler executes the most important measures through personal trustees, table-companions, or court-servants. Their commissions and powers are not precisely delimited and are temporarily called into being for each case.

II. The principles of *office hierarchy* and of channels of appeal (*Instanzenzug*) stipulate a clearly established system of super- and subordination in which there is a supervision of the lower offices by the higher ones. Such a system offers the governed the possibility of appealing, in a precisely regulated manner, the decision of a lower office to the corresponding superior authority. With the full development of the bureaucratic type, the office hierarchy is *monocratically* organized. The principle of hierarchical office authority is found in all bureaucratic structures: in state and ecclesiastical structures as well as in large party organizations and private enterprises. It does not matter for the character of bureaucracy whether its authority is called "private" or "public."

When the principle of jurisdictional "competency" is fully carried through, hierarchical subordination—at least in public office—does not mean that the "higher" authority is authorized simply to take over the business of the "lower." Indeed, the opposite is the rule; once an office has been set up, a new incumbent will always be appointed if a vacancy occurs.

III. The management of the modern office is based upon written documents (the "files"), which are preserved in their original or draft form, and upon a staff of subaltern officials and scribes of all sorts. The body of officials working in an agency along with the respective apparatus of material implements and the files makes up a *bureau* (in private enterprises often called the "counting house," *Kontor*).

In principle, the modern organization of the civil service separates the bureau from the

SOURCE: Excerpts from Max Weber's *Economy and Society,* 2 vols. Translated and edited by Guenther Roth and Claus Wittich. Copyright © 1978 the Regents of the University of California. Published by the University of California Press.

private domicile of the official and, in general, segregates official activity from the sphere of private life. Public monies and equipment are divorced from the private property of the official. This condition is everywhere the product of a long development. Nowadays, it is found in public as well as in private enterprises; in the latter, the principle extends even to the entrepreneur at the top. In principle, the *Kontor* (office) is separated from the household, business from private correspondence, and business assets from private wealth. The more consistently the modern type of business management has been carried through, the more are these separations the case. The beginnings of this process are to be found as early as the Middle Ages.

It is the peculiarity of the modern entrepreneur that he conducts himself as the "first official" of his enterprise, in the very same way in which the ruler of a specifically modern bureaucratic state [Frederick II of Prussia] spoke of himself as "the first servant" of the state. The idea that the bureau activities of the state are intrinsically different in character from the management of private offices is a continental European notion and, by way of contrast, is totally foreign to the American way.

IV. Office management, at least all specialized office management—and such management is distinctly modern—usually presupposes thorough training in a field of specialization. This, too, holds increasingly for the modern executive and employee of a private enterprise, just as it does for the state officials.

V. When the office is fully developed, official activity demands the *full working capacity* of the official, irrespective of the fact that the length of his obligatory working hours in the bureau may be limited. In the normal case, this too is only the product of a long development, in the public as well as in the private office. Formerly the normal state of affairs was the reverse: Official business was discharged as a secondary activity.

VI. The management of the office follows *general rules*, which are more or less stable, more or less exhaustive, and which can be learned. Knowledge of these rules represents a special technical expertise which the officials possess. It involves jurisprudence, administrative or business management.

The reduction of modern office management to rules is deeply embedded in its very nature. The theory of modern public administration, for instance, assumes that the authority to order certain matters by decree—which has been legally granted to an agency—does not entitle the agency to regulate the matter by individual commands given for each case, but only to regulate the matter abstractly. This stands in extreme contrast to the regulation of all relationships through individual privileges and bestowals of favor, which, as we shall see, is absolutely dominant in patrimonialism, at least in so far as such relationships are not fixed by sacred tradition.

The Position of the Official Within and Outside of Bureaucracy

All this results in the following for the internal and external position of the official:

I. Office Holding as a Vocation

That the office is a "vocation" (*Beruf*) finds expression, first, in the requirement of a prescribed course of training, which demands the entire working capacity for a long period of time, and in generally prescribed special examinations as prerequisites of employment. Furthermore, it finds expression in that the position of the official is in the nature of a "duty" (*Pflicht*). This determines the character of his relations in the following manner: Legally and actually, office holding is not considered ownership of a source of income, to be exploited for rents or emoluments in exchange for the rendering of certain services, as was normally the case during the Middle Ages and frequently up to the threshold of recent times, nor is office holding considered a common exchange of services, as in the case of free employment contracts. Rather, entrance into an office, including one in the private economy, is considered an acceptance of a specific duty of fealty to the purpose of the office (*Amtstreue*) in return for the grant of a secure existence. It is decisive for the modern loyalty to an office that, in the pure type, it does not establish a relationship to a *person*, like the vassal's or disciple's faith under feudal or patrimonial authority, but rather is devoted to *impersonal* and *functional* purposes. These purposes, of course, frequently gain an ideological halo from cultural values, such as state, church, community, party or enterprise, which appear as surrogates for a this-worldly or other-worldly personal master and which are embodied by a given group.

The political official—at least in the fully developed modern state—is not considered the personal servant of a ruler. Likewise, the bishop, the priest and the preacher are in fact no longer, as in early Christian times, carriers of a purely personal charisma, which offers other-worldly sacred values under the personal mandate of a master, and in principle responsible only to him, to everybody who appears worthy of them and asks for them. In spite of the partial survival of the old theory, they have become officials in the service of a functional purpose, a purpose which in the present-day "church" appears at once impersonalized and ideologically sanctified.

II. The Social Position of the Official

A. Social esteem and status convention.
Whether he is in a private office or a public bureau, the modern official, too, always strives for and usually attains a distinctly elevated *social esteem* vis-à-vis the governed. His social position is protected by prescription about rank order and, for the political official, by special prohibitions of the criminal code against "insults to the office" and "contempt" of state and church authorities.

The social position of the official is normally highest where, as in old civilized countries, the following conditions prevail: a strong demand for administration by trained experts; a strong and stable social differentiation, where the official predominantly comes from socially and economically privileged strata because of the social distribution of power or the costliness of the required training and of status conventions. The possession of educational certificates or patents . . . is usually linked with qualification for office; naturally, this enhances the "status element" in the social position of the official.

Sometimes the status factor is explicitly acknowledged; for example, in the prescription that the acceptance of an aspirant to an office career depends upon the consent ("election") by the members of the official body. . . .

Usually the social esteem of the officials is especially low where the demand for expert administration and the hold of status conventions are weak. This is often the case in new settlements by virtue of the great economic opportunities and the great instability of their social stratification: witness the United States.

B. Appointment versus election: Consequences for expertise.
Typically, the bureaucratic official is appointed by a superior authority. An official elected by the governed is no longer a purely bureaucratic figure. Of course, a formal election may hide an appointment—in politics especially by party bosses. This does not depend upon legal statutes, but upon the way in which the party mechanism functions. Once firmly organized, the parties can turn a formally free election into the mere acclamation of a candidate designated by the party chief, or at least into a contest, conducted according to certain rules, for the election of one of two designated candidates.

In all circumstances, the designation of officials by means of an election modifies the rigidity of hierarchical subordination. In principle, an official who is elected has an autonomous position vis-à-vis his superiors, for he does not derive his position "from above" but "from below," or at least not from a superior authority of the official hierarchy but from powerful party men ("bosses"), who also determine his further career. The career of the elected official is not primarily dependent upon his chief in the administration. The official who is not elected, but appointed by a master, normally functions, from a technical point of view, more accurately because it is more likely that purely functional points of consideration and qualities will determine his selection and career. As laymen, the governed can evaluate the expert qualifications of a candidate for office only in terms of experience, and hence only after his service. Moreover, if political parties are involved in any sort of selection of officials by election, they quite naturally tend to give decisive weight not to technical competence but to the services a follower renders to the party boss. This holds for the designation of otherwise freely elected officials by party bosses when they determine the slate of candidates as well as for the free appointment of officials by a chief who has himself been elected. The contrast, however, is relative: substantially similar conditions hold where legitimate monarchs and their subordinates appoint officials, except that partisan influences are then less controllable.

Where the demand for administration by trained experts is considerable, and the party faithful have to take into account an intellectually developed, educated, and free "public opinion," the use of unqualified officials redounds upon the party in power at the next election. Naturally, this is more likely to happen when the officials are appointed by the chief. The demand for a trained administration now exists in the United States, but wherever, as in the large cities, immigrant votes are "corralled," there is, of course, no effective public opinion. Therefore, popular election not only of the administrative chief but also of his subordinate officials usually endangers, at least in very large administrative bodies which are difficult to supervise, the expert qualification of the officials as well as the precise functioning of the bureaucratic mechanism, besides weakening the dependence of the officials upon the hierarchy. The superior qualification and integrity of Federal judges appointed by the president, as over and against elected judges, in the United States is well known, although both types of officials are selected primarily in terms of party considerations. The great changes in American metropolitan administrations demanded by reformers have been effected essentially by elected mayors working with an apparatus of officials who were appointed by them. These reforms have thus come about in a "caesarist" fashion. Viewed technically, as an organized form of domination, the efficiency of "caesarism," which often grows out of democracy, rests in general upon the position of the "caesar" as a free trustee of the masses (of the army or of the citizenry), who is unfettered by tradition. The "caesar" is thus the unrestrained master of a body of highly qualified military officers and officials whom he selects freely and personally without regard to tradition or to any other impediments. Such "rule of the personal genius," however, stands in conflict with the formally "democratic" principle of a generally elected officialdom.

C. Tenure and the inverse relationship between judicial independence and social prestige. Normally, the position of the official is held for life, at least in public bureaucracies, and this is increasingly the case for all similar structures. As a factual rule, *tenure for life* is presupposed even where notice can be given or periodic reappointment occurs. In a private enterprise, the fact of such tenure normally differentiates the official from the worker. Such legal or actual life-tenure, however, is not viewed as a proprietary right of the official to the possession of office as was the case in many structures of authority of the past. Wherever legal guarantees against discretionary dismissal or transfer are developed, as in Germany for all judicial and increasingly also for administrative officials, they merely serve the purpose of guaranteeing a strictly impersonal discharge of specific office duties. . . .

D. Rank as the basis of regular salary. The official as a rule receives a *monetary* compensation in the form of a *salary*, normally fixed, and the old age security provided by a pension. The salary is not measured like a wage in terms of work done, but according to "status," that is, according to the kind of function (the "rank") and, possibly, according to the length of service. The relatively great security of the official's income, as well as the rewards of social esteem, make the office a sought-after position, especially in countries which no longer provide opportunities for colonial profits. In such countries, this situation permits relatively low salaries for officials.

E. Fixed career lines and status rigidity. The official is set for a "career" within the hierarchical order of the public service. He expects to move from the lower, less important and less well paid, to the higher positions. The average official naturally desires a mechanical fixing of the conditions of promotion: if not of the offices, at least of the salary levels. He wants these conditions fixed in terms of "seniority," or possibly according to grades achieved in a system of examinations. Here and there, such grades actually form a *character indelebilis* of the official and have lifelong effects on his career. To this is joined the desire to reinforce the right to office and to increase status group closure and economic security. All of this makes for a tendency to consider the offices as "prebends" of those qualified by educational certificates. The necessity of weighing general personal and intellectual qualifications without concern for the often subaltern character of such patents of specialized education, has brought it about that the highest political offices, especially the "ministerial" positions, are as a rule filled without reference to such certificates. . . .

The Technical Superiority of Bureaucratic Organization Over Administration by Notables

The decisive reason for the advance of bureaucratic organization has always been its purely *technical* superiority over any other form of organization. The fully developed bureaucratic apparatus compares with other organizations exactly as does the machine with the non-mechanical modes of production. Precision, speed, unambiguity, knowledge of the files, continuity, discretion, unity, strict subordination, reduction of friction and of material and personal costs—these are raised to the optimum point in the strictly bureaucratic administration, and especially in its monocratic form. As compared with all collegiate, honorific, and avocational forms of administration, trained bureaucracy is superior on all these points. And as far as complicated tasks are concerned, paid bureaucratic work is not only more precise but, in the last analysis, it is often cheaper than even formally unremunerated honorific service. . . .

Today, it is primarily the capitalist market economy which demands that the official business of public administration be discharged precisely, unambiguously, continuously, and with as much speed as possible. Normally, the very large modern capitalist enterprises are themselves unequalled models of strict bureaucratic organization. Business management throughout rests on increasing precision, steadiness, and, above all, speed of operations. This, in turn, is determined by the peculiar nature of the modern means of communication, including, among other things, the news service of the press. The extraordinary increase in the speed by which public announcements, as well as economic and political facts, are transmitted exerts a steady and sharp pressure in the direction of speeding up the tempo of administrative reaction towards various situations. The optimum of such reaction time is normally attained only by a strictly bureaucratic organization. (The fact that the bureaucratic apparatus also can, and indeed does, create certain definite impediments for the discharge of business in a manner best adapted to the individuality of each case does not belong in the present context.)

Bureaucratization offers above all the optimum possibility for carrying through the principle of specializing administrative functions according to purely objective considerations. Individual performances are allocated to functionaries who have specialized training and who by constant practice increase their expertise. "Objective" discharge of business primarily means a discharge of business according to *calculable rules* and "without regard for persons."

"Without regard for persons," however, is also the watchword of the market and, in general, of all pursuits of naked economic interests. Consistent bureaucratic domination means the leveling of "status honor." Hence, if the principle of the free market is not at the same time restricted, it means the universal domination of the "class situation." That this consequence of bureaucratic domination has not set in everywhere proportional to the extent of bureaucratization is due to the differences between possible principles by which polities may supply their requirements. However, the second element mentioned, calculable rules, is the most important one for modern bureaucracy. The peculiarity of modern culture, and specifically of its technical and economic basis, demands this very "calculability" of results. When fully developed, bureaucracy also stands, in a specific sense, under the principle of *sine ira ac studio*. Bureaucracy develops the more perfectly, the more it is "dehumanized," the more completely it succeeds in eliminating from official business love, hatred, and all purely personal, irrational, and emotional elements which escape calculation. This is appraised as its special virtue by capitalism.

The more complicated and specialized modern culture becomes, the more its external supporting apparatus demands the personally detached and strictly objective *expert*, in lieu of the lord of older social structures who was moved by personal sympathy and favor, by grace and gratitude. Bureaucracy offers the attitudes demanded by the external apparatus of modern culture in the most favorable combination. In particular, only bureaucracy has established the foundation for the administration of a rational law conceptually systematized on the basis of "statutes," such as the later Roman Empire first created with a high degree of technical perfection. During the Middle Ages, the reception of this [Roman] law coincided with the bureaucratization of legal administration: The advance of the rationally trained expert displaced the old trial procedure which was bound to tradition or to irrational presuppositions. . . .

THE LEVELING OF SOCIAL DIFFERENCES

In spite of its indubitable technical superiority, bureaucracy has everywhere been a relatively late development. A number of obstacles have contributed to this, and only under certain social and political conditions have they definitely receded into the background.

A. Administrative Democratization

Bureaucratic organization has usually come into power on the basis of a leveling of economic and social differences. This leveling has been at least relative, and has concerned the significance of social and economic differences for the assumption of administrative functions.

Bureaucracy inevitably accompanies modern *mass democracy*, in contrast to the democratic self-government of small homogeneous units. This results from its characteristic principle: the abstract regularity of the exercise of authority, which is a result of the demand for "equality before the law" in the personal and functional sense—hence, of the horror of "privilege," and the principled rejection of doing business "from case to case." Such regularity also follows from the social pre-conditions of its origin. Any non-bureaucratic administration of a large social structure rests in some way upon the fact that existing social, material, or honorific preferences and ranks are connected with administrative functions and duties. This usually means that an economic or a social exploitation of position, which every sort of administrative activity provides to its bearers, is the compensation for the assumption of administrative functions.

Bureaucratization and democratization within the administration of the state therefore signify an increase of the cash expenditures of the public treasury, in spite of the fact that bureaucratic administration is usually more "economical" in character than other forms. Until recent times— at least from the point of view of the treasury— the cheapest way of satisfying the need for administration was to leave almost the entire local administration and lower judicature to the landlords of Eastern Prussia. The same is true of the administration by justices of the peace in England. Mass democracy which makes a clean sweep of the feudal, patrimonial, and—at least in intent—the plutocratic privileges in administration unavoidably has to put paid professional labor in place of the historically inherited "avocational" administration by notables.

B. Mass Parties and the Bureaucratic Consequences of Democratization

This applies not only to the state. For it is no accident that in their own organizations the democratic mass parties have completely broken with traditional rule by notables based upon personal relationships and personal esteem. Such personal structures still persist among many old conservative as well as old liberal parties, but democratic mass parties are bureaucratically organized under the leadership of party officials, professional party and trade union secretaries, etc. In Germany, for instance, this has happened in the Social Democratic party and in the agrarian mass-movement; in England earliest in the caucus democracy of Gladstone and Chamberlain which spread from Birmingham in the 1870's. In the United States, both parties since Jackson's administration have developed bureaucratically. In France, however, attempts to organize disciplined political parties on the basis of an election system that would compel bureaucratic organization have repeatedly failed. The resistance of local circles of notables against the otherwise unavoidable bureaucratization of the parties, which would encompass the entire country and break their influence, could not be overcome. Every advance of simple election techniques based on numbers alone as, for instance, the system of proportional representation, means a strict and inter-local bureaucratic organization of the parties and therewith an increasing domination of party bureaucracy and discipline, as well as the elimination of the local circles of notables—at least this holds for large states.

The progress of bureaucratization within the state administration itself is a phenomenon paralleling the development of democracy, as is quite obvious in France, North America, and now in England. Of course, one must always remember that the term "democratization" can be misleading. The *demos*, itself, in the sense of a shapeless mass, never "governs" larger associations, but rather is governed. What changes is only the way in which the executive leaders are selected and the measure of influence which the *demos*, or better, which social circles from its midst are able to exert upon the content and the direction of administrative activities by means of "public opinion." "Democratization," in the sense here intended, does not necessarily mean an increasingly active share of the subjects in government. This may be a result of democratization, but it is not necessarily the case.

We must expressly recall at this point that the political concept of democracy, deduced from the "equal rights" of the governed, includes these further postulates: (1) prevention of the development of a closed status group of officials in the interest of a universal accessibility of office, and (2) minimization of the authority of officialdom in the interest of expanding the sphere of influence of "public opinion" as far as practicable. Hence, wherever possible, political democracy strives to shorten the term of office through election and recall, and to be relieved from a limitation to candidates with special expert qualifications. Thereby democracy inevitably comes into conflict with the bureaucratic tendencies which have been produced by its very fight against the notables. The loose term "democratization" cannot be used here, in so far as it is understood to mean the minimization of the civil servants' power in favor of the greatest possible "direct" rule of the *demos*, which in practice means the respective party leaders of the *demos*. The decisive aspect here—indeed it is rather exclusively so—is *the leveling of the governed* in face of the governing and bureaucratically articulated group, which in its turn may occupy a quite autocratic position, both in fact and in form. . . .

THE OBJECTIVE AND SUBJECTIVE BASES OF BUREAUCRATIC PERPETUITY

Once fully established, bureaucracy is among those social structures which are the hardest to destroy. Bureaucracy is *the* means of transforming social action into rationally organized action. Therefore, as an instrument of rationally organizing authority relations, bureaucracy was and is a power instrument of the first order for one who controls the bureaucratic apparatus. Under otherwise equal conditions, rationally organized and directed action (*Gesellschaftshandeln*) is superior to every kind of collective behavior (*Massenhandeln*) and also social action (*Gemein-schaftshandeln*) opposing it. Where administration has been completely bureaucratized, the resulting system of domination is practically indestructible.

The individual bureaucrat cannot squirm out of the apparatus into which he has been harnessed. In contrast to the "notable" performing administrative tasks as a honorific duty or as a subsidiary occupation (avocation), the professional bureaucrat is chained to his activity in his entire economic and ideological existence. In the great majority of cases he is only a small cog in a ceaselessly moving mechanism which prescribes to him an essentially fixed route of march. The official is entrusted with specialized tasks, and normally the mechanism cannot be put into motion or arrested by him, but only from the very top. The individual bureaucrat is, above all, forged to the common interest of all the functionaries in the perpetuation of the apparatus and the persistence of its rationally organized domination.

The ruled, for their part, cannot dispense with or replace the bureaucratic apparatus once it exists, for it rests upon expert training, a functional specialization of work, and an attitude set on habitual virtuosity in the mastery of single yet methodically integrated functions. If the apparatus stops working, or if its work is interrupted by force, chaos results, which it is difficult to master by improvised replacements from among the governed. This holds for public administration as well as for private economic management. Increasingly the material fate of the masses depends upon the continuous and correct functioning of the ever more bureaucratic organizations of private capitalism, and the idea of eliminating them becomes more and more utopian.

Increasingly, all order in public and private organizations is dependent on the system of files and the discipline of officialdom, that means, its habit of painstaking obedience within its wonted sphere of action. The latter is the more decisive element, however important in practice the files are. The naive idea of Bakuninism of destroying the basis of "acquired rights" together with "domination" by destroying the public documents overlooks that the settled orientation of *man* for observing the accustomed rules and regulations will survive independently of the documents. Every reorganization of defeated or scattered army units, as well as every restoration of an administrative order destroyed by revolts, panics, or other catastrophes, is effected by an appeal to this conditioned orientation, bred both in the officials and in the subjects, of obedient adjustment to such [social and political] orders. If the appeal is successful it brings, as it were, the disturbed mechanism to "snap into gear" again.

The objective indispensability of the once-existing apparatus, in connection with its peculiarly "impersonal" character, means that the mechanism—in contrast to the feudal order based upon personal loyalty—is easily made to work for anybody who knows how to gain control over it. A rationally ordered officialdom .continues to function smoothly after the enemy has occupied the territory; he merely needs to change the top officials. It continues to operate because it is to the vital interest of everyone concerned, including above all the enemy. After Bismarck had, during the long course of his years in power, brought his ministerial colleagues into unconditional bureaucratic dependence by eliminating all independent statesmen, he saw to his surprise that upon his resignation they continued to administer their offices unconcernedly and undismayedly, as if it had not been the ingenious lord and very creator of these tools who had left, but merely some individual figure in the bureaucratic machine which had been exchanged for some other figure. In spite of all the changes of masters in France since the time of the First Empire, the power apparatus remained essentially the same.

Such an apparatus makes "revolution," in the sense of the forceful creation of entirely new formations of authority, more and more impossible—technically, because of its control over the modern means of communication (telegraph etc.), and also because of its increasingly rationalized inner structure. The place of "revolutions" is under this process taken by *coups d'état*, as again France demonstrates in the classical manner since all successful transformations there have been of this nature. . . .

Discussion Questions

1. How can the rise of "new age" movements, extreme sports, religious fundamentalisms, and spiritual healers be explained in light of Weber's discussion of rationalization and the "disenchantment of the world"?

2. What are some of the essential differences between Weber's view of religion and Durkheim's? Which view better explains the role of religion in contemporary life?

3. In developing his ideal type of bureaucracy, Weber highlights the rational aspects of this organizational form. In what ways might bureaucracies exhibit "irrational" or inefficient features? How have bureaucracies "dehumanized" social life, transforming modernity into an iron cage?

4. Given Weber's three types of legitimate domination, the political system in the United States is best characterized as based on legal authority. What elements of the other types of authority can, nevertheless, still be found? How might political controversies result from the "illegitimate mixing" of different types of authority? Provide examples.

5. Following Weber's argument in the *Protestant Ethic and Spirit of Capitalism*, what role did the "calling" and outward signs of grace play in the development of capitalism? When capitalism was firmly established, what effect did it have on the religiously based ideas?

6. In what way(s) is Weber's analysis of class, status, and party different from Marx's understanding of class and social stratification? What are the implications of the difference(s) for designating the proletariat a revolutionary force for social change and for understanding the exercise of power in the United States?

5 CHARLOTTE PERKINS GILMAN (1860–1935)

Key Concepts

■ Gender inequality

■ Women's economic independence

The labor of women in the house, certainly, enables men to produce more than they otherwise could; and in this way women are economic factors in society. But so are horses.

(Gilman 1898/1998:7)

In 1980, the United Nations summed up the burden of **gender inequality**: women comprised half the world's population, did two-thirds of the world's work, earned one-tenth of the world's income, and owned one-hundredth of the world's property. In the past 30 years, not much has changed. Even in one of the most developed nations in the world, the United States, significant economic gender inequities continue: according to the United States Department of Labor, in 2000 American women earned approximately 77 percent of what men with similar educational and other qualifications earned. Imagine, then, what life was like for women in the *nineteenth* century, when, legally speaking, women were analogous to children, without the right either to own property or to vote.

In the nineteenth and early twentieth centuries, American and European women were not only legally prevented from owning property or voting, but they also were denied access to higher education. Yet, despite this absence of equality, since at least the 1800s a number of self-educated women have done remarkable scholarly work.[1] In this chapter, we consider one of these extraordinary women: Charlotte Perkins Gilman (1860–1935). Gilman was an accomplished writer, feminist, and sociologist—though in her own day she was not widely recognized as a sociologist. However, despite her lack of institutional credentials, there are a number of reasons for regarding Gilman as a "sociologist." As Deegan (1997:11) points out, Gilman

1. identified herself as a sociologist;

2. was identified by others as a sociologist;

3. taught sociology courses (though she declined an academic appointment in sociology, she was a frequent freelance lecturer on college campuses);

4. wrote sociological books and articles, some of which were published in the influential *American Journal of Sociology*; and

5. was a charter member of the American Sociological Society and remained a member for 25 years.

In addition, Gilman is a pivotal feminist theorist: she was one of the first to seek to explain *how* women and men came to have their respective societal roles and *why* societies developed gender inequalities. Significantly, the three main theoretical traditions from which Gilman drew—social Darwinism, symbolic interactionism, and Marxist theory (and their offshoots)—are the main traditions around which feminist social theories still revolve today. It is for these reasons, then, that we consider Gilman a core classical theorist.

▐▌ A BIOGRAPHICAL SKETCH

Charlotte Perkins was born on July 3, 1860, to Mary Wescott Perkins and Frederic Beecher Perkins. She was the third of four children, although the Perkins's first child died at birth and their fourth died in infancy. Charlotte's mother, Mary, was said to be an attractive woman who had had many suitors. Charlotte's father, Frederic, was part of the distinguished New England Beecher clan. His grandfather was the influential theologian Lyman Beecher; his uncle was the famous clergyman and abolitionist, Henry Ward Beecher; and his aunt was the famous abolitionist and author of *Uncle Tom's Cabin*, Harriet Beecher Stowe. Frederic Beecher Perkins himself was a librarian and writer of some distinction, but he abandoned the family soon after Charlotte was born. Though Frederic provided some financial support to his family, Charlotte and her brother and mother suffered financially.

The Perkins family moved 19 times in 18 years; 14 of the moves were from one city to another. Though a lack of money prevented Charlotte from receiving much in the way of a

---

[1] These women include Harriet Martineau (1802–1876, see Significant Others box below), Jane Addams (1860–1953), Anna Julia Cooper (1858–1964, see Significant Others box, pages 328–329, Chapter 7), Marianne Weber (1870–1954), and Beatrice Potter Webb (1858–1943). In addition to this brief list of "first wave" (i.e., nineteenth and early twentieth centuries) feminists who wrote sociologically about the origins and dynamics of gender inequality, are the many feminist activists devoted to remedying gender inequality, e.g., Sojourner Truth (1797–1883), Elizabeth Cady Stanton (1815–1902), Susan B. Anthony (1820–1906), and Ida B. Wells-Barnett (1862–1931).

Significant Others

Harriet Martineau (1802–76): The First Woman Sociologist

Harriet Martineau was born the sixth of eight children in a well-to-do English family. Her father was a successful textile manufacturer and a devout Unitarian whose relatively progressive views allowed Harriet to pursue academic subjects normally reserved for men. Despite the comforts her father's career afforded the family, Martineau's life was far from idyllic. By the age of 12, she was deaf, and she did not possess the senses of smell or taste. Her father died in 1826, and shortly thereafter the family business closed. During this period, her fiancé died as well, and Martineau was left without financial support.

As it would turn out, Martineau's misfortunes thrust her into the world of professional writing as a means of earning a living. In keeping with her upbringing, she began her career by publishing articles for the Unitarian journal, the *Monthly Repository*, as well as writing religious books. However, she quickly expanded her range by publishing works on political economy, sociology, and English history; novels; travel books; and children's stories. Moreover, she worked as a journalist for the *Daily News* where over the course of some 15 years she contributed more than 1,500 articles. In much of her writing and public life, she proved to be a staunch advocate for women's rights and the abolition of slavery. As for her more sociological writings, she examined many of the issues that would later occupy the attention of Marx, Durkheim, and Weber, including class relations, suicide, religion, and social science methodology. Her analyses of these subjects are particularly noteworthy, given that the discipline of sociology had not yet been born at the time of her writing. Of her works, two stand out for their impact on the field: *The Positive Philosophy of Auguste Comte* (1853), which is a translation and condensing of Comte's six-volume opus, and *Society in America* (1837). Considered by many to be a masterpiece study of American life, *Society in America* was based on Martineau's two-year stay in the United States. In it, she compared the democratic principles of equality and freedom on which the young nation was founded with its actual institutionalized practices. Certainly, were it not for the sexism of the Victorian era and its lingering effects in academia, Martineau's position in the discipline would be more secure.

Note: This account is based largely on Valerie K. Pichanick's *Harriet Martineau: The Woman and Her Work, 1802–76* (Ann Arbor: University of Michigan Press, 1980).

---

formal education, she attended the Rhode Island School of Design in 1878–79. The training she received there enabled her to earn money as a commercial artist. But most of Charlotte's education was gained through her voracious reading, some of which was overseen by her father. Although he was only an "occasional visitor," he sent Charlotte books and catalogues of books, as well as lists of books to read (Degler 1966:viii–ix; Gilman 1935/1972:5).

Charlotte was known as a very "willful" child. She had little interest in the pursuits deemed proper for a girl, and, as she matured, spurned the traditional roles assigned to women. She refused to play the part of the precious and frail coquette, choosing instead to exercise vigorously and develop her physical strength (Degler 1966:x).

Charlotte cherished her independence and vowed never to marry. However, a young artist named Charles Walter Stetson fell in love with Charlotte and proposed to her. Despite some initial reluctance, Charlotte agreed to marry him in 1884. But her reservations about marriage proved to be well founded. In the course of the first year of their marriage, she became increasingly and inexplicably despondent. The birth of a daughter, Katharine, just 10 and a half months after her wedding, seemed to make her depression even more severe. At the time, the prescribed remedy for women's melancholia was rest and no intellectual activity, which nearly drove the young, spirited, intellectually curious Perkins into madness. In a now-famous passage in her autobiography, Perkins described the "rest cure" that nearly did her in: "Live as domestic a life as possible. Have your child with you all the time. . . . Lie down an hour after each meal. Have but two hours' intellectual life a day. And never touch pen, brush or pencil again" (Gilman 1935/1972:96).

A trip away from her daughter and her husband made Perkins recognize that it was not rest but activity she needed. In 1890, Charlotte decided to separate from Stetson and move to California: "Better for that dear child [Katharine] to have separated parents than a lunatic mother," she wrote in her autobiography (ibid.:97).

Life in California was very difficult for Perkins. She undertook what work she could find—lecturing on women's rights, writing for small periodicals, and even managing a boarding house after her destitute mother came to live with her in Oakland (Degler 1966:xii). Yet, during this time, Perkins published several important works. In 1892, "The Yellow Wallpaper" was published. This poignant, semiautobiographical account of her experience with depression would become one of her most well known and highly acclaimed stories. The following year, Perkins's *In This Our World* was published. This sensational book of poems "enjoyed a near cult following in the United States and England" (Golden and Zangrando 2000:11).

In 1894, Charlotte Perkins and Charles Walter Stetson were divorced, and Stetson married one of Charlotte's closest friends, Grace Channing. Charlotte freely gave the couple her blessing, and she permitted her daughter, Katharine, to live with Stetson and his new wife. This was a flagrant departure from the accepted attitude of the day, and Perkins was pilloried in the newspapers for giving away both her husband and her child to another woman. Nevertheless, she remained on close terms with Stetson and Channing (Degler 1966:xi–xii).

With her daughter no longer her direct responsibility, Perkins expanded her lecturing tours from California to the rest of the nation. She became well known in women's suffrage circles for her provocative ideas about women's rights as well as other social issues. After the publication of *Women and Economics* in 1898, her reputation as a feminist was secured. In this book, which was translated into seven languages, Perkins denounced women's economic dependence on men and advocated for public day care and cooperative kitchens.

Not surprisingly, Perkins was allied with progressive political movements. Most noteworthy was her commitment to Fabian socialism, a brand of socialism that called for the collective ownership and democratic control of resources. While advocating equality for all, Fabian socialists, nevertheless, rejected the classical Marxist theory of revolutionary class struggle in favor of peaceful and gradual social change.

In conjunction with Fabian socialism, Perkins emphasized how the insular, nuclear family was dysfunctional for women. Gilman believed in communal kitchens and child care, rather than individual mothers "doing it all"—cooking, cleaning, and childrearing—while isolated in their own homes. Incidentally, this same inefficient individualism is readily apparent today, for instance, in the gridlock of SUVs in suburban school parking lots as mothers (and some fathers) arrive en masse—each in her (or his) own car—to drop off and pick up children.

In 1900, at the age of 40, Perkins married George Houghton Gilman, a man seven years her junior. Between her second marriage and 1914 Gilman reached the peak of her public activity and fame. She published nine books during this period, including *Concerning Children* (1900), *The Home* (1903), *Human Work* (1904), and *The Man-Made World* (1911). She also founded the journal *Forerunner* (1909–16), in which she published feminist stories and articles. In 1932, Gilman was diagnosed with breast cancer. She continued to write and lecture for a few more years, but, unfortunately, the cancer could not be arrested. On August 17, 1935, Charlotte Perkins Gilman took her own life, writing in her suicide note that she "preferred chloroform to cancer" (Degler 1966:xvii). Gilman's autobiography, *The Living of Charlotte Perkins Gilman: An Autobiography* (Gilman 1935/1972), was published shortly after her death.

INTELLECTUAL INFLUENCES AND CORE IDEAS

As noted above, Gilman's approach to gender reflects the basic building blocks that inform feminist theory to this day: She drew from a variety of theoretical wells including Marxism, symbolic interactionism, and social Darwinism. In this section, we outline her indebtedness to each of these frameworks. As you will see, her multidimensional theory of gender inequality combines (1) a Marxist emphasis on the economic and political basis for gender inequality, (2) a symbolic interactionist emphasis on how these gender differences are reinforced and institutionalized through the process of socialization, and (3) a sociobiological emphasis on the evolutionary advantages or roots of gender differences.

First, following the Marxist tradition, Gilman analyzed the political and economic factors that produce and reproduce gender inequality. Gilman sought to show that the division of labor of the traditional family (breadwinner husband/stay-at-home wife) was inherently problematic because it makes women economically dependent on men. To be sure, Gilman did not focus on the evils of capitalism as did Marx. However, just as Marx considered the system of capitalism inherently exploitative because workers do not own the means of production, Gilman considered the traditional family structure inherently exploitative because the economic compensation of women bears absolutely no relation to her labor. Regardless of how much work she actually does (or doesn't do) in the home, the housewife's social and economic standing comes from her husband. Thus, her labor belongs to her husband, not to her. This is the major point of the selection from *Women and Economics* that you will read below.

Second, in conjunction with the Chicago School and the symbolic interactionist tradition, Gilman emphasized how differential socialization leads to and sustains gender inequality. In doing so, she challenged the longstanding assumption that inherent biological differences precluded men and women from effectively pursuing overlapping social activities (Degler 1966:xii–xxiii). Instead, Gilman maintained that from the earliest age young girls were encouraged, if not forced, to act, think, look, and talk differently from boys, though their interests and capabilities at that age might be identical. For instance, Gilman states,

> [o]ne of the first things we force upon the child's dawning consciousness is the fact that he is a boy or that she is a girl, and that, therefore, each must regard everything from a different point of view. They must be dressed differently, not on account of their personal needs, which are exactly similar at this period, but so that neither they, nor any one beholding them, may for a moment forget the distinction of sex. (Gilman 1898/1998:28)

Ironically, differential gender socialization is not only evident today as in Gilman's time—in some ways, it is even *more* apparent. Because of consumerism and marketing, as well as the fact that parents often find out the sex of their child even before he or she is born, there is a huge array of gendered baby paraphernalia (up to and including pink and blue diapers). Today, children are subjected to extensive marketing campaigns, most notably by fast food restaurants and toy manufacturers, which from a very early age "force upon the child" a sense of being a "boy" or a "girl." All one need do is stroll down the toy aisle at a local department store to see how children are not taught to view themselves (and each other) as *human beings* but rather are taught to view themselves (and each other) as "boys" and "girls."

However, despite her emphasis on differential socialization, Gilman did not deny that biological differences exist between men and women. On the contrary, Gilman borrowed theoretically not only from Fabian socialism (with its origins in Marxism) and symbolic interactionism, but also from social Darwinism. Social Darwinists, such as Herbert Spencer

Photo 5.1   One of the Most Gendered Sites Today: The Toy Store

in England and William Graham Sumner in America, applied Darwin's theory of evolution to human societies and maintained that human existence was based on "survival of the fittest." Like many social Darwinists, Gilman was fascinated by the animal world. She used animal analogies to explain the human condition as well as biological and behavioral differences between the sexes. Specifically, Gilman contended that women and men, in general, have different biological "principles" to which they adhere. She maintained that women's unique capabilities—particularly their love and concern for others—have tremendous social value, though they are grossly underappreciated.

Indeed, Gilman went so far as to assert the natural superiority of the female sex. She enthusiastically endorsed the "scientifically" based gynocentric theory promoted by the Harvard sociologist Lester Ward, suggesting that that the civilizing capacities of women could compensate for the destructive combativeness of men. As Gilman notes,

> [t]he innate underlying difference [between the sexes] is one of principle. On the one hand, the principle of struggle, conflict, and competition. . . . On the other, the principle of growth, of culture, of applying services and nourishment in order to produce improvement. (As cited in Hill 1989:45)

Thus, Gilman maintained that, in contrast to men, women did not want to fight, to take, to oppress, but rather to love. Women exhibited "the growing altruism of work, founded in mother love, in the antiselfish instinct of reproduction" (ibid.).

However, as the acclaimed Gilman historian Mary Hill (1989:45) notes, Gilman's view of women as "saintly givers" and men as "warring beasts" is problematic: when she glorifies the female "instincts" of love and service, her radical feminist theory dissolves into a "sentimental worship of the status quo." Her insistence on the "giving" nature of women as compared to the "combative" nature of men seems to indicate that women must be and should be the primary caretakers of children, and that there is only so far a man can go in his role as nurturer. In short, Gilman's biological determinism seems quite antiquated today. Many contemporary fathers are far more nurturing toward their children than were their fathers, though they have the same basic biological constitution.

Despite significant shortcomings in her biological arguments, however, Gilman raised interesting issues that today are being explored by brain researchers. Contemporary neuro-scientists, geneticists, evolutionary psychologists, and others are breathing new life into some of the same gender differences that Gilman noted more than a century ago. For instance, researchers today find that females tend to be more highly sensitive to touch, sound, and smell than males. In one study of day-old infants, researchers found that although both male and female infants reacted most intensely to the sound of another's trouble (as opposed to other sounds), infant females reacted more strongly than did males. The researchers suggested that the infant girls were more finely attuned to an empathetic response, and that this sensitivity would run like an "underground stream" throughout their entire lives (Blum 1997:66–67).

Most importantly, in contrast to the static "nature-versus-nurture" debates of the past, contemporary neurological research illuminates the interconnectedness of nature and nur-ture. Researchers now are investigating the complex ways in which the environment sparks significant neurological or chemical changes and developments, and vice versa. Nature is understood to be more malleable and more of a process than social Darwinists ever imag-ined. For example, even genetic instructions that are often thought to be determinate, such as height, are known to be influenced by the environment. A baby whose genes blueprint him to grow six feet tall but who does not receive adequate nutrition, will fall short by an inch or more. Some researchers even suggest that stress in childhood can interfere with height by suppressing growth hormones (ibid.:21–22). To be sure, it is outside the scope of this chapter to delve into these highly complex and contentious neurological arguments, but the point is that the questions about nature versus nurture that Gilman raised continue to be at the heart of gender theory and research today.

Far more problematic than her biological arguments about gender were Gilman's biolog-ical arguments about race. Drawing not only on social Darwinist theories of "survival of the fittest," but also on the "commonsense" notions of "manifest destiny" and the "white man's burden" dominant in her day (see Chapter 7; and Edles 2002:9), Gilman made patently racist remarks. For instance, she maintained that some races

> combine well, making a good blend, [but] some do not. We are perfectly familiar in this country with the various blends of black and white, and the wisest of both races prefer the pure stock. The Eurasian mixture is generally considered unfortunate by most observers. (Gilman 1923; see also Scharnhorst 2000:69)

Less insidiously, but no less problematic, Gilman also implicitly assumed she was speak-ing about all women when she was really referring to white women. Certainly, Gilman did not discuss or consider the resources or situations of nonwhite women (though, to be sure, this implicit privileging of "white lives" was typical of *all* of sociology's white classical figures).

For instance, in her discussion of the isolation of traditional housewives, Gilman ignored the fact that African American women may have had traditions of support and community that white women did not (see Stack 1974). And, of course, the entire notion of the isolated stay-at-home mom is out of sync with the fact that paid work has long been "an important and valued dimension of Afrocentric definitions of Black motherhood" (Collins 1987:5). Yet, despite her racist assumptions, Gilman often made pointedly antiracist comments too. She decried slavery and the continued oppression of African Americans, and she spoke out against the genocide and ill treatment of Native Americans and Native Hawaiians.

▪▪ GILMAN'S THEORETICAL ORIENTATION

Given that Gilman melds the distinct traditions of neo-Marxism, symbolic interactionism, and social Darwinism, it should come as no surprise that Gilman's approach to gender is theoretically multidimensional. Specifically, as shown in Figure 5.2, Gilman highlights differential gender socialization (i.e., how boys and girls are taught to *behave* differently) as well as the distinct sex "principles" with which they are born. These concerns reflect a nonrational orientation to action primarily at the individual level. At the collective level, she highlights both (rationalist) political and economic structures—most importantly, those that prohibit **women's economic independence**—and the (nonrationalist) normative, symbolic structures or codes that ensure differential socialization.

Nevertheless, in her sociological work, especially *Women and Economics*, it is the structural, institutional basis of inequality with which Gilman is most concerned. Hence, as shown in Figure 5.1, we situate her at the more rational/collective end of the theoretical continuum. As indicated previously, Figure 5.1 and the related figures in other chapters reflect our perception of each major theorist's *overall* or most basic theoretical orientation. The point is not that each theorist is situated in a particular "box." Rather, each of these figures is a heuristic device with which to compare and contrast the theoretical orientation of each sociologist discussed in this book (or any other theorist). These positions can and should be discussed and contested; that is why there are no fixed points in these figures.

In any case, Gilman's particular theoretical approach (as well as her similarities with Marx) is readily apparent in the following passage on women's corsets, which can be considered a metaphor for the general constraints placed on women:

Figure 5.1   Gilman's Basic Theoretical Orientation

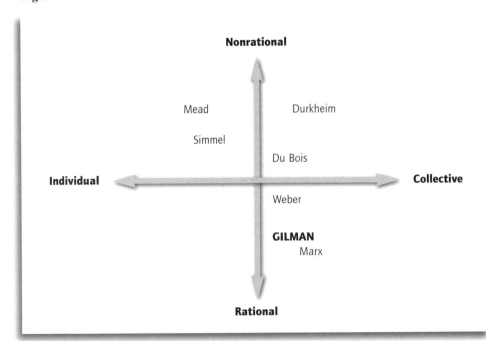

Figure 5.2 Gilman's Multidimensional Explanation of Gender Inequality

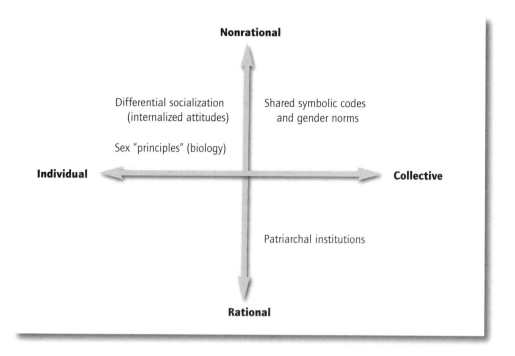

Put a corset, even a loose one, on a vigorous man or woman who never wore one, and there is intense discomfort, and a vivid consciousness thereof. The healthy muscles of the trunk resent the pressure, the action of the whole body is checked in the middle, the stomach is choked, the process of digestion is interfered with; and the victim says, "how can you bear such a thing?" (Gilman 1898/1998:40)

Just as the corset "chokes" the stomach, so, too, do the traditional institutional features of the family "choke" women. Just as "healthy muscles" resent the "pressure," so, too, do healthy women. In short, the metaphor of the corset reflects that the constraints placed on women originate outside her. As such, Gilman views these as external pressures, thus pointing to the rationalist aspect of her theory.

Yet, at the same time, Gilman argues that women learn to accept and internalize such pressures. In fact, women are so indoctrinated that they resist their own "freedom." As Gilman goes on to note,

[b]ut the person habitually wearing a corset does not feel these evils. They exist, assuredly, the facts are there, the body is not deceived; but the nerves have become accustomed to these disagreeable sensations, and no longer respond to them. The person "does not feel it." In fact, the wearer becomes so used to the sensations that when they are removed,—with the corset,—there is a distinct sense of loss and discomfort. (Gilman 1898/1998:40)

Thus, Gilman's metaphor of the corset is similar to Marx's notion of false consciousness. In both cases, "the facts are there"—the inequality is there—but the person "does not feel

Photo 5.2 In this famous scene from the 1939 film *Gone with the Wind*, Scarlett O'Hara (played by Vivien Leigh) insists that her corset be pulled even tighter to achieve her enviable 18-inch waist. The juxtaposition of white Scarlett O'Hara and "her" black "Mammy" (played by Hattie McDaniel) reflects the racial as well as class dimensions of this standard of beauty.

it"; he does not see or know of it. She has internalized the pressures and constraints as her own. This view reflects Gilman's incorporation of a more nonrationalist theoretical position. In addition, Gilman acknowledges that women may be resistant to developing a "true" consciousness because of the safety and familiarity provided by false consciousness. Gilman argues that women not only accept but also believe in the legitimacy of the traditional division of labor; thus, it should be no surprise that they feel discomfited by any other reality.

In terms of order, Gilman acknowledges that social patterns exist prior to any individual; that is, the traditional gendered division of labor is a long-established social structure. However, she also emphasizes that it is in individual interactions that this social structure is maintained. On the other hand, it is at the individual level of "free will" that such social structures can be resisted and changed. Yet, the exertion of individual will against existing conditions or the forces of "natural law" is not a common trait. Instead, it takes an "advanced" individual to see through the taken-for-granted symbolic codes that legitimate social hierarchies and oppression. As Gilman remarks,

[i]n the course of social evolution there are developed individuals so constituted as not to fit existing conditions, but to be organically adapted to more advanced conditions. These advanced individuals respond in sharp and painful consciousness to existing conditions, and cry out against them according to their lights. The history of religion, of political and social reform, is full of familiar instances of this. The heretic, the reformer, the agitator, these feel what their compeers do not, see what they do not, and naturally, say what they do not. The mass of the people are invariably displeased by the outcry of these uneasy

spirits. In simple primitive periods they were promptly put to death. (Gilman 1898/1998:41)

The autobiographical bent of this passage is readily apparent. Much more so than other social scientists who challenged existing dogma, feminists such as Gilman challenged and threatened core, sacred "family values." As indicated previously, Gilman was perceived as "giving away" both her husband and her daughter, and Gilman was publicly skewered for her actions—not only for her beliefs. No doubt Gilman felt very much like the isolated heretic.

Nevertheless, theoretically, Gilman's assumption that social change rests entirely at the level of the individual is problematic. Gilman did not recognize that it was not solely individual "free will" or social "advancement," but the examination and study of alternative social and cultural schemas (e.g., Fabian socialism) that allowed her to see through and cast off the "corsets" of her day. Moreover, because intellectuals are no different from anyone else, except that they may have more intellectual "wells" from which to draw, Gilman fails to acknowledge that there might be a wealth of other social "corsets" that the intellectual herself never perceives and consequently never rejects. For instance, as we have seen, in some of her essays that draw on social Darwinism, Gilman reinforces some of the racist assumptions common in her day. In sum, *all* individuals are necessarily affected by their environment, not just those who are not "advanced." But social change can arise when people—intellectuals or not—begin to see their lives in a new way. This process occurs not just because of revelations at the level of the individual, but because of new social and cultural conditions. The virtue of social theorists, and intellectuals and social activists more generally, is that they can produce and disseminate new ideas and symbolic schemes that potentially can lead to positive social change.

Readings

In this section, you will read excerpts from Gilman's most influential fiction as well as nonfiction. We begin with Gilman's pivotal semi-autobiographical story, "The Yellow Wallpaper," first published in 1892. The story is followed by a brief statement that Gilman wrote in 1913, in response to the many queries she fielded about why she wrote "The Yellow Wallpaper." We then turn to Gilman's major nonfiction sociological treatise, *Women and Economics: A Study of the Economic Relation Between Men and Women as a Factor in Social Evolution* (1898).

Introduction to "The Yellow Wallpaper"

"The Yellow Wallpaper" is a chilling, semiautobiographical story about a woman's descent into madness. It was first published in *The New England Magazine* in 1892, after having been rejected by numerous other publishers. In the 1890s, the notion that literature should be morally uplifting was dominant, and editors such as A. E. Scudder of the *Atlantic Monthly* found the story entirely too depressing for publication (Hedges 1973:40). Indeed, as you will see, the poignancy of "The Yellow Wallpaper" lies in its directness. The reader experiences the protagonist's mental breakdown from the inside out.

In the late nineteenth and early twentieth centuries, white middle-class women were discouraged from developing either their minds or their bodies. Physically as well as intellectually, they were cherished for being "childlike" and "fragile." Gilman's main point in "The Yellow Wallpaper" is that these constraints placed on women simply because they are women drive healthy, independent women to insanity.

Though it initially received mixed reviews, "The Yellow Wallpaper" soon became widely read and known, and it was republished numerous times. In the 1970s, "The Yellow Wallpaper" received a "second wave" of acclaim as it became a keystone in newly developing women's studies programs.

Ironically, however, when it was first printed no one seems to have made the connection between the main character's descent into madness and society's gender roles. As Hedges (1973:39) notes, "The Yellow Wallpaper" is one of the rare pieces of literature in the nineteenth century to directly confront the sexual politics of male-female and husband-wife relationships. Yet it is extremely significant that, on its debut, this linkage between the "personal" and the "political" was overlooked. In this sense, this story was ahead of its time. It was not until Gilman's work was rediscovered in the 1960s and 1970s that Gilman's eloquent insights into the intertwined nature of the personal and the political would be completely appreciated.

As you will shortly see, the main character in "The Yellow Wallpaper" is prohibited from intellectual activity and kept socially isolated in order to cure her of her "hysteria." This is one of the critical "semiautobiographical" aspects of this piece. Gilman wrote the story shortly after recovering from a severe case of postpartum depression—a depression she overcame precisely by *disobeying* the advice of her well-known and respected physician, Dr. S. Weir Mitchell, who had prescribed for her the "rest cure" (Wagner-Martin 1989:52). Mitchell's treatment for Gilman included seclusion, immobility, and overfeeding. Isolated for up to six weeks, Gilman was prohibited from using her mind toward any intellectual pursuits of any sort. In her autobiography, Gilman attests to how the rest cure drove her to the brink of insanity.

The "rest cure"—at the heart of both Gilman's personal descent into madness and that of her protagonist in "The Yellow Wallpaper"—was the most widely accepted treatment for "female" ailments in the Victorian era. In this story (as well as in her other works), Gilman poignantly demonstrates that this social isolation has dire mental consequences for women. Interestingly, Gilman's main point—that social bonds are absolutely essential to mental health—coincides with that made by Durkheim in *Suicide* (see Chapter 3). Recall that Durkheim maintained that social and moral isolation could in severe cases result in suicide. Recall, too, that Durkheim found that women had lower rates of suicide than men because they were more likely to have tighter social bonds—through both formal religion and informal social relationships. Yet, ironically, American doctors in the nineteenth and early twentieth centuries attempted to cure "hysterical" women by disallowing social intimacy.

A second, vital, interrelated feminist point contained in "The Yellow Wallpaper" concerns masculinist logic. In the story, truth is the domain both of science and of men. The protagonist in the story is at the mercy of her husband not only because he is "The Man," but also because he is "The Doctor." This dual legitimacy means that it is *his*—and only his—assessment of the protagonist's health and treatment that counts. The undergirding assumption is that women's finely tuned emotions and sensibility prevent her from rational, scientific thought. As Gilman (1892/1973:10) states,

> If a physician of high standing, and one's own husband, assures friends and relatives that there is really nothing the matter with one but temporary nervous depression—a slight hysterical tendency—what is one to do?

> My brother is also a physician, and also of high standing, and he says the same thing.

> So I take phosphates or phospites—whichever it is, and tonics, and journeys, and air, and exercise, and am absolutely forbidden to "work" until I am well again.

> Personally, I disagree with their ideas.

Personally, I believe that congenial work, with excitement and change, would do me good.

But what is one to do?

Most important, the assumed intellectual inferiority of women places them in a childlike relationship with their husbands. This childlike status is revealed in the above quote (as well as in numerous other places in the story) with the phrase "So I take phosphates or phospites—*whichever it is*" (emphasis added). Like the "good" child who dutifully carries out her parents' wishes, the woman in the story is supposed to do what she's told by her doctor/husband, regardless of whether or not she agrees—or understands—at all. In other words, the problem with patriarchy is not only that, at the rational level, women are not able to *do* as they please. The even more profound problem is at the level of the *nonrational*: women are not encouraged to *think.* Nowhere is this interrelation between the rationalist and nonrationalist levels of oppression more readily apparent than in suffrage. The dominant nineteenth-century attitude toward women's suffrage was simply that it was not necessary. There was simply no need for women to vote, because each household would "best" be represented by the vote of the husband/father.

Yet, one of the most insidious aspects of patriarchy is that, despite the assumption that women were "incapable" of rational, scientific, logical thought, women were not valued for their "feminine" ways of knowing either. Rational, logical, scientific thought was (is?) deemed *superior* to intuition, emotion, or feeling, and the latter "ways of knowing" were (are?) roundly eschewed. Of course, since the 1960s, feminists have forcefully challenged this assumption. Thus, for instance, the medium of the fictional, semiautobiographical short story became valued as a source of "truth." This brings us to the issue of why this *fictional* short story is included in this *theory* reader. By including this story in this volume, we are aligning ourselves with the idea that formally nontheoretical material can have theoretical relevance. In addition, however, we include this story in this volume for an important historical reason. Women were largely prohibited, both formally and informally, from becoming scientists and sociologists in the nineteenth and early twentieth centuries. As a result, many women with provocative sociological theories turned to other means (e.g., the short story) for expressing their ideas. Thus, in order to gain access to their ideas, we need to include such stories as "The Yellow Wallpaper." We cannot leave theory to the institutionally acknowledged "scientists" if we want to uncover nineteenth- and early twentieth–century women's thoughts, even if these ideas are expressed in modes that seem contrary to "standard" sociological theorizing.

"The Yellow Wallpaper" (1892)

Charlotte Perkins Gilman

It is very seldom that mere ordinary people like John and myself secure ancestral halls for the summer.

A colonial mansion, a hereditary estate, I would say a haunted house and reach the height of romantic felicity—but that would be asking too much of fate!

Still I will proudly declare that there is something queer about it.

Else, why should it be let so cheaply? And why have stood so long untenanted?

John laughs at me, of course, but one expects that.

John is practical in the extreme. He has no patience with faith, an intense horror of superstition, and he scoffs openly at any talk of things not to be felt and seen and put down in figures.

SOURCE: From *The Yellow Wallpaper* by Charlotte Perkins Gilman. Originally appeared in the January 1892 issue of *The New England Magazine*. First published in book form in 1899 by The Feminist Press.

John is a physician, and *perhaps*—(I would not say it to a living soul, of course, but this is dead paper and a great relief to my mind)—*perhaps* that is one reason I do not get well faster.

You see, he does not believe I am sick! And what can one do?

If a physician of high standing, and one's own husband, assures friends and relatives that there is really nothing the matter with one but temporary nervous depression—a slight hysterical tendency—what is one to do?

My brother is also a physician, and also of high standing, and he says the same thing.

So I take phosphates or phospites—whichever it is—and tonics, and air and exercise, and journeys, and am absolutely forbidden to "work" until I am well again.

Personally, I disagree with their ideas.

Personally, I believe that congenial work, with excitement and change, would do me good.

But what is one to do?

I did write for a while in spite of them; but it *does* exhaust me a good deal—having to be so sly about it, or else meet with heavy opposition.

I sometimes fancy that in my condition, if I had less opposition and more society and stimulus—but John says the very worst thing I can do is to think about my condition, and I confess it always makes me feel bad.

So I will let it alone and talk about the house.

The most beautiful place! It is quite alone, standing well back from the road, quite three miles from the village. It makes me think of English places that you read about, for there are hedges and walls and gates that lock, and lots of separate little houses for the gardeners and people.

There is a *delicious* garden! I never saw such a garden—large and shady, full of box-bordered paths, and lined with long grape-covered arbors with seats under them.

There were greenhouses, but they are all broken now.

There was some legal trouble, I believe, something about the heirs and coheirs; anyhow, the place has been empty for years.

That spoils my ghostliness, I am afraid, but I don't care—there is something strange about the house—I can feel it.

I even said so to John one moonlight evening, but he said what I felt was a *draught*, and shut the window.

I get unreasonably angry with John sometimes. I'm sure I never used to be so sensitive. I think it is due to this nervous condition.

But John says if I feel so I shall neglect proper self-control; so I take pains to control myself—before him, at least, and that makes me very tired.

I don't like our room a bit. I wanted one downstairs that opened onto the piazza and had roses all over the window, and such pretty old-fashioned chintz hangings! but John would not hear of it.

He said there was only one window and not room for two beds, and no near room for him if he took another.

He is very careful and loving, and hardly lets me stir without special direction.

I have a schedule prescription for each hour in the day; he takes all care from me, and so I feel basely ungrateful not to value it more.

He said he came here solely on my account, that I was to have perfect rest and all the air I could get. "Your exercise depends on your strength, my dear," said he, "and your food somewhat on your appetite; but air you can absorb all the time." So we took the nursery at the top of the house.

It is a big, airy room, the whole floor nearly, with windows that look all ways, and air and sunshine galore. It was nursery first, and then playroom and gymnasium, I should judge, for the windows are barred for little children, and there are rings and things in the walls.

The paint and paper look as if a boys' school had used it. It is stripped off—the paper—in great patches all around the head of my bed, about as far as I can reach, and in a great place on the other side of the room low down. I never saw a worse paper in my life.

One of those sprawling, flamboyant patterns committing every artistic sin.

It is dull enough to confuse the eye in following, pronounced enough constantly to irritate and provoke study, and when you follow the lame uncertain curves for a little distance they suddenly commit suicide—plunge off at outrageous angles, destroy themselves in unheard-of contradictions.

The color is repellent, almost revolting: a smouldering unclean yellow, strangely faded by the slow-turning sunlight.

It is a dull yet lurid orange in some places, a sickly sulphur tint in others.

No wonder the children hated it! I should hate it myself if I had to live in this room long.

There comes John, and I must put this away—he hates to have me write a word.

We have been here two weeks, and I haven't felt like writing before, since that first day.

I am sitting by the window now, up in this atrocious nursery, and there is nothing to hinder my writing as much as I please, save lack of strength.

John is away all day, and even some nights when his cases are serious.

I am glad my case is not serious!

But these nervous troubles are dreadfully depressing.

John does not know how much I really suffer. He knows there is no *reason* to suffer, and that satisfies him.

Of course it is only nervousness. It does weigh on me so not to do my duty in any way!

I meant to be such a help to John, such a real rest and comfort, and here I am a comparative burden already!

Nobody would believe what an effort it is to do what little I am able—to dress and entertain, and order things.

It is fortunate Mary is so good with the baby. Such a dear baby!

And yet I *cannot* be with him, it makes me so nervous.

I suppose John never was nervous in his life. He laughs at me so about this wallpaper!

At first he meant to repaper the room, but afterward he said that I was letting it get the better of me, and that nothing was worse for a nervous patient than to give way to such fancies.

He said that after the wallpaper was changed it would be the heavy bedstead, and then the barred windows, and then that gate at the head of the stairs, and so on.

"You know the place is doing you good," he said, "and really, dear, I don't care to renovate the house just for a three months' rental."

"Then do let us go downstairs," I said. "There are such pretty rooms there."

Then he took me in his arms and called me a blessed little goose, and said he would go down cellar, if I wished, and have it whitewashed into the bargain.

But he is right enough about the beds and windows and things.

It is as airy and comfortable a room as anyone need wish, and, of course, I would not be so silly as to make him uncomfortable just for a whim.

I'm really getting quite fond of the big room, all but that horrid paper.

Out of one window I can see the garden—those mysterious deep-shaded arbors, the riotous old-fashioned flowers, and bushes and gnarly trees.

Out of another I get a lovely view of the bay and a little private wharf belonging to the estate. There is a beautiful shaded lane that runs down there from the house. I always fancy I see people walking in these numerous paths and arbors, but John has cautioned me not to give way to fancy in the least. He says that with my imaginative power and habit of story-making, a nervous weakness like mine is sure to lead to all manner of excited fancies, and that I ought to use my will and good sense to check the tendency. So I try.

I think sometimes that if I were only well enough to write a little it would relieve the press of ideas and rest me.

But I find I get pretty tired when I try.

It is so discouraging not to have any advice and companionship about my work. When I get really well, John says we will ask Cousin Henry and Julia down for a long visit; but he says he would as soon put fireworks in my pillow-case as to let me have those stimulating people about now.

I wish I could get well faster.

But I must not think about that. This paper looks to me as if it *knew* what a vicious influence it had!

There is a recurrent spot where the pattern lolls like a broken neck and two bulbous eyes stare at you upside down.

I get positively angry with the impertinence of it and the everlastingness. Up and down and sideways they crawl, and those absurd unblinking eyes are everywhere. There is one place where two breaths didn't match, and the eyes go all up and down the line, one a little higher than the other.

I never saw so much expression in an inanimate thing before, and we all know how much expression they have! I used to lie awake as a child and get more entertainment and terror out of blank walls and plain furniture than most children could find in a toy-store.

I remember what a kindly wink the knobs of our big, old bureau used to have, and there was one chair that always seemed like a strong friend.

I used to feel that if any of the other things looked too fierce I could always hop into that chair and be safe.

The furniture in this room is no worse than inharmonious, however, for we had to bring it all from downstairs. I suppose when this was used as a playroom they had to take the nursery things out, and no wonder! I never saw such ravages as the children have made here.

The wallpaper, as I said before, is torn off in spots, and it sticketh closer than a brother—they must have had perseverance as well as hatred.

Then the floor is scratched and gouged and splintered, the plaster itself is dug out here and there, and this great heavy bed, which is all we found in the room, looks as if it had been through the wars.

But I don't mind it a bit—only the paper.

There comes John's sister. Such a dear girl as she is, and so careful of me! I must not let her find me writing.

She is a perfect and enthusiastic housekeeper, and hopes for no better profession. I verily believe she thinks it is the writing which made me sick!

But I can write when she is out, and see her a long way off from these windows.

There is one that commands the road, a lovely shaded winding road, one that just looks off over the country. A lovely country, too, full of great elms and velvet meadows.

This wallpaper has a kind of sub-pattern in a different shade, a particularly irritating one, for you can only see it in certain lights, and not clearly then.

But in the places where it isn't faded and where the sun is just so—I can see a strange, provoking, formless sort of figure, that seems to skulk about behind that silly and conspicuous front design.

There's sister on the stairs!

Well, the Fourth of July is over! The people are all gone, and I am tired out. John thought it might do me good to see a little company, so we just had Mother and Nellie and the children down for a week.

Of course I didn't do a thing. Jennie sees to everything now.

But it tired me all the same.

John says if I don't pick up faster he shall send me to Weir Mitchell in the fall.

But I don't want to go there at all. I had a friend who was in his hands once, and she says he is just like John and my brother, only more so!

Besides, it is such an undertaking to go so far.

I don't feel as if it was worthwhile to turn my hand over for anything, and I'm getting dreadfully fretful and querulous.

I cry at nothing, and cry most of the time.

Of course I don't when John is here, or anybody else, but when I am alone.

And I am alone a good deal just now. John is kept in town very often by serious cases, and Jennie is good and lets me alone when I want her to.

So I walk a little in the garden or down that lovely lane, sit on the porch under the roses, and lie down up here a good deal.

I'm getting really fond of the room in spite of the wallpaper. Perhaps *because* of the wallpaper.

It dwells in my mind so!

I lie here on this great immovable bed—it is nailed down, I believe—and follow that pattern about by the hour. It is as good as gymnastics, I assure you. I start, we'll say, at the bottom, down in the corner over there where it has not been touched, and I determine for the thousandth time that I *will* follow that pointless pattern to some sort of a conclusion.

I know a little of the principle of design, and I know this thing was not arranged on any laws of radiation, or alternation, or repetition, or symmetry, or anything else that I ever heard of.

It is repeated, of course, by the breadths, but not otherwise.

Looked at in one way, each breadth stands alone; the bloated curves and flourishes—a kind of "debased Romanesque" with *delirium tremens*—go waddling up and down in isolated columns of fatuity.

But, on the other hand, they connect diagonally, and the sprawling outlines run off in great slanting waves of optic horror, like a lot of wallowing seaweeds in full chase.

The whole thing goes horizontally, too, at least it seems so, and I exhaust myself trying to distinguish the order of its going in that direction.

They have used a horizontal breadth for a frieze, and that adds wonderfully to the confusion.

There is one end of the room where it is almost intact, and there, when the crosslights fade and the low sun shines directly upon it,

I can almost fancy radiation after all—the interminable grotesque seems to form around a common center and rush off in headlong plunges of equal distraction.

It makes me tired to follow it. I will take a nap, I guess.

I don't know why I should write this.

I don't want to.

I don't feel able.

And I know John would think it absurd. But I *must* say what I feel and think in some way—it is such a relief!

But the effort is getting to be greater than the relief.

Half the time now I am awfully lazy, and lie down ever so much.

John says I mustn't lose my strength, and has me take cod liver oil and lots of tonics and things, to say nothing of ale and wine and rare meat.

Dear John! He loves me very dearly, and hates to have me sick. I tried to have a real earnest reasonable talk with him the other day, and tell him how I wish he would let me go and make a visit to Cousin Henry and Julia.

But he said I wasn't able to go, nor able to stand it after I got there; and I did not make out a very good case for myself, for I was crying before I had finished.

It is getting to be a great effort for me to think straight. Just this nervous weakness, I suppose.

And dear John gathered me up in his arms, and just carried me upstairs and laid me on the bed, and sat by me and read to me till it tired my head.

He said I was his darling and his comfort and all he had, and that I must take care of myself for his sake, and keep well.

He says no one but myself can help me out of it, that I must use my will and self-control and not let any silly fancies run away with me.

There's one comfort—the baby is well and happy, and does not have to occupy this nursery with the horrid wallpaper.

If we had not used it, that blessed child would have! What a fortunate escape! Why, I wouldn't have a child of mine, an impressionable little thing, live in such a room for worlds.

I never thought of it before, but it is lucky that John kept me here after all; I can stand it so much easier than a baby, you see.

Of course I never mention it to them any more—I am too wise—but I keep watch of it all the same.

There are things in that wallpaper that nobody knows about but me, or ever will.

Behind that outside pattern the dim shapes get clearer every day.

It is always the same shape, only very numerous.

And it is like a woman stooping down and creeping about behind that pattern. I don't like it a bit. I wonder—I begin to think—I wish John would take me away from here!

It is so hard to talk with John about my case, because he is so wise, and because he loves me so.

But I tried it last night.

It was moonlight. The moon shines in all around just as the sun does.

I hate to see it sometimes, it creeps so slowly, and always comes in by one window or another.

John was asleep and I hated to waken him, so I kept still and watched the moonlight on that undulating wallpaper till I felt creepy.

The faint figure behind seemed to shake the pattern, just as if she wanted to get out.

I got up softly and went to feel and see if the paper *did* move, and when I came back John was awake.

"What is it, little girl?" he said. "Don't go walking about like that—you'll get cold."

I thought it was a good time to talk, so I told him that I really was not gaining here, and that I wished he would take me away.

"Why, darling!" said he. "Our lease will be up in three weeks, and I can't see how to leave before.

"The repairs are not done at home, and I cannot possibly leave town just now. Of course, if you were in any danger, I could and would, but you really are better, dear, whether you can see it or not. I am a doctor, dear, and I know. You are gaining flesh and color, your appetite is better, I feel really much easier about you."

"I don't weigh a bit more," said I, "nor as much; and my appetite may be better in the evening when you are here but it is worse in the morning when you are away!"

"Bless her little heart!" said he with a big hug. "She shall be as sick as she pleases! But now let's improve the shining hours by going to sleep, and talk about it in the morning!"

"And you won't go away?" I asked gloomily.

"Why, how can I, dear? It is only three weeks more and then we will take a nice little trip of a few days while Jennie is getting the house ready. Really, dear, you are better!"

"Better in body perhaps—" I began, and stopped short, for he sat up straight and looked at me with such a stern, reproachful look that I could not say another word.

"My darling," said he, "I beg of you, for my sake and for our child's sake, as well as for your own, that you will never for one instant let that idea enter your mind! There is nothing so dangerous, so fascinating, to a temperament like yours. It is a false and foolish fancy. Can you not trust me as a physician when I tell you so?"

So of course I said no more on that score, and we went to sleep before long. He thought I was asleep first, but I wasn't, and lay there for hours trying to decide whether that front pattern and the back pattern really did move together or separately.

On a pattern like this, by daylight, there is a lack of sequence, a defiance of law, that is a constant irritant to a normal mind.

The color is hideous enough, and unreliable enough, and infuriating enough, but the pattern is torturing.

You think you have mastered it, but just as you get well under way in following, it turns a back-somersault and there you are. It slaps you in the face, knocks you down, and tramples upon you. It is like a bad dream.

The outside pattern is a florid arabesque, reminding one of a fungus. If you can imagine a toadstool in joints, an interminable string of toadstools, budding and sprouting in endless convolutions—why, that is something like it.

That is, sometimes!

There is one marked peculiarity about this paper, a thing nobody seems to notice but myself, and that is that it changes as the light changes.

When the sun shoots in through the east window—I always watch for that first long, straight ray—it changes so quickly that I never can quite believe it.

That is why I watch it always.

By moonlight—the moon shines in all night when there, is a moon—I wouldn't know it was the same paper.

At night in any kind of light, in twilight, candlelight, lamplight, and worst of all by moonlight, it becomes bars! The outside pattern, I mean, and the woman behind it is as plain as can be.

I didn't realize for a long time what the thing was that showed behind, that dim sub-pattern, but now I am quite sure it is a woman.

By daylight she is subdued, quiet. I fancy it is the pattern that keeps her so still. It is so puzzling. It keeps me quiet by the hour.

I lie down ever so much now. John says it is good for me, and to sleep all I can.

Indeed he started the habit by making me lie down for an hour after each meal.

It is a very bad habit, I am convinced, for you see, I don't sleep.

And that cultivates deceit, for I don't tell them I'm awake—oh, no!

The fact is I am getting a little afraid of John.

He seems very queer sometimes, and even Jennie has an inexplicable look.

It strikes me occasionally, just as a scientific hypothesis,— that perhaps it is the paper!

I have watched John when he did not know I was looking, and come into the room suddenly on the most innocent excuses, and I've caught him several times *looking at the paper!* And Jennie too. I caught Jennie with her hand on it once.

She didn't know I was in the room, and when I asked her in a quiet, a very quiet voice, with the most restrained manner possible, what she was doing with the paper—she turned around as if she had been caught stealing, and looked quite angry—asked me why I should frighten her so!

Then she said that the paper stained everything it touched, that she had found yellow smooches on all my clothes and John's, and she wished we would be more careful!

Did not that sound innocent? But I know she was studying that pattern, and I am determined that nobody shall find it out but myself!

Life is very much more exciting now than it used to be. You see, I have something more to expect, to look forward to, to watch. I really do eat better, and am more quiet than I was.

John is so pleased to see me improve! He laughed a little the other day, and said I seemed to be flourishing in spite of my wallpaper.

I turned it off with a laugh. I had no intention of telling him it was *because* of the wallpaper—he would make fun of me. He might even want to take me away.

I don't want to leave now until I have found it out. There is a week more, and I think that will be enough.

I'm feeling ever so much better! I don't sleep much at night, for it is so interesting to watch developments; but I sleep a good deal during the daytime.

In the daytime it is tiresome and perplexing.

There are always new shoots on the fungus, and new shades of yellow all over it. I cannot keep count of them, though I have tried conscientiously.

It is the strangest yellow, that wallpaper! It makes me think of all the yellow things I ever saw—not beautiful ones like buttercups, but old, foul, bad yellow things.

But there is something else about that paper—the smell! I noticed it the moment we came into the room, but with so much air and sun it was not bad. Now we have had a week of fog and rain, and whether the windows are open or not, the smell is here.

It creeps all over the house.

I find it hovering in the dining-room, skulking in the parlor, hiding in the hall, lying in wait for me on the stairs.

It gets into my hair.

Even when I go to ride, if I turn my head suddenly and surprise it—there is that smell!

Such a peculiar odor, too! I have spent hours in trying to analyze it, to find what it smelled like.

It is not bad—at first—and very gentle, but quite the subtlest, most enduring odor I ever met.

In this damp weather it is awful. I wake up in the night and find it hanging over me.

It used to disturb me at first. I thought seriously of burning the house—to reach the smell.

But now I am used to it. The only thing I can think of that it is like is the *color* of the paper! A yellow smell.

There is a very funny mark on this wall, low down, near the mopboard. A streak that runs round the room. It goes behind every piece of furniture, except the bed, a long, straight, even *smooch,* as if it had been rubbed over and over.

I wonder how it was done and who did it, and what they did it for. Round and round and round—round and round and round—it makes me dizzy!

I really have discovered something at last.

Through watching so much at night, when it changes so, I have finally found out.

The front pattern *does* move—and no wonder! The woman behind shakes it!

Sometimes I think there are a great many women behind, and sometimes only one, and she crawls around fast, and her crawling shakes it all over.

Then in the very bright spots she keeps still, and in the very shady spots she just takes hold of the bars and shakes them hard.

And she is all the time trying to climb through. But nobody could climb through that pattern—it strangles so; I think that is why it has so many heads.

They get through, and then the pattern strangles them off and turns them upside down, and makes their eyes white!

If those heads were covered or taken off it would not be half so bad.

I think that woman gets out in the daytime!

And I'll tell you why—privately—I've seen her!

I can see her out of every one of my windows!

It is the same woman, I know, for she is always creeping, and most women do not creep by daylight.

I see her on that long road under the trees, creeping along, and when a carriage comes she hides under the blackberry vines.

I don't blame her a bit. It must be very humiliating to be caught creeping by daylight!

I always lock the door when I creep by daylight. I can't do it at night, for I know John would suspect something at once.

And John is so queer now, that I don't want to irritate him. I wish he would take another room! Besides, I don't want anybody to get that woman out at night but myself.

I often wonder if I could see her out of all the windows at once.

But, turn as fast as I can, I can only see out of one at one time.

And though I always see her, she *may* be able to creep faster than I can turn!

I have watched her sometimes away off in the open country, creeping as fast as a cloud shadow in a wind.

If only that top pattern could be gotten off from the under one! I mean to try it, little by little.

I have found out another funny thing, but I shan't tell it this time! It does not do to trust people too much.

There are only two more days to get this paper off, and I believe John is beginning to notice. I don't like the look in his eyes.

And I heard him ask Jennie a lot of professional questions about me. She had a very good report to give.

She said I slept a good deal in the daytime.

John knows I don't sleep very well at night, for all I'm so quiet!

He asked me all sorts of questions, too, and pretended to be very loving and kind.

As if I couldn't see through him!

Still, I don't wonder he acts so, sleeping under this paper for three months.

It only interests me, but I feel sure John and Jennie are affected by it.

Hurrah! This is the last day, but it is enough. John is to stay in town over night, and won't be out until this evening.

Jennie wanted to sleep with me—the sly thing! but I told her I should undoubtedly rest better for a night all alone.

That was clever, for really I wasn't alone a bit! As soon as it was moonlight and that poor thing began to crawl and shake the pattern, I got up and ran to help her.

I pulled and she shook. I shook and she pulled, and before morning we had peeled off yards of that paper.

A strip about as high as my head and half around the room.

And then when the sun came and that awful pattern began to laugh at me, I declared I would finish it today!

We go away tomorrow, and they are moving all my furniture down again to leave things as they were before.

Jennie looked at the wall in amazement, but I told her merrily that I did it out of pure spite at the vicious thing.

She laughed and said she wouldn't mind doing it herself, but I must not get tired.

How she betrayed herself that time!

But I am here, and no person touches this paper but me—not *alive!*

She tried to get me out of the room—it was too patent! But I said it was so quiet and empty and clean now that I believed I would lie down again and sleep all I could; and not to wake me even for dinner—I would call when I woke.

So now she is gone, and the servants are gone, and the things are gone, and there is nothing left but that great bedstead nailed down, with the canvas mattress we found on it.

We shall sleep downstairs tonight, and take the boat home tomorrow.

I quite enjoy the room, now it is bare again.

How those children did tear about here!

This bedstead is fairly gnawed!

But I must get to work.

I have locked the door and thrown the key down into the front path.

I don't want to go out, and I don't want to have anybody come in, till John comes.

I want to astonish him.

I've got a rope up here that even Jennie did not find. If that woman does get out, and tries to get away, I can tie her!

But I forgot I could not reach far without anything to stand on!

"This bed will *not* move!"

I tried to lift and push it until I was lame, and then I got so angry I bit off a little piece at one corner—but it hurt my teeth.

Then I peeled off all the paper I could reach standing on the floor. It sticks horribly and the pattern just enjoys it! All those strangled heads and bulbous eyes and waddling fungus growths just shriek with derision!

I am getting angry enough to do something desperate. To jump out of the window would be admirable exercise, but the bars are too strong even to try.

Besides I wouldn't do it. Of course not. I know well enough that a step like that is improper and might be misconstrued.

I don't like to *look* out of the windows even—there are so many of those creeping women, and they creep so fast.

I wonder if they all come out of that wallpaper as I did?

But I am securely fastened now by my well-hidden rope—you don't get *me* out in the road there!

I suppose I shall have to get back behind the pattern when it comes night, and that is hard!

It is so pleasant to be out in this great room and creep around as I please!

I don't want to go outside. I won't, even if Jennie asks me to.

For outside you have to creep on the ground, and everything is green instead of yellow.

But here I can creep smoothly on the floor, and my shoulder just fits in that long smooch around the wall, so I cannot lose my way.

Why, there's John at the door!

It is no use, young man, you can't open it!

How he does call and pound!

Now he's crying for an axe.

It would be a shame to break down that beautiful door!

"John, dear!" said I in the gentlest voice, "the key is down by the front steps, under a plantain leaf!"

That silenced him for a few moments.

Then he said—very quietly indeed, "Open the door, my darling!"

"I can't," said I. "The key is down by the front door under a plantain leaf!"

And then I said it again, several times, very gently and slowly, and said it so often that he had to go and see, and he got it of course, and came in. He stopped short by the door.

"What is the matter?" he cried. "For God's sake, what are you doing!"

I kept on creeping just the same, but I looked at him over my shoulder.

"I've got out at last," said I, "in spite of you and Jane. And I've pulled off most of the paper, so you can't put me back!"

Now why should that man have fainted? But he did, and right across my path by the wall, so that I had to creep over him every time!

"Why I Wrote 'The Yellow Wallpaper'"

Many and many a reader has asked that. When the story first came out, in the *New England Magazine* about 1891, a Boston physician made protest in *The Transcript*. Such a story ought not to be written, he said; it was enough to drive anyone mad to read it.

Another physician, in Kansas I think, wrote to say that it was the best description of incipient insanity he had ever seen, and—begging my pardon—had I been there?

Now the story of the story is this:

For many years I suffered from a severe and continuous nervous breakdown tending to melancholia—and beyond. During about the third year of this trouble I went, in devout faith and some faint stir of hope, to a noted specialist in nervous diseases, the best known in the country. This wise man put me to bed and applied the rest cure, to which a still-good physique responded so promptly that he concluded there was nothing much the matter with me, and sent me home with solemn advice to "live as domestic a life as far as possible," to "have but two hours' intellectual life a day," and "never to touch pen, brush, or pencil again" as long as I lived. This was in 1887.

I went home and obeyed those directions for some three months, and came so near the borderline of utter mental ruin that I could see over.

Then, using the remnants of intelligence that remained, and helped by a wise friend, I cast the noted specialist's advice to the winds and went to work again—work, the normal life of every human being; work, in which is joy and growth and service, without which one is a pauper and a parasite—ultimately recovering some measure of power.

Being naturally moved to rejoicing by this narrow escape, I wrote "The Yellow Wallpaper," with its embellishments and additions, to carry out the ideal (I never had hallucinations or objections to my mural decorations) and sent a copy to the physician who so nearly drove me mad. He never acknowledged it.

The little book is valued by alienists and as a good specimen of one kind of literature. It has, to my knowledge, saved one woman from a similar fate—so terrifying her family that they let her out into normal activity and she recovered.

But the best result is this. Many years later I was told that the great specialist had admitted to friends of his that he had altered his treatment of neurasthenia since reading "The Yellow Wallpaper."

It was not intended to drive people crazy, but to save people from being driven crazy, and it worked.

SOURCE: "Why I Wrote 'The Yellow Wallpaper'" appeared in the October 1913 issue of *The Forerunner*.

Introduction to *Women and Economics*

Women and Economics is Charlotte Perkins Gilman's most sociological, as well as most theoretical, work of nonfiction. In this highly acclaimed book, Gilman seeks to show that the
traditional division of labor (breadwinner husband/stay-at-home wife) is inherently problematic. Contrary to the (still prevalent) commonsense rhetoric, which holds that in the traditional division of labor women are equal partners to men (or that both the man and the
woman are dependent on each other), Gilman maintains that the traditional division of labor
renders women economically dependent on men and, hence, necessarily strips women of
their freedom. In this arrangement, the woman receives both her social status and her economic viability not through her own labor, but through that of her husband. This makes her
labor not her "own," but a property of the male. Indeed, as the opening quote in this chapter reflects, rather than viewing the woman in the traditional family as an "equal partner" to
her husband, Gilman compares the traditional position of the woman to the domesticated
horse: neither the horse nor the woman is "free." Specifically, Gilman argues that if women
were actually compensated for their work in the home (and not "given" the status of their
husband), poor women with lots of children would get the most money (for they are doing
the most work), while women with no children and those who do no work in the home (i.e.,
those who have nannies, maids, etc.) would get no compensation. But, of course, the fact is
that poor women (i.e., women married to unemployed or working-class men) do the most
amount of work and receive the least amount of money. They work long and hard, cleaning,
cooking, and raising children. Meanwhile, rich women (women married to wealthy men) do
the least amount of work and have the most money to spend, because these women have
domestic help, servants, and nannies who perform the household and childrearing labor for
them. As Gilman states,

> Whatever the economic value of the domestic industry of women is, they do not get
> it. The women who do the most work get the least money, and the women who have
> the most money do the least work. Their labor is neither given nor taken as a factor in
> economic exchange. It is held to be their duty as women to do this work; and their eco-
> nomic status bears no relation to their domestic labors, unless an inverse one. (Gilman
> 1898/1998:8)

For those who argue that "a woman's place is in the home" because of her *child-bearing*
responsibilities, Gilman argues that "women's work" is actually mostly *house* service (cooking, cleaning, mending, etc.), not *child* service (bearing children, breastfeeding, etc.). Thus,
Gilman contends that the traditional division of labor is not biologically driven. On this
point, Gilman asserts,

> The poor man's wife has far too much of other work to do to spend all her time in wait-
> ing on her children. The rich man's wife could do it, but does not, partly because she
> hires some one to do it for her, and partly because she, too, has other duties to occupy
> her time. (ibid.:94).

Most provocatively, however, Gilman maintains that her economic dependency makes the
woman more akin to a horse than an equal partner in traditional marriage. As Gilman states,

> The horse, in his free natural condition, is economically independent. He gets his liv-
> ing by his own exertions irrespective of any other creature. The horse, in his present
> condition of slavery, is economically dependent. He gets his living at the hands of his
> master; and his exertions, though strenuous, bear no direct relation to his living. . . . The

horse works, it is true; but what he gets to eat depends on the power and will of his master. His living comes through another. He is economically dependent. (ibid.:4)

Translated into the human condition, Gilman remarks,

From the day laborer to the millionaire, the wife's worn dress or flashing jewels, her low roof or her lordly one, her weary feet or her rich equipage,—these speak of the economic ability of the husband. The comfort, the luxury, the necessities of life itself, which the woman receives, are obtained by the husband and given her by him. And, when the woman, left alone with no man to "support" her, tries to meet her own economic necessities, the difficulties which confront her prove conclusively what the general economic status of the woman is. (ibid.:5)

In short, like a horse, women are subject to the "power and will of another" because their domestic labor, for which no wages are received in return, belongs not to themselves but to their husbands. Women are thus rendered economically dependent.

Consequently, Gilman argues, rather than develop her own capabilities, women reduce themselves to attracting a viable life partner. Economically, this makes sense for women, because "their profit comes through the power of sex-attraction," not through their own talents (ibid.:33). As evidence for this state of affairs, Gilman remarks that

when we honestly care as much for motherhood as we pretend, we shall train the woman for her duty, not the girl for her guileless maneuvers to secure a husband. We talk about the noble duties of the mother, but our maidens are educated for economically successful marriage. (ibid.:100)

Thus, the problem with women's economic dependence on men is that their energies are focused on "catching" a man rather than on being productive citizens. Gilman saw it as a tragic waste that women were forced to spend their time and energy on grooming and "finding a man" rather than on intellectual concerns. Moreover, in denying her capabilities, she reduces herself to being, literally, the "weaker sex." As Gilman states,

The degree of feebleness and clumsiness common to women, the comparative inability to stand, walk, run, jump, climb, and perform other race-functions common to both sexes, is an excessive sex-distinction; and the ensuing transmission of this relative feebleness to their children, boys and girls alike, retards human development. . . . The relative weakness of women is a sex-distinction. It is apparent in her to a degree that injures motherhood, that injures wifehood, that injures the individual. (ibid.:24)

This brings us back to the issue of socialization with which we began this section. Women (especially middle- and upper-class women) are encouraged not to use, but to *deny*, their talents and capabilities:

The daughters and wives of the rich fail to perform even the domestic service expected of the women of poorer families. They are from birth to death absolutely nonproductive in goods of labor of economic value, and consumers of such goods and labor to an extent limited only by the purchasing power of their male relatives. (ibid.:85)

This is the *sociobiological* tragedy that Gilman perceives: She contends that women are not "underdeveloped men, but the feminine half of humanity in undeveloped form." Women are "oversexed," there is too much emphasis on their sex distinction. Rather than a healthy "survival of the fittest" in which women would be taught to be strong and productive, bourgeois women are mandated to be soft and weak, dependent, emotional, and frail.

It is this emphasis on economic dependency that distinguishes Gilman's perspective from that of Marx. Marx implied that bourgeois women are privileged because of their economic status, but both Marx's coauthor Friedrich Engels and Gilman see bourgeois women as economically dependent and, therefore, also oppressed (though both Engels and Gilman also fully recognize that bourgeois women are economically privileged in comparison to poor women). Specifically, in *The Origin of Family, Private Property and the State* (1884; see pp. 82–93 of this volume) Engels maintained that in traditional marriage, the man "is the bourgeois; the wife represents the proletariat." Gilman similarly suggests that just as the proletariat are exploited by the bourgeoisie, so, too, women are exploited by men.

Despite significant institutional advances in educational and professional opportunities, legal rights, and other spheres, some of the social and cultural gender inequities that Gilman discussed are still readily apparent today. Television shows such as "Who Wants To Marry a Millionaire" implicitly affirm that "catching" a (rich) man is a sensible career path for a woman, and that a woman's most important asset is her body. From this point of view, a woman would be better off focusing on her looks than on what Gilman considered "matters of real importance." Similarly, the popularity of silicone breast implants and the "Girls Gone Wild" culture industry seem to reflect that women continue to be "oversexed" in Gilman's terms (although certainly one might argue that men's bodies are often objectified as well).

Of course, sadly, there also are many contemporary societies where women still lack basic legal and civil rights, in addition to enduring major cultural and social inequality. For instance, M. Steven Fish (2003) finds that while the average literacy gap between the sexes in most non-Muslim countries is 7 percentage points; it is 20 percentage points in Iran, 23 in Egypt, and 28 in Syria. Many women in these countries, in addition to restricted educational opportunities, are confronted with significantly inferior health care, while being denied basic legal rights.

From *Women and Economics* (1898)

Charlotte Perkins Gilman

PREFACE

This book is written to offer a simple and natural explanation of one of the most common and most perplexing problems of human life,—a problem which presents itself to almost every individual for practical solution, and which demands the most serious attention of the moralist, the physician, and the sociologist—

To show how some of the worst evils under which we suffer, evils long supposed to be inherent and ineradicable in our natures, are but the result of certain arbitrary conditions of our own adoption, and how, by removing those conditions, we may remove the evil resultant—

To point out how far we have already gone in the path of improvement, and how irresistibly the social forces of to-day are compelling us further, even without our knowledge and against our violent opposition,—an advance which may be greatly quickened by our recognition and assistance—

To reach in especial the thinking women of to-day, and urge upon them a new sense, not only of their social responsibility as individuals, but of their measureless racial importance as makers of men.

It is hoped also that the theory advanced will prove sufficiently suggestive to give rise to such further study and discussion as shall prove its error or establish its truth.

Charlotte Perkins Stetson

SOURCE: From *Women and Economics* by Charlotte Perkins Gilman. 1898/1998. New York: Dover.

I

Since we have learned to study the development of human life as we study the evolution of species throughout the animal kingdom, some peculiar phenomena which have puzzled the philosopher and moralist for so long, begin to show themselves in a new light. We begin to see that, so far from being inscrutable problems, requiring another life to explain, these sorrows and perplexities of our lives are but the natural results of natural causes, and that, as soon as we ascertain the causes, we can do much to remove them.

In spite of the power of the individual will to struggle against conditions, to resist them for a while, and sometimes to overcome them, it remains true that the human creature is affected by his environment, as is every other living thing. The power of the individual will to resist natural law is well proven by the life and death of the ascetic. In any one of those suicidal martyrs may be seen the will, misdirected by the ill-informed intelligence, forcing the body to defy every natural impulse,—even to the door of death, and through it.

But, while these exceptions show what the human will can do, the general course of life shows the inexorable effect of conditions upon humanity. Of these conditions we share with other living things the environment of the material universe. We are affected by climate and locality, by physical, chemical, electrical forces, as are all animals and plants. With the animals, we farther share the effect of our own activity, the reactionary force of exercise. What we do, as well as what is done to us, makes us what we are. But, beyond these forces, we come under the effect of a third set of conditions peculiar to our human status; namely, social conditions. In the organic interchanges which constitute social life, we are affected by each other to a degree beyond what is found even among the most gregarious of animals. This third factor, the social environment, is of enormous force as a modifier of human life. Throughout all these environing conditions, those which affect us through our economic necessities are most marked in their influence.

Without touching yet upon the influence of the social factors, treating the human being merely as an individual animal, we see that he is modified most by his economic conditions, as is every other animal. Differ as they may in color and size, in strength and speed, in minor adaptation to minor conditions, all animals that live on grass have distinctive traits in common, and all animals that eat flesh have distinctive traits in common,—so distinctive and so common that it is by teeth, by nutritive apparatus in general, that they are classified, rather than by means of defence or locomotion. The food supply of the animal is the largest passive factor in his development; the processes by which he obtains his food supply, the largest active factor in his development. It is these activities, the incessant repetition of his exertions by which he is fed, which most modify his structure and develope his functions. The sheep, the cow, the deer, differ in their adaptation to the weather, their locomotive ability, their means of defence; but they agree in main characteristics, because of their common method of nutrition.

The human animal is no exception to this rule. Climate affects him, weather affects him, enemies affect him; but most of all he is affected, like every other living creature, by what he does for his living. Under all the influence of his later and wider life, all the reactive effect of social institutions, the individual is still inexorably modified by his means of livelihood: "the hand of the dyer is subdued to what he works in." As one clear, world-known instance of the effect of economic conditions upon the human creature, note the marked race-modification of the Hebrew people under the enforced restrictions of the last two thousand years. Here is a people rising to national prominence, first as a pastoral, and then as an agricultural nation; only partially commercial through race affinity with the Phoenicians, the pioneer traders of the world. Under the social power of a united Christendom—united at least in this most unchristian deed—the Jew was forced to get his livelihood by commercial methods solely. Many effects can be traced in him to the fierce pressure of the social conditions to which he was subjected: the intense family devotion of a people who had no country, no king, no room for joy and pride except the family; the reduced size and tremendous vitality and endurance of the pitilessly selected survivors of the Ghetto; the repeated bursts of erratic genius from the human spirit so inhumanly restrained. But more patent still is the effect of the economic conditions,—the artificial development of a race of traders and dealers in money, from the lowest pawnbroker to the house of Rothschild; a special kind of people, bred of the economic environment in which they were compelled to live.

One rough but familiar instance of the same effect, from the same cause, we can all see in the marked distinction between the pastoral, the agricultural, and the manufacturing classes in any nation, though their other conditions be the same. On the clear line of argument that functions and organs are developed by use, that what we use most is developed most, and that the daily processes of supplying economic needs are the processes that we most use, it follows that, when we find special economic conditions affecting any special class of people, we may look for special results, and find them.

In view of these facts, attention is now called to a certain marked and peculiar economic condition affecting the human race, and unparalleled in the organic world. We are the only animal species in which the female depends on the male for food, the only animal species in which the sex-relation is also an economic relation. With us an entire sex lives in a relation of economic dependence upon the other sex, and the economic relation is combined with the sex-relation. The economic status of the human female is relative to the sex-relation.

It is commonly assumed that this condition also obtains among other animals, but such is not the case. There are many birds among which, during the nesting season, the male helps the female feed the young, and partially feeds her; and, with certain of the higher carnivora, the male helps the female feed the young, and partially feeds her. In no case does she depend on him absolutely, even during this season, save in that of the hornbill, where the female, sitting on her nest in a hollow tree, is walled in with clay by the male, so that only her beak projects; and then he feeds her while the eggs are developing. But even the female hornbill does not expect to be fed at any other time. The female bee and ant are economically dependent, but not on the male. The workers are females, too, specialized to economic functions solely. And with the carnivora, if the young are to lose one parent, it might far better be the father: the mother is quite competent to take care of them herself. With many species, as in the case of the common cat, she not only feeds herself and her young, but has to defend the young against the male as well. In no case is the female throughout her life supported by the male.

In the human species the condition is permanent and general, though there are exceptions, and though the present century is witnessing the beginnings of a great change in this respect. We have not been accustomed to face this fact beyond our loose generalization that it was "natural," and that other animals did so, too.

To many this view will not seem clear at first; and the case of working peasant women or females of savage tribes, and the general household industry of women, will be instanced against it. Some careful and honest discrimination is needed to make plain to ourselves the essential facts of the relation, even in these cases. The horse, in his free natural condition, is economically independent. He gets his living by his own exertions, irrespective of any other creature. The horse, in his present condition of slavery, is economically dependent. He gets his living at the hands of his master; and his exertions, though strenuous, bear no direct relation to his living. In fact, the horses who are the best fed and cared for and the horses who are the hardest worked are quite different animals. The horse works, it is true; but what he gets to eat depends on the power and will of his master. His living comes through another. He is economically dependent. So with the hard-worked savage or peasant women. Their labor is the property of another: they work under another will; and what they receive depends not on their labor, but on the power and will of another. They are economically dependent. This is true of the human female both individually and collectively.

In studying the economic position of the sexes collectively, the difference is most marked. As a social animal, the economic status of man rests on the combined and exchanged services of vast numbers of progressively specialized individuals. The economic progress of the race, its maintenance at any period, its continued advance, involve the collective activities of all the trades, crafts, arts, manufactures, inventions, discoveries, and all the civil and military institutions that go to maintain them. The economic status of any race at any time, with its involved effect on all the constituent individuals, depends on their world-wide labors and their free exchange. Economic progress, however, is almost exclusively masculine. Such economic processes as women have been allowed to exercise are of the earliest and most primitive kind. Were men to perform no economic services save such as are still performed by women, our racial status in economics would be reduced to most painful limitations.

To take from any community its male workers would paralyze it economically to a far greater degree than to remove its female workers. The labor now performed by the women could be performed by the men, requiring only the setting back of many advanced workers into earlier forms of industry; but the labor now performed by the men could not be performed by the women without generations of effort and adaptation. Men can cook, clean, and sew as well as women; but the making and managing of the great engines of modern industry, the threading of earth and sea in our vast systems of transportation, the handling of our elaborate machinery of trade, commerce, government,—these things could not be done so well by women in their present degree of economic development.

This is not owing to lack of the essential human faculties necessary to such achievements, nor to any inherent disability of sex, but to the present condition of woman, forbidding the development of this degree of economic ability. The male human being is thousands of years in advance of the female in economic status. Speaking collectively, men produce and distribute wealth; and women receive it at their hands. As men hunt, fish, keep cattle, or raise corn, so do women eat game, fish, beef, or corn. As men go down to the sea in ships, and bring coffee and spices and silks and gems from far away, so do women partake of the coffee and spices and silks and gems the men bring.

The economic status of the human race in any nation, at any time, is governed mainly by the activities of the male: the female obtains her share in the racial advance only through him.

Studied individually, the facts are even more plainly visible, more open and familiar. From the day laborer to the millionaire, the wife's worn dress or flashing jewels, her low roof or her lordly one, her weary feet or her rich equipage,—these speak of the economic ability of the husband. The comfort, the luxury, the necessities of life itself, which the woman receives, are obtained by the husband, and given her by him. And, when the woman, left alone with no man to "support" her, tries to meet her own economic necessities, the difficulties which confront her prove conclusively what the general economic status of the woman is. None can deny these patent facts,—that the economic status of women generally depends upon that of men generally, and that the economic status of women individually depends upon that of men individually, those men to whom they are related. But we are instantly confronted by the commonly received opinion that, although it must be admitted that men make and distribute the wealth of the world, yet women earn their share of it as wives. This assumes either that the husband is in the position of employer and the wife as employee, or that marriage is a "partnership," and the wife an equal factor with the husband in producing wealth.

Economic independence is a relative condition at best. In the broadest sense, all living things are economically dependent upon others,—the animals upon the vegetables, and man upon both. In a narrower sense, all social life is economically interdependent, man producing collectively what he could by no possibility produce separately. But, in the closest interpretation, individual economic independence among human beings means that the individual pays for what he gets, works for what he gets, gives to the other an equivalent for what the other gives him. I depend on the shoemaker for shoes, and the tailor for coats; but, if I give the shoemaker and the tailor enough of my own labor as a house-builder to pay for the shoes and coats they give me, I retain my personal independence. I have not taken of their product, and given nothing of mine. As long as what I get is obtained by what I give, I am economically independent.

Women consume economic goods. What economic product do they give in exchange for what they consume? The claim that marriage is a partnership, in which the two persons married produce wealth which neither of them, separately, could produce, will not bear examination. A man happy and comfortable can produce more than one unhappy and uncomfortable, but this is as true of a father or son as of a husband. To take from a man any of the conditions which make him happy and strong is to cripple his industry, generally speaking. But those relatives who make him happy are not therefore his business partners, and entitled to share his income.

Grateful return for happiness conferred is not the method of exchange in a partnership. The comfort a man takes with his wife is not in the nature of a business partnership, nor are her frugality and industry. A housekeeper, in her place, might be as frugal, as industrious, but would not therefore be a partner. Man and wife are partners truly in their mutual obligation to their children,—their common love, duty, and service. But a manufacturer who marries, or a doctor, or a lawyer, does not take a partner in his business, when he takes a partner in parenthood, unless his wife is also a manufacturer, a doctor, or a lawyer. In his business, she cannot even advise wisely without training and experience. To love her husband, the composer, does not enable her to compose; and the loss of a man's wife, though it may break his heart, does not cripple his business, unless his mind is affected by grief. She is in no sense a business partner, unless she contributes capital or experience or labor, as a man would in like relation. Most men would hesitate very seriously before entering a business partnership with any woman, wife or not.

If the wife is not, then, truly a business partner, in what way does she earn from her husband the food, clothing, and shelter she receives at his hands? By house service, it will be instantly replied. This is the general misty idea upon the subject,—that women earn all they get, and more, by house service. Here we come to a very practical and definite economic ground. Although not producers of wealth, women serve in the final processes of preparation and distribution. Their labor in the household has a genuine economic value.

For a certain percentage of persons to serve other persons, in order that the ones so served may produce more, is a contribution not to be overlooked. The labor of women in the house, certainly, enables men to produce more wealth than they otherwise could; and in this way

women are economic factors in society. But so are horses. The labor of horses enables men to produce more wealth than they otherwise could. The horse is an economic factor in society. But the horse is not economically independent, nor is the woman. If a man plus a valet can perform more useful service than he could minus a valet, then the valet is performing useful service. But, if the valet is the property of the man, is obliged to perform this service, and is not paid for it, he is not economically independent.

The labor which the wife performs in the household is given as part of her functional duty, not as employment. The wife of the poor man, who works hard in a small house, doing all the work for the family, or the wife of the rich man, who wisely and gracefully manages a large house and administers its functions, each is entitled to fair pay for services rendered.

To take this ground and hold it honestly, wives, as earners through domestic service, are entitled to the wages of cooks, housemaids, nursemaids, seamstresses, or housekeepers, and to no more. This would of course reduce the spending money of the wives of the rich, and put it out of the power of the poor man to "support" a wife at all, unless, indeed, the poor man faced the situation fully, paid his wife her wages as house servant, and then she and he combined their funds in the support of their children. He would be keeping a servant: she would be helping keep the family. But nowhere on earth would there be "a rich woman" by these means. Even the highest class of private housekeeper, useful as her services are, does not accumulate a fortune. She does not buy diamonds and sables and keep a carriage. Things like these are not earned by house service.

But the salient fact in this discussion is that, whatever the economic value of the domestic industry of women is, they do not get it. The women who do the most work get the least money, and the women who have the most money do the least work. Their labor is neither given nor taken as a factor in economic exchange. It is held to be their duty as women to do this work; and their economic status bears no relation to their domestic labors, unless an inverse one. Moreover, if they were thus fairly paid,—given what they earned, and no more,— all women working in this way would be reduced to the economic status of the house servant. Few women—or men either—care to face this condition. The ground that women earn their living by domestic labor is instantly forsaken, and we are told that they obtain their livelihood as mothers. This is a peculiar position. We speak of it commonly enough, and often with deep feeling, but without due analysis.

In treating of an economic exchange, asking what return in goods or labor women make for the goods and labor given them,—either to the race collectively or to their husbands individually,—what payment women make for their clothes and shoes and furniture and food and shelter, we are told that the duties and services of the mother entitle her to support.

If this is so, if motherhood is an exchangeable commodity given by women in payment for clothes and food, then we must of course find some relation between the quantity or quality of the motherhood and the quantity and quality of the pay. This being true, then the women who are not mothers have no economic status at all; and the economic status of those who are must be shown to be relative to their motherhood. This is obviously absurd. The childless wife has as much money as the mother of many,—more; for the children of the latter consume what would otherwise be hers; and the inefficient mother is no less provided for than the efficient one. Visibly, and upon the face of it, women are not maintained in economic prosperity proportioned to their motherhood. Motherhood bears no relation to their economic status. Among primitive races, it is true,—in the patriarchal period, for instance,—there was some truth in this position. Women being of no value whatever save as bearers of children, their favor and indulgence did bear direct relation to maternity; and they had reason to exult on more grounds than one when they could boast a son. To-day, however, the maintenance of the woman is not conditioned upon this. A man is not allowed to discard his wife because she is barren. The claim of motherhood as a factor in economic exchange is false to-day. But suppose it were true. Are we willing to hold this ground, even in theory? Are we willing to consider motherhood as a business, a form of commercial exchange? Are the cares and duties of the mother, her travail and her love, commodities to be exchanged for bread?

It is revolting so to consider them; and, if we dare face our own thoughts, and force them to their logical conclusion, we shall see that nothing could be more repugnant to human feeling, or more socially and individually injurious, than to make motherhood a trade. Driven off these alleged grounds of women's economic independence; shown that women, as a class, neither produce nor distribute wealth; that women, as individuals, labor mainly as house servants, are not paid as such, and would not be satisfied with such an economic status if they were so paid; that wives are not business partners or co-producers of wealth with their husbands, unless they actually practise the same profession; that they are not salaried as mothers, and that it would be unspeakably degrading if they were,—what remains to those who deny that women are supported by men? This (and a most amusing position it is),—that the function of maternity unfits a woman for economic production, and, therefore, it is right that she should be supported by her husband.

The ground is taken that the human female is not economically independent, that she is fed by the male of her species. In denial of this, it is first alleged that she is economically independent,—that she does support herself by her own industry in the house. It being shown that there is no relation between the economic status of woman and the labor she performs in the home, it is then alleged that not as house servant, but as mother, does woman earn her living. It being shown that the economic status of woman bears no relation to her motherhood, either in quantity or quality, it is then alleged that motherhood renders a woman unfit for economic production, and that, therefore, it is right that she be supported by her husband. Before going farther, let us seize upon this admission,—that she *is* supported by her husband.

Without going into either the ethics or the necessities of the case, we have reached so much common ground: the female of genus homo is supported by the male. Whereas, in other species of animals, male and female alike graze and browse, hunt and kill, climb, swim, dig, run, and fly for their livings, in our species the female does not seek her own living in the specific activities of our race, but is fed by the male.

Now as to the alleged necessity. Because of her maternal duties, the human female is said to be unable to get her own living. As the maternal duties of other females do not unfit them for getting their own living and also the livings of their young, it would seem that the human maternal duties require the segregation of the entire energies of the mother to the service of the child during her entire adult life, or so large a proportion of them that not enough remains to devote to the individual interests of the mother.

Such a condition, did it exist, would of course excuse and justify the pitiful dependence of the human female, and her support by the male. As the queen bee, modified entirely to maternity, is supported, not by the male, to be sure, but by her co-workers, the "old maids," the barren working bees, who labor so patiently and lovingly in their branch of the maternal duties of the hive, so would the human female, modified entirely to maternity, become unfit for any other exertion, and a helpless dependant.

Is this the condition of human motherhood? Does the human mother, by her motherhood, thereby lose control of brain and body, lose power and skill and desire for any other work? Do we see before us the human race, with all its females segregated entirely to the uses of motherhood, consecrated, set apart, specially developed, spending every power of their nature on the service of their children?

We do not. We see the human mother worked far harder than a mare, laboring her life long in the service, not of her children only, but of men; husbands, brothers, fathers, whatever male relatives she has; for mother and sister also; for the church a little, if she is allowed; for society, if she is able; for charity and education and reform,—working in many ways that are not the ways of motherhood.

It is not motherhood that keeps the housewife on her feet from dawn till dark; it is house service, not child service. Women work longer and harder than most men, and not solely in maternal duties. The savage mother carries the burdens, and does all menial service for the family tribe. The peasant mother toils in the fields, and the workingman's wife in the home. Many mothers, even now, are wage-earners for the family, as well as bearers and rearers of it. And the women who are not so occupied, the women who belong to rich men,—here perhaps is the exhaustive devotion to maternity which is supposed to justify an admitted economic dependence. But we

do not find it even among these. Women of ease and wealth provide for their children better care than the poor woman can; but they do not spend more time upon it themselves, nor more care and effort. They have other occupation.

In spite of her supposed segregation to maternal duties, the human female, the world over, works at extra-maternal duties for hours enough to provide her with an independent living, and then is denied independence on the ground that motherhood prevents her working!

If this ground were tenable, we should find a world full of women who never lifted a finger save in the service of their children, and of men who did *all* the work besides, and waited on the women whom motherhood prevented from waiting on themselves. The ground is not tenable. A human female, healthy, sound, has twenty-five years of life before she is a mother, and should have twenty-five years more after the period of such maternal service as is expected of her has been given. The duties of grandmotherhood are surely not alleged as preventing economic independence.

The working power of the mother has always been a prominent factor in human life. She is the worker *par excellence,* but her work is not such as to affect her economic status. Her living, all that she gets,—food, clothing, ornaments, amusements, luxuries,—these bear no relation to her power to produce wealth, to her services in the house, or to her motherhood. These things bear relation only to the man she marries, the man she depends on,—to how much he has and how much he is willing to give her. The women whose splendid extravagance dazzles the world, whose economic goods are the greatest, are often neither houseworkers nor mothers, but simply the women who hold most power over the men who have the most money. The female of genus homo is economically dependent on the male. He is her food supply.

II

Knowing how important a factor in the evolution of species is the economic relation, and finding in the human species an economic relation so peculiar, we may naturally look to find effects peculiar to our race. We may expect to find phenomena in the sex-relation and in the economic relation of humanity of a unique character,—phenomena not traceable to human superiority, but singularly derogatory to that superiority; phenomena so marked, so morbid, as to give rise to much speculation as to their cause. Are these natural inferences fulfilled? Are these peculiarities in the sex-relation and in the economic relation manifested in human life? Indisputably these are,—so plain, so prominent, so imperiously demanding attention, that human thought has been occupied from its first consciousness in trying some way to account for them. To explain and relate these phenomena, separating what is due to normal race-development from what is due to this abnormal sexuo-economic relation, is the purpose of the line of study here suggested.

As the racial distinction of humanity lies in its social relation, so we find the distinctive gains and losses of humanity to lie also in its social relation. We are more affected by our relation to each other than by our physical environment.

Disadvantages of climate, deficiencies in food supply, competition from other species,—all these conditions society, in its organic strength, is easily able to overcome or to adjust. But in our inter-human relations we are not so successful. The serious dangers and troubles of human life arise from difficulties of adjustment with our social environment, and not with our physical environment. These difficulties, so far, have acted as a continual check to social progress. The more absolutely a nation has triumphed over physical conditions, the more successful it has become in its conquest of physical enemies and obstacles, the more it has given rein to the action of social forces which have ultimately destroyed the nation, and left the long ascent to be begun again by others.

There is the moral of all human tales:
'Tis but the same rehearsal of the past,—
First Freedom, and then Glory; when that fails,
Wealth, Vice, Corruption,—barbarism at last.
And History, with all her volumes vast,
Hath but *one* page.[i]

The path of history is strewn with fossils and faint relics of extinct races,—races which died

of what the sociologist would call internal diseases rather than natural causes. This, too, has been clear to the observer in all ages. It has been easily seen that there was something in our own behavior which did us more harm than any external difficulty; but what we have not seen is the natural cause of our unnatural conduct, and how most easily to alter it.

Rudely classifying the principal fields of human difficulty, we find one large proportion lies in the sex-relation, and another in the economic relation, between the individual constituents of society. To speak broadly, the troubles of life as we find them are mainly traceable to the heart or the purse. The other horror of our lives—disease—comes back often to these causes,—to something wrong either in economic relation or in sex-relation. To be ill-fed or ill-bred, or both, is largely what makes us the sickly race we are. In this wrong breeding, this maladjustment of the sex-relation in humanity, what are the principal features? We see in social evolution two main lines of action in this department of life. One is a gradual orderly development of monogamous marriage, as the form of sex-union best calculated to advance the interests of the individual and of society. It should be clearly understood that this is a natural development, inevitable in the course of social progress; not an artificial condition, enforced by laws of our making. Monogamy is found among birds and mammals: it is just as natural a condition as polygamy or promiscuity or any other form of sex-union; and its permanence and integrity are introduced and increased by the needs of the young and the advantage to the race, just as any other form of reproduction was introduced. Our moral concepts rest primarily on facts. The moral quality of monogamous marriage depends on its true advantage to the individual and to society. If it were not the best form of marriage for our racial good, it would not be right. All the way up, from the promiscuous horde of savages, with their miscellaneous matings, to the lifelong devotion of romantic love, social life has been evolving a type of sex-union best suited to develope and improve the individual and the race. This is an orderly process, and a pleasant one, involving only such comparative pain and difficulty as

[i]*Childe Harold's Pilgrimage,* Canto IV., cviii.

always attend the assumption of new processes and the extinction of the old; but accompanied by far more joy than pain.

But with the natural process of social advancement has gone an unnatural process,—an erratic and morbid action, making the sex-relation of humanity a frightful source of evil. So prominent have been these morbid actions and evil results that hasty thinkers of all ages have assumed that the whole thing was wrong, and that celibacy was the highest virtue. Without the power of complete analysis, without knowledge of the sociological data essential to such analysis, we have sweepingly condemned as a whole what we could easily see was so allied with pain and loss. But, like all natural phenomena, the phenomena of sex may be studied, both the normal and the abnormal, the physiological and the pathological; and we are quite capable of understanding why we are in such evil case, and how we may attain more healthful conditions.

So far, the study of this subject has rested on the assumption that man must be just as we find him, that man behaves just as he chooses, and that, if he does not choose to behave as he does, he can stop. Therefore, when we discovered that human behavior in the sex-relation was productive of evil, we exhorted the human creature to stop so behaving, and have continued so to exhort for many centuries. By law and religion, by education and custom, we have sought to enforce upon the human individual the kind of behavior which our social sense so clearly showed was right.

But always there has remained the morbid action. Whatever the external form of sex-union to which we have given social sanction, however Bible and Koran and Vedas have offered instruction, some hidden cause has operated continuously against the true course of social evolution, to pervert the natural trend toward a higher and more advantageous sex-relation; and to maintain lower forms, and erratic phases, of a most disadvantageous character.

Every other animal works out the kind of sex-union best adapted to the reproduction of his species, and peacefully practises it. We have worked out the kind that is best for us,—best for the individuals concerned, for the young resultant, and for society as a whole; but we do not peacefully practise it. So palpable is this fact that we have commonly accepted it, and

taken it for granted that this relation must be a continuous source of trouble to humanity. "Marriage is a lottery," is a common saying among us. "The course of true love never did run smooth." And we quote with unction *Punch*'s advice to those about to marry,—"Don't!" That peculiar sub-relation which has dragged along with us all the time that monogamous marriage has been growing to be the accepted form of sex-union—prostitution—we have accepted, and called a "social necessity." We also call it "the social evil." We have tacitly admitted that this relation in the human race must be more or less uncomfortable and wrong, that it is part of our nature to have it so.

Now let use examine the case fairly and calmly, and see whether it is as inscrutable and immutable as hitherto believed. What are the conditions? What are the natural and what the unnatural features of the case? To distinguish these involves a little study of the evolution of the processes of reproduction.

Very early in the development of species it was ascertained by nature's slow but sure experiments that the establishment of two sexes in separate organisms, and their differentiation, was to the advantage of the species. Therefore, out of the mere protoplasmic masses, the floating cells, the amorphous early forms of life, grew into use the distinction of the sexes,—the gradual development of masculine and feminine organs and functions in two distinct organisms. Developed and increased by use, the distinction of sex increased in the evolution of species. As the distinction increased, the attraction increased, until we have in all the higher races two markedly different sexes, strongly drawn together by the attraction of sex, and fulfilling their use in the reproduction of species. These are the natural features of sex-distinction and sex-union, and they are found in the human species as in others. The unnatural feature by which our race holds an unenviable distinction consists mainly in this,—a morbid excess in the exercise of this function.

It is this excess, whether in marriage or out, which makes the health and happiness of humanity in this relation so precarious. It is this excess, always easily seen, which law and religion have mainly striven to check. Excessive sex-indulgence is the distinctive feature of humanity in this relation.

To define "excess" in this connection is not difficult. All natural functions that require our conscious co-operation for their fulfilment are urged upon our notice by an imperative desire. We do not have to desire to breathe or to digest or to circulate the blood, because that is done without our volition; but we do have to desire to eat and drink, because the stomach cannot obtain its supplies without in some way spurring the whole organism to secure them. So hunger is given us as an essential factor in our process of nutrition. In the same manner sex-attraction is an essential factor in the fulfilment of our processes of reproduction. In a normal condition the amount of hunger we feel is exactly proportioned to the amount of food we need. It tells us when to eat and when to stop. In some diseased conditions "an unnatural appetite" sets in; and we are impelled to eat far beyond the capacity of the stomach to digest, of the body to assimilate. This is an excessive hunger.

We, as a race, manifest an excessive sex-attraction, followed by its excessive indulgence, and the inevitable evil consequence. It urges us to a degree of indulgence which bears no relation to the original needs of the organism, and which is even so absurdly exaggerated as to react unfavorably on the incidental gratification involved; an excess which tends to pervert and exhaust desire as well as to injure reproduction.

The human animal manifests an excess in sex-attraction which not only injures the race through its morbid action on the natural processes of reproduction, but which injures the happiness of the individual through its morbid reaction on his own desires.

What is the cause of this excessive sex-attraction in the human species? The immediately acting cause of sex-attraction is sex-distinction. The more widely the sexes are differentiated, the more forcibly they are attracted to each other. The more highly developed becomes the distinction of sex in either organism, the more intense is its attraction for the other. In the human species we find sex-distinction carried to an excessive degree. Sex-distinction in humanity is so marked as to retard and confuse race-distinction, to check individual distinction, seriously to injure the race. Accustomed as we are simply to accept the facts of life as we find them, to consider people as permanent types instead of seeing them and the whole race in continual change according to the action of many forces, it seems strange at first to differentiate between familiar manifestations of sex-distinction, and to say: "This is normal, and should not be disturbed. This is abnormal, and should be removed." But that is precisely what must be done.

Normal sex-distinction manifests itself in all species in what are called primary and secondary sex-characteristics. The primary are those organs and functions essential to reproduction; the secondary, those modifications of structure and function which subserve the uses of reproduction ultimately, but are not directly essential,—such as the horns of the stag, of use in sex-combat; the plumage of the peacock, of use in sex-competition. All the minor characteristics of beard or mane, comb, wattles, spurs, gorgeous color or superior size, which distinguish the male from the female,—these are distinctions of sex. These distinctions are of use to the species through reproduction only, the processes of race-preservation. They are not of use in self-preservation. The creature is not profited personally by his mane or crest or tail-feathers: they do not help him get his dinner or kill his enemies.

On the contrary, they react unfavorably upon his personal gains, if, through too great development, they interfere with his activity or render him a conspicuous mark for enemies. Such development would constitute excessive sex-distinction, and this is precisely the condition of the human race. Our distinctions of sex are carried to such a degree as to be disadvantageous to our progress as individuals and as a race. The sexes in our species are differentiated not only enough to perform their primal functions; not only enough to manifest all sufficient secondary sexual characteristics and fulfil their use in giving rise to sufficient sex-attraction; but so much as seriously to interfere with the processes of self-preservation on the one hand; and, more conspicuous still, so much as to react unfavorably upon the very processes of race-preservation which they are meant to serve. Our excessive sex-distinction, manifesting the characteristics of sex to an abnormal degree, has given rise to a degree of attraction which demands a degree of indulgence that directly injures motherhood and fatherhood. We are not better as parents, nor better as people, for our existing degree of sex-distinction, but visibly worse. To what conditions are we to look for the developing cause of these phenomena?

Let us first examine the balance of forces by which these two great processes, self-preservation and race-preservation, are conducted in the world. Self-preservation involves the expenditure of energy in those acts, and their ensuing modifications of structure and function, which tend to the maintenance of the individual life. Race-preservation involves the expenditure of energy in those acts, and their ensuing modifications of structure and function, which tend to the maintenance of the racial life, even to the complete sacrifice of the individual. This primal distinction should be clearly held in mind. Self-preservation and race-preservation are in no way identical processes, and are often directly opposed. In the line of self-preservation, natural selection, acting on the individual, develops those characteristics which enable it to succeed in "the struggle for existence," increasing by use those organs and functions by which it directly profits. In the line of race-preservation, sexual selection, acting on the individual, develops those characteristics which enable it to succeed in what Drummond has called "the struggle for the existence of others," increasing by use those organs and functions by which its young are to profit, directly or indirectly. The individual has been not only modified to its environment, under natural selection, but modified to its mate, under sexual selection, each sex developing the qualities desired by the other by the simple process of choice, those best sexed being first chosen, and transmitting their sex-development as well as their racial development.

The order mammalia is the resultant of a primary sex-distinction developed by natural selection; but the gorgeous plumage of the peacock's tail is a secondary sex-distinction developed by sexual selection. If the peacock's tail were to increase in size and splendor till it shone like the sun and covered an acre,—if it tended so to increase, we will say,—such excessive sex-distinction would be so inimical to the personal prosperity of that peacock that he would die, and his tail-tendency would perish with him. If the pea-hen, conversely, whose sex-distinction attracts in the opposite direction, not by being large and splendid, but small and dull,—if she should grow so small and dull as to fail to keep herself and her young fed and defended, then she would die; and there would be another check to excessive sex-distinction. In herds of deer and cattle the male is larger and stronger, the female smaller and weaker; but, unless the latter is large and strong enough to keep up with the male in the search for food or the flight from foes, one is taken and the other left, and there is no more of that kind of animal. Differ as they may in sex, they must remain alike in species, equal in race-development, else destruction overtakes them. The force of natural selection, demanding and producing identical race-qualities, acts as a check on sexual selection, with its production of different sex-qualities. As sexes, they perform different functions, and therefore tend to develope differently. As species, they perform the same functions, and therefore tend to develope equally.

And as sex-functions are only used occasionally, and race-functions are used all the time,—as they mate but yearly or tri-monthly, but eat daily and hourly,—the processes of obtaining food or of opposing constant enemies act more steadily than the processes of reproduction, and produce greater effect.

We find the order mammalia accordingly producing and suckling its young in the same manner through a wide variety of species which obtain their living in a different manner. The calf and colt and cub and kitten are produced by the same process; but the cow and horse, the bear and cat, are produced by different processes. And, though cow and bull, mare and stallion, differ as to sex, they are alike in species; and the likeness in species is greater than the difference in sex. Cow, mare, and cat are all females of the order mammalia, and so far alike; but how much more different they are than similar!

Natural selection developes race. Sexual selection developes sex. Sex-development is one throughout its varied forms, tending only to reproduce what is. But race-development rises ever in higher and higher manifestation of energy. As sexes, we share our distinction with the animal kingdom almost to the beginning of life, and with the vegetable world as well. As races, we differ in ascending degree; and the human race stands highest in the scale of life so far.

When, then, it can be shown that sex-distinction in the human race is so excessive as not only to affect injuriously its own purposes, but to check and pervert the progress of the race, it becomes a matter for most serious consideration. Nothing could be more inevitable, however, under our sexuo-economic relation. By the economic dependence of the human female upon the male, the balance of forces is altered. Natural selection no longer checks the action of sexual selection, but co-operates with it. Where both sexes obtain their food through the same exertions, from the same sources, under the same conditions, both sexes are acted upon alike, and developed alike by their environment. Where the two sexes obtain their food under different conditions, and where that difference consists in one of them being fed by the other, then the feeding sex becomes the environment of the fed. Man, in supporting woman, has become her economic environment. Under natural selection, every creature is modified to its environment, developing perforce the qualities needed to obtain its livelihood under that environment. Man, as the feeder of woman, becomes the strongest modifying force in her economic condition. Under sexual selection the human creature is of course modified to its mate, as with all creatures. When the mate becomes also the master, when economic necessity is added to sex-attraction, we have the two great evolutionary forces acting together to the same end; namely, to develope sex-distinction in the human female. For, in her position of economic dependence in the sex-relation, sex-distinction is with her not only a means of attracting a mate, as with all creatures, but a means of getting her livelihood, as is the case with no other creature under heaven. Because of the economic dependence of the human female on her mate, she is modified to sex to an excessive degree. This excessive modification she transmits to her children; and so is steadily implanted in the human constitution the morbid tendency to excess in this relation, which has acted so universally upon us in all ages, in spite of our best efforts to restrain it. It is not the normal sex-tendency, common to all creatures, but an abnormal sex-tendency, produced and maintained by the abnormal economic relation which makes one sex get its living from the other by the exercise of sex-functions. This is the immediate effect upon individuals of the peculiar sexuo-economic relation which obtains among us.

III

In establishing the claim of excessive sex-distinction in the human race, much needs to be said to make clear to the general reader what is meant by the term. To the popular mind, both the coarsely familiar and the over-refined, "sexual" is thought to mean "sensual"; and the charge of excessive sex-distinction seems to be a reproach. This should be at once dismissed, as merely showing ignorance of the terms used. A man does not object to being called "masculine," nor a woman to being called "feminine." Yet whatever is masculine or feminine is sexual. To be distinguished by femininity is to be distinguished by sex. To be over-feminine is to be over-sexed. To manifest in excess any of the distinctions of sex, primary or secondary, is to be over-sexed. Our hypothetical peacock, with his too large and splendid tail, would be over-sexed, and no offence to his moral character!

The primary sex-distinctions in our race as in others consist merely in the essential organs and functions of reproduction. The secondary distinctions, and this is where we are to look for our largest excess—consist in all those differences in organ and function, in look and action, in habit, manner, method, occupation, behavior, which distinguish men from women. In a troop of horses, seen at a distance, the sexes are indistinguishable. In a herd of deer the males are distinguishable because of their antlers. The male lion is distinguished by his mane, the male cat only by a somewhat heavier build. In certain species of insects the male and female differ so widely in appearance that even naturalists have supposed them to belong to separate species. Beyond these distinctions lies that of conduct. Certain psychic attributes are manifested by either sex. The intensity of the maternal passion is a sex-distinction as much as the lion's mane or the stag's horns. The belligerence and dominance of the male is a sex-distinction: the modesty and timidity of the female is a sex-distinction. The tendency to "sit" is a sex-distinction of the hen: the tendency to strut is a sex-distinction of the cock. The tendency to fight is a sex-distinction of males in general: the tendency to protect and provide for is a sex-distinction of females in general.

With the human race, whose chief activities are social, the initial tendency to sex-distinction is carried out in many varied functions. We have differentiated our industries, our responsibilities, our very virtues, along sex lines. It will therefore be clear that the claim of excessive sex-distinction in humanity, and especially in woman, does not carry with it any specific "moral" reproach, though it does in the larger sense prove a decided evil in its effect on human progress.

In primary distinctions our excess is not so marked as in the farther and subtler development; yet, even here, we have plain proof of it. Sex-energy in its primal manifestation is exhibited in the male of the human species to a degree far greater than is necessary for the processes of reproduction,—enough, indeed, to subvert and injure those processes. The direct injury to reproduction from the excessive indulgence of the male, and the indirect injury through its debilitating effect upon the female, together with the enormous evil to society produced by extra-marital indulgence,—these are facts quite generally known. We have recognized them for centuries; and sought to check the evil action by law, civil, social, moral. But we have treated it always as a field of voluntary action, not as a condition of morbid development. We have held it as right that man should be so, but wrong that man should do so. Nature does not work in that way. What it is right to be, it is right to do. What it is wrong to do, it is wrong to be. This inordinate demand in the human male is an excessive sex-distinction. In this, in a certain over-coarseness and hardness, a too great belligerence and pride, a too great subservience to the power of sex-attraction, we find the main marks of excessive sex-distinction in men. It has been always checked and offset in them by the healthful activities of racial life. Their energies have been called out and their faculties developed along all the lines of human progress. In the growth of industry, commerce, science, manufacture, government, art, religion, the male of our species has become human, far more than male. Strong as this passion is in him, inordinate as is his indulgence, he is a far more normal animal than the female of his species,—far less over-sexed. To him this field of special activity is but part of life,—an incident. The whole world remains besides. To her it is the world. This has been well stated in the familiar epigram of Madame de Staël,—"Love with man is an episode, with woman a history." It is in woman that we find

most fully expressed the excessive sex-distinction of the human species,—physical, psychical, social. See first the physical manifestation.

To make clear by an instance the difference between normal and abnormal sex-distinction, look at the relative condition of a wild cow and a "milk cow," such as we have made. The wild cow is a female. She has healthy calves, and milk enough for them; and that is all the femininity she needs. Otherwise than that she is bovine rather than feminine. She is a light, strong, swift, sinewy creature, able to run, jump, and fight, if necessary. We, for economic uses, have artificially developed the cow's capacity for producing milk. She has become a walking milk-machine, bred and tended to that express end, her value measured in quarts. The secretion of milk is a maternal function,—a sex-function. The cow is over-sexed. Turn her loose in natural conditions, and, if she survive the change, she would revert in a very few generations to the plain cow, with her energies used in the general activities of her race, and not all running to milk.

Physically, woman belongs to a tall, vigorous, beautiful animal species, capable of great and varied exertion. In every race and time when she has opportunity for racial activity, she developes accordingly, and is no less a woman for being a healthy human creature. In every race and time where she is denied this opportunity,—and few, indeed, have been her years of freedom,—she has developed in the lines of action to which she was confined; and those were always lines of sex-activity. In consequence the body of woman, speaking in the largest generalization, manifests sex-distinction predominantly.

Woman's femininity—and "the eternal feminine" means simply the eternal sexual—is more apparent in proportion to her humanity than the femininity of other animals in proportion to their caninity or felinity or equinity. "A feminine hand" or "a feminine foot" is distinguishable anywhere. We do not hear of "a feminine paw" or "a feminine hoof." A hand is an organ of prehension, a foot an organ of locomotion: they are not secondary sexual characteristics. The comparative smallness and feebleness of woman is a sex-distinction. We have carried it to such an excess that women are commonly known as "the weaker sex." There is no such glaring difference between male and female in other advanced species. In the long migrations of birds, in the ceaseless motion of the grazing herds that used to swing up and down over the continent each year, in the wild, steep journeys of the breeding salmon, nothing is heard of the weaker sex. And among the higher carnivora, where longer maintenance of the young brings their condition nearer ours, the hunter dreads the attack of the female more than that of the male. The disproportionate weakness is an excessive sex-distinction. Its injurious effect may be broadly shown in the Oriental nations, where the female in curtained harems is confined most exclusively to sex-functions and denied most fully the exercise of race-functions. In such peoples the weakness, the tendency to small bones and adipose tissue of the over-sexed female, is transmitted to the male, with a retarding effect on the development of the race. Conversely, in early Germanic tribes the comparatively free and humanly developed women—tall, strong, and brave—transmitted to their sons a greater proportion of human power and much less of morbid sex-tendency.

The degree of feebleness and clumsiness common to women, the comparative inability to stand, walk, run, jump, climb, and perform other race-functions common to both sexes, is an excessive sex-distinction; and the ensuing transmission of this relative feebleness to their children, boys and girls alike, retards human development. Strong, free, active women, the sturdy, field-working peasant, the burden- bearing savage, are no less good mothers for their human strength. But our civilized "feminine delicacy," which appears somewhat less delicate when recognized as an expression of sexuality in excess,—makes us no better mothers, but worse. The relative weakness of women is a sex-distinction. It is apparent in her to a degree that injures motherhood, that injures wifehood, that injures the individual. The sex-usefulness and the human usefulness of women, their general duty to their kind, are greatly injured by this degree of distinction. In every way the over-sexed condition of the human female reacts unfavorably upon herself, her husband, her children, and the race.

In its psychic manifestation this intense sex-distinction is equally apparent. The primal instinct of sex-attraction has developed under social forces into a conscious passion of enormous power, a deep and lifelong devotion, overwhelming in its force. This is excessive in both sexes, but more so in women than in men,—not so commonly in its simple physical form, but in the unreasoning intensity of emotion that refuses all guidance, and drives those possessed by it to risk every other good for this one end. It is not at first sight easy, and it may seem an irreverent and thankless task, to discriminate here between what is good in the "master passion" and what is evil, and especially to claim for one sex more of this feeling than for the other; but such discrimination can be made.

It is good for the individual and for the race to have developed such a degree of passionate and permanent love as shall best promote the happiness of individuals and the reproduction of species. It is not good for the race or for the individual that his feeling should have become so intense as to override all other human faculties, to make a mock of the accumulated wisdom of the ages, the stored power of the will; to drive the individual—against his own plain conviction—into a union sure to result in evil, or to hold the individual helpless in such an evil union, when made.

Such is the condition of humanity, involving most evil results to its offspring and to its own happiness. And, while in men the immediate dominating force of the passion may be more conspicuous, it is in women that it holds more universal sway. For the man has other powers and faculties in full use, whereby to break loose from the force of this; and the woman, specially modified to sex and denied racial activity, pours her whole life into her love, and, if injured here, she is injured irretrievably. With him it is frequently light and transient, and, when most intense, often most transient. With her it is a deep, all-absorbing force, under the action of which she will renounce all that life offers, take any risk, face any hardships, bear any pain. It is maintained in her in the face of a lifetime of neglect and abuse. The common instance of the police court trials—the woman cruelly abused who will not testify against her husband—shows this. This devotion, carried to such a degree as to lead to the mismating of individuals with its personal and social injury, is an excessive sex-distinction.

But it is in our common social relations that the predominance of sex-distinction in women is made most manifest. The fact that, speaking broadly, women have, from the very beginning, been spoken of expressively enough as "the sex," demonstrates clearly that this is the main impression which they have made upon observers and recorders. Here one need attempt no farther proof than to turn the mind of the reader to an unbroken record of facts and feelings perfectly patent to every one, but not hitherto looked at as other than perfectly natural and right. So utterly has the status of woman been accepted as a sexual one that it has remained for the woman's movement of the nineteenth century to devote much contention to the claim that women are persons! That women are persons as well as females,—an unheard of proposition!

In a "Handbook of Proverbs of All Nations," a collection comprising many thousands, these facts are to be observed: first, that the proverbs concerning women are an insignificant minority compared to those concerning men; second, that the proverbs concerning women almost invariably apply to them in general,—to the sex. Those concerning men qualify, limit, describe, specialize. It is "a lazy man," "a violent man," "a man in his cups." Qualities and actions are predicated of man individually, and not as a sex, unless he is flatly contrasted with woman, as in "A man of straw is worth a woman of gold," "Men are deeds, women are words," or "Man, woman, and the devil are the three degrees of comparison." But of woman it is always and only "a woman," meaning simply a female, and recognizing no personal distinction: "As much pity to see a woman weep as to see a goose go barefoot." "He that hath an eel by the tail and a woman by her word hath a slippery handle." "A woman, a spaniel, and a walnut-tree,—the more you beat 'em, the better they be." Occasionally a distinction is made between "a fair woman" and "a black woman"; and Solomon's "virtuous woman," who commanded such a high price, is familiar to us all. But in common thought it is simply "a woman" always. The boast of the profligate that he knows "the sex," so recently expressed by a new poet,—"The things you will learn from the Yellow and Brown, they'll 'elp you an' 'eap with the White"; the complaint of the angry rejected that "all women are just alike!"—the consensus of public opinion of all time goes to show that the characteristics common to the sex have predominated over the characteristics distinctive of the individual,—a marked excess in sex-distinction.

From the time our children are born, we use every means known to accentuate sex-distinction in both boy and girl; and the reason that the boy is not so hopelessly marked by it as the girl is that he has the whole field of human expression open to him besides. In our steady insistence on proclaiming sex-distinction we have grown to consider most human attributes as masculine attributes, for the simple reason that they were allowed to men and forbidden to women.

A clear and definite understanding of the difference between race-attributes and sex-attributes should be established. Life consists of action. The action of a living thing is along two main lines,—self-preservation and race-preservation. The processes that keep the individual alive, from the involuntary action of his internal organs to the voluntary action of his external organs,—every act, from breathing to hunting his food, which contributes to the maintenance of the individual life,—these are the processes of self-preservation. Whatever activities tend to keep the race alive, to reproduce the individual, from the involuntary action of the internal organs to the voluntary action of the external organs; every act from the development of germ-cells to the taking care of children, which contributes to the maintenance of the racial life,—these are the processes of race-preservation. In race-preservation, male and female have distinctive organs, distinctive functions, distinctive lines of action. In self-preservation, male and female have the same organs, the same functions, the same lines of action. In the human species our processes of race-preservation have reached a certain degree of elaboration; but our processes of self-preservation have gone farther, much farther.

All the varied activities of economic production and distribution, all our arts and industries, crafts and trades, all our growth in science, discovery, government, religion,—these are along the line of self-preservation: these are, or should be, common to both sexes. To teach, to rule, to make, to decorate, to distribute,—these are not sex-functions: they are race-functions. Yet so inordinate is the sex-distinction of the human race that the whole field of human progress has been considered a masculine prerogative. What could more absolutely prove the excessive sex-distinction of the human race? That this difference should surge over all its natural boundaries and blazon itself across every act of life, so that every step of the human creature is marked "male" or "female,"—surely, this is enough to show our over-sexed condition.

Little by little, very slowly, and with most unjust and cruel opposition, at cost of all life holds most dear, it is being gradually established by many martyrdoms that human work is woman's as well as man's. Harriet Martineau must conceal her writing under her sewing when callers came, because "to sew" was a feminine verb, and "to write" a masculine one. Mary Somerville must struggle to hide her work from even relatives, because mathematics was a "masculine" pursuit. Sex has been made to dominate the whole human world,—all the main avenues of life marked "male," and the female left to be a female, and nothing else.

But while with the male the things he fondly imagined to be "masculine" were merely human, and very good for him, with the female the few things marked "feminine" were feminine, indeed; and her ceaseless reiterance of one short song, however sweet, has given it a conspicuous monotony. In garments whose main purpose is unmistakably to announce her sex; with a tendency to ornament which marks exuberance of sex-energy, with a body so modified to sex as to be grievously deprived of its natural activities; with a manner and behavior wholly attuned to sex-advantage, and frequently most disadvantageous to any human gain; with a field of action most rigidly confined to sex-relations; with her overcharged sensibility, her prominent modesty, her "eternal femininity,"—the female of genus homo is undeniably over-sexed.

This excessive distinction shows itself again in a marked precocity of development. Our little children, our very babies, show signs of it when the young of other creatures are serenely asexual in general appearance and habit. We eagerly note this precocity. We are proud of it. We carefully encourage it by precept and example, taking pains to develope the sex-instinct in little children, and think no harm. One of the first things we force upon the child's dawning consciousness is the fact that he is a boy or that she is a girl, and that, therefore, each must regard everything from a different point of view. They must be dressed differently, not on account of their personal needs, which are exactly similar at this period, but so that neither they, nor any one beholding them, may for a moment forget the distinction of sex.

Our peculiar inversion of the usual habit of species, in which the male carries ornament and the female is dark and plain, is not so much a proof of excess indeed, as a proof of the peculiar reversal of our position in the matter of sex-selection. With the other species the males compete in ornament, and the females select. With us the females compete in ornament, and the males select. If this theory of sex-ornament is disregarded, and we prefer rather to see in masculine decoration merely a form of exuberant sex-energy, expending itself in non-productive excess, then, indeed, the fact that with us the females manifest such a display of gorgeous adornment is another sign of excessive sex-distinction. In either case the forcing upon girl-children of an elaborate ornamentation which interferes with their physical activity and unconscious freedom, and fosters a premature sex-consciousness, is as clear and menacing a proof of our condition as could be mentioned. That the girl-child should be so dressed as to require a difference in care and behavior, resting wholly on the fact that she is a girl,—a fact not otherwise present to her thought at that age,—is a precocious insistence upon sex-distinction, most unwholesome in its results. Boys and girls are expected, also, to behave differently to each other, and to people in general,—a behavior to be briefly described in two words. To the boy we say, "Do"; to the girl, "Don't." The little boy must "take care" of the little girl, even if she is larger than he is. "Why?" he asks. Because he is a boy. Because of sex. Surely, if she is the stronger, she ought to take care of him, especially as the protective instinct is purely feminine in a normal race. It is not long before the boy learns his lesson. He is a boy, going to be a man; and that means all. "I thank the Lord that I was not born a woman," runs the Hebrew prayer. She is a girl, "only a girl," "nothing but a girl," and going to be a woman,—only a woman. Boys are encouraged from the beginning to show the feelings supposed to be proper to their sex. When our infant son bangs about, roars, and smashes things, we say proudly that he is "a regular boy!" When our infant daughter coquettes with visitors, or wails in maternal agony because her brother has broken her doll, whose sawdust remains she nurses with piteous care, we say proudly that "she is a perfect little mother already!" What business has a little girl with the instincts of maternity? No more than the little boy should have with the instincts of paternity. They are sex-instincts, and should not appear till the period of adolescence.

The most normal girl is the "tom-boy,"—whose numbers increase among us in these wiser days,—a healthy young creature, who is human through and through, not feminine till it is time to be. The most normal boy has calmness and gentleness as well as vigor and courage. He is a human creature as well as a male creature, and not aggressively masculine till it is time to be. Childhood is not the period for these marked manifestations of sex. That we exhibit them, that we admire and encourage them, shows our over-sexed condition.

IV

Having seen the disproportionate degree of sex-distinction in humanity and its greater manifestation in the female than in the male, and having seen also the unique position of the human female as an economic dependant on the male of her species, it is not difficult to establish a relation between these two facts. The general law acting to produce this condition of exaggerated sex-development was briefly referred to in the second chapter. It is as follows: the natural tendency of any function to increase in power by use causes sex-activity to increase under the action of sexual selection. This tendency is checked in most species by the force of natural selection, which diverts the energies into other channels and developes race-activities. Where the female finds her economic environment in the male, and her economic advantage is directly conditioned upon the sex-relation, the force of natural selection is added to the force of sexual selection, and both together operate to develope sex-activity. In any animal species, free from any other condition, such a relation would have inevitably developed sex to an inordinate degree, as may be readily seen in the comparatively similar cases of those insects where the female, losing economic activity and modified entirely to sex, becomes a mere egg-sac, an organism with no powers of self-preservation, only those of race-preservation. With these insects the only race-problem is to maintain and reproduce the species, and such a condition is not necessarily evil; but with a race like ours, whose development as human creatures is but comparatively begun, it is evil because of its check to individual and racial progress. There

are other purposes before us besides mere maintenance and reproduction.

It should be clear to any one accustomed to the working of biological laws that all the tendencies of a living organism are progressive in their development, and are held in check by the interaction of their several forces. Each living form, with its dominant characteristics, represents a balance of power, a sort of compromise. The size of earth's primeval monsters was limited by the tensile strength of their material. Sea monsters can be bigger, because the medium in which they move offers more support. Birds must be smaller for the opposite reason. The cow requires many stomachs of a liberal size, because her food is of low nutritive value; and she must eat large quantities to keep her machine going. The size of arboreal animals, such as monkeys or squirrels, is limited by the nature of their habitat: creatures that live in trees cannot be so big as creatures that live on the ground. Every quality of every creature is relative to its condition, and tends to increase or decrease accordingly; and each quality tends to increase in proportion to its use, and to decrease in proportion to its disuse. Primitive man and his female were animals, like other animals. They were strong, fierce, lively beasts; and she was as nimble and ferocious as he, save for the added belligerence of the males in their sex-competition. In this competition, he, like the other male creatures, fought savagely with his hairy rivals; and she, like the other female creatures, complacently viewed their struggles, and mated with the victor. At other times she ran about in the forest, and helped herself to what there was to eat as freely as he did.

There seems to have come a time when it occurred to the dawning intelligence of this amiable savage that it was cheaper and easier to fight a little female, and have it done with, than to fight a big male every time. So he instituted the custom of enslaving the female; and she, losing freedom, could no longer get her own food nor that of her young. The mother ape, with her maternal function well fulfilled, flees leaping through the forest,—plucks her fruit and nuts, keeps up with the movement of the tribe, her young one on her back or held in one strong arm. But the mother woman, enslaved, could not do this. Then man, the father, found that slavery had its obligations: he must care for what he forbade to care for itself, else it died on his hands. So he slowly and reluctantly shouldered the duties of his new position. He began to feed her, and not only that, but to express in his own person the thwarted uses of maternity: he had to feed the children, too. It seems a simple arrangement. When we have thought of it at all, we have thought of it with admiration. The naturalist defends it on the ground of advantage to the species through the freeing of the mother from all other cares and confining her unreservedly to the duties of maternity. The poet and novelist, the painter and sculptor, the priest and teacher, have all extolled this lovely relation. It remains for the sociologist, from a biological point of view, to note its effects on the constitution of the human race, both in the individual and in society.

When man began to feed and defend women, she ceased proportionately to feed and defend herself. When he stood between her and her physical environment, she ceased proportionately to feel the influence of that environment and respond to it. When he became her immediate and all-important environment, she began proportionately to respond to this new influence, and to be modified accordingly. In a free state, speed was of as great advantage to the female as to the male, both in enabling her to catch prey and in preventing her from being caught by enemies; but, in her new condition, speed was a disadvantage. She was not allowed to do the catching, and it profited her to be caught by her new master. Free creatures, getting their own food and maintaining their own lives, develope an active capacity for attaining their ends. Parasitic creatures, whose living is obtained by the exertions of others, develope powers of absorption and of tenacity,—the powers by which they profit most. The human female was cut off from the direct action of natural selection, that mighty force which heretofore had acted on male and female alike with inexorable and beneficial effect, developing strength, developing skill, developing endurance, developing courage,—in a word, developing species. She now met the influence of natural selection acting indirectly through the male, and developing, of course, the faculties required to secure and obtain a hold on him. Needless to state that these faculties were those of sex-attraction, the one power that has made him cheerfully maintain, in what luxury he

could, the being in whom he delighted. For many, many centuries she had no other hold, no other assurance of being fed. The young girl had a prospective value, and was maintained for what should follow; but the old woman, in more primitive times, had but a poor hold on life. She who could best please her lord was the favorite slave or favorite wife, and she obtained the best economic conditions.

With the growth of civilization, we have gradually crystallized into law the visible necessity for feeding the helpless female; and even old women are maintained by their male relatives with a comfortable assurance. But to this day—save, indeed, for the increasing army of women wage-earners, who are changing the face of the world by their steady advance toward economic independence—the personal profit of women bears but too close a relation to their power to win and hold the other sex. From the odalisque with the most bracelets to the débutante with the most bouquets, the relation still holds good,—woman's economic profit comes through the power of sex-attraction.

When we confront this fact boldly and plainly in the open market of vice, we are sick with horror. When we see the same economic relation made permanent, established by law, sanctioned and sanctified by religion, covered with flowers and incense and all accumulated sentiment, we think it innocent, lovely, and right. The transient trade we think evil. The bargain for life we think good. But the biological effect remains the same. In both cases the female gets her food from the male by virtue of her sex-relationship to him. In both cases, perhaps even more in marriage because of its perfect acceptance of the situation, the female of genus homo, still living under natural law, is inexorably modified to sex in an increasing degree.

Followed in specific detail, the action of the changed environment upon women has been in given instances as follows: In the matter of mere passive surroundings she has been immediately restricted in her range. This one factor has an immense effect on man and animal alike. An absolutely uniform environment, one shape, one size, one color, one sound, would render life, if any life could be, one helpless, changeless thing. As the environment increases and varies, the development of the creature must increase and vary with it; for he acquires knowledge and power, as the material for knowledge and the need for power appear. In migratory species the female is free to acquire the same knowledge as the male by the same means, the same development by the same experiences. The human female has been restricted in range from the earliest beginning. Even among savages, she has a much more restricted knowledge of the land she lives in. She moves with the camp, of course, and follows her primitive industries in its vicinity; but the war-path and the hunt are the man's. He has a far larger habitat. The life of the female savage is freedom itself, however, compared with the increasing constriction of custom closing in upon the woman, as civilization advanced, like the iron torture chamber of romance. Its culmination is expressed in the proverb: "A woman should leave her home but three times,—when she is christened, when she is married, and when she is buried." Or this: "The woman, the cat, and the chimney should never leave the house." The absolutely stationary female and the wide-ranging male are distinctly human institutions, after we leave behind us such low forms of life as the gypsy moth, whose female seldom moves more than a few feet from the pupa moth. She has aborted wings, and cannot fly. She waits humbly for the winged male, lays her myriad eggs, and dies,—a fine instance of modification to sex.

To reduce so largely the mere area of environment is a great check to race-development; but it is not to be compared in its effects with the reduction in voluntary activity to which the human female has been subjected. Her restricted impression, her confinement to the four walls of the home, have done great execution, of course, in limiting her ideas, her information, her thought-processes, and power of judgment; and in giving a disproportionate prominence and intensity to the few things she knows about; but this is innocent in action compared with her restricted expression, the denial of freedom to act. A living organism is modified far less through the action of external circumstances upon it and its reaction thereto, than through the effect of its own exertions. Skin may be thickened gradually by exposure to the weather; but it is thickened far more quickly by being rubbed against something, as the handle of an oar or of a broom. To be surrounded by beautiful things has much influence upon the human creature: to make beautiful things has more. To live among

beautiful surroundings and make ugly things is more directly lowering than to live among ugly surroundings and make beautiful things. What we do modifies us more than what is done to us. The freedom of expression has been more restricted in women than the freedom of impression, if that be possible. Something of the world she lived in she has seen from her barred windows. Some air has come through the purdah's folds, some knowledge has filtered to her eager ears from the talk of men. Desdemona learned somewhat of Othello. Had she known more, she might have lived longer. But in the ever-growing human impulse to create, the power and will to make, to do, to express one's new spirit in new forms,—here she has been utterly debarred. She might work as she had worked from the beginning,—at the primitive labors of the household; but in the inevitable expansion of even those industries to professional levels we have striven to hold her back. To work with her own hands, for nothing, in direct body-service to her own family,—this has been permitted,—yes, compelled. But to be and do anything further from this she has been forbidden. Her labor has not only been limited in kinds, but in degree. Whatever she has been allowed to do must be done in private and alone, the first-hand industries of savage times.

Our growth in industry has been not only in kind, but in class. The baker is not in the same industrial grade with the house-cook, though both make bread. To specialize any form of labor is a step up: to organize it is another step. Specialization and organization are the basis of human progress, the organic methods of social life. They have been forbidden to women almost absolutely. The greatest and most beneficent change of this century is the progress of women in these two lines of advance. The effect of this check in industrial development, accompanied as it was by the constant inheritance of increased racial power, has been to intensify the sensations and emotions of women, and to develope great activity in the lines allowed. The nervous energy that up to present memory has impelled women to labor incessantly at something, be it the veriest folly of fancy work, is one mark of this effect.

In religious development the same dead-line has held back the growth of women through all the races and ages. In dim early times she was sharer in the mysteries and rites; but, as religion developed, her place receded, until Paul commanded her to be silent in the churches. And she has been silent until to-day. Even now, with all the ground gained, we have but the beginnings—the slowly forced and disapproved beginnings—of religious equality for the sexes. In some nations, religion is held to be a masculine attribute exclusively, it being even questioned whether women have souls. An early Christian council settled that important question by vote, fortunately deciding that they had. In a church whose main strength has always been derived from the adherence of women, it would have been an uncomfortable reflection not to have allowed them souls. Ancient family worship ran in the male line. It was the son who kept the sacred grandfathers in due respect, and poured libations to their shades. When the woman married, she changed her ancestors, and had to worship her husband's progenitors instead of her own. This is why the Hindu and the Chinaman and many others of like stamp must have a son to keep them in countenance,—a deep-seated sex-prejudice, coming to slow extinction as women rise in economic importance.

It is painfully interesting to trace the gradual cumulative effect of these conditions upon women: first, the action of large natural laws, acting on her as they would act on any other animal; then the evolution of social customs and laws (with her position as the active cause), following the direction of mere physical forces, and adding heavily to them; then, with increasing civilization, the unbroken accumulation of precedent, burnt into each generation by the growing force of education, made lovely by art, holy by religion, desirable by habit; and, steadily acting from beneath, the unswerving pressure of economic necessity upon which the whole structure rested. These are strong modifying conditions, indeed.

The process would have been even more effective and far less painful but for one important circumstance. Heredity has no Salic law. Each girl-child inherits from her father a certain increasing percentage of human development, human power, human tendency; and each boy as well inherits from his mother the increasing percentage of sex-development, sex-power, sex-tendency. The action of heredity has been to equalize what every tendency of environment and education made to differ. This has saved us from such a female as the gypsy moth. It has

held up the woman, and held down the man. It has set iron bounds to our absurd effort to make a race with one sex a million years behind the other. But it has added terribly to the pain and difficulty of human life,—a difficulty and a pain that should have taught us long since that we were living on wrong lines. Each woman born, re-humanized by the current of race activity carried on by her father and re-womanized by her traditional position, has had to live over again in her own person the same process of restriction, repression, denial; the smothering "no" which crushed down all her human desires to create, to discover, to learn, to express, to advance. Each woman has had, on the other hand, the same single avenue of expression and attainment; the same one way in which alone she might do what she could, get what she might. All other doors were shut, and this one always open; and the whole pressure of advancing humanity was upon her. No wonder that young Daniel in the apocryphal tale proclaimed: "The king is strong! Wine is strong! But women are stronger!"

To the young man confronting life the world lies wide. Such powers as he has he may use, must use. If he chooses wrong at first, he may choose again, and yet again. Not effective or successful in one channel, he may do better in another. The growing, varied needs of all mankind call on him for the varied service in which he finds his growth. What he wants to be, he may strive to be. What he wants to get, he may strive to get. Wealth, power, social distinction, fame,—what he wants he can try for.

To the young woman confronting life there is the same world beyond, there are the same human energies and human desires and ambition within. But all that she may wish to have, all that she may wish to do, must come through a single channel and a single choice. Wealth, power, social distinction, frame,—not only these, but home and happiness, reputation, ease and pleasure, her bread and butter,—all must come to her through a small gold ring. This is a heavy pressure. It has accumulated behind her through heredity, and continued about her through environment. It has been subtly trained into her through education, till she herself has come to think it a right condition, and pours its influence upon her daughter with increasing impetus. Is it any wonder that women are over-sexed? But for the constant inheritance from the more human male, we should have been queen bees, indeed, long before this. But the daughter of the soldier and the sailor, of the artist, the inventor, the great merchant, has inherited in body and brain her share of his development in each generation, and so stayed somewhat human for all her femininity.

All morbid conditions tend to extinction. One check has always existed to our inordinate sex-development,—nature's ready relief, death. Carried to its furthest excess, the individual has died, the family has become extinct, the nation itself has perished, like Sodom and Gomorrah. Where one function is carried to unnatural excess, others are weakened, and the organism perishes. We are familiar with this in individual cases,—at least, the physician is. We can see it somewhat in the history of nations. From younger races, nearer savagery, nearer the healthful equality of pre-human creatures, has come each new start in history. Persia was older than Greece, and its highly differentiated sexuality had produced the inevitable result of enfeebling the racial qualities. The Greek commander stripped the rich robes and jewels from his Persian captives, and showed their unmanly feebleness to his men. "You have such bodies as these to fight for such plunder as this," he said. In the country, among peasant classes, there is much less sex-distinction than in cities, where wealth enables the women to live in absolute idleness; and even the men manifest the same characteristics. It is from the country and the lower classes that the fresh blood pours into the cities, to be weakened in its turn by the influence of this unnatural distinction until there is none left to replenish the nation.

The inevitable trend of human life is toward higher civilization; but, while that civilization is confined to one sex, it inevitably exaggerates sex-distinction, until the increasing evil of this condition is stronger than all the good of the civilization attained, and the nation falls. Civilization, be it understood, does not consist in the acquisition of luxuries. Social development is an organic development. A civilized State is one in which the citizens live in organic industrial relation. The more full, free, subtle, and easy that relation; the more perfect the differentiation of labor and exchange of product, with their correlative institutions,—the more perfect is that civilization. To eat, drink, sleep,

and keep warm,—these are common to all animals, whether the animal couches in a bed of leaves or one of eiderdown, sleeps in the sun to avoid the wind or builds a furnace-heated house, lies in wait for game or orders a dinner at a hotel. These are but individual animal processes. Whether one lays an egg or a million eggs, whether one bears a cub, a kitten, or a baby, whether one broods its chickens, guards its litter, or tends a nursery full of children, these are but individual animal processes. But to serve each other more and more widely; to live only by such service; to develope special functions, so that we depend for our living on society's return for services that can be of no direct use to ourselves,—this is civilization, our human glory and race-distinction.

All this human progress has been accomplished by men. Women have been left behind, outside, below, having no social relation whatever, merely the sex-relation, whereby they lived. Let us bear in mind that all the tender ties of family are ties of blood, of sex-relationship. A friend, a comrade, a partner,—this is a human relative. Father, mother, son, daughter, sister, brother, husband, wife,—these are sex-relatives. Blood is thicker than water, we say. True. But ties of blood are not those that ring the world with the succeeding waves of progressive religion, art, science, commerce, education, all that makes us human. Man is the human creature. Woman has been checked, starved, aborted in human growth; and the swelling forces of race-development have been driven back in each generation to work in her through sex-functions alone.

This is the way in which the sexuo-economic relation has operated in our species, checking race-development in half of us, and stimulating sex-development in both.

V

The facts stated in the foregoing chapters are familiar and undeniable, the argument seems clear; yet the mind reacts violently from the conclusions it is forced to admit, and tries to find relief in the commonplace conditions of every-day life. From this looming phantom of the over-sexed female of genius homo we fly back in satisfaction to familiar acquaintances and relatives,—to Mrs. John Smith and Miss Imogene Jones, to mothers and sisters and daughters and sweethearts and wives. We feel that such a dreadful state of things cannot be true, or we should surely have noticed it. We may even perform that acrobatic feat so easy to most minds,—admit that the statement may be theoretically true, but practically false!

Two simple laws of brain action are responsible for the difficulty of convincing the human race of any large general truths concerning itself. One is common to all brains, to all nerve sensations indeed, and is cheerfully admitted to have nothing to do with the sexuo-economic relation. It is this simple fact, in popular phrase,—that what we are used to we do not notice. This rests on the law of adaptation, the steady, ceaseless pressure that tends to fit the organism to the environment. A nerve touched for the first time with a certain impression feels this first impression far more than the hundredth or thousandth, though the thousandth be far more violent than the first. If an impression be constant and regular, we become utterly insensitive to it, and only respond under some special condition, as the ticking of a clock, the noise of running water or waves on the beach, even the clatter of railroad trains, grows imperceptible to those who hear it constantly. It is perfectly possible for an individual to become accustomed to the most disadvantageous conditions, and fail to notice them.

It is equally possible for a race, a nation, a class, to become accustomed to most disadvantageous conditions, and fail to notice them. Take, as an individual instance, the wearing of corsets by women. Put a corset, even a loose one, on a vigorous man or woman who never wore one, and there is intense discomfort, and a vivid consciousness thereof. The healthy muscles of the trunk resent the pressure, the action of the whole body is checked in the middle, the stomach is choked, the process of digestion interfered with; and the victim says, "How can you bear such a thing?"

But the person habitually wearing a corset does not feel these evils. They exist, assuredly, the facts are there, the body is not deceived; but the nerves have become accustomed to these disagreeable sensations, and no longer respond to them. The person "does not feel it." In fact, the wearer becomes so used to the sensations that,

when they are removed,—with the corset,—there is a distinct sense of loss and discomfort. The heavy folds of the cravat, stock, and neckcloth of earlier men's fashions, the heavy horse-hair peruke, the stiff high collar of to-day, the kind of shoes we wear,—these are perfectly familiar instances of the force of habit in the individual.

This is equally true of racial habits. That a king should rule because he was born, passed unquestioned for thousands of years. That the eldest son should inherit the titles and estates was a similar phenomenon as little questioned. That a debtor should be imprisoned, and so entirely prevented from paying his debts, was common law. So glaring an evil as chattel slavery was an unchallenged social institution from earliest history to our own day among the most civilized nations of the earth. Christ himself let it pass unnoticed. The hideous injustice of Christianity to the Jew attracted no attention through many centuries. That the serf went with the soil, and was owned by the lord thereof, was one of the foundations of society in the Middle Ages.

Social conditions, like individual conditions, become familiar by use, and cease to be observed. This is the reason why it is so much easier to criticise the customs of other persons or other nations than our own. It is also the reason why we so naturally deny and resent the charges of the critic. It is not necessarily because of any injustice on the one side or dishonesty on the other, but because of a simple and useful law of nature. The Englishman coming to America is much struck by America's political corruption; and, in the earnest desire to serve his brother, he tells us all about it. That which he has at home he does not observe, because he is used to it. The American in England finds also something to object to, and omits to balance his criticism by memories of home.

When a condition exists among us which began in those unrecorded ages back of tradition even, which obtains in varying degree among every people on earth, and which begins to act upon the individual at birth, it would be a miracle past all belief if people should notice it. The sexuo-economic relation is such a condition. It began in primeval savagery. It exists in all nations. Each boy and girl is born into it, trained into it, and has to live in it. The world's progress in matters like these is attained by a slow and painful process, but one which works to good ends.

In the course of social evolution there are developed individuals so constituted as not to fit existing conditions, but to be organically adapted to more advanced conditions. These advanced individuals respond in sharp and painful consciousness to existing conditions, and cry out against them according to their lights. The history of religion, of political and social reform, is full of familiar instances of this. The heretic, the reformer, the agitator, these feel what their compeers do not, see what they do not, and, naturally, say what they do not. The mass of the people are invariably displeased by the outcry of these uneasy spirits. In simple primitive periods they were promptly put to death. Progress was slow and difficult in those days. But this severe process of elimination developed the kind of progressive person known

as a martyr; and this remarkable sociological law was manifested: that the strength of a current of social force is increased by the sacrifice of individuals who are willing to die in the effort to promote it. "The blood of the martyrs is the seed of the church." This is so commonly known to-day, though not formulated, that power hesitates to persecute, lest it intensify the undesirable heresy. A policy of "free speech" is found to let pass most of the uneasy pushes and spurts of these stirring forces, and lead to more orderly action. Our great anti-slavery agitation, the heroic efforts of the "women's rights" supporters, are fresh and recent proofs of these plain facts: that the mass of the people do not notice existing conditions, and that they are not pleased with those who do. This is one strong reason why the sexuo-economic relation passes unobserved among us, and why any statement of it will be so offensive to many. . . .

Discussion Questions

1. One of Gilman's main points is that women cannot be equal to men unless they are economically independent. Do you agree or disagree? Why or why not? Use concrete examples to explain and support your point of view.

2. For Gilman, "women's work" is lonely and demeaning. Do you think "parentwork" and housework has to be this way? Why or why not? What specific measures, institutions, or practices might help prevent or combat these tendencies?

3. Discuss the specific advances in gender equality that have occurred since Gilman's day. What specific issues highlighted by Gilman do you consider still problematic, and which problems do you consider "eradicated" (at least in the United States)?

4. Compare and contrast Gilman's discussion of social bonds and mental health with

that of Durkheim. Does each theorist conceive of social bonds as working in the same way? How so or why not? Discuss how each theorist would construe the social bonds between (1) mothers and their children, and (2) husbands and wives.

5. Compare and contrast Gilman's theory as to the oppression of women in patriarchy with Marx's theory as to how and why workers are oppressed under capitalism. What similarities do you see in their arguments? What are the differences in these two theories of oppression?

6. Compare and contrast Gilman's perspective on the family, gender roles, and the division of labor in *Women and Economics* with that of Engels in *The Origin of the Family, Private Property and the State*. To what extent do Engels and Gilman's historical trajectories of the family and perceptions about monogamy cohere with one another?

6 GEORG SIMMEL (1858–1918)

Key Concepts

- Duality
- Form
- Content
- Types
- Tragedy of culture
- Blasé attitude

Just as the universe needs "love and hate," that is, attractive and repulsive forces, in order to have any form at all, so society, too, in order to attain a determinate shape, needs some quantitative ratio of harmony and disharmony, of association and competition, of favorable and unfavorable tendencies.

(Simmel 1908b/1971:72)

As the quotation above suggests, Georg Simmel's work is informed by a combining of opposites into a whole, where social life is based on seeming contradictions. Consider, for instance, one example from Simmel's work that expresses this dynamic: fashion. Sociologically, the fascinating thing about fashion is that whether you wear baggy jeans, toe rings and tattoos, or Armani suits, fashion signals both your individuality *and* your attachment to specific social groups. Whether age-related (e.g., youth), hobby-related (e.g., skaters), or attitude-related (e.g., hip, urban), fashion is at once a process of conforming to some groups, while distancing yourself from others. Moreover, even within a specific attachment, you strive for your "own" look, that is, you don't want to look identical to your friends. Fashion, then, is not a singular, "pure" expression, but is built on two opposing forces: differentiation/individuality and conformity/imitation.

Simmel's analyses of social life have produced a unique legacy among the classical theorists. Unlike Marx, Durkheim, Weber, and Mead (see Chapter 8), no definable theoretical school has coalesced around his work. Nevertheless, it should not be inferred from this that

Simmel's work has proved less fertile for sociologists. On the contrary, his impact on the discipline has been far reaching. The influence of Simmel's insights is found in the work of his contemporaries (most notably Weber and Mead) as well as in that of succeeding generations of sociologists. Indeed, the names of those who have drawn inspiration from Simmel reads as a veritable list of "who's who" in sociology. Theorists have drawn on his work in their explorations of topics ranging from small group dynamics, networks of interpersonal relationships, processes of exchange behavior, and the nature of social conflict, to the character of the urban environment and its effects on the individual and social life, the consequences of modernization for culture and individual personality, and the patterns of apprehension through which we experience our world.

In a sense, Simmel himself predicted this outcome:

I know that I shall die without intellectual heirs, and that is as it should be. My legacy will be, as it were, in cash, distributed to many heirs, each transforming his part into use conformed to *his* nature: a use which will reveal no longer its indebtedness to this heritage. (Levine 1971:xii)

■■ A Biographical Sketch

Georg Simmel was born on March 1, 1858, in Berlin, the urban center of Germany. His father, a successful businessman and part owner of a Berlin chocolate factory, died when Georg was an infant. However, Georg's financial future was secured when a friend of the family, the owner of a profitable music publishing house, was appointed his guardian. The guardianship provided Georg with a comfortable childhood and, upon adulthood, allowed him to pursue his intellectual interests with relatively few monetary worries.

As fate would have it, Simmel's inheritance proved more beneficial than might appear at first sight. Though his reputation as a scholar drew international attention, and his skills as a lecturer earned him a degree of popularity among students and Berlin's cultural elite, Simmel was unable to obtain a permanent academic position throughout most of his career. This was due, in part, to the seemingly fragmentary nature of his work. Simmel wrote on a wide range of subjects that crossed disciplinary boundaries. For instance, he wrote books on the artists Rembrandt, Goethe, and Rodin, and published essays, many of which appeared in nonacademic journals, on unusual topics such as the sociology of smell and the sociology of secrecy. Such eclecticism was frowned on in the conservative climate of German universities. Instead, "real" scholars were those who committed themselves to sustained analyses of a limited set of questions. Moreover, the discipline of sociology, itself only recently born, was met with significant resistance within the university establishment. Thus, appointments to full-time positions in sociology departments were far less available relative to the other disciplines.

However, in addition to the more academic obstacles, there is another reason for Simmel's lack of professional success. The German university establishment—and, for that matter, much of Europe more generally—was tainted by anti-Semitism. As a Jew, Simmel encountered discrimination first hand. In 1908, he was recommended for the chair of philosophy at the University of Heidelberg. His candidacy was rejected, however, when the minister of education in Baden requested an evaluation of Simmel's qualifications from Professor Dietrich Schaefer, a prominent historian at the University of Berlin. Schaefer wrote of Simmel,

[h]e is . . . a dyed-in-the-wool Israelite, in his outward appearance, in his bearing, and in his manner of thinking. . . . He spices his words with clever sayings. And the audience he recruits is composed accordingly. The ladies constitute a very large portion. . . . For the rest, there [appears at his lectures] an extraordinarily numerous contingent of the oriental world. (Quoted in Coser 1977:209)

Simmel was thus "guilty" on three counts: (1) he was Jewish, (2) he was "clever" (i.e., his analyses were superficial), and (3) his lectures, though well attended, attracted intellectually inferior "Orientals" (i.e., foreigners) and women.

Anti-Semitism together with his own refusal to pursue a conventional intellectual track left Simmel confined to the margins of academia for most of his career. In 1885, he was appointed a *Privatdozent*, or unpaid lecturer, at the University of Berlin. After 16 years, he was granted the purely honorary title of *Außerordentlicher Professor*, or Extraordinary Professor, a position that still precluded him from fully participating in departmental and university affairs. Finally, in 1914, at the age of 56, Simmel was awarded a full professorship at the University of Strasbourg. His time at Strasbourg was far from ideal, however. Shortly after his arrival, the dormitories and lecture halls were turned over to the military for use as barracks and hospitals; World War I had just erupted. Simmel would die four years later of liver cancer.

Simmel's marginal status within the university establishment did not prevent him from playing a pivotal role in the development of sociology, however. What Durkheim was to sociology in France, Simmel was in Germany. He taught some of the first sociology courses in Germany (Ashley and Orenstein 1998:309) and, together with Max Weber and Ferdinand Tönnies (1855–1936), founded the German Society for Sociology. Indeed, it was through the efforts of these three that the emerging discipline gained a foothold within the German university system. In addition, Simmel was an active figure in a number of Berlin's intellectual circles. Through the salons, he developed friendships with writers, poets, journalists, and artists. He was also friends with a number of leading academics including Weber, the philosopher and founder of phenomenology Edmund Husserl, and the philosopher Heinrich Rickert. Each of these men lent their continued support to and endorsement of Simmel as he sought to win a full-time university appointment. Finally, lack of recognition from the university establishment did not adversely affect Simmel's productivity. During his life, he published more than 200 articles and 30 books, some of which were translated into five languages (Ashley and Orenstein 1998:311; Coser 1977:195).

INTELLECTUAL INFLUENCES AND CORE IDEAS

Outlining the core ideas that make up Simmel's work is a task made particularly difficult for a number of reasons. First, while he saw himself first and foremost as a philosopher (Frisby 1984:25), Simmel's intellectual interests spanned three disciplines: philosophy, history, and sociology (Levine 1971:xxi). Second, even within these three fields, his intellectual pursuits led him into a number of directions that were not necessarily related. In addition to his more sociological writings, varied as they are, Simmel published works on aesthetics, ethics, religion, the philosophy of history, the philosophies of Nietzsche and Schopenhauer, and the metaphysics of individuality. Third, Simmel, unlike, say, Marx or Mead, did not set out to construct a coherent theoretical scheme, nor did he explicitly aim to develop a systematic critique of or to build on a specific theoretical paradigm. As a result, his work perhaps is best seen as a "collection of insights" (Collins and Makowsky 1998:160) that provides for the sociologist a unique and subtle understanding of social interaction.[1]

In this section, we restrict ourselves to presenting an overview of Simmel's central sociological ideas. To this end, we touch on the following issues: (1) Simmel's image of society, (2) his view of sociology as a discipline, and (3) the plight of the individual in modern society.

---

[1]There is yet another reason why tracing the connections between Simmel's ideas and those articulated by other scholars is a complicated endeavor: he refused to put footnotes or citations in his publications. His avoidance of the professionally expected practice was one more reason for the disdain he evoked from many within the academic establishment.

Society

Simmel's conception of society stood in contrast both to the organic view developed by Comte, Spencer, and Durkheim, and to the purely abstract view articulated within German idealist philosophy. From the former vantage point, society is seen as having a reality outside or independent of the existence of the interacting individuals who compose it. Thus, society is seen as working *down* on individuals as it shapes, or even determines, their behavior, attitudes, and interests. On the other hand, idealists reject the notion of society as a reality sui generis, arguing instead that "society" is only an abstract label. Despite whatever practical uses the label may have, society is not a real object or thing. From this latter perspective, then, what "really" exist are unique individuals and the ideas that motivate their actions.

Simmel attempted to carve out a middle ground with respect to these competing visions. For him, the essence of society lies in the *interactions* that take place between individuals and groups. Society, then, is not a system of overarching institutions or symbolic codes nor is it merely an abstract idea used to describe a collection of individuals atomistically pursuing lines of conduct. Thus, Simmel defined society as a "number of individuals connected by interaction. . . . It is not a 'substance,' nothing concrete, but an *event*: It is the function of receiving and affecting the fate and development of one individual by the other" (Simmel 1917/1950:10,11; emphasis in original). The centrality of interaction for Simmel's view of society is expressed further when he states, "the significance of interactions among men [sic] lies in the fact that it is because of them that the individuals . . . form a unity, that is, a society" (Simmel 1908d/1971:23). In short, society is the array of interactions engaged in by individuals.

But what of those large-scale institutions and organizations—corporations, governments, schools, advocacy groups—that often come to mind when talking about society? Indeed, are not such supraindividual systems often equated with society itself? For Simmel, they "are nothing but immediate interactions that . . . have become crystallized as permanent fields, as autonomous phenomena" (Simmel 1917/1950:10). Here, he acknowledges that while such institutions develop under their own laws and logic and, thus, confront the individual as a seeming outside force, society is a process "constantly being realized" (ibid.). Society is something individuals *do* as they influence and are influenced by each other.

Thus, according to Simmel, society and the individuals that compose it constitute an interdependent **duality**. In other words, the existence of one presupposes the existence of the other. Moreover, this duality has a profound effect on the nature of individuality. For while who you are as an individual is in an important sense defined and made possible by the groups to which you belong, preserving your individuality demands that your identity not be completely submerged into or engulfed by group membership. Otherwise, you have no self that you can call your own.[2] As Simmel remarks, society is

a structure which consists of beings who stand inside and outside of it at the same time. This fact forms the basis for one of the most important sociological phenomena, namely, that between a society and its component individuals a relation may exist as if between

---

[2]Simmel's emphasis on the dualistic relationship between the individual and society is shaped in large measure by the philosophy of Immanuel Kant (1724–1804). In asking, "How is nature possible?," Kant argued that a priori (preexisting) "categories" or concepts that exist within the human mind structure our experience of the external world. The never-ending streams of sensory impressions that confront the individual have no underlying unity or pattern. It is only through the mind's a priori categories (particularly our conceptions of time, space, and causality) that the raw data that enter our consciousness are given a form, and hence meaning. However, without sensory impressions, the categories have no effect on human existence.

two parties. . . . [T]he individual can never stay within a unit which he does not at the same time stay outside of, that he is not incorporated into any order without also confronting it. (Simmel 1908e/1971:14,15)

In other words, your self-directed efforts to express and satisfy your interests and desires require that you engage in interaction with others. Yet interaction, in turn, shapes which aspects of your self can be expressed and how they are expressed. In an important sense, then, your "individuality" is created out of a synthesis of two seemingly contradictory forces: you are at the same time both an autonomous being with a unique disposition and history, and a product of society.

Sociology

Simmel's view of society provides, not surprisingly, an important glimpse into his understanding of sociology as a discipline.[3] His emphasis on the duality existing between society and the individual led him to define sociology as the study of social interaction or, as he often called it, "sociation." But it was not interaction per se that interested Simmel. Rather, he sought to analyze the **forms** in which interaction takes place. For instance, understanding the specific **content** of interactions that take place between an employer and employee—what they talk about and why—is not of central concern to sociologists. What is of sociological significance, however, is determining the uniformities or commonalities that such interactions have with those between, say, a husband and wife in a patriarchal society or between a ruler and his subjects. Recognizing the commonalities leads to the realization that each is based on a reciprocal relationship of domination and subordination. The main task of sociology, then, is to uncover the basic forms of interaction through which individuals pursue their interests or satisfy their desires, for it is only in relating to others, acting both with them and against them, that we are able to satisfy our ambitions.

Let's further clarify Simmel's notion of content and form. The content of interaction refers to the drives, purposes, interests, or inclinations that individuals have for interacting with another. In themselves, such motivations are not social, but rather are simply isolated psychological or biological impulses. What is social, however, are the actions that we take in concert with others in order to fulfill our drives or realize our interests. Moreover, when joined with others, our actions take on an identifiable, though not necessarily stable, form. Thus, our interactions may take on the form of conflict or cooperation, for example, or perhaps domination or equality (Simmel 1908d/1971:24).

Simmel also noted that while, on the one hand, the most dissimilar of contents (individual motivations) may be realized in an identical form of sociation, on the other hand, the same motivations or interests can be expressed in a range of forms of interaction. For instance, interaction within and between families, gangs, corporations, political organizations, and governments may all take the form of conflict. Thus, despite the varied interests or purposes that led to interaction in each of these cases, the individuals involved all may find themselves facing an opposing party that hinders the realization of their impulses or desires. Conversely, the same drive or interest can be expressed through a number of forms. Attempts to gain economic advantage, for instance, can be asserted through cooperative agreements among parties as well as through forcing submission from others.

Simmel's distinction between the forms and contents of interaction led him to draw a parallel between the subject matter of sociology and that of geometry. Both sciences develop

---

[3]Bear in mind that at this time sociology was confronted with the task of justifying its existence as an academic discipline. This required, among other things, carving out a unique subject matter and methodology to distinguish it from other disciplines.

general principles from determining the regularities that exist among diverse materials. Simmel thus saw himself devising a geometry of social life:

> Geometric abstraction investigates only the spatial forms of bodies, although empirically, these forms are given merely as the forms of some material content. Similarly, if society is conceived as interaction among individuals, the description of the forms of this inter-action is the task of the science of society in its strictest and most essential sense. (Simmel 1917/1950:21,22)

To take a simple example, while a car tire, clock face, and nickel are different things or "con-tents" that serve different purposes (you wouldn't look at your tire to find out what time it is), to the mathematician they all take the form of a three-dimensional circle and thus share spatial properties. On the other hand, although giving a wedding present, paying for music lessons, and volunteering at the food co-op are motivated by different intentions or "contents," to the soci-ologist they all take the form of an exchange relation and thus share interaction properties.

The Individual in Modern Society

Nowhere is the duality between individual identity and the "web of association" expressed more vividly than in Simmel's discussion of the nature of modern society. For Simmel, modern, urban societies allow individuals to cultivate their unique talents and inter-ests, but at the same time also lead to a "tragic" leveling of the human spirit.

Underlying this process is the changing nature of group life in modern societies. In con-trast to premodern societies, modern societies are characterized by an increasingly special-ized division of labor and expanding circles of affiliations. No longer is an individual's total personality absorbed into a particular group or controlled by a particular leader or landlord as tribal life or feudal relations once demanded. Unlike the peasant of feudal times who was economically, socially, and legally dependent on the landlord, the modern individual is not bound so extensively to any specific person. Thus, for example, employers today do not establish or adjudicate your legal status, nor do they determine whom you can associate with or marry. The demands of group membership in modern society are not totalizing. Modern, functionally specific organizations require only a "part" of the self in the service of its aims.

Echoing Durkheim's earlier argument, Simmel notes how occupational specialization and membership in an array of social circles has allowed individuals to differentiate them-selves from others while developing their unique personalities. The modern individual has separate relationships that tie her to work, family, community, creative pursuits, or hobbies, as well as to her religious and ethnic identity (see Figure 6.1). Moreover, in extending the network of their affiliations, modern individuals become less dependent on any one group for meeting their needs. Because individuals are not completely immersed in any specific group, they are freed from the dominating control of group life that characterized premod-ern societies. Expanding group memberships thus brings with it liberation and the potential to develop one's individuality.

This freedom not only prevents the individual's entire identity from being absorbed by a single set of relations, but also enables a person to refine his skills, talents, and personal preferences. Individuality is enhanced because as the number of affiliations increases, the less likely it becomes that the whole of an individual's multiple memberships is identical to another person's. Because individual identity is forged through group affiliations, occupy-ing a distinct position in the space of possible affiliations allows for the development of a unique personality (Simmel 1908a/1955).

However, while the evolution from homogeneous, premodern ways of life to heteroge-neous, modern societies has created an increased potential for freedom and the cultivation of individuality, it has also produced new types of constraints. Reminiscent of Marx's discussion

Figure 6.1 Simmel's Web of Group Affiliations

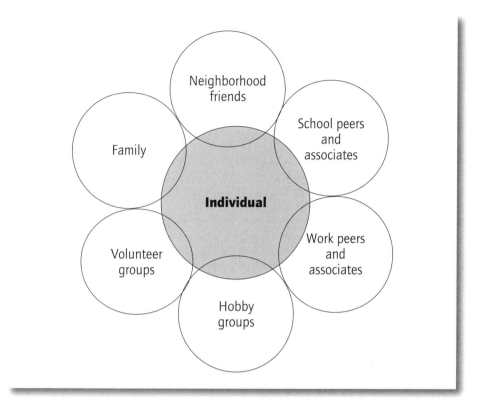

of alienation and commodity fetishism, Simmel speaks of the **tragedy of culture** where "objective culture"—the ideas and products of human creativity—comes to dominate individual will and self-development or "subjective culture." In short, the cultivation of individual potential (subjective culture) is inextricably bound up with the cultivation of products that are the tangible, objectified expression of individual creativity (objective culture). While Marx centered his critique on the oppression inherent in the capitalist mode of production, Simmel, like Weber, saw advances in science, technology, and the arts, as well as in the economy, bringing in their wake an increasing rationalization and routinization of many domains of life.

The alienating effects are an inevitable consequence of modernization. As the complexities of modern societies and its extensive division of labor create a need for the specialization of skills, they also sow the seeds for individual creativity and innovation. The freedom to cultivate one's desires and talents that accompanies individualization leads to an ever-increasing number of cultural objects—things such as new goods to consume, new scientific discoveries, and new forms of technology in production, transportation, and communication. However, these products of individual invention become reified—that is, they take on a life independent of their creators. As a result, what was initially the expression of individual growth and creativity (subjective culture) later confronts individuals as an autonomous force that requires submission to its own internal logic, thus compromising the self-development that these creations were intended to foster. This is one reason, for instance, that politicians and military specialists feel compelled to produce ever-more-lethal weapons as a way to prevent war. Mutual annihilation made possible by scientific discoveries has promoted a logic that guarantees peace through the fear of destruction. This is also why objects that did not exist 20 years ago (e.g., the cell phone) seem indispensable today. Indeed, our homes have become bigger and bigger in order to contain all the stuff we think we cannot live without. In the past century, the average square footage of the American home has almost tripled, going from 800 square feet in 1900 to 983 square feet in 1950 to 2,265 square feet in 2003 (National Association of Home Builders, as cited in the *Los Angeles Times*, September 14, 2003). And that home is itself part of the vast suburbanization of the country that, while offering the "American Dream" to many, has also led to homogeneous neighborhoods, insular families, and the loss of a sense of community (Putnam 2001). Meanwhile, the network of highways that has made suburban sprawl possible is converted daily into a veritable parking lot with average speeds of less than 30 miles per hour commonplace during rush hour commutes in cities such as Los Angeles.

Yet, the issue here is not whether such objects and developments represent real advances, but rather that they are fetishized. For example, the car is not only a means of transportation that, among other things, has allowed individuals to be liberated from the once-confining aspects of distance and thereby experience new people and places: it is also an object that is infused with expressive meaning. Seen as an extension of one's self, cars are invested with a "personality" though they are in reality an assemblage of metal and wires. Much like Weber's iron cage, such is the tragedy of culture in which the very cultural objects (developments in science, technology, politics, religion, the production of goods, and the arts) that were once the expression of individual ingenuity have imprisoned us. We are dependent on them not only for meeting our everyday needs, but also because they are, paradoxically, the alienating mediums through which we continue to express our individual creativity and contribute to subjective culture. As Simmel notes, in modern society,

> [t]hings become more perfected, more intellectual, and to some degree more controlled by an internal, objective logic tied to their instrumentality; but the supreme cultivation, that of subjects, does not increase proportionately. . . . The dissonance in modern life— in particular manifested in the improvement of technique in every area and the simultaneous deep dissatisfaction with technical progress—is caused in large part by the fact that things are becoming more and more cultivated, while men are less able to gain from the perfection of objects a perfection of subjective life. (Simmel 1908f/1971:234)

Significant Others

Ferdinand Tönnies (1855–1936): Gemeinschaft and Gesellschaft

A contemporary of both Georg Simmel and Max Weber, Tönnies crafted a sociological framework that shared much with the theories of his compatriots as well as with the evolutionary social theories offered by Herbert Spencer and Émile Durkheim. In his most influential work, *Gemeinschaft und Gesellschaft* (*Community and Society*), published in 1887, he set out a basic dichotomy describing the foundation of both individual and social action, as well as the nature of the broader social order. The ideal types that served as the polar ends of this dichotomy are "Gemeinschaft" ("community") and "Gesellschaft" ("society"), and each is rooted in a particular "will" that possesses "binding force within or determine[s] the individual wills" (Tönnies 1887/1957:205). According to Tönnies, Gemeinschaft is the principle that structures agricultural and handicraft life within families, villages, and towns. It gives rise to a "natural will" or mentality that inclines individuals and groups to associate with one another on the basis of concord (harmony), folkways, and religious beliefs. These likewise form the core of the normative order as it is manifested in both morality and laws.

In contrast to the warmth and emotional depth nurtured by natural will, Gesellschaft is rooted in a "rational will" that promotes "cold" instrumental relationships in which individuals are viewed as "tools" or as means to further one's ends. Rational self-interest, not feelings of sympathy, ties individuals to one another. This outlook predominates in the affairs of industry and commerce that make up city, national, and cosmopolitan (international) life. It is not concord, folkways, and religion that structure actions and attitudes, morality and law, but rather conventions, state-sponsored legislation, and scientifically informed public opinion. Social life is regulated according to legally sanctioned contracts, not long-standing customs, while "the intellectual attitude of the individual becomes gradually less and less influenced by religion and more and more influenced by science" (ibid.:226).

Like Durkheim, Tönnies viewed societies transitioning from a form of solidarity based on a relatively simple, common way of life and a widely shared set of normative and moral codes, to a more complex social bond based on individualism and interdependence. Interestingly, however, unlike Durkheim, Tönnies described the evolution from Gemeinschaft to Gesellschaft as a shift from an organic to a mechanical form of social life. As he characterized the two forms, "*Gemeinschaft . . .* is the lasting and genuine form of living together. In contrast to Gemeinschaft, Gesellschaft is transitory and superficial. Accordingly, Gemeinschaft should be understood as a living organism, Gesellschaft as a mechanical aggregate and artifact" (ibid.:35). Gemeinschaft is "intimate," "real and organic life" as opposed to the "imaginary and mechanical structure" of the public Gesellschaft based on the "mere coexistence of people independent of each other" (ibid.:33,34).

From these remarks, it is apparent that Tönnies did not hail the transformation to Gesellschaft as an indisputable step toward progress. In fact, even more than Weber and Simmel he harbored a pronounced pessimism with regard to modernity where "money and capital are unlimited and almighty" (ibid.:228). However, he was troubled not only by the increasing rationalization of modern life and the alienating effects of the money economy: like Marx, he saw in capitalism a source of conflict and exploitation that could only be overcome through revolutionary class struggle and the destruction of Gesellschaft. As a social order in which "only the upper strata, the rich and the cultured, are really active and alive . . . city life and Gesellschaft down the common people to decay and death" (ibid.:227,230). Only a return to Gemeinschaft can rescue humanity from this "decaying" society where "individuals remain in isolation and veiled hostility toward each other so that only fear of clever retaliation restrains them from attacking one another, and, therefore, even peaceful and neighborly relations are in reality based upon a warlike situation . . . and underlying mutual fear" (ibid.:224).

The Individual and Money

Perhaps the most profound expression of the "tragedy" inherent in modern society is money. While money, as a medium of exchange, allows us to enter into a wide range of relationships as we seek to satisfy our individual needs, it does so at the cost of transforming the nature of our ties to others. It expands the freedom to pursue self-fulfillment and self-expression as it trivializes the personal qualities we share with others and ushers in superficial, impersonal, and fragmented social relationships. A money economy, then, represents in particularly clear fashion the alienating duality between objective and subjective culture.

Just as clearly revealed in his discussion of money is Simmel's own ambivalence regarding the "advances" offered by modern society.

What are the properties of money that produce these effects on the individual and social relationships more generally? First, money is an abstract, general standard for measuring the value of goods and services. As such, it can be used as a medium of exchange in a vast number of interactions. Few things cannot be bought or sold for money. It can pay for food, entertainment, education, shelter, sex, transportation—the list is virtually endless. The flexibility of money opens up greater opportunity for individuals to satisfy an expanded range of desires and needs (Simmel 1900a/1978:212,213).

Second, to the extent that we are dependent on money to meet our needs, we become less dependent on others. As the locus of dependency shifts from people to money, the possibilities for freedom of expression are enhanced. In modern societies, the pursuit of interests and fulfillment of desires are limited more by the amount of money you possess than by the demands or whims of others or the constraints imposed by tradition or custom. Consider, for instance, the teenager's summer job that often makes possible her first foray into independence. More pressingly, the fight for women's equality has always hinged on the ability not only to enter into a full range of occupations, but also to earn equal pay for equal work.

Third, because money retains the same value regardless of who possesses it, it can cancel out social inequalities. Thus, one's position in society is related more to the objective, impersonal issue of how much money one has than to tradition, personal qualities, or even ethnicity and gender. In a money economy, group membership, whether it is membership in a neighborhood, a college, a profession, or a gardening club, is tied more to the *quantity* of money an individual has than to his personal *qualities*. For instance, African American "bad boy" Dennis Rodman owns both a home and a restaurant in Newport Beach, California, a wealthy, predominantly white community. Though his personal qualities and antics (wild parties, public drunkenness, etc.) continually infuriate his neighbors, Rodman's money prevents his neighbors from keeping him out.

Fourth, the impersonal quality of money allows for a qualitative and quantitative expansion of one's social network. As more needs and desires are satisfied through monetary transactions with others, money promotes contacts with a greater range of people. This leads to an expansion of avenues for the cultivation of personal freedom, because individuality can be expressed only in relationships with others. Money buys us membership in groups (fraternities, sororities, professional associations) or access to activities (education, travel, athletics, pottery classes). With the widening of social circles that money makes possible comes the ability to pursue self-expressive interests and thus further individuation (Simmel 1900a/1978:344,345).

However, while money offers a number of advantages for individuals, Simmel also notes the far less beneficial consequences it brings about. Most important, in modern societies in which money serves as a nexus for social integration, relations with others become less personal and intimate. Without significant emotional attachment or investment, relations become more and more standardized, like money itself. The impersonality and generalizability of money as a medium of exchange transform the nature of social interaction. Money breeds relationships based on rational calculation and instrumental purposes as opposed to personal ties of family or attraction to another's individual, subjective qualities. Coworkers come into and go out of our lives with little notice and only passing concern. We pick doctors based on the convenience of their office location and hours. Dating services, for a fee, will "match" individuals by way of quantitative surveys. Sharing an affinity with Weber's analysis of Western societies, Simmel argues that money plays an integral role in the trend toward the rationalization of social life. As money becomes the standard by which the value of all objects and relations are quantified, life is emptied of emotional connections and the enchantment of traditions. Personal loyalties and emotional attachment have been replaced with indifference (ibid.:297–303).

Moreover, the expanding social relations and involvements made possible through a money economy leave individuals able only to partially invest their self in any one activity. Thus, the self becomes atomized or fragmented, while others with whom the individual interacts "know" but a small part of his self (ibid.:342–345).

SIMMEL'S THEORETICAL ORIENTATION

Turning to our metatheoretical framework, Simmel's work is predominantly individualistic and nonrationalist in orientation (Figure 6.2). Let's consider first the individualistic dimension of his theory. Simmel's view of the social order—the routines and patterns of behavior that comprise everyday life—emphasizes the importance of *interaction*. As we noted above, Simmel did not see society as a "thing" or reality that existed independently of the individuals who compose it. Instead, Simmel saw society as an "event"—that is, the ongoing interactions that individuals engage in as they move through their daily lives seeking to satisfy their needs and realize their desires. The regularities or unity that characterize social life stems from the basic *forms* of interaction or structure of relationships, not from some overarching system of institutional arrangements that determines one's interests, behavior, or consciousness.

Figure 6.2 Simmel's Basic Theoretical Orientation

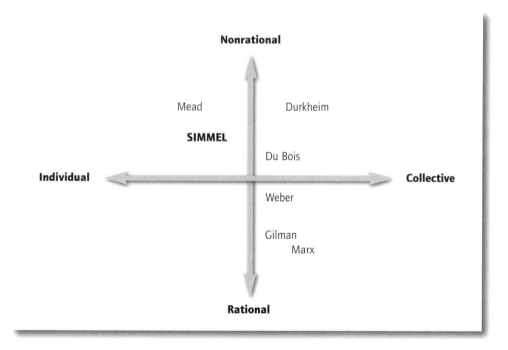

To be sure, although Simmel viewed society primarily as an ongoing process of interaction in which individuals mutually influence one another, his perspective incorporates collectivist elements as well. In particular, as discussed previously, Simmel saw "objective culture" taking on a seemingly autonomous presence in modern, urban societies. Here, the increasingly complex division of labor promotes the specialization of individual skills and, with it, a greater cultivation of individuality and personal expression. However, the sense of liberation is countered by external objects that confront individuals as dominating forces.

Such alien objects are in reality the products of individual innovation that have become reified. Advances in science, industry, technology, and the political and legal administration of the populace are often experienced as oppressive systems that compel conformity. Individuals' lives move to the impersonal rhythms of a money economy and the institutional demands around which urban society is organized. This duality between "objective" and "subjective" culture—and the individualist/nonrationalist versus collectivist/rationalist presuppositions—is illustrated in Table 6.1.

Table 6.1 Simmel's Duality Regarding Subjective/Objective Culture

		ORDER	
		Individual	**Collective**
ACTION	**Nonrational**	*Subjective culture:* Ongoing interactions that individuals engage in as they move through their daily lives seeking to satisfy their needs and realize their desires.	
	Rational		*Objective culture:* Reification; external objects confront individuals as oppressive systems that compel conformity.

With regard to his view on the nature of action, Simmel adopts a predominantly (though not exclusively) nonrationalist approach. A variety of conscious and unconscious motivations can impel individuals to interact with one another. As Simmel states,

> interaction always arises on the basis of certain drives or for the sake of certain purposes. Erotic, religious, or merely associative impulses; and purposes of defense, attack, play, gain, aid, or instruction—these and countless others cause man to live with other men, to act for them, with them, against them, and thus to correlate his condition with theirs. (Simmel 1908e/1971:23)

Thus, while Simmel acknowledges that action can be based on calculative attempts to minimize costs and maximize gains, by no means does he restrict motivations to such a rationalist basis. Indeed, Simmel points out that we often are led to interact simply for the pleasure we derive from being in the company of others.

Readings

As we noted previously, Simmel saw in social life a synthesis of contradictions. This emphasis on the dualities underlying modern society and interaction is found in all his writings. Within this general tendency, however, three topics are most relevant. In the selections below, you will read Simmel's discussions on (1) forms of social interaction, namely "exchange," "conflict," and "sociability"; (2) social types, particularly "the stranger"; and (3) the relationship between the individual and the broader social structure as evident in fashion and the metropolis.

Forms of Interaction

Introduction to "Exchange"

As we noted above, Simmel saw in exchange "the purest and most concentrated form of all human interactions in which serious interests are at stake" (Simmel 1900b/1971:43). Indeed, for Simmel every interaction (a performance, a conversation, or even a romantic affair) could be understood as a form of exchange in which each participant gives the other "more than he had himself possessed" (ibid.:44). In the reading that follows, you will find Simmel focusing his attention on a specific form of exchange, *economic* exchange, particularly as it relates to the creation of value. Here, you also will see Simmel engaged in an important debate with Marx's view on the source of economic value.

What separates specifically economic forms of exchange from more general exchange interactions is *sacrifice*. For Simmel, the measure of sacrifice necessary to attain goods or goals is the source of their economic value. Unlike Marx, who argued that the value of goods is equal to the amount of labor power invested in their production, Simmel found in sacrifice—the giving up of one's money, time, services, possessions—"the condition of all value" (ibid.:49). Hence, there can be no universal, objective standard by which value can be established (see Table 6.2). For instance, while an avid collector of comic books may be willing to buy a rare comic "valued" at $200, most of us would not, instead preferring to sacrifice our money on other goods such as clothes or CDs. As a result, the value of the comic book, the ratio of desire to sacrifice, would be far less than $200 for most of us.

Table 6.2 Comparison of Simmel and Marx on the Issue of Economic Value

		ORDER	
		Individual	**Collective**
ACTION	**Nonrational**	*Simmel:* Value of goods is determined in *interaction*, as actors weigh their *desire* for goods against *sacrifice* required to attain them.	
	Rational		*Marx:* Value of goods is equal to the amount of labor invested in *production*.

Value, then, is always subjective and relative. It is determined by the interaction at hand in which actors weigh their desire for the goods in question against the amount of sacrifice required to attain them. Moreover, without having to endure obstacles or some form of self-denial, not even the most intensely felt desire for an object will make it valuable. This is because value is created out of the "distance" that separates desire from its satisfaction and the willingness to sacrifice something in order to overcome that distance. This is precisely the lesson that some parents try to teach their children when they assign chores. By having children earn money in exchange for sacrificing their time and energies, the toys they purchase for themselves will acquire value and thus won't be so freely lost or discarded—at least in theory!

From "Exchange" (1907)

Georg Simmel

Most relationships among men can be considered under the category of exchange. Exchange is the purest and most concentrated form of all human interactions in which serious interests are at stake.

Many actions which at first glance appear to consist of mere unilateral process in fact involve reciprocal effects. The speaker before an audience, the teacher before a class, the journalist writing to his public—each appears to be the sole source of influence in such situations, whereas each of them is really acting in response to demands and directions that emanate from apparently passive, ineffectual groups. The saying "I am their leader, therefore I must follow them" holds good for politicians the world over. Even in hypnosis, which is manifestly the most clear-cut case where one person exercises influence and the other shows total passivity, reciprocity still obtains. As an outstanding hypnotist has recently stressed, the hypnotic effect would not be realized were it not for a certain ineffable reaction of the person hypnotized back on the hypnotist himself.

INTERACTIONS AS EXCHANGE

Now every interaction is properly viewed as a kind of exchange. This is true of every conversation, every love (even when requited unfavorably), every game, every act of looking one another over. It might seem that the two categories are dissimilar, in that in interaction one gives something one does not have, whereas in exchange one gives only what one does have, but this distinction does not really hold. What one expends in interaction can only be one's own energy, the transmission of one's own substance. Conversely, exchange takes place not for the sake of an object previously possessed by another person, but rather for the sake of one's own feeling about an object, a feeling which the other previously did not possess. The meaning

of exchange, moreover, is that the sum of values is greater afterward than it was before, and this implies that each party gives the other more than he had himself possessed.

Interaction is, to be sure, the broader concept, exchange the narrower one. In human relations, however, interaction generally appears in forms which lend themselves to being viewed as exchange. The ordinary vicissitudes of daily life produce a continuous alternation of profit and loss, an ebbing and flowing of the contents of life. Exchange has the effect of rationalizing these vicissitudes, through the conscious act of setting the one *for* the other. The same synthetic process of mind that from the mere juxtaposition of things creates a with-another and for-another—the same ego which, permeated by sense data, informs them with its own unified character—has through the category of exchange seized that naturally given rhythm of our existence and organized its elements into a meaningful nexus.

THE NATURE OF ECONOMIC EXCHANGE

Of all kinds of exchange, the exchange of economic values is the least free of some tinge of sacrifice. When we exchange love for love, we release an inner energy we would otherwise not know what to do with. Insofar as we surrender it, we sacrifice no real utility (apart from what may be the external consequences of involvement). When we communicate intellectual matters in conversation, these are not thereby diminished. When we reveal a picture of our personality in the course of taking in that of others, this exchange in no way decreases our possession of ourselves. In all these exchanges the increase of value does not occur through the calculation of profit and loss. Either the contribution of each party stands beyond such a consideration, or else simply to be allowed to contribute is itself a gain—in which case we perceive the response of the other, despite our own offering, as an unearned gift. In contrast, economic exchange—whether it involves substances, labor, or labor power invested in substances—always entails the sacrifice of some good that has other potential uses, even though utilitarian gain may prevail in the final analysis.

The idea that all economic action is interaction, in the specific sense of exchange that involves sacrifice, may be met with the same objection which has been raised against the doctrine that equates all economic value with exchange value. The point has been made that the totally isolated economic man, who neither buys nor sells, would still have to evaluate his products and means of production—would therefore have to construct a concept of value independent of all exchange—if his expenditures and results were to stand in proper relation to one another. This fact, however, proves exactly what it is supposed to disprove, for all consideration whether a certain product is worth enough to justify a certain expenditure of labor or other goods is, for the economic agent, precisely the same as the appraisal which takes place in connection with exchange.

In dealing with the concept of exchange there is frequently a confusion of thought which leads one to speak of a relationship as though it were something external to the elements between which it occurs. Exchange means, however, only a condition of or a change within each of these elements, nothing that is *between* them in the sense of an object separated in space between the two other objects. When we subsume the two acts or changes of condition which occur in reality under the concept "exchange," it is tempting to think that with the exchange something has happened in addition to or beyond that which took place in each of the contracting parties.

This is just like being misled by the substantive concept of "the kiss" (which to be sure is also "exchanged") into thinking that a kiss is something that lies outside of the two pairs of lips, outside of their movements and sensations. Considered with reference to its immediate content, exchange is nothing more than the causally connected repetition of the fact that an actor now has something which he previously did not have, and for that has lost something which he previously did have.

That being the case, the isolated economic man, who surely must make certain sacrifices in order to gain certain fruits, behaves exactly like the one who makes exchanges. The only difference is that the party with whom he contracts is not a second free agent, but the natural order and regularity of things, which no more satisfy our desires without a sacrifice on our part than would another person. His calculations of value, in accordance with which he governs his actions, are generally the same as in exchange. For the economic actor as such it is surely quite immaterial whether the substances or labor capacities which he possesses are sunk into the ground or given to another man, if what he gains from the sacrifice is exactly the same in both cases.

This subjective process of sacrifice and gain within the individual psyche is by no means something secondary or imitative in relation to interindividual exchange. On the contrary, the give-and-take between sacrifice and attainment within the individual is the fundamental presupposition and, as it were, the essence of every two-sided exchange. The latter is only a subspecies of the former; that is, it is the sort in which the sacrifice is occasioned by the demand of another individual, whereas the sacrifice can be occasioned by things and their natural properties with the same sort of consequences for the actor.

It is extremely important to carry through this reduction of the economic process to that which takes place in *actuality,* that is, within the psyche of every economic actor. We should not let ourselves be misled because in exchange this process is reciprocal, conditioned by a similar process within another party. The natural and "solipsistic" economic transaction goes back to the same fundamental form as the two-sided exchange: to the process of balancing two subjective events within an individual. This is basically unaffected by the secondary question whether the process is instigated by the nature of things or the nature of man, whether it is a matter of purely natural economy or exchange economy. All feelings of value, in other words, which are set free by producible objects are in general to be gained only by foregoing other values. Such self-denial consists not only in that indirect labor for ourselves which appears as labor for others, but frequently enough in direct labor on behalf of our own personal ends. . . .

THE SIGNIFICANCE OF SACRIFICE

The fact that value is the issue of a process of sacrifice discloses the infinity of riches for which our life is indebted to this basic form. Because we strive to minimize sacrifice and perceive it as painful, we tend to suppose that only with its complete disappearance would life attain its highest level of value. But this notion overlooks the fact that sacrifice is by no means always an external barrier to our goals. It is rather the *inner* condition of the goal and of the way to it. Because we dissect the problematic unity of our practical relations to things into the categories of sacrifice and profit, of obstacle and attainment, and because these categories are frequently separated into differentiated temporal stages, we forget that if a goal were granted to us without the interposition of obstacles it would no longer be the same goal.

The resistance which has to be eliminated is what gives our powers the possibility of proving themselves. Sin, after whose conquest the soul ascends to salvation, is what assures that special "joy in heaven" which those who were upright from the outset do not possess there. Every synthesis requires at the same time an effective analytic principle, which actually negates it (for without this it would be an absolute unity rather than a synthesis of several elements). By the same token every analysis requires a synthesis, in the dissolution of which it consists (for analysis demands always a certain coherence of elements if it is not to amount to a mere congeries without relations). The most bitter enmity is still more of a connection than simple indifference, indifference still more than not even knowing of one another. In short: the inhibiting countermovement, the diversion of which signifies sacrifice, is often—perhaps, seen from the point of view of elementary processes, even always—the positive presupposition of the goal itself. Sacrifice by no means belongs in the category of the undesirable, though superficiality and greed might portray it as such. It is not only the condition of individual values but, in what concerns us here, the economic realm, sacrifice is the condition of all value; not only the price to be paid for individual values that are already established, but that through which alone values can come into being.

Exchange occurs in two forms, which I shall discuss here in connection with the value of labor. All labor is indisputably a sacrifice if it is accompanied by a desire for leisure, for the mere self-satisfying play of skills, or for the avoidance of strenuous exertion. In addition to such desires, however, there exists a quantum of latent work energy which either we do not know what to do with or which presents itself as a drive to carry out voluntary labor, labor called forth neither by necessity nor by ethical motives. The expenditure of this energy is in itself no sacrifice, yet for this quantum of energy there compete a number of demands all of which it cannot satisfy. For every expenditure of the energy in question one or more possible and desirable alternative uses of it must be sacrificed. Could we not usefully spend the energy with which we accomplish task A also on task B, then the first would not entail any sacrifice; the same would hold for B in the event we chose it rather than A. In this utilitarian loss what is sacrificed is not labor, but *nonlabor*. What we pay for A is not the sacrifice of labor—for our assumption here is that the latter in itself poses not the slightest hardship on us—but the giving up of task B.

The sacrifice which we make of labor in exchange is therefore of two sorts, of an absolute and a relative sort. The discomfort we accept is in the one case directly bound up with the labor itself, because the labor is annoying and troublesome. In the case where the labor itself is of eudaemonistic irrelevance or even of positive value, and when we can attain one object only at the cost of denying ourselves another, the frustration is indirect. The instances of happily done labor are thereby reduced to the form of exchange entailing renunciation, the form which characterizes all aspects of economic life. . . .

THE SOURCE OF VALUE

If we regard economic activity as a special case of the universal life-form of exchange, as a sacrifice in return for a gain, we shall from the very beginning intuit something of what takes place within this form, namely, that the value of the gain is not, so to speak, brought with it, readymade, but accrues to the desired object, in part or even entirely through the measure of the sacrifice demanded in acquiring it. These cases, which are as frequent as they are important for the theory of value, seem, to be sure, to harbor an inner contradiction: they have us making a sacrifice of a value for things which in themselves are worthless.

No one in his right mind would forego value without receiving for it at least an equal value; that, on the contrary, an end should receive its value only through the price that we must give for it could be the case only in an absurd world. Yet common sense can readily see why this is so.

The value which an actor surrenders for another value can never be greater, for the subject himself under the actual circumstances of the moment, than that for which it is given. All contrary appearances rest on the confusion of the value actually estimated by the actor with the value which the object of exchange in question usually has or has by virtue of some apparently objective assessment. Thus if someone at the point of death from hunger gives away a jewel for a piece of bread, he does so because the latter is worth more to him under the circumstances than the former. Some particular circumstances, however, are *always* involved when one attaches a feeling of value to an object. Every such feeling of value is lodged in a whole complex system of our feelings which is in constant flux, adaptation, and reconstruction. Whether these circumstances are exceptional or relatively constant is obviously in principle immaterial. Through the fact that the starving man gives away his jewel he shows unambiguously that the bread is worth more to him.

There can thus be no doubt that in the moment of the exchange, of the making of the sacrifice, the value of the exchanged object forms the limit which is the highest point to which the value of the object being given away can rise. Quite independent of this is the question whence that former object derives its exigent value, and whether it may not come from the sacrifices to be offered for it, such that the equivalence between gain and cost would be established a posteriori, so to speak, and by virtue of the latter. We will presently see how frequently value comes into being psychologically in this apparently illogical manner.

Given the existence of the value, however, it is psychologically necessary to regard it, no less than values constituted in every other way, as a positive good at least as great as the negative of what has been sacrificed for it. There is in fact a whole range of cases known to the untrained psychological observer in which sacrifice not only heightens the value of the goal, but even generates it by itself. What comes to expression in this process is the desire to prove one's strength, to overcome difficulties, indeed often to oppose for the sheer joy of opposition. The detour required to attain certain things is often the occasion, often the cause as well, of perceiving them as values. In human relationships, most frequently and clearly in erotic relations, we notice how reserve, indifference, or rejection inflames the most passionate desire to prevail over these obstacles, and spurs us to efforts and sacrifices which, without these obstacles, would surely seem to us excessive. For many people the aesthetic gain from climbing the high Alps would not be considered worth further notice if it did not demand the price of extraordinary exertion and dangers and thereby acquire character, appeal, and consecration.

The charm of antiques and curios is frequently of the same sort. Even if antiques possess no intrinsic aesthetic or historical interest, a substitute for this is furnished by the mere difficulty of acquiring them: they are worth as much as they cost. It then comes to appear that they cost what they are worth. Furthermore, all ethical merit signifies that for the sake of the morally desirable deed contrary drives and wishes must be combatted and given up. If the act occurs without any conquest, as the direct issue of uninhibited impulses, its content may be objectively desirable, but it is not accorded a subjective moral value in the same sense. Only through the sacrifice of the lower and yet so seductive goods does one reach the height of ethical merit; and the more tempting the seductions and the more profound their sacrifice, the loftier the height. If we observe which human achievements attain to the highest honors and evaluations, we find them always to be those which manifest, or at least appear to manifest, the most depth, the most exertion, the most persistent concentration of the whole being—which is to say the most self-denial, sacrifice of all that is subsidiary, and devotion of the subjective to the objective ideal.

And if, in contrast with all this, aesthetic production and everything sweet and light, flowing from the naturalness of impulse, unfolds an incomparable charm, this charm derives its special quality from feelings associated with the burdens and sacrifices which are ordinarily required to gain such things. The liability and inexhaustible richness of combination of the contents of our minds frequently transform the significance of a connection into its exact converse, somewhat as the association between two ideas follows equally whether they are asserted or denied of each other. We perceive the specific value of something obtained without difficulty as a gift of fortune only on the grounds of the significance which things have for us that are hard to come by and measured by sacrifice. It is the same value, but with the negative sign; and the latter is the primary from which the former may be derived—but not vice versa.

We may be speaking here of course of exaggerated or exceptional cases. To find their counterpart in the whole realm of the economy it seems necessary, first of all, to make an analytic distinction between the universal substance of value, and economic activity as a differentiated form thereof. If for the moment we take value as something given, then in accord with our foregoing discussion the following proposition is established beyond doubt: *Economic value as such does not inhere in an object in its isolated self-existence, but comes to an object only through the expenditure of another object which is given for it.* Wild fruit picked without effort, and not given in exchange, but immediately consumed, is no economic good. It can at most count as such only when its consumption saves some other economic expense. If, however, all of life's requirements were to be satisfied in this manner, so that at no point was sacrifice involved, men would simply not have *economic* activity, any more than do birds or fish or the denizens of fairyland. Whatever the way two objects, A and B, became values, A becomes an *economic* value only because I must give B for it, B only because I can obtain A for it. As mentioned above, it is in principle immaterial here whether the sacrifice takes place by transferring a value to another person, that is, through

interindividual exchange, or within the circle of the individual's own interests, through a balancing of efforts and results. In articles of commerce there is simply nothing else to be found other than the meaning each one directly or indirectly has for our consumption needs and the exchange which takes place between them. Since, as we have seen, the former does not of itself suffice to make a given object an object of economic activity, it follows that the latter alone can supply to it the specific difference which we call economic.

This distinction between value and its economic form, is, however, an artificial one. If at first economy appears to be a mere form, in the sense that it presumes values as its contents, in order to be able to draw them into the process of balancing between sacrifice and profit, in reality the same process which forms the presumed values into an economy can be shown to be the creator of the economic *values themselves*. This will now be demonstrated.

The economic form of the value stands between two boundaries: on the one hand, *desire* for the object, connected with the anticipated feeling of satisfaction from its possession and enjoyment; on the other hand, this *enjoyment* itself which, strictly speaking, is not an economic act. That is, as soon as one concedes, as was shown above, that the immediate consumption of wild fruit is not an economic act and therefore the fruit itself is not an economic value (except insofar as it saves the production of economic values), then the consumption of real economic values is no longer economic, for the act of consumption in the latter case is not distinguishable from that in the former. Whether the fruit someone eats has been accidentally found, stolen, home-grown, or bought makes not the slightest difference in the act of eating and its direct consequences for the eater. . . .

The qualities of objects which account for their subjective desirability cannot, consequently, be credited with producing an absolute amount of value. It is always the relation of desires to one another, realized in exchange, which turns their objects into economic values. With respect to scarcity, the other element supposed to constitute value, this consideration is more directly apparent. Exchange is, indeed, nothing other than the interindividual attempt to improve an unfavorable situation arising out of a shortage of goods; that is, to reduce as much as possible the amount of subjective abstinence by the mode of distributing the available supply. Thereupon follows immediately a universal correlation between what is called scarcity-value (a term justly criticized) and what is called exchange-value.

For us, however, the connection is more important in the reverse direction. As I have already emphasized, the fact that goods are scarce would not lead us to value them unless we could not somehow modify that scarcity. It is modifiable in only two ways: by expending labor to increase the supply of goods, or by giving up objects already possessed in order to make whatever items an individual most desires less scarce for him. One can accordingly say that the scarcity of goods in relation to the desires directed to them objectively conditions exchange, but that it is exchange alone that makes scarcity a factor in value. It is a mistake of many theories of value to assume that, when utility and scarcity are given, economic value—that is, the exchange process—is something to be taken for granted, a conceptually necessary consequence of those premises. In this they are by no means correct. If, for instance, those conditions are accompanied by ascetic resignation, or if they instigated only combat or robbery—which, to be sure, is indeed true often enough—no economic value and no economic life would emerge. . . .

The Process of Value Formation: Creating Objects Through Exchange

Now an object is not a value so long as it remains a mere emotional stimulus enmeshed in the subjective process—a natural part of our sensibility, as it were. It must first be separated from this subjective sensibility for it to attain the peculiar significance which we call value. For not only is it certain that desire in and of itself could not establish any value if it did not encounter obstacles—trade in economic values could never have arisen if every desire was satisfied without struggle or exertion—but even desire itself would never have ascended to such a considerable height if it could be satisfied

without further ado. It is only the postponement of satisfaction through impediment, the anxiety that the object may escape, the tension of struggle for it, that brings about the cumulation of desires to a point of intensified volition and continuous striving.

If, however, even the highest pitch of desire were generated wholly from within, we still would not confer value on the object which satisfies it if the object were available to us in unlimited abundance. The important thing in that case would be the total enjoyment, the existence of which guarantees to us the satisfaction of our wishes, but not that particular quantum which we actually take possession of, since this could be replaced quite as easily by another. Even that totality would acquire some sense of value only by virtue of the thought of its possible shortage. Our consciousness would in this case be filled simply with the rhythm of subjective desires and satisfactions, without attaching any attention to the mediating object. Neither need nor enjoyment contains in itself value or economic process. These are actualized simultaneously through exchange between two subjects, each of whom requires some self-denial by the other as a condition of feeling satisfied, or through the counterpart of this process in the solipsistic economy. Through exchange, economic process and economic values emerge simultaneously, because exchange is what sustains or produces the distance between subject and object which transmutes the subjective state of feeling into objective valuation.

Kant once summarized his theory of knowledge in the proposition: "The conditions of experience are at the same time the conditions of the objects of experience." By this he meant that the process we call experience and the concepts which constitute its contents or objects are subject to the same laws of reason. The objects can enter into our experience, that is, can be experienced by us, because they exist as concepts within us, and the same energy which forms and defines the experience manifests itself in the formation of those concepts. In the same spirit we may say here that the possibility of economy is at the same time the possibility of the objects of economy. The very transaction between two possessors of objects (substances, labor energies, rights) which brings them into the so-called economic relation, namely, reciprocal sacrifice, at the same time elevates each of these objects into the category of value. The logical difficulty raised by the argument that values must first exist, and exist as values, in order to enter into the form and process of economic action, is now removed. It is removed thanks to the significance we have perceived in that psychic relationship which we designated as the distance between us and things. This distance differentiates the original subjective state of feeling into (1) a desiring subject, anticipating feelings, and (2) counterposed to him, an object that is now imbued with value; while the distance, on its side, is produced in the economic realm by exchange, that is, by the two-sided operation of barriers, restraint, and self-denial. Economic *values* thus emerge through the same reciprocity and relativity in which the *economic condition* of values consists.

Exchange is not merely the addition of the two processes of giving and receiving. It is, rather, something new. Exchange constitutes a third process, something that emerges when each of those two processes is simultaneously the cause and the effect of the other. Through this process, the value which the necessity of self-denial for an object imparts to it becomes an economic value. If it is true that value arises in general in the interval which obstacles, renunciations, and sacrifices interpose between desire and its satisfaction, and if the process of exchange consists in that reciprocally conditioned taking and giving, there is no need to invoke a prior process of valuation which makes a value of an isolated object for an isolated subject. What is required for this valuation takes place in the very act of exchange itself. . . .

VALUE AND PRICE

If value is, as it were, the offspring of price, then it seems logical to assert that their amounts must be the same. I refer now to what has been established above, that in each individual case no contrasting party pays a price which to him under the given circumstances is too high for the thing obtained. If, in the poem of Chamisso, the highwayman with pistol drawn compels the victim to sell his watch and ring for three

coppers, the fact is that under the circumstances, since the victim could not otherwise save his life, the thing obtained in exchange is actually worth the price. No one would work for starvation wages if, in the situation in which he actually found himself, he did not prefer this wage to not working. The appearance of paradox in the assertion that value and price are equivalent in every individual case arises from the fact that certain conceptions of other kinds of equivalence of value and price are brought into our estimate.

Two kinds of considerations bring this about: (1) the relative stability of the relations which determine the majority of exchange transactions, and (2) the analogies which set still uncertain value-relations according to the norms of those that already exist. Together these produce the notion that if for a certain object this and that other object were exchange equivalents, then these two objects, or the circle of objects which they define, would have the same position in the scale of values. They also give rise to the related notion that if abnormal circumstances caused us to exchange this object for values that lie higher or lower in the scale, price and value would become discrepant—although in each individual case, considering *its* circumstances, we would find them actually to coincide. We should not forget that the objective and just equivalence of value and price, which we make the norm of the actual and individual case, holds good only under very specific historical and technical conditions; and that, with the change of these conditions, the equivalence vanishes at once. Between the norm itself and the cases which it defines as either exceptional or standard there is no difference of kind: there is, so to speak, only a quantitative difference. This is somewhat like when we say of an extraordinarily elevated or degraded individual, "He is really no longer a man." The fact is that this idea of man is only an average. It would lose its normative character at the moment a majority of men ascended or descended to that level of character, which would then pass for the generically "human."

To perceive this requires an energetic effort to disentangle two deeply rooted conceptions of value which have substantial practical justification. In relations that are somewhat evolved these conceptions are lodged in two superimposed levels. One kind of standard is formed from the traditions of society, from the majority of experiences, from demands that seem to be purely logical; the other, from individual constellations, from demands of the moment, from the constraints of a capricious environment. Looking at the rapid changes which take place within the latter sphere, we lose sight of the slow evolution of the former and its development out of the sublimation of the latter; and the former seems suitably justified as the expression of an objective proportion. In an exchange that takes place under such circumstances, when the feelings of loss and gain at least balance each other (for otherwise no actor who made any comparisons at all would consummate the exchange) yet when these same feelings of value are discrepant when measured by those general standards, one speaks of a divergence between value and price. This occurs most conspicuously under two conditions, which almost always go together: (1) when a single value-quality is counted as the economic value and two objects consequently are adjudged equal in value only insofar as the same quantum of that fundamental value is present in them, and (2) when a certain proportion between two values is expected not only in an objective sense but also as a moral imperative.

The conception, for example, that the real value-element in all values is the socially required labor time objectified in them has been applied in both of these ways, and provides a standard, directly or indirectly applicable, which makes value fluctuate positively and negatively with respect to price. The fact of that single *standard of value* in no way establishes how labor power comes to be a value in the first place. It could hardly have done so if the labor power had not, by acting on various materials and fashioning various products, created the possibility of exchange, or if the use of the labor power were not perceived as a sacrifice which one makes for the sake of its fruits. Labor energy also, then, is aligned with the category of value only through the possibility and reality of exchange, irrespective of the fact that subsequently *within* this category of value labor may itself provide a standard for the remaining contents. If the labor power therefore is also the content of every value, it receives its form as value in the first place only because it enters into the relations between sacrifice and gain, or profit and value (here in the narrower sense).

In the cases of discrepancy between price and value the one contracting party would, according to this theory, give a certain amount of immediately objectified labor power for a lesser amount of the same. Other factors, not involving labor power, would then lead the party to complete the exchange, factors such as the satisfaction of a terribly urgent need, amateurish fancy, fraud, monopoly, and so on. In the wider and subjective sense, therefore, the equivalence of value and countervalue holds fast in these cases, and the single norm, labor power, which makes the discrepancy possible, does not cease to derive the genesis of its character as a value from exchange. . . .

From all the foregoing it appears that exchange is a sociological structure sui generis, a primary form and function of interindividual life. By no means does it follow logically from those qualitative and quantitative properties of things which we call utility and scarcity. On the contrary, both these properties derive their significance as generators of value only under the presupposition of exchange. Where exchange, offering a sacrifice for the sake of a gain, is impossible for any reason, no degree of scarcity of a desired object can convert it to an economic value until the possibility of that relation reappears.

The meaning that an object has for an individual always rests solely in its desirability. For whatever an object is to accomplish for us, its qualitative character is decisive. When we possess it, it is a matter of indifference whether in addition there exist many, few, or no other specimens of its kind. (I do not distinguish here those cases in which scarcity itself is a kind of qualitative property which makes the object desirable to us, such as old postage stamps, curiosities, antiques without aesthetic or historical value, etc.) The sense of difference, incidentally, important for enjoyment in the narrower sense of the word, may be everywhere conditioned by a scarcity of the object, that is, by the fact that it is not enjoyed everywhere and at all times. This inner psychological condition of enjoyment, however, is not a practical factor, because it would have to lead not to the overcoming of scarcity but to its conservation, its increase even—which is patently not the case. The only relevant question apart from the direct enjoyment of things for their qualities is the question of the way to it. As soon as this way is a long and difficult one, involving sacrifices in patience, disappointment, toil, inconvenience, feats of self-denial, and so on, we call the object "scarce." One can express this directly: things are not difficult to obtain because they are scarce, but they are scarce because they are difficult to obtain. The inflexible external fact that the supply of certain goods is too small to satisfy all our desires for them would be in itself insignificant. There are many objectively scarce things which are not scarce in the economic sense of the term. Whether they are scarce in this sense depends entirely upon what measure of energy, patience, and devotion is necessary for their acquisition—sacrifices which naturally presume the desirability of the object.

The difficulty of attainment, that is, the magnitude of the sacrifice involved in exchange, is thus the element that peculiarly constitutes value. Scarcity constitutes only the outer appearance of this element, only its objectification in the form of quantity. One often fails to observe that scarcity, purely as such, is only a negative property, an existence characterized by nonexistence. The nonexistent, however, cannot be operative. Every positive consequence must be the issue of a positive property and force, of which that negative property is only the shadow. These concrete forces are, however, manifestly the only ingredients of exchange. The aspect of concreteness is in no wise reduced because we are not dealing here with individuals as such. Relativity among things has a peculiar property: it involves reaching out beyond the individual, it subsists only within a plurality, and yet it does not constitute a mere conceptual generalization and abstraction.

Herewith is expressed the profound relation between relativity and society, which is the most immediate demonstration of relativity in regard to the material of humanity: society is the supra-singular structure which is nonetheless not abstract. Through this concept historical life is spared the alternatives of having to run either in mere individuals or in abstract generalities. Society is the generality that has, simultaneously, concrete vitality. From this can be seen the unique meaning which exchange, as the economic realization of the relativity of things, has for society. It lifts the individual thing and its significance for the individual man out of their singularity, not into the sphere of the abstract but into the liveliness of interaction, which is, so to speak, the body of economic value. We may examine an object ever so closely with respect to its self-sufficient properties, but we shall not find its economic value. For this consists exclusively in the *reciprocal relationship* which comes into being among several objects on the basis of these properties, each determining the other and each returning to the other the significance it has received therefrom.

— ▓▓ —

Introduction to "Conflict"

In this essay, Simmel maintains that conflict is an inevitable—and in many ways beneficial—feature of social life. The development of a sense of self as well as the creation of group unity depends on conflict or antagonism. In other words, just as our individuality is forged in opposition to the demands placed on us by other individuals and groups, so too we require someone or something to stand *against* in order to realize our essence. As Simmel remarks, "Our opposition makes us feel that we are not completely victims of circumstances. It allows us to prove our strength consciously and only thus gives vitality and reciprocity to conditions from which, without such corrective, we would withdraw at any cost" (Simmel 1908b/1971:75). Moreover, it is often the case that competition or conflict with others spurs us to increase our efforts in the pursuit of goals, whether it be an athlete honing her physical skills or an entrepreneur devising a more efficient business enterprise to outpace competitors.

With regard to group life, conflict generates a clearer distinction between those who belong to a group and those who do not, thereby breeding an intensified sense of group membership. For instance, nothing stokes the flames of school unity like the existence of a rival school. Antagonism directed toward the rival school promotes cohesion or integration within the group. So, too, patriotism is greater in times of war when there is a clear enemy toward which to direct our antagonisms. Group cohesion can also be sparked when a community rallies together in the wake of a natural disaster. In short, Simmel maintains that, without having to overcome a common crisis or attain a common goal in the face of obstacles (in short, without some measure of conflict), there would be no basis for cooperation, group feelings, or "harmony of interest"—the unifying forces that make society possible.

From "Conflict" (1908)

Georg Simmel

The sociological significance of conflict (*Kampf*) has in principle never been disputed. Conflict is admitted to cause or modify interest groups, unifications, organizations. On the other hand, it may sound paradoxical in the common view if one asks whether irrespective of any phenomena that result from conflict or that accompany it, it itself is a form of sociation. At first glance, this sounds like a rhetorical question. If every interaction among men is a sociation, conflict—after all one of the most vivid interactions, which, furthermore, cannot possibly be carried on by one individual alone—must certainly be considered as sociation. And in fact, dissociating factors—hate, envy, need, desire—are the *causes* of conflict; it breaks out because of them. Conflict is thus designed to resolve divergent dualisms; it is a way of achieving some kind of unity, even if it be through the annihilation of one of the conflicting parties. This is roughly parallel to the fact that it is the most violent symptoms of a disease which represent the effort of the organism to free itself of disturbances and damages caused by them.

But this phenomenon means much more than the trivial "si vis pacem para bellum" [if you want peace, prepare for war]; it is something quite general, of which this maxim only describes a special case. Conflict itself resolves the tension between contrasts. The fact that it aims at peace is only one, an especially obvious, expression of its nature: the synthesis of elements that work both against and for one another. This nature appears more clearly when it is realized that both forms of relation—the antithetical and the convergent—are fundamentally distinguished from the mere indifference of two or more individuals or groups. Whether it implies the rejection or the termination of sociation, indifference is purely negative. In contrast to such pure negativity, conflict contains something positive. Its positive and negative aspects, however, are integrated; they can be separated conceptually, but not empirically.

THE SOCIOLOGICAL RELEVANCE OF CONFLICT

Social phenomena appear in a new light when seen from the angle of this sociologically positive character of conflict. It is at once evident then that if the relations among men (rather than what the individual is to himself and in his relations to objects) constitute the subject matter of a special science, sociology, then the traditional topics of that science cover only a subdivision of it: it is more comprehensive and is truly defined by a principle. At one time it appeared as if there were only two consistent subject matters of the science of man: the individual unit and the unit of individuals (society); any third seemed logically excluded. In this conception, conflict itself—irrespective of its contributions to these immediate social units—found no place for study. It was a phenomenon of its own, and its subsumption under the concept of unity would have been arbitrary as well as useless, since conflict meant the negation of unity.

A more comprehensive classification of the science of the relations of men should distinguish, it would appear, those relations which constitute a unit, that is, social relations in the strict sense, from those which counteract unity. It must be realized, however, that both relations can usually be found in every historically real situation. The individual does not attain the unity of his personality exclusively by an exhaustive harmonization, according to logical, objective, religious, or ethical norms, of the contents of his personality. On the contrary, contradiction and conflict not only precede this unity but are operative in it at every moment of its existence. Just so, there probably exists no social unit in which convergent and divergent currents among its members are not inseparably interwoven. An absolutely centripetal and harmonious group, a pure "unification" *("Vereinigung")*, not only is empirically unreal, it could show no real life process. The society of saints which Dante sees in the Rose of Paradise may be like such a group, but it is without any change and development; whereas the holy assembly of Church Fathers in Raphael's *Disputa* shows if not actual conflict, at leaset a considerable differentiation of moods and directions of thought, whence flow all the vitality and the really organic structure of that group. Just as the universe needs "love and hate," that is, attractive and repulsive forces, in order to have any form at all, so society, too, in order to attain a determinate shape, needs some quantitative ratio of harmony and disharmony, of association and competition, of favorable and unfavorable tendencies. But these discords are by no means mere sociological liabilities or negative instances. Definite, actual society does not result only from other social forces which are positive, and only to the extent that the negative factors do not hinder them. This common conception is quite superficial. Society, as we know it, is the result of both categories of interaction, which thus both manifest themselves as wholly positive.[i]

---

[i]This is the sociological instance of a contrast between two much more general conceptions of life. According to the common view, life always shows two parties in opposition. One of them represents the positive aspect of life, its content proper, if not its substance, while the very meaning of the other is non-being, which must be subtracted from the positive elements before they can constitute life. This is the common view of the relation between happiness and suffering, virtue and vice, strength and inadequacy, success and failure—between all possible contents and interruptions of the course of life. The highest conception indicated in respect to these contrasting pairs appears to me different: we must conceive of all these polar differentiations as of *one* life; we must sense the pulse of a central vitality even in that which, if seen from the standpoint of a particular ideal, ought not to be at all and is merely something negative; we must allow the total meaning of our existence to grow out of *both* parties. In the most comprehensive context of life, even that which as a single element is disturbing and destructive, is wholly positive; it is not a gap but the fulfillment of a role reserved for it alone. Perhaps it is not given to us to attain, much less always to maintain, the height from which all phenomena can be felt as making up the unity of life, even though from an objective or value standpoint, they appear to oppose one another as pluses and minuses, contradictions, and mutual eliminations. We are too inclined to think and feel that our essential being, our true, ultimate significance, is identical with one of these factions. According to our optimistic or pessimistic feeling of life, one of them appears to us as surface or accident, as something to be eliminated or subtracted, in order for the true and intrinsically consistent life to emerge. We are everywhere enmeshed in this dualism (which will presently be discussed in more detail in the text above)—in the most intimate as in the most comprehensive provinces of life, personal, objective, and social. We think we have, or are, a whole or unit which is composed of two logically and objectively opposed parties, and we identify this totality of ours with one of them, while we feel the other to be something alien which does not properly belong and which denies our central and comprehensive being. Life constantly moves between these two tendencies. The one has just been described. The other lets the whole really *be* the whole. It makes the unity, which after all comprises both contrasts, alive in each of these contrasts and in their juncture. It is all the more necessary to assert the right of this second tendency in respect to the sociological phenomenon of conflict, because conflict impresses us with its socially destructive force as with an apparently indisputable fact.

Unity and Discord

There is a misunderstanding according to which one of these two kinds of interaction tears down what the other builds up, and what is eventually left standing is the result of the subtraction of the two (while in reality it must rather be designated as the result of their addition). The misunderstanding probably derives from the twofold meaning of the concept of unity. We designate as "unity" the consensus and concord of interacting individuals, as against their discords, separations, and disharmonies. But we also call "unity" the total group-synthesis of persons, energies, and forms, that is, the ultimate wholeness of that group, a wholeness which covers both strictly speaking unitary relations and dualistic relations. We thus account for the group phenomenon which we feel to be "unitary" in terms of functional components considered *specifically* unitary; and in so doing, we disregard the other, larger meaning of the term.

This imprecision is increased by the corresponding twofold meaning of "discord" or "opposition." Since discord unfolds its negative, destructive character between particular individuals, we naïvely conclude that it must have the same effect on the total group. In reality, however, something which is negative and damaging between individuals if it is considered in isolation and as aiming in a particular direction, does not necessarily have the same effect within the total relationship of these individuals. For a very different picture emerges when we view the conflict in conjunction with other interactions not affected by it. The negative and dualistic elements play an entirely positive role in this more comprehensive picture, despite the destruction they may work on particular relations. All this is very obvious in the competition of individuals within an economic unit.

Conflict as an Integrative Force in the Group

Here, among the more complex cases, there are two opposite types. First, we have small groups,

such as the marital couple, which nevertheless involve an unlimited number of vital relations among their members. A certain amount of discord, inner divergence and outer controversy, is organically tied up with the very elements that ultimately hold the group together; it cannot be separated from the unity of the sociological structure. This is true not only in cases of evident marital failure but also in marriages characterized by a *modus vivendi* which is bearable or at least borne. Such marriages are not "less" marriages by the amount of conflict they contain; rather, out of so many elements, among which there is that inseparable quantity of conflict, they have developed into the definite and characteristic units which they are. Secondly, the positive and integrating role of antagonism is shown in structures which stand out by the sharpness and carefully preserved purity of their social divisions and gradations. Thus, the Hindu social system rests not only on the hierarchy, but also directly on the mutual repulsion, of the castes. Hostilities not only prevent boundaries within the group from gradually disappearing, so that these hostilities are often consciously cultivated to guarantee existing conditions. Beyond this, they also are of direct sociological fertility: often they provide classes and individuals with reciprocal positions which they would not find, or not find in the same way, if the causes of hostility were not accompanied by the *feeling* and the expression of hostility—even if the same objective causes of hostility were in operation.

The disappearance of repulsive (and, considered in isolation, destructive) energies does by no means always result in a richer and fuller social life (as the disappearance of liabilities results in larger property) but in as different and unrealizable a phenomenon as if the group were deprived of the forces of cooperation, affection, mutual aid, and harmony of interest. This is not only true for competition generally, which determines the form of the group, the reciprocal positions of its participants, and the distances between them, and which does so purely as a formal matrix of tensions, quite irrespective of its objective *results*. It is true also where the group is based on the attitudes of its members. For instance, the opposition of a member to an associate is no purely negative social factor, if only because such opposition is often the only means for making life with actually unbearable people at least possible. If we did not even have the power and the right to rebel against tyranny, arbitrariness, moodiness, tactlessness, we could not bear to have any relation to people from whose characters we thus suffer. We would feel pushed to take desperate steps—and these, indeed, would end the relation but do *not,* perhaps, constitute "conflict." Not only because of the fact (though it is not essential here) that oppression usually increases if it is suffered calmly and without protest, but also because opposition gives us inner satisfaction, distraction, relief, just as do humility and patience under different psychological conditions. Our opposition makes us feel that we are not completely victims of the circumstances. It allows us to prove our strength consciously and only thus gives vitality and reciprocity to conditions from which, without such corrective, we would withdraw at any cost.

Opposition achieves this aim even where it has no noticeable success, where it does not become manifest but remains purely covert. Yet while it has hardly any practical effect, it may yet achieve an inner balance (sometimes even on the part of *both* partners to the relation), may exert a quieting influence, produce a feeling of virtual power, and thus save relationships whose continuation often puzzles the observer. In such cases, opposition is an element in the relation itself; it is intrinsically interwoven with the other reasons for the relation's existence. It is not only a *means* for preserving the relation but one of the concrete functions which actually constitute it. Where relations are purely external and at the same time of little practical significance, this function can be satisfied by conflict in its *latent* form, that is, by aversion and feelings of mutual alienness and repulsion which upon more intimate contact, no matter how occasioned, immediately change into positive hatred and fight.

Without such aversion, we could not imagine what form modern urban life, which every day brings everybody in contact with innumerable others, might possibly take. The whole inner organization of urban interaction is based on an extremely complex hierarchy of sympathies, indifferences, and aversions of both the most short-lived and the most enduring kind. And in this complex, the sphere of indifference is relatively limited. Our psychological activity responds to almost every impression that comes from another person with a certain determinate feeling. The subconscious, fleeting, changeful nature of this feeling only *seems* to reduce it to indifference. Actually, such indifference would be as unnatural to us as the vague character of innumerable contradictory stimuli would be unbearable. We are protected against both of these typical dangers of the city by antipathy, which is the preparatory phase of concrete antagonism and which engenders the distances and aversions without which we could not lead the urban life at all. The extent and combination of antipathy, the rhythm of its appearance and disappearance, the forms in which it is satisfied, all these, along with the more literally unifying elements, produce the metropolitan form of life in its irresolvable totality; and what at first glance appears in it as dissociation, actually is one of its elementary forms of association. . . .

CONFLICT IN INTIMATE RELATIONS

At the highest level of spiritual cultivation it is possible to avoid this, for it is characteristic of this level to combine complete mutual devotion with complete mutual differentiation. Whereas undifferentiated passion involves the totality of the individual in the excitement of a part or an element of it, the cultivated person allows no such part or element to transcend its proper, clearly circumscribed domain. Cultivation thus gives relations between harmonious persons the advantage that they become aware, precisely on the occasion of conflict, of its trifling nature in comparison with the magnitude of the forces that unify them.

Furthermore, the refined discriminatory sense, especially of deeply sensitive persons, makes attractions and antipathies more passionate if these feelings contrast with those of the past. This is true in the case of unique, irrevocable decisions concerning a given relationship, and it must be sharply distinguished from the everyday vacillations within a mutual belongingness which is felt, on the whole, to be unquestionable. Sometimes between men and women a fundamental aversion, even a feeling of hatred—not in regard to certain particulars, but the reciprocal repulsion of the total person—is the first stage of a relation whose second phase is passionate love. One might entertain the paradoxical suspicion that when individuals are destined to the closest mutual emotional relationship, the emergence of the intimate phase is guided by an instinctive pragmatism so that the eventual feeling attains its most passionate intensification and awareness of what it has achieved by means of an opposite prelude—a step back before running, as it were.

The inverse phenomenon shows the same form: the deepest hatred grows out of broken love. Here, however, not only the sense of discrimination is probably decisive but also the denial of one's own past—a denial involved in such change of feeling. To have to recognize that a deep love—and not only a sexual love—was an error, a failure of intuition (*Instinkt*); so compromises us before ourselves, so splits the security and unity of our self-conception, that we unavoidably make the object of this intolerable feeling pay for it. We cover our secret awareness of our own responsibility for it by hatred which makes it easy for us to pass all responsibility on to the other.

This particular bitterness which characterizes conflicts within relationships whose nature would seem to entail harmony is a sort of positive intensification of the platitude that relations show their closeness and strength in the absence of differences. But this platitude is by no means true without exception. That very intimate groups, such as marital couples, which dominate, or at least touch on, the whole content of life, should contain no occasions for conflict is quite out of the question. It is by no means the sign of

the most genuine and deep affection never to yield to those occasions but instead to prevent them in far-ranging anticipation and to cut them short immediately by mutual yielding. On the contrary, this behavior often characterizes attitudes which though affectionate, moral, and loyal, nevertheless lack the ultimate, unconditional emotional devotion. Conscious of this lack, the individual is all the more anxious to keep the relation free from any shadow and to compensate his partner for that lack through the utmost friendliness, self-control, and consideration. But another function of this behavior is to soothe one's own consciousness in regard to its more or less evident untruthfulness which even the most sincere or even the most passionate will cannot change into truthfulness—because feelings are involved which are not accessible to the will but, like fate itself, exist or do not exist.

The felt insecurity concerning the basis of such relations often moves us, who desire to maintain the relation at all cost, to acts of exaggerated selflessness, to the almost mechanical insurance of the relationship through the avoidance, on principle, of every possibility of conflict. Where on the other hand we are certain of the irrevocability and unreservedness of our feeling, such peace at any price is not necessary. We know that no crisis can penetrate to the foundation of the relationship—we can always find the other again on this foundation. The strongest love can stand a blow most easily, and hence it does not even occur to it, as is characteristic of a weaker one, to fear that the consequences of such a blow cannot be faced, and it must therefore be avoided by all means. Thus, although conflict among intimates can have more tragic results than among less intimate persons, in the light of the circumstances discussed, precisely the most firmly grounded relation may take a chance at discord, whereas good and moral but less deeply rooted relationships apparently follow a much more harmonious and conflictless course.

Introduction to "Sociability"

Thus far, we have seen that people often enter into interactions because they have a specific goal to attain. For instance, we might be motivated to engage in exchange or conflict relations in order to advance our material position or status, or perhaps to fulfill a self-protective need to resist those in authority. However, in his essay "Sociability," Simmel points out that we do not always engage in interactions for strategic or objective purposes. Sometimes we find ourselves interacting with others simply for the sake of the connection itself. Simmel called this form of interaction "sociability," or the "play-form of association."

At its most basic and commonplace, sociability is the "stuff" of conversations. Sociable conversations have no significance or ulterior motive outside the encounter itself; "in sociability talking is an end in itself (Simmel 1910/1971:136). We talk about movies or concerts we've seen, classes we've taken, or the latest news. The impetus for such conversations lies not in proving one's point or in the advantages gained from the interaction, but in the pure pleasure of conversing and satisfying the impulse to associate with others. As soon as the truthfulness of the conversation's content or the striving for personal rewards or goals is made the focus, the encounter loses its playfulness. We are all familiar with the shift in the form of interaction that occurs when a conversation that begins with carefree talk about recent movies seen or sports teams turns into a heated discussion to be won as attempts are made to prove that one's subjective view on the issues is right; or when casual chitchat at a party is diverted by the other person—who you just learned lives an hour away—asking for a ride home. Simmel's point is not that, in order to be sociable, conversations must avoid discussions about serious topics. Instead, his point is that no matter the subject matter, sociability finds "its justification, its place, and its purpose only in the functional play of conversation as such" (ibid.).

One important element of sociability that contributes to its frictionless quality, and thus its appeal, is its "democratic" nature. To the extent that it is a form of interaction freed from the conflicts and pressures that make up "real" life, sociability establishes an "artificial" world, a world without friction or inequalities. In this self-contained, temporary escape, actors leave their personal ambitions and burdens tied to real life behind, thus encouraging the treatment of others simply as "human beings," acting as if all are equal as individuals. Simmel puts it like this:

> Inasmuch as sociability is the abstraction of association—an abstraction of the character of art or of play—it demands the purest, most transparent, most engaging kind of interaction—that among *equals*. It must, because of its very nature, posit beings who give up so much of their objective content, who are so modified in both their outward and their inner significance, that they are sociably equal, and every one of them can win sociability values for himself only under the condition that the others, interacting with him, can also win them. It is a game in which one "acts" as though all were equal, as though he especially esteemed everyone. (ibid.:133, emphasis in the original)

In addition, Simmel suggests that there is a particular kind of sociability that epitomizes the duality between conformity and differentiation discussed previously: flirtation or coquetry. As depicted in Figure 6.3, flirtation is a type of erotic play in which an actor continuously alternates between consent and denial. The idea is "to draw the man on without letting matters come to a decision, to rebuff him without making him lose all hope" (ibid.:134).

Underlying flirtation is the fundamental antithesis arising from a playful unwillingness to surrender oneself to another that otherwise could lead to submission, and a giving up of oneself, behind which lies the threat of refusal or withdrawal. However, once her decision is revealed, resolving the tension between these two polar opposites, the "play" is over. She has either ended consenting to the man's desire to continue the performance (behind which sexual pleasures exist only as a distant possibility) or has made real her attraction for him.

Figure 6.3 Duality of Flirtation

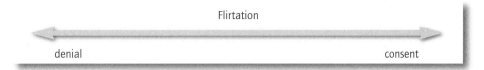

Flirtation

denial   consent

Photo 6.1   Flirtation: It's All in the Look—Maybe Yes . . . or Maybe No

---

From "Sociability" (1910)

Georg Simmel

There is an old conflict over the nature of society. One side mystically exaggerates its significance, contending that only through society is human life endowed with reality. The other regards it as a mere abstract concept by means of which the observer draws the realities, which are individual human beings, into a whole, as one calls trees and brooks, houses and meadows, a "landscape." However one decides this conflict, he must allow society to be a reality in a double sense. On the one hand are the individuals in their directly perceptible existence, the bearers of the processes of association, who are united by these processes into the higher unity which one calls "society"; on the other hand, the interests which, living in the individuals, motivate such union: economic and ideal interests, warlike and erotic, religious and charitable. To satisfy such urges and to attain such purposes arise the innumerable forms of social life, all the with-one-another, for-one-another, in-one-another, against-one-another, and through-one-another, in state and commune, in church and economic associations, in family and clubs. The energy effects of atoms upon each other bring matter into the innumerable forms which we see as "things." Just so the impulses and interests which a man experiences in himself and which push him out toward other men bring about all the forms of association by which a mere sum of separate individuals are made into a "society."

Within this constellation called society, or out of it, there develops a special sociological structure corresponding to those of art and play, which draw their form from these realities but nevertheless leave their reality behind them. It may be an open question whether the concept of a play impulse or an artistic impulse possesses explanatory value; at least it directs attention to the fact that in every play or artistic activity there is contained a common element not affected by their differences of content. Some residue of satisfaction lies in gymnastics, as in card-playing, in music, and in plastic art, something which has nothing to do with the peculiarities of music or plastic art as such but only with the fact that both of the latter are art and both of the former are play. A common element, a likeness of psychological reaction and need, is found in all these various things—something easily distinguishable from the special interest which gives each its distinction. In the same sense one may speak of an impulse to sociability in man. To be sure, it is for the sake of special needs and interests that men unite in economic associations or blood fraternities, in cult societies or robber bands. But above and beyond their special content, all these associations are accompanied by a feeling for, by a satisfaction in, the very fact that one is associated with others and that the solitariness of the individual is resolved into togetherness, a union with others. Of course, this feeling can, in individual cases, be nullified by contrary psychological factors; association can be felt as a mere burden, endured for the sake of our objective aims. But typically there is involved in all effective motives for association a feeling of the worth of association as such, a drive which presses toward this form of existence and often only later calls forth that objective content which carries the particular association along. And as that which I have called artistic impulse draws its form from the complexes of perceivable things and builds this form into a special structure corresponding to the artistic impulse, so also the impulse to sociability distils, as it were, out of the realities of social life the pure essence of association, of the associative process as a value and a satisfaction. It thereby constitutes what we call sociability in the narrower sense. It is no mere accident of language that all sociability, even the purely spontaneous, if it is to have meaning and stability, lays such great value on form, on good form. For "good form" is mutual self-definition, interaction of the elements, through which a unity is made; and since in sociability the concrete motives bound up with life-goals fall away, so must the pure form, the free-playing, interacting interdependence of individuals stand out so much the more strongly and operate with so much the greater effect.

And what joins art with play now appears in the likeness of both to sociability. From the realities of life play draws its great, essential themes: the chase and cunning; the proving of physical and mental powers, the contest and reliance on chance and the favor of forces which one cannot influence. Freed of substance, through which these activities make up the seriousness of life, play gets its cheerfulness but also that symbolic significance which distinguishes it from pure pastime. And just this will show itself more and more as the essence of sociability; that it makes up its substance from numerous fundamental forms of serious relationships among men, a substance, however, spared the frictional relations of real life; but out of its formal relations to real life, sociability (and the more so as it approaches pure sociability) takes on a symbolically playing fulness of life and a significance which a superficial rationalism always seeks only in the content. Rationalism, finding no content there, seeks to do away with sociability as empty idleness, as did the savant who asked concerning a work of art, "What does that prove?" It is nevertheless not without significance that in many, perhaps in all, European languages, the word "society" [*Gesellschaft*] designates a sociable gathering. The political, the economic, the purposive society of any sort is, to be sure, always "society." But only the sociable gathering is "society" without qualifying adjectives, because it alone presents the pure, abstract play of form, all the specific contents of the one-sided and qualified societies being dissolved away.[i]

Sociability is, then, the *play-form of association*. It is related to the content-determined concreteness of association as art is related to reality. Now the great problem of association comes to a solution possible only in sociability. The problem is that of the measure of significance and accent which belongs to the individual as such in and as against the social milieu. Since sociability in its pure form has no ulterior end, no content, and no result outside itself, it is oriented completely about personalities. Since nothing but the satisfaction of the impulse to sociability—although with a resonance left over—is to be gained, the process remains, in its conditions as in its results, strictly limited to its personal bearers; the personal traits of amiability, breeding, cordiality, and attractiveness of all kinds determine the character of purely sociable association. But precisely because all is oriented about them, the personalities must not emphasize themselves too individually. Where real interests, co-operating or clashing, determine the social form, they provide of themselves that the individual shall not present his peculiarities and individuality with too much abandon and aggressiveness. But where this restraint is wanting, if association is to be possible at all, there must prevail another restriction of personal pushing, a restriction springing solely out of the form of the association. It is for this reason that the sense of tact is of such special significance in society, for it guides the self-regulation of the individual in his personal relations to others where no outer or directly egoistic interests provide regulation. And perhaps it is the specific function of tact to mark out for individual impulsiveness, for the ego and for outward demands, those limits which the rights of other require. A

very remarkable sociological structure appears at this point. In sociability, whatever the personality has of objective importance, of features which have their orientation toward something outside the circle, must not interfere. Riches and social position, learning and fame, exceptional capacities and merits of the individual have no role in sociability or, at most, as a slight nuance of that immateriality with which alone reality dares penetrate into the artificial structure of sociability. As these objective qualities which gather about the personality, so also must the most purely and deeply personal qualities be excluded from sociability. The most personal things—character, mood, and fate—have thus no place in it. It is tactless to bring in personal humor, good or ill, excitement and depression, the light and shadow of one's inner life. Where a connection, begun on the sociable level—and not necessarily a superficial or conventional one—finally comes to center about personal values, it loses the essential quality of sociability and becomes an association determined by a content—not unlike a business or religious relation, for which contact, exchange, and speech are but instruments for ulterior ends, while for sociability they are the whole meaning and content of the social processes. This exclusion of the personal reaches into even the most external matters; a lady would not want to appear in such extreme *décolletage* in a really personal, intimately friendly situation with one or two men as she would in a large company without any embarrassment. In the latter she would not feel herself personally involved in the same measure and could therefore abandon herself to the impersonal freedom of the mask. For she is, in the larger company, herself, to be sure, but not quite completely herself, since she is only an element in a formally constituted gathering.

A man, taken as a whole, is, so to speak, a somewhat unformed complex of contents, powers, potentialities; only according to the motivations and relationships of a changing existence is he articulated into a differentiated, defined structure. As an economic and political agent, as a member of a family or of a profession, he is, so to speak, an *ad hoc* construction; his

---

[i]The point is more striking in German, where the word *Gesellschaft* means both "society" and "party" (in the sense of a sociable gathering).—Ed.

life-material is ever determined by a special idea, poured into a special mold, whose relatively independent life is, to be sure, nourished from the common but somewhat undefinable source of energy, the ego. In this sense, the man, as a social creature, is also a unique structure, occurring in no other connection. On the one hand, he has removed all the objective qualities of the personality and entered into the structure of sociability with nothing but the capacities, attractions, and interests of his pure humanity. On the other hand, this structure stops short of the purely subjective and inward parts of his personality. That discretion which is one's first demand upon others in sociability is also required of one's own ego, because a breach of it in either direction causes the sociological artifact of sociability to break down into a sociological naturalism. One can therefore speak of an upper and a lower sociability threshold for the individual. At the moment when people direct their association toward objective content and purpose, as well as at the moment when the absolutely personal and subjective matters of the individual enter freely into the phenomenon, sociability is no longer the central and controlling principle but at most a formalistic and outwardly instrumental principle.

From this negative definition of the nature of sociability through boundaries and thresholds, however, one can perhaps find the positive motif. Kant set it up as the principle of law that everyone should have that measure of freedom which could exist along with the freedom of every other person. If one stands by the sociability impulse as the source or also as the substance of sociability, the following is the principle according to which it is constituted: everyone should have as much satisfaction of this impulse as is consonant with the satisfaction of the impulse for all others. If one expresses this not in terms of the impulse but rather in terms of success, the principle of sociability may be formulated thus: everyone should guarantee to the other that maximum of sociable values (joy, relief, vivacity) which is consonant with the maximum of values he himself receives. As justice upon the Kantian basis is thoroughly democratic, so likewise this principle shows the democratic structure of all sociability, which to be sure every social stratum can realize only within itself, and which so often

makes sociability between members of different social classes burdensome and painful. But even among social equals the democracy of their sociability is a play. Sociability creates, if one will, an ideal sociological world, for in it—so say the enunciated principles—the pleasure of the individual is always contingent upon the joy of others; here, by definition, no one can have his satisfaction at the cost of contrary experiences on the part of others. In other forms of association such lack of reciprocity is excluded only by the ethical imperative which govern them but not by their own immanent nature.

This world of sociability, the only one in which a democracy of equals is possible without friction, is an *artificial* world, made up of beings who have renounced both the objective and the purely personal features of the intensity and extensiveness of life in order to bring about among themselves a pure interaction, free of any disturbing material accent. If we now have the conception that we enter into sociability purely as "human beings," as that which we really are, lacking all the burdens, the agitations, the inequalities with which real life disturbs the purity of our picture, it is because modern life is overburdened with objective content and material demands. Ridding ourselves of this burden in sociable circles, we believe we return to our natural-personal being and overlook the fact that this personal aspect also does not consist in its full uniqueness and natural completeness, but only in a certain reserve and stylizing of the sociable man. In earlier epochs, when a man did not depend so much upon the purposive, objective content of his associations, his "formal personality" stood out more clearly against his personal existence: hence personal bearing in the society of earlier times was much more ceremonially, rigidly, and impersonally regulated than now. This reduction of the personal periphery, of the measure of significance which homogeneous interaction with others allowed the individual, has been followed by a swing to the opposite extreme; today one may even find in society that courtesy by which the strong, outstanding person not only places himself on a level with the weaker but goes so far as to assume the attitude that the weaker is the more worthy and superior. If association itself is interaction, it appears in its purest and most stylized form when it goes on among equals, just as symmetry and balance are the most outstanding forms of artistic stylizing of visible elements. Inasmuch as sociability is the abstraction of association—an abstraction of the character of art or of play—it demands the purest, most transparent, most engaging kind of interaction—that among *equals.* It must, because of its very nature, posit beings who give up so much of their objective content, who are so modified in both their outward and their inner significance, that they are sociably equal, and every one of them can win sociability values for himself only under the condition that the others, interacting with him, can also win them. It is a game in which one "acts" as though all were equal, as though he especially esteemed everyone. This is just as far from being a lie as is play or art in all their departures from reality. But the instant the intentions and events of practical reality enter into the speech and behavior of sociability, it does become a lie—just as a painting does when it attempts, panorama fashion, to be taken for reality. That which is right and proper within the self-contained life of sociability, concerned only with the immediate play of its forms, becomes a lie when this is mere pretense, which in reality is guided by purposes of quite another sort than the sociable or is used to conceal such purposes—and indeed sociability may easily get entangled with real life.

It is an obvious corollary that everything may be subsumed under sociability which one can call sociological play-form; above all, play itself, which assumes a large place in the sociability of all epochs. The expression "social game" is significant in the deeper sense which I have indicated. The entire interactional or associational complex among men: the desire to gain advantage, trade, formation of parties and the desire to win from another, the movement between opposition and co-operation, outwitting and revenge—all this, fraught with purposive content in the serious affairs of reality, in play leads a life carried along only and completely by the stimulus of these functions. For even when play turns about a money prize, it is not the prize, which indeed could be won in many other ways, which is the specific point of the play; but the attraction for the true sportsman lies in the dynamics and in the chances of

that sociologically significant form of activity itself. The social game has a deeper double meaning—that it is played not only *in* a society as its outward bearer but that with its help people actually "play" "society."

Further, in the sociology of the sexes, eroticism has elaborated a form of play: *coquetry,* which finds in sociability its lightest, most playful, and yet its widest realization. If the erotic question between the sexes turns about consent or denial (whose objects are naturally of endless variety and degree and by no means only of strictly physiological nature), so is it the essence of feminine coquetry to play hinted consent and hinted denial against each other to draw the man on without letting matters come to a decision, to rebuff him without making him lose all hope. The coquette brings her attractiveness to its climax by letting the man hang on the verge of getting what he wants without letting it become too serious for herself; her conduct swings between yes and no, without stopping at one or the other. She thus playfully shows the simple and pure form of erotic decision and can bring its polar opposites together in a quite integrated behavior, since the decisive and fateful content, which would bring it to one of the two decisions, by definition does not enter into coquetry. And this freedom from all the weight of firm content and residual reality gives coquetry that character of vacillation, of distance, of the ideal, which allows one to speak with some right of the "art"—not of the "arts"—of coquetry. In order, however, for coquetry to spread as so natural a growth on the soil of sociability, as experience shows it to be, it must be countered by a special attitude on the part of men. So long as the man denies himself the stimulation of coquetry, or so long as he is—on the contrary—merely a victim who is involuntarily carried along by her vacillations from a half-yes to a half-no—so long does coquetry lack the adequate structure of sociability. It lacks that free interaction and equivalence of the elements which is the fundamental condition of sociability. The latter appears only when the man desires nothing more than this free moving play, in which something definitively erotic lurks only as a remote symbol, and when he does not get his pleasure in these gestures and preliminaries from erotic desire or fear of it. Coquetry, as it unfolds its grace on the heights of sociable cultivation, has left behind the reality of erotic desire, of consent or denial, and becomes a play of shadow pictures of these serious matters. Where the latter enter or lurk, the whole process becomes a private affair of the two persons, played out on the level of reality; under the sociological sign of sociability, however, in which the essential orientation of the person to the fulness of life does not enter, coquetry is the teasing or even ironic play with which eroticism has distilled the pure essence of its interaction out from its substantive or individual content. As sociability plays at the forms of society, so coquetry plays out the forms of eroticism.

In what measure sociability realizes to the full the abstraction of the forms of sociological interaction otherwise significant because of their content and gives them—now turning about themselves, so to speak—a shadow body is revealed finally in that most extensive instrument of all human common life, *conversation*. The decisive point is expressed in the quite banal experience that in the serious affairs of life men talk for the sake of the content which they wish to impart or about which they want to come to an understanding—in sociability talking is an end in itself; in purely sociable conversation the content is merely the indispensable carrier of the stimulation, which the lively exchange of talk as such unfolds. All the forms with which this exchange develops: argument and the appeals to the norms recognized by both parties; the conclusion of peace through compromise and the discovery of common convictions; the thankful acceptance of the new and the parrying-off of that on which no understanding is to be hoped for—all these forms of conversational interaction, otherwise in the service of innumerable contents and purposes of human intercourse, here have their meaning in themselves; that is to say, in the excitement of the play of relations which they establish between individuals, binding and loosening, conquering and being vanquished, giving and taking. In order that this play may retain its self-sufficiency at the level of pure form, the content must receive no weight on its own account; as soon as the discussion gets business-like, it is no longer sociable; it turns its compass point around as soon as the verification of a truth becomes its purpose. Its character as sociable converse is disturbed just as when it turns into a serious argument. The form of the common search of the truth, the form of the argument, may occur; but it must not permit the seriousness of the momentary content to become its substance any more than one may put a piece of three-dimensional reality into the perspective of a painting. Not that the content of sociable conversation is a matter of indifference; it must be interesting, gripping, even significant—only it is not the purpose of the conversation that these qualities should square with objective results, which stand by definition outside the conversation. Outwardly, therefore, two conversations may run a similar course, but only that one of them is sociable in which the subject matter, with all its value and stimulation, finds its justification, its place, and its purpose only in the functional play of conversation as such, in the form of repartee with its special unique significance. It therefore inheres in the nature of sociable conversation that its object matter can change lightly and quickly; for, since the matter is only the means, it has an entirely interchangeable and accidental character which inheres in means as against fixed purposes. Thus sociability offers, as was said, perhaps the only case in which talk is a legitimate end in itself. For by the fact that it is two-sided—indeed with the possible exception of looking-each-other-over the purest and most sublimated form of mutuality among all sociological phenomena—it becomes the most adequate fulfillment of a relation, which is, so to speak, nothing but relationship, in which even that which is otherwise pure form of interaction is its own self-sufficient content. It results from this whole complex that also the telling of tales, witticisms, anecdotes, although often a stopgap and evidence of conversational poverty, still can show a fine tact in which all the motives of sociability are apparent. For, in the first place, the conversation is by this means kept above all individual intimacy, beyond everything purely personal which would not fit into the categories of sociability. This objective element is brought in not for the sake of its content but in the interest of sociability; that something is said and accepted is not an end in itself but a mere means to maintain the liveliness, the mutual understanding, the common consciousness of the group. Not only thereby is it given a content which all can share but it is a gift of the individual to the whole, behind which the giver can remain invisible; the finest sociably told story is that in which the narrator allows his own person to remain completely in the background; the most effective story holds itself in the happy balance of the sociable ethic, in which the subjectively individual as well as the objectively substantive have dissolved themselves completely in the service of pure sociability. . . .

Social Types

Introduction to "The Stranger"

Simmel juxtaposed his analysis of the forms of social interaction with a discussion of social **types**. Social types derive not from qualities intrinsic to the individual in question, nor from an individual's choice to be one "type" or another. Rather, being assigned or identified as a type of individual is a product of one's *relationship* to others. Simmel identified a number of social types, including "the poor," "the nobility," "the miser," "the spendthrift," and "the adventurer," but it is his analysis of "the stranger" that has become most well known.

The relationship of the stranger to the group is rooted in a unique synthesis of opposites: "nearness" and "remoteness" (see Figure 6.4). As a distinct social type, the stranger is near or close to us insofar as we share with him general, impersonal qualities, such as nationality, gender, or race. But because such similarities connect us with so many others, the stranger is indistinct or "remote." No unique or specific qualities are shared with him that could in turn form the basis of a personal relationship. As a result, the stranger is seen not as an individual, but, rather, as a "type" of person whose *particular* characteristics make him fundamentally different from the group. Yet, this unique position of the stranger relative to the group allows him to provide services that are otherwise unattainable or "unfit" for the in-group to perform. Often this attachment makes the stranger an indispensable element of the group, though his positive contributions are dependent on his outsider status. Though he is part of the group, the stranger, then, exists outside of, and is thus confronted by the group.

For Simmel, the classical example of the stranger was the European Jew who served as a trader. The trader is a "middleman" who makes possible the exchange of goods with people who live beyond the boundaries of the group. Jews became traders because they were denied many of the legal, political, and property rights granted to ordinary citizens.[1] As a result, European Jews were often restricted in their professional activities to "mobile" occupations such as trading and finance. Simmel argued that it is the mobility of the stranger within a group—he is "no landowner"—that makes the position a "synthesis of nearness and remoteness." As Simmel remarked, "The purely mobile person comes incidentally into

Figure 6.4 Duality of the "Stranger"

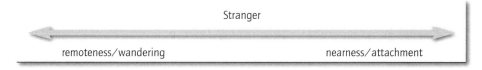

Stranger

remoteness/wandering   nearness/attachment

---

[1] As you will read in the next chapter, there are many interesting parallels between Simmel's notion of the "stranger" and Du Bois's notion of "double consciousness" and the "place" of African Americans in the United States.

contact with *every* single element [of a group] but is not bound up organically, through established ties of kinship, locality, or occupation, with any single one" (Simmel 1908c/1971:145, emphasis in the original).

In addition to these occupational consequences, the unique relation of the stranger to the larger group allows the stranger to adopt an objective attitude toward internal conflicts. Nonpartisanship grants the stranger a position of objectivity in efforts to resolve disputes. In other words, the stranger is not likely to be committed in advance to any one party should disagreements arise between individuals who possess full group membership. Furthermore, the remoteness and freedom from prejudiced understanding that objectivity entails can also make the stranger a valued confidant. Certainly, in modern society it is not uncommon to be willing to share otherwise unaired, intimate details about one's life and relationships with a complete stranger. Indeed, now such strangers—professional therapists and counselors—are paid.

Significantly, then, the stranger is not the same as the complete "outcast." The stranger has elements of nearness *and* remoteness—he is attached, but not completely—while the social outcast is only remote. However, despite the services that strangers are able to provide to a community, nonetheless we should be careful not to romanticize the position of this social type. Strangers often are exceptionally vulnerable to discrimination, if not violence.[2]

"The Stranger" (1908)

Georg Simmel

If Wandering, considered as a state of detachment from every given point in space, is the conceptual opposite of attachment to any point, then the sociological form of "the stranger" presents the synthesis, as it were, of both of these properties. (This is another indication that spatial relations not only are determining conditions of relationships among men, but are also symbolic of those relationships.) The stranger will thus not be considered here in the usual sense of the term, as the wanderer who comes today and goes tomorrow, but rather as the man who comes today and stays tomorrow—the potential wanderer, so to speak, who, although he has gone no further, has not quite got over the freedom of coming and going. He is fixed within a certain spatial circle—or within a group whose boundaries

are analogous to spatial boundaries—but his position within it is fundamentally affected by the fact that he does not belong in it initially and that he brings qualities into it that are not, and cannot be, indigenous to it.

In the case of the stranger, the union of closeness and remoteness involved in every human relationship is patterned in a way that may be succinctly formulated as follows: the distance within this relation indicates that one who is close by is remote, but his strangeness indicates that one who is remote is near. The state of being a stranger is of course a completely positive relation; it is a specific form of interaction. The inhabitants of Sirius are not exactly strangers to us, at least not in the sociological sense of the word as we are considering it. In that sense they

[2]While Jews historically have been strangers throughout Europe, America has its own history of "strangers." African Americans first were enslaved, then, after abolition, were kept "second-class" citizens with the establishment of Jim Crow laws. During the late nineteenth and early twentieth centuries, immigrants from southern and eastern Europe faced intense hatred and discrimination. Most ironic, however, is the plight suffered by Native Americans and Mexican Americans, who have always been treated as strangers in their own land as a result of colonization. All the while, they have played important roles by risking their lives in the armed forces, building the nation's infrastructure, and performing jobs that are considered "beneath" the dominant groups.

do not exist for us at all; they are beyond being far and near. The stranger is an element of the group itself, not unlike the poor and sundry "inner enemies"—an element whose membership within the group involves both being outside it and confronting it.

The following statements about the stranger are intended to suggest how factors of repulsion and distance work to create a form of being together, a form of union based on interaction.

In the whole history of economic activity the stranger makes his appearance everywhere as a trader, and the trader makes his as a stranger. As long as production for one's own needs is the general rule, or products are exchanged within a relatively small circle, there is no need for a middleman within the group. A trader is required only for goods produced outside the group. Unless there are people who wander out into foreign lands to buy these necessities, in which case they are themselves "strange" merchants in this other region, the trader *must* be a stranger; there is no opportunity for anyone else to make a living at it.

This position of the stranger stands out more sharply if, instead of leaving the place of his activity, he settles down there. In innumerable cases even this is possible only if he can live by trade as a middleman. Any closed economic group where land and handicrafts have been apportioned in a way that satisfies local demands will still support a livelihood for the trader. For trade alone makes possible unlimited combinations, and through it intelligence is constantly extended and applied in new areas, something that is much harder for the primary producer with his more limited mobility and his dependence on a circle of customers that can be expanded only very slowly. Trade can always absorb more men than can primary production. It is therefore the most suitable activity for the stranger, who intrudes as a supernumerary, so to speak, into a group in which all the economic positions are already occupied. The classic example of this is the history of European Jews. The stranger is by his very nature no owner of land—land not only in the physical sense but also metaphorically as a vital substance which is fixed, if not in space, then at least in an ideal position within the social environment.

Although in the sphere of intimate personal relations the stranger may be attractive and meaningful in many ways, so long as he is regarded as a stranger he is no "landowner" in the eye of the other. Restriction to intermediary trade and often (although sublimated from it) to pure finance gives the stranger the specific character of *mobility*. The appearance of this mobility within a bounded group occasions that synthesis of nearness and remoteness which constitutes the formal position of the stranger. The purely mobile person comes incidentally into contact with *every* single element but is not bound up organically, through established ties of kinship, locality, or occupation, with any single one.

Another expression of this constellation is to be found in the objectivity of the stranger. Because he is not bound by roots to the particular constituents and partisan dispositions of the group, he confronts all of these with a distinctly "objective" attitude, an attitude that does not signify mere detachment and nonparticipation but is a distinct structure composed of remoteness and nearness, indifference and involvement. I refer to my analysis of the dominating positions gained by aliens, in the discussion of superordination and subordination, typified by the practice in certain Italian cities of recruiting their judges from outside, because no native was free from entanglement in family interest and factionalism.

Connected with the characteristic of objectivity is a phenomenon that is found chiefly, though not exclusively, in the stranger who moves on. This is that he often receives the most surprising revelations and confidences, at times reminiscent of a confessional about matters which are kept carefully hidden from everybody with whom one is close. Objectivity is by no means nonparticipation, a condition that is altogether outside the distinction between subjective and objective orientations. It is rather a positive and definite kind of participation, in the same way that the objectivity of theoretical observation clearly does not mean that the mind is a passive tabula rasa on which things inscribe their qualities, but rather signifies the full activity of a mind working according to its own laws, under conditions that exclude accidental distortions and emphases whose individual and subjective differences would produce quite different pictures of the same object.

Objectivity can also be defined as freedom. The objective man is not bound by ties which could prejudice his perception, his understanding, and his assessment of data. This freedom, which permits the stranger to experience and treat even his close relationships as though from a bird's-eye view, contains many dangerous possibilities. From earliest times, in uprisings of all sorts the attacked party has claimed that there has been incitement from the outside, by foreign emissaries and agitators. Insofar as this has happened, it represents an exaggeration of the specific role of the stranger: he is the freer man, practically and theoretically; he examines conditions with less prejudice; he assesses them against standards that are more general and more objective; and his actions are not confined by custom, piety, or precedent.

Finally, the proportion of nearness and remoteness which gives the stranger the character of objectivity also finds practical expression in the more *abstract* nature of the relation to him. That is, with the stranger one has only certain *more general* qualities in common, whereas the relation with organically connected persons is based on the similarity of just those specific traits which differentiate them from the merely universal. In fact, all personal relations whatsoever can be analyzed in terms of this scheme. They are not determined only by the existence of certain common characteristics which the individuals share in addition to their individual differences, which either influence the relationship or remain outside of it. Rather, the kind of effect which that commonality has on the relation essentially depends on whether it exists only among the participants themselves, and thus, although general within the relation, is specific and incomparable with respect to all those on the outside, or whether the participants feel that what they have in common is so only because it is common to a group, a type, or mankind in general. In the latter case, the effect of the common features becomes attenuated in proportion to the size of the group bearing the same characteristics. The commonality provides a basis for unifying the members, to be sure; but it does not specifically direct *these* particular persons to one another. A similarity so widely shared could just as easily unite each person with every possible other. This, too, is evidently a way in which a relationship includes both nearness and remoteness simultaneously. To the extent to which the similarities assume a universal nature, the warmth of the connection based on them will acquire an element of coolness, a sense of the contingent nature of precisely *this* relation—the connecting forces have lost their specific, centripetal character.

In relation to the stranger, it seems to me, this constellation assumes an extraordinary preponderance in principle over the individual elements peculiar to the relation in question. The stranger is close to us insofar as we feel between him and ourselves similarities of nationality or social position, of occupation or of general human nature. He is far from us insofar as these similarities extend beyond him and us, and connect us only because they connect a great many people.

A trace of strangeness in this sense easily enters even the most intimate relationships. In the stage of first passion, erotic relations strongly reject any thought of generalization. A love such as this has never existed before; there is nothing to compare either with the person one loves or with our feelings for that person. An estrangement is wont to set in (whether as cause or effect is hard to decide) at the moment when this feeling of uniqueness disappears from the relationship. A skepticism regarding the intrinsic value of the relationship and its value for us adheres to the very thought that in this relation, after all, one is only fulfilling a general human destiny, that one has had an experience that has occurred a thousand times before, and that, if one had not accidentally met this precise person, someone else would have acquired the same meaning for us.

Something of this feeling is probably not absent in any relation, be it ever so close, because that which is common to two is perhaps never common *only* to them but belongs to a general conception which includes much else besides, many *possibilities* of similarities. No matter how few of these possibilities are realized and how often we may forget about them, here and there, nevertheless, they crowd in like shadows between men, like a mist eluding every designation, which must congeal into solid corporeality for it to be called jealousy. Perhaps this is in many cases a more general, at least more insurmountable, strangeness than that due to differences and obscurities. It is strangeness caused by the fact that similarity, harmony, and closeness are accompanied by the feeling that they are actually not the exclusive property of this particular relation, but stem from a more general one—a relation that potentially includes us and an indeterminate number of others, and therefore prevents that relation which alone was experienced from having an inner and exclusive necessity.

On the other hand, there is a sort of "strangeness" in which this very connection on the basis of a general quality embracing the parties is precluded. The relation of the Greeks to the barbarians is a typical example; so are all the cases in which the general characteristics one takes as peculiarly and merely human are disallowed to the other. But here the expression "the stranger" no longer has any positive meaning. The relation with him is a non-relation; he is not what we have been discussing here: the stranger as a member of the group itself.

As such, the stranger is near and far *at the same time,* as in any relationship based on merely universal human similarities. Between these two factors of nearness and distance, however, a peculiar tension arises, since the consciousness of having only the absolutely general in common has exactly the effect of putting a special emphasis on that which is not common. For a stranger to the country, the city, the race, and so on, what is stressed is again nothing individual, but alien origin, a quality which he has, or could have, in common with many other strangers. For this reason strangers are not really perceived as individuals, but as strangers of a certain type. Their remoteness is no less general than their nearness.

This form appears, for example, in so special a case as the tax levied on Jews in Frankfurt and elsewhere during the Middle Ages. Whereas the tax paid by Christian citizens varied according to their wealth at any given time, for every single Jew the tax was fixed once and for all. This amount was fixed because the Jew had his social position as a *Jew,* not as the bearer of certain objective contents. With respect to taxes every other citizen was regarded as possessor of a certain amount of wealth, and his tax could follow the fluctuations of his fortune. But the Jew as taxpayer was first of all a Jew, and thus his fiscal position contained an invariable element. This appears most forcefully, of course, once the differing circumstances of individual Jews are no longer considered, limited though this consideration is by fixed assessments, and all strangers pay exactly the same head tax.

Despite his being inorganically appended to it, the stranger is still an organic member of the group. Its unified life includes the specific conditioning of this element. Only we do not know how to designate the characteristic unity of this position otherwise than by saying that it is put together of certain amounts of nearness and of remoteness. Although both these qualities are found to some extent in all relationships, a special proportion and reciprocal tension between them produce the specific form of the relation to the "stranger."

Individuality and Social Structure

Introduction to "Fashion"

As the example at the opening of this chapter indicated, Simmel explores the world of fashion as yet another aspect of social life built on the coupling of opposites. Whether it be taste in music or in cars, the design of clothes or furniture, Simmel sees fashion as an expression of individualization, an attempt to cultivate one's distinctiveness, on the one hand, and an attempt to express imitation and conformity, on the other (see Figure 6.5). As Simmel contends,

Figure 6.5 Duality of Fashion

Fashion

eccentricity
(adopted by very few)

conformity/imitation
(widespread acceptance)

> From the fact that fashion as such can never be generally in vogue, the individual derives the satisfaction of knowing that as adopted by him it still represents something special and striking, while at the same time he feels inwardly supported by a set of persons who are . . . actually doing the same thing. (Simmel 1904/1971:304)

As a result, adopting a particular fashion allows the individual to cultivate his uniqueness and sense of self-identity with the security of knowing that, should the trend be met with reproach, he is not responsible for creating it. After all, he is merely following the latest fashion.

The passage quoted above suggests an interesting question: why can't a fashion ever be generally in vogue? To this, Simmel offers an answer that, not surprisingly, speaks to the paradoxical nature of social life. Simply put, fashions remain fashionable only to the extent that the general population does not adopt them. Once fashions become widely disseminated and take on an air of universality or permanence, they become a common fact of life and thus are no longer able to convey distinctions between individuals and groups. Turning again to Simmel,

> The very character of fashion demands that it should be exercised at one time only by a portion of the given group, the great majority merely being on the road to adopting it. As soon as . . . anything that was originally done only by few has really come to be practiced by all . . . we no longer speak of fashion. As fashion spreads, it gradually goes to its doom. (ibid.:302)

The capacity for a particular fashion to create a sense of distinction for the individuals who first adopt it is destroyed as more and more people practice it; as the fashionable difference is transformed into a commonplace standard. With the destruction of its very purpose—to cultivate individuality—the fashion dies, only to be replaced by a new trend that, through its inevitable spread, will also face its equally inevitable death. And so the cycle continues with the introduction of styles that have no functional usefulness: wide or narrow jacket lapels, bell-bottomed or straight-legged pants, thick- or thin-striped shirts, high-heeled shoes, etc. etc.

The ebb and flow of fashion trends raises an important issue. Aside from an individual's quest to be fashionable, what group forces shape the rise and demise of trends? In addressing this question, Simmel looks to the dynamics of class relations. Indeed, the cycle described above is in large measure a consequence of the upper classes within a society attempting to distance themselves from the lower classes. Fashion is a visible and easily identifiable sign of class position, making it a domain well suited for publicly demonstrating one's place in the class hierarchy. (The overt connection between cars and class position clearly indicates this.) However, as the lower classes set out to imitate those above them in the "externals of life," the upper classes necessarily must seek out an alternative form of fashion in order to retain and express their distinctiveness.

The pace at which styles change is quickened in modern societies. To the extent that the lower classes in advanced societies possess greater wealth relative to those in less-developed or premodern societies, they have an advantage in chasing the fashion trends established by the upper classes. With the mass production of goods, the costs for manufacturing them

Photo 6.2 "Punk" Fashion: Looking the Same, Only Different

decreases, in turn making them more affordable to the lower classes. More purchasing power coupled with cheaper products and increased supply shortens the life span of fashions first adopted by the upper classes in their attempt to differentiate themselves from the masses. Finally, societal development has produced yet another Simmelian irony: insofar as we try to express our uniqueness or individuality through fashion, we often turn to buying mass-produced, standardized goods.

From "Fashion" (1904)

Georg Simmel

Fashion is the imitation of a given example and satisfies the demand for social adaptation; it leads the individual upon the road which all travel, it furnishes a general condition, which resolves the conduct of every individual into a mere example. At the same time it satisfies in no less degree the need of differentiation, the tendency towards dissimilarity, the desire for change and contrast, on the one hand by a constant change of contents, which gives to the fashion of today an individual stamp as opposed to that of yesterday and of to-morrow, on the other hand because fashions differ for different classes—the fashions of the upper stratum of society are never identical with those of the lower; in fact, they are abandoned by the former as soon as the latter prepares to appropriate them. Thus fashion represents nothing more than one of the many forms of life by the aid of which we seek to combine in uniform spheres of activity the tendency towards social equalization with the desire for individual differentiation and change. Every phase of the conflicting pair strives visibly beyond the degree of satisfaction that any fashion offers to an absolute control of the sphere of life in question. If we should study the history of fashions (which hitherto have been examined only from the view-point of the development of their contents) in connection with their importance for the form of the social process, we should find that it reflects the history of the attempts to adjust the satisfaction of the two counter-tendencies more and more perfectly to the condition of the existing individual and social culture. The various psychological elements in fashion all conform to this fundamental principle.

Fashion, as noted above, is a product of class distinction and operates like a number of other forms, honor especially, the double function of which consists in revolving within a given circle and at the same time emphasizing it as separate from others. Just as the frame of a picture characterizes the work of art inwardly as a coherent, homogeneous, independent entity and at the same time outwardly severs all direct relations with the surrounding space, just as the uniform energy of such forms cannot be expressed unless we determine the double effect, both inward and outward, so honor owes its character, and above all its moral rights, to the fact that the individual in his personal honor at the same time represents and maintains that of his social circle and his class. These moral rights, however, are frequently considered unjust by those without the pale. Thus fashion on the one hand signifies union with those in the same class, the uniformity of a circle characterized by it, and, *uno actu,* the exclusion of all other groups.

Union and segregation are the two fundamental functions which are here inseparably united, and one of which, although or because it forms a logical contrast to the other, becomes the condition of its realization. Fashion is merely a product of social demands, even though the individual object which it creates or recreates may represent a more or less individual need. This is clearly proved by the fact that very frequently not the slightest reason can be found for the creations of fashion from the standpoint of an objective, aesthetic, or other expediency. While

in general our wearing apparel is really adapted to our needs, there is not a trace of expediency in the method by which fashion dictates, for example, whether wide or narrow trousers, colored or black scarfs shall be worn. As a rule the material justification for an action coincides with its general adoption, but in the case of fashion there is a complete separation of the two elements, and there remains for the individual only this general acceptance as the deciding motive to appropriate it. Judging from the ugly and repugnant things that are sometimes in vogue, it would seem as though fashion were desirous of exhibiting its power by getting us to adopt the most atrocious things for its sake alone. The absolute indifference of fashion to the material standards of life is well illustrated by the way in which it recommends something appropriate in one instance, something abstruse in another, and something materially and aesthetically quite indifferent in a third. The only motivations with which fashion is concerned are formal social ones. The reason why even aesthetically impossible styles seem *distingué,* elegant, and artistically tolerable when affected by persons who carry them to the extreme, is that the persons who do this are generally the most elegant and pay the greatest attention to their personal appearance, so that under any circumstances we would get the impression of something *distingué* and aesthetically cultivated. This impression we credit to the questionable element of fashion, the latter appealing to our consciousness as the new and consequently most conspicuous feature of the *tout ensemble.*

Fashion occasionally will affect objectively determined subjects such as religious faith, scientific interests, even socialism and individualism; but it does not become operative as fashion until these subjects can be considered independent of the deeper human motives from which they have risen. For this reason the rule of fashion becomes in such fields unendurable. We therefore see that there is good reason why externals—clothing, social conduct, amusements—constitute the specific field of fashion, for here no dependence is placed on really vital motives of human action. It is the field which we can most easily relinquish to the bent towards imitation, which it would be a sin to follow in important questions. We encounter here a close connection between the consciousness of personality and that of the material forms of life, a connection that runs all through history. The more objective our view of life has become in the last centuries, the more it has stripped the picture of nature of all subjective and anthropomorphic elements, and the more sharply has the conception of individual personality become defined. The social regulation of our inner and outer life is a sort of embryo condition, in which the contrasts of the purely personal and the purely objective are differentiated, the action being synchronous and reciprocal. Therefore wherever man appears essentially as a social being we observe neither strict objectivity in the view of life nor absorption and independence in the consciousness of personality.

Social forms, apparel, aesthetic judgment, the whole style of human expression, are constantly transformed by fashion, in such a way, however, that fashion—*i.e.,* the latest fashion—in all these things affects only the upper classes. Just as soon as the lower classes begin to copy their style, thereby crossing the line of demarcation the upper classes have drawn and destroying the uniformity of their coherence, the upper classes turn away from this style and adopt a new one, which in its turn differentiates them from the masses; and thus the game goes merrily on. Naturally the lower classes look and strive towards the upper, and they encounter the least resistance in those fields which are subject to the whims of fashion; for it is here that mere external imitation is most readily applied. The same process is at work as between the different sets within the upper classes, although it is not always as visible here as it is, for example, between mistress and maid. Indeed, we may often observe that the more nearly one set has approached another, the more frantic becomes the desire for imitation from below and the seeking for the new from above. The increase of wealth is bound to hasten the process considerably and render it visible, because the objects of fashion, embracing as they do the externals of life, are most accessible to the mere call of money, and conformity to the higher set is more easily acquired here than in fields which demand an individual test that gold and silver cannot affect. . . .

Fashion plays a more conspicuous *rôle* in modern times, because the differences in our standards of life have become so much more strongly accentuated, for the more numerous and the more sharply drawn these differences are, the greater the opportunities for emphasizing them at every turn. In innumerable instances this cannot be accomplished by passive inactivity, but only by the development of forms established by fashion; and this has become all the more pronounced since legal restrictions prescribing various forms of apparel and modes of life for different classes have been removed. . . .

The very character of fashion demands that it should be exercised at one time only by a portion of the given group, the great majority being merely on the road to adopting it. As soon as an example has been universally adopted, that is, as soon as anything that was originally done only by a few has really come to be practiced by all—as is the case in certain portions of our apparel and in various forms of social conduct—we no longer speak of fashion. As fashion spreads, it gradually goes to its doom. The distinctiveness which in the early stages of a set fashion assures for it a certain distribution is destroyed as the fashion spreads, and as this element wanes, the fashion also is bound to die. By reason of this peculiar play between the tendency towards universal acceptance and the destruction of its very purpose to which this general adoption leads, fashion includes a peculiar attraction of limitation, the attraction of a simultaneous beginning and end, the charm of novelty coupled to that of transitoriness. The attractions of both poles of the phenomena meet in fashion, and show also here that they belong together unconditionally, although, or rather because, they are contradictory in their very nature. Fashion always occupies the dividing-line between the past and the future, and consequently conveys a stronger feeling of the present, at least while it is at its height, than most other phenomena. What we call the present is usually nothing more than a combination of a fragment of the past with a fragment of the future. Attention is called to the present less often than colloquial usage, which is rather liberal in its employment of the word, would lead us to believe.

Few phenomena of social life possess such a pointed curve of consciousness as does fashion. As soon as the social consciousness attains to the highest point designated by fashion, it marks the beginning of the end for the latter. This transitory character of fashion, however, does not on the whole degrade it, but adds a new element of attraction. At all events an object does not suffer degradation by being called fashionable, unless we reject it with disgust or wish to debase it for other, material reasons, in which case, of course, fashion becomes an idea of value. In the practice of life anything else similarly new and suddenly disseminated is not called fashion, when we are convinced of its continuance and its material justification. If, on the other hand, we feel certain that the fact will vanish as rapidly as it came, then we call it fashion. We can discover one of the reasons why in these latter days fashion exercises such a powerful influence on our consciousness in the circumstance that the great, permanent, unquestionable convictions are continually losing strength, as a consequence of which the transitory and vacillating elements of life acquire more room for the display of their activity. The break with the past, which, for more than a century, civilized mankind has been laboring unceasingly to bring about, makes the consciousness turn more and more to the present. This accentuation of the present evidently at the same time emphasizes the element of change, and a class will turn to fashion in all fields, by no means only in that of apparel, in proportion to the degree in which it supports the given civilizing tendency. It may almost be considered a sign of the increased power of fashion, that it has overstepped the bounds of its original domain, which comprised only personal externals, and has acquired an increasing influence over taste, over theoretical convictions, and even over the moral foundations of life.

From the fact that fashion as such can never be generally in vogue, the individual derives the satisfaction of knowing that as adopted by him it still represents something special and striking, while at the same time he feels inwardly supported by a set of persons who are striving for the same thing, not as in the case of other social satisfactions, by a set actually doing the same thing. The fashionable person is regarded with mingled feelings of approval and envy; we envy him as an individual, but approve of him as a member of a set or group. Yet even this envy has a peculiar coloring. There is a shade of envy

which includes a species of ideal participation in the envied object itself. An instructive example of this is furnished by the conduct of the poor man who gets a glimpse of the feast of his rich neighbor. The moment we envy an object or a person, we are no longer absolutely excluded from it; some relation or other has been established—between both the same psychic content now exists—although in entirely different categories and forms of sensations. This quiet personal usurpation of the envied property contains a kind of antidote, which occasionally counter-acts the evil effects of this feeling of envy. The contents of fashion afford an especially good chance of the development of this conciliatory shade of envy, which also gives to the envied person a better conscience because of his satisfaction over his good fortune. This is due to the fact that these contents are not, as many other psychic contents are, denied absolutely to any one, for a change of fortune, which is never entirely out of the question, may play them into the hands of an individual who had previously been confined to the state of envy.

From all this we see that fashion furnishes an ideal field for individuals with dependent natures, whose self-consciousness, however, requires a certain amount of prominence, attention, and singularity. Fashion raises even the unimportant individual by making him the representative of a class, the embodiment of a joint spirit. And here again we observe the curious intermixture of antagonistic values. Speaking broadly, it is characteristic of a standard set by a general body, that its acceptance by any one individual does not call attention to him; in other words, a positive adoption of a given norm signifies nothing. Whoever keeps the laws the breaking of which is punished by the penal code, whoever lives up to the social forms prescribed by his class, gains no conspicuousness or notoriety. The slightest infraction or opposition, however, is immediately noticed and places the individual in an exceptional position by calling the attention of the public to his action. All such norms do not assume positive importance for the individual until he begins to depart from them. It is peculiarly characteristic of fashion that it renders possible a social obedience, which at the same time is a form of individual differentiation. Fashion does this because in its very nature it represents a standard that can never be accepted by all. While fashion postulates a certain amount of general acceptance, it nevertheless is not without significance in the characterization of the individual, for it emphasizes his personality not only through omission but also through observance. In the dude the social demands of fashion appear exaggerated to such a degree that they completely assume an individualistic and peculiar character. It is characteristic of the dude that he carries the elements of a particular fashion to an extreme; when pointed shoes are in style, he wears shoes that resemble the prow of a ship; when high collars are all the rage, he wears collars that come up to his ears; when scientific lectures are fashionable, you cannot find him anywhere else, etc., etc. Thus he represents something distinctly individual, which consists in the quantitative intensification of such elements as are qualitatively common property of the given set of class. He leads the way, but all travel the same road. Representing as he does the most recently conquered heights of public taste, he seems to be marching at the head of the general procession. In reality, however, what is so frequently true of the relation between individuals and groups applies also to him: As a matter of fact, the leader allows himself to be led. . . .

Inasmuch as we are dealing here not with the importance of a single fact or a single satisfaction, but rather with the play between two contents and their mutual distinction, it becomes evident that the same combination which extreme obedience to fashion acquires can be won also by opposition to it. Whoever consciously avoids following the fashion does not attain the consequent sensation of individualization through any real individual qualification, but rather through mere negation of the social example. If obedience to fashion consists in imitation of such an example, conscious neglect of fashion represents similar imitation, but under an inverse sign. The latter, however, furnishes just as fair testimony of the power of the social tendency, which demands our dependence in some positive or negative manner. The man who consciously pays no heed to fashion accepts its forms just as much as the dude does, only he embodies it in another category, the former in that of exaggeration, the latter in that of negation. Indeed, it occasionally happens that it becomes fashionable in whole bodies of a large class to depart altogether from the standards set by fashion. This constitutes a most curious social-psychological complication, in which the tendency towards individual conspicuousness primarily rests content with a mere inversion of the social imitation and secondly draws in strength from approximation to a similarly characterized narrower circle. If the club-haters organized themselves into a club, it would not be logically more impossible and psychologically more possible than the above case. Similarly atheism has been made into a religion, embodying the same fanaticism, the same intolerance, the same satisfying of the needs of the soul that are embraced in religion proper. Freedom, likewise, after having put a stop to tyranny, frequently becomes no less tyrannical and arbitrary. So the phenomenon of conscious departure from fashion illustrates how ready the fundamental forms of human character are to accept the total antithesis of contents and to show their strength and their attraction in the negation of the very thing to whose acceptance they seemed a moment before irrevocably committed. It is often absolutely impossible to tell whether the elements of personal strength or of personal weakness preponderate in the group of causes that lead to such a departure from fashion. It may result from a desire not to make common cause with the mass, a desire that has at its basis not independence of the mass, to be sure, but yet an inherently sovereign position with respect to the latter. However, it may be due to a delicate sensibility, which causes the individual to fear that he will be unable to maintain his individuality in case he adopts the forms, the tastes, and the customs of the general public. Such opposition is by no means always a sign of personal strength. . . .

We have seen that in fashion the different dimensions of life, so to speak, acquire a peculiar convergence, that fashion is a complex structure in which all the leading antithetical tendencies of the soul are represented in one way or another. This will make clear that the total rhythm in which the individuals and the groups move will exert an important influence also upon their relation to fashion, that the various strata of a group, altogether aside from their different contents of life and external possibilities, will bear different relations to fashion simply because their contents of life are evolved either in conservative or in rapidly varying form. On the one hand the lower classes are difficult to put in motion and they develop slowly. A very clear and instructive example of this may be found in the attitude of the lower classes in England towards the Danish and the Norman conquests. On the whole the changes brought about affected the upper classes only; in the lower classes we find such a degree of fidelity to arrangements and forms of life that the whole continuity of English life which was retained through all those national vicissitudes rests entirely upon the persistence and immovable conservatism of the lower classes. The upper classes, however, were most intensely affected and transformed by new influences, just as the upper branches of a tree are most responsive to the movements of the air. The highest classes, as everyone knows, are the most conservative, and frequently enough they are even archaic. They dread every motion and change, not because they have an antipathy for the contents or because the latter are injurious to them, but simply because it is change and because they regard every modification of the whole as suspicious and dangerous. No change can bring them additional power, and every change can give them something to fear, but nothing to hope for. The real variability of historical life is therefore vested in the middle classes, and for this reason the history of social and cultural movements has fallen into an entirely different pace since the *tiers état* assumed control. For this reason fashion, which represents the variable and contrasting forms of life, has since then become much broader and more animated, and also because of the transformation in the immediate political life, for man requires an ephemeral tyrant the moment he has rid himself of the absolute and permanent one. The frequent change of fashion represents a tremendous subjugation of the individual and in that respect forms one of the essential complements of the increased social and political freedom. A form of life, for the contents of which the moment of acquired

height marks the beginning of decline, belongs to a class which is inherently much more variable, much more restless in its rhythms than the lowest classes with their dull, unconscious conservatism, and the highest classes with their consciously desired conservatism. Classes and individuals who demand constant change, because the rapidity of their development gives them the advantage over others, find in fashion something that keeps pace with their own soul-movements. Social advance above all is favorable to the rapid change of fashion, for it capacitates lower classes so much for imitation of upper ones, and thus the process characterized above, according to which every higher set throws aside a fashion the moment a lower set adopts it, has acquired a breadth and activity never dreamed of before.

This fact has important bearing on the content of fashion. Above all else it brings in its train a reduction in the cost and extravagance of fashions. In earlier times there was a compensation for the costliness of the first acquisition or the difficulties in transforming conduct and taste in the longer duration of their sway. The more an article becomes subject to rapid changes of fashion, the greater the demand for *cheap* products of its kind, not only because the larger and therefore poorer classes nevertheless have enough purchasing power to regulate industry and demand objects, which at least bear the outward semblance of style, but also because even the higher circles of society could not afford to adopt the rapid changes in fashion forced upon them by the imitation of the lower circles, if the objects were not relatively cheap. The rapidity of the development is of such importance in actual articles of fashion that it even withdraws them from certain advances of economy gradually won in other fields. It has been noticed, especially in the older branches of modern productive industry, that the speculative element gradually ceases to play an influential *rôle*. The movements of the market can be better overlooked, requirements can be better foreseen and production can be more accurately regulated than before, so that the rationalization of production makes greater and greater inroads on chance conjunctures, on the aimless vacillation of supply and demand. Only pure articles of fashion seem to prove an exception. The polar oscillations, which modern economics in many instances knows how to avoid and from which it is visibly striving towards entirely new economic orders and forms, still hold sway in the field immediately subject to fashion. The element of feverish change is so essential here that fashion stands, as it were, in a logical contrast to the tendencies for development in modern economics.

In contrast to this characteristic, however, fashion possesses this peculiar quality, that every individual type to a certain extent makes its appearance as though it intended to live forever. When we furnish a house these days, intending the articles to last a quarter of a century, we invariably invest in furniture designed according to the very latest patterns and do not even consider articles in vogue two years before. Yet it is evident that the attraction of fashion will desert the present article just as it left the earlier one, and satisfaction or dissatisfaction with both forms is determined by other material criterions. A peculiar psychological process seems to be at work here in addition to the mere bias of the moment. Some fashion always exists and fashion *per se* is indeed immortal, which fact seems to affect in some manner or other each of its manifestations, although the very nature of each individual fashion stamps it as being transitory. The fact that change itself does not change in this instance endows each of the objects which it affects with a psychological appearance of duration.

This apparent duration becomes real for the different fashion-contents within the change itself in the following special manner. Fashion, to be sure, is concerned only with change, yet like all phenomena it tends to conserve energy; it endeavors to attain its objects as completely as possible, but nevertheless with the relatively most economical means. For this very reason, fashion repeatedly returns to old forms, as is illustrated particularly in wearing-apparel; and the course of fashion has been likened to a circle. As soon as an earlier fashion has partially been forgotten there is no reason why it should not be allowed to return to favor and why the charm of difference, which constitutes its very essence, should not be permitted to exercise an influence similar to that which it exerted conversely some time before.

▓

Introduction to "The Metropolis and Mental Life"

"The Metropolis and Mental Life" addresses several key themes that we have already discussed. For instance, you will find Simmel examining the nature of the struggle for individuality in modern societies as well as the relationship between objective culture and subjective or "individual" culture. You will also read Simmel's views on money and its psychological effects on the individual and her relationships with others.

This essay also contains an important theme not discussed previously: the intensity of stimuli created by the urban environment and its consequences for the psychology of the city dweller. Unlike the slower tempo and rhythms of small town life, city life is characterized by a "swift and continuous shift of external and internal stimuli . . . the rapid telescoping of changing images, pronounced differences within what is grasped at a single glance" (Simmel 1903/1971:325). While the slower tempo and limited social contacts within small towns fosters the development of emotional bonds that tie its inhabitants together, the metropolis, with its unceasing fluctuations of stimuli and expansiveness of interpersonal contacts, is antithetical to nurturing a rich emotional life. Indeed, it is impossible for the city dweller to absorb or become emotionally invested in all the happenings and encounters that make up his daily life. Attempting to do so would lead to an overstimulation of the senses that would in turn produce a virtual psychological and emotional paralysis. This leaves the urbanite to react to her world "primarily in a rational manner. . . . Thus the reaction of the metropolitan person to those events is moved to a sphere of mental activity which is least sensitive and which is furthest removed from the depths of personality" (ibid.:326).

Photo 6.3   Simmel's Metropolis: Berlin, Circa 1900

Photo 6.4   The Quintessential Modern Metropolis: New York City, Circa 2000

Bombarded with sensory impressions, the metropolitan person, in order to shield himself from the onslaught of stimuli and disruptions, adopts out of necessity an intellectualized approach to life. This psychological disposition, or **blasé attitude**, produces a dulling of differences, an emotional "graying" of reactions, which protects the individual from becoming overwhelmed by the sensory intensity of city life. As a result, the metropolitan personality experiences "quality" and differences as meaningless: "They appear to the blasé person in a homogenous, flat and gray color with no one of them worthy of being preferred to another" (ibid.:330). The blasé attitude, while an adaptive outlook, is coupled with a money economy that further hinders the development of an emotionally meaningful life. As we discussed previously, money is a standardized, impersonal measure of value. It levels all subjective qualitative distinctions into objective differences of quantity. The emphasis on exactness and calculability required by the urban, capitalist economy finds its expression in the life of the individual to the extent that he likewise becomes indifferent to the qualitative distinctions in his surroundings and in his relationships. The more money mediates our relationships and serves as the medium for self-expression, the more life itself takes on a quantitative quality.

Like Marx and Weber, Simmel's analysis of urban life , in part, was intended as a critique of modernity and its corruption of individuality and the human spirit. In the end, however, his project was more an intellectual journey whose goal was to further our insight into the duality that exists between the individual and the "sovereign powers of society." "To the extent that such forces have been integrated . . . into the root as well as the crown of the totality of historical life to which we belong—it is our task not to complain or condone but only to understand" (ibid.:339).

Photo 6.5a During the late nineteenth and early twentieth centuries, Impressionist painters often depicted the distinctly metropolitan blasé attitude in their art, as here in Edouard Manet's *Bar at Folies Bergère.*

Photo 6.5b Today, this metropolitan attitude often finds its expression in our impersonal indifference to the plight of others—particularly those with whom we would prefer not to come into contact.

———— "The Metropolis and Mental Life" (1903) ————

Georg Simmel

The deepest problems of modern life flow from the attempt of the individual to maintain the independence and individuality of his existence against the sovereign powers of society, against the weight of the historical heritage and the external culture and technique of life. This antagonism represents the most modern form of the conflict which primitive man must carry on with nature for his own bodily existence. The eighteenth century may have called for liberation from all the ties which grew up historically in politics, in religion, in morality and in economics in order to permit the original natural virtue of man, which is equal in everyone, to develop without inhibition; the nineteenth century may have sought to promote, in addition to man's freedom, his individuality (which is connected with the division of labor) and his achievements which make him unique and indispensable but which at the same time make him so much the more dependent on the complementary activity of others; Nietzsche may have seen the relentless struggle of the individual as the prerequisite for his full development, while Socialism found the same thing in the suppression of all competition—but in each of these the same fundamental motive was at work, namely the resistance of the individual to being levelled, swallowed up in the social-technological mechanism. When one inquires about the products of the specifically modern aspects of contemporary life with reference to their inner meaning—when, so to speak, one examines the body of culture with reference to the soul, as I am to do concerning the metropolis today— the answer will require the investigation of the relationship which such a social structure promotes between the individual aspects of life and those which transcend the existence of single individuals. It will require the investigation of the adaptations made by the personality in its adjustment to the forces that lie outside of it.

The psychological foundation, upon which the metropolitan individuality is erected, is the intensification of emotional life due to the swift and continuous shift of external and internal stimuli. Man is a creature whose existence is dependent on differences, i.e., his mind is stimulated by the difference between present impressions and those which have preceded. Lasting impressions, the slightness in their differences, the habituated regularity of their course and contrasts between them, consume, so to speak, less mental energy than the rapid telescoping of changing images, pronounced differences within what is grasped at a single glance, and the unexpectedness of violent stimuli. To the extent that the metropolis creates these psychological conditions—with every crossing of the street, with the tempo and multiplicity of economic, occupational and social life—it creates in the sensory foundations of mental life, and in the degree of awareness necessitated by our organization as creatures dependent on differences, a deep contrast with the slower, more habitual, more smoothly flowing rhythm of the sensory-mental phase of small town and rural existence. Thereby the essentially intellectualistic character of the mental life of the metropolis becomes intelligible as over against that of the small town which rests more on feelings and emotional relationships. These latter are rooted in the unconscious levels of the mind and develop most readily in the steady equilibrium of unbroken customs. The locus of reason, on the other hand, is in the lucid, conscious upper strata of the mind and it is the most adaptable of our inner forces. In order to adjust itself to the shifts and contradictions in events, it does not require the disturbances and inner upheavals which are the only means whereby more conservative personalities are able to adapt themselves to the same rhythm of events. Thus the metropolitan type—which naturally takes on a thousand individual modifications—creates a protective organ for itself against the profound disruption with which the fluctuations and discontinuities of the external milieu threaten it. Instead of reacting emotionally, the metropolitan type reacts primarily in a rational manner, thus creating a mental predominance through the intensification of consciousness, which in turn is caused by it. Thus the reaction of the metropolitan person to those events is moved to a sphere of mental activity which is least sensitive and which is furthest removed from the depths of the personality.

This intellectualistic quality which is thus recognized as a protection of the inner life against the domination of the metropolis, becomes ramified into numerous specific phenomena. The metropolis has always been the seat of money economy because the many-sidedness and concentration of commercial activity have given the medium of exchange an importance which it could not have acquired in the commercial aspects of rural life. But money economy and the domination of the intellect stand in the closest relationship to one another. They have in common a purely matter-of-fact attitude in the treatment of persons and things in which a formal justice is often combined with an unrelenting hardness. The purely intellectualistic person is indifferent to all things personal because, out of them, relationships and reactions develop which are not to be completely understood by purely rational methods—just as the unique element in events never enters into the principle of money. Money is concerned only with what is common to all, i.e., with the exchange value which reduces all quality and individuality to a purely quantitative level. All emotional relationships between persons rest on their individuality, whereas intellectual relationships deal with persons as with numbers, that is, as with elements which, in themselves, are indifferent, but which are of interest only insofar as they offer something objectively perceivable. It is in this very manner that the inhabitant of the metropolis reckons with his merchant, his customer, and with his servant, and frequently with the persons with whom he is thrown into obligatory association. These relationships stand in distinct contrast with the nature of the smaller circle in which the inevitable knowledge of individual characteristics produces, with an equal inevitability, an emotional tone in conduct, a sphere which is beyond the mere objective weighting of tasks performed and payments made. What is essential here as regards the economic-psychological aspect of the problem is that in less advanced cultures production was for the customer who ordered the product so that the producer and the purchaser knew one another. The modern city, however, is supplied almost exclusively by production for the market, that is, for entirely unknown purchasers who never appear in the actual field of vision of the producers themselves. Thereby, the interests of each party acquire a relentless matter-of-factness, and its rationally calculated economic egoism need not fear any divergence from its set path because of the imponderability of personal relationships. This is all the more the case in the money economy which dominates the metropolis in which the last remnants of domestic production and direct barter of goods have been eradicated and in which the amount of production on direct personal order is reduced daily. Furthermore, this psychological intellectualistic attitude and the money economy are in such close integration that no one is able to say whether it was the former that effected the latter or *vice versa*. What is certain is only that the form of life in the metropolis is the soil which nourishes this interaction most fruitfully, a point which I shall attempt to demonstrate only with the statement of the most outstanding English constitutional historian to the effect that through the entire course of English history London has never acted as the heart of England but often as its intellect and always as its money bag.

In certain apparently insignificant characters or traits of the most external aspects of life are to be found a number of characteristic mental tendencies. The modern mind has become more and more a calculating one. The calculating exactness of practical life which has resulted from a money economy corresponds to the ideal of natural science, namely that of transforming the world into an arithmetical problem and of fixing every one of its parts in a mathematical formula. It has been money economy which has thus filled the daily life of so many people with weighing, calculating, enumerating and the reduction of qualitative values to quantitative terms. Because of the character of calculability which money has there has come into the relationships of the elements of life a precision and a degree of certainty in the definition of the equalities and inequalities and an unambiguousness in agreements and arrangements, just as externally this precision has been brought about through the general diffusion of pocket watches. It is, however, the conditions of the metropolis which are cause as well as effect for this essential characteristic. The relationships and concerns of the typical metropolitan resident are so manifold and complex that, especially as a result of the agglomeration of so many persons with such differentiated interests, their relationships and activities intertwine with one another into a many-membered organism. In view of this fact, the lack of the most exact punctuality in promises and performances would cause the whole to break down into an inextricable chaos. If all the watches in Berlin suddenly went wrong in different ways even only as much as an hour, its entire economic and commercial life would be derailed for some time. Even though this may seem more superficial in its significance, it transpires that the magnitude of distances results in making all waiting and the breaking of appointments an ill-afforded waste of time. For this reason the technique of metropolitan life in general is not conceivable without all of its activities and reciprocal relationships being organized and coordinated in the most punctual way into a firmly fixed framework of time which transcends all subjective elements. But here too there emerge those conclusions which are in general the whole task of this discussion, namely, that every event, however restricted to this superficial level it may appear, comes immediately into contact with the depths of the soul, and that the most banal externalities are, in the last analysis, bound up with the final decisions concerning the meaning and the style of life. Punctuality, calculability, and exactness, which are required by the complications and extensiveness of metropolitan life are not only most intimately connected with its capitalistic and intellectualistic character but also color the content of life and are conductive to the

exclusion of those irrational, instinctive, sovereign human traits and impulses which originally seek to determine the form of life from within instead of receiving it from the outside in a general, schematically precise form. Even though those lives which are autonomous and characterised by these vital impulses are not entirely impossible in the city, they are, none the less, opposed to it *in abstracto*. It is in the light of this that we can explain the passionate hatred of personalities like Ruskin and Nietzsche for the metropolis—personalities who found the value of life only in unschematized individual expressions which cannot be reduced to exact equivalents and in whom, on that account, there flowed from the same source as did that hatred, the hatred of the money economy and of the intellectualism of existence.

The same factors which, in the exactness and the minute precision of the form of life, have coalesced into a structure of the highest impersonality, have, on the other hand, an influence in a highly personal direction. There is perhaps no psychic phenomenon which is so unconditionally reserved to the city as the blasé outlook. It is at first the consequence of those rapidly shifting stimulations of the nerves which are thrown together in all their contrasts and from which it seems to us the intensification of metropolitan intellectuality seems to be derived. On that account it is not likely that stupid persons who have been hitherto intellectually dead will be blasé. Just as an immoderately sensuous life makes one blasé because it stimulates the nerves to their utmost reactivity until they finally can no longer produce any reaction at all, so, less harmful stimuli, through the rapidity and the contradictoriness of their shifts, force the nerves to make such violent responses, tear them about so brutally that they exhaust their last reserves of strength and, remaining in the same milieu, do not have time for new reserves to form. This incapacity to react to new stimulations with the required amount of energy constitutes in fact that blasé attitude which every child of a large city evinces when compared with the products of the more peaceful and more stable milieu.

Combined with this physiological source of the blasé metropolitan attitude there is another which derives from a money economy. The essence of the blasé attitude is an indifference toward the distinctions between things. Not in the sense that they are not perceived, as is the case of mental dullness, but rather that the meaning and the value of the distinctions between things, and therewith of the things themselves, are experienced as meaningless. They appear to the blasé person in a homogeneous, flat and gray color with no one of them worthy of being preferred to another. This psychic mood is the correct subjective reflection of a complete money economy to the extent that money takes the place of all the manifoldness of things and expresses all qualitative distinctions between them in the distinction of "how much." To the extent that money, with its colorlessness and its indifferent quality, can become a common denominator of all values it becomes the frightful leveler—it hollows out the core of things, their peculiarities, their specific values and their uniqueness and incomparability in a way which is beyond repair. They all float with the same specific gravity in the constantly moving stream of money. They all rest on the same level and are distinguished only by their amounts. In individual cases this coloring, or rather this de-coloring of things, through their equation with money, may be imperceptibly small. In the relationship, however, which the wealthy person has to objects which can be bought for money, perhaps indeed in the total character which, for this reason, public opinion now recognizes in these objects, it takes on very considerable proportions. This is why the metropolis is the seat of commerce and it is in it that the purchasability of things appears in quite a different aspect than in simpler economies. It is also the peculiar seat of the blasé attitude. In it is brought to a peak, in a certain way, that achievement in the concentration of purchasable things which stimulates the individual to the highest degree of nervous energy. Through the mere quantitative intensification of the same conditions this achievement is transformed into its opposite, into this peculiar adaptive phenomenon—the blasé attitude—in which the nerves reveal their final possibility of adjusting themselves to the content and the form of metropolitan life by renouncing the response to them. We see that the self-preservation of certain types of personalities is obtained at the cost of devaluing the entire objective world, ending inevitably in dragging the personality downward into a feeling of its own valuelessness.

Whereas the subject of this form of existence must come to terms with it for himself, his self-preservation in the face of the great city requires of him a no less negative type of social conduct. The mental attitude of the people of the metropolis to one another may be designated formally as one of reserve. If the unceasing external contact of numbers of persons in the city should be met by the same number of inner reactions as in the small town, in which one knows almost every person he meets and to each of whom he has a positive relationship, one would be completely atomized internally and would fall into an unthinkable mental condition. Partly this psychological circumstance and partly the privilege of suspicion which we have in the face of the elements of metropolitan life (which are constantly touching one another in fleeting contact) necessitates in us that reserve, in consequence of which we do not know by sight neighbors of years standing and which permits us to appear to small-town folk so often as cold and uncongenial. Indeed, if I am not mistaken, the inner side of this external reserve is not only indifference but more frequently than we believe, it is a slight aversion, a mutual strangeness and repulsion which, in a close contact which has arisen any way whatever, can break out into hatred and conflict. The entire inner organization of such a type of extended commercial life rests on an extremely varied structure of sympathies, indifferences and aversions of the briefest as well as of the most enduring sort. This sphere of indifference is, for this reason, not as great as it seems superficially. Our minds respond, with some definite feeling, to almost every impression emanating from another person. The unconsciousness, the transitoriness and the shift of these feelings seem to raise them only into indifference. Actually this latter would be as unnatural to us as immersion into a chaos of unwished-for suggestions would be unbearable. From these two typical dangers of metropolitan life we are saved by antipathy which is the latent adumbration of actual antagonism since it brings about the sort of distanciation and deflection without which this type of life could not be carried on at all. Its extent and its mixture, the rhythm of its emergence and disappearance, the forms in which it is adequate—these constitute, with the simplified motives (in the narrower sense) an inseparable totality of the form of metropolitan life. What appears here directly as dissociation is in reality only one of the elementary forms of socialization.

This reserve with its overtone of concealed aversion appears once more, however, as the form or the wrappings of a much more general psychic trait of the metropolis. It assures the individual of a type and degree of personal freedom to which there is no analogy in other circumstances. It has its roots in one of the great developmental tendencies of social life as a whole; in one of the few for which an approximately exhaustive formula can be discovered. The most elementary stage of social organization which is to be found historically, as well as in the present, is this: a relatively small circle almost entirely closed against neighboring foreign or otherwise antagonistic groups but which has however within itself such a narrow cohesion that the individual member has only a very slight area for the development of his own qualities and for free activity for which he himself is responsible. Political and familial groups began in this way as do political and religious communities; the self-preservation of very young associations requires a rigorous setting of boundaries and a centripetal unity and for that reason it cannot give room to freedom and the peculiarities of inner and external development of the individual. From this stage social evolution proceeds simultaneously in two divergent but none the less corresponding directions. In the measure that the group grows numerically, spatially, and in the meaningful content of life, its immediate inner unity and the definiteness of its original demarcation against others are weakened and rendered mild by reciprocal interactions and interconnections. And at the same time the individual gains a freedom of movement far beyond the first jealous delimitation, and gains also a peculiarity and individuality to which the division of labor in groups, which have become larger, gives both occasion and necessity. However much the particular conditions and forces of the individual situation might modify

the general scheme, the state and Christianity, guilds and political parties and innumerable other groups have developed in accord with this formula. This tendency seems, to me, however to be quite clearly recognizable also in the development of individuality within the framework of city life. Small town life in antiquity as well as in the Middle Ages imposed such limits upon the movements of the individual in his relationships with the outside world and on his inner independence and differentiation that the modern person could not even breathe under such conditions. Even today the city dweller who is placed in a small town feels a type of narrowness which is very similar. The smaller the circle which forms our environment and the more limited the relationships which have the possibility of transcending the boundaries, the more anxiously the narrow community watches over the deeds, the conduct of life and the attitudes of the individual and the more will a quantitative and qualitative individuality tend to pass beyond the boundaries of such a community.

The ancient *polis* seems in this regard to have had a character of a small town. The incessant threat against its existence by enemies from near and far brought about that stern cohesion in political and military matters, that supervision of the citizen by other citizens, and that jealousy of the whole toward the individual whose own private life was repressed to such an extent that he could compensate himself only by acting as a despot in his own household. The tremendous agitation and excitement, and the unique colorfulness of Athenian life is perhaps explained by the fact that a people of incomparably individualized personalities were in constant struggle against the incessant inner and external oppression of a de-individualizing small town. This created an atmosphere of tension in which the weaker were held down and the stronger were impelled to the most passionate type of self-protection. And with this there blossomed in Athens, what, without being able to define it exactly, must be designated as "the general human character" in the intellectual development of our species. For the correlation, the factual as well as the historical validity of which we are here maintaining, is that the broadest and the most general contents and forms of life are intimately bound up with the most individual ones. Both have a common prehistory and also common enemies in the narrow formations and groupings, whose striving for self-preservation set them in conflict with the broad and general on the outside, as well as the freely mobile and individual on the inside. Just as in feudal times the "free" man was he who stood under the law of the land, that is, under the law of the largest social unit, but he was unfree who derived his legal rights only from the narrow circle of a feudal community—so today in an intellectualized and refined sense the citizen of the metropolis is "free" in contrast with the trivialities and prejudices which bind the small town person. The mutual reserve and indifference, and the intellectual conditions of life in large social units are never more sharply appreciated in their significance for the independence of the individual than in the dense crowds of the metropolis because the bodily closeness and lack of space make intellectual distance really perceivable for the first time. It is obviously only the obverse of this freedom that, under certain circumstances, one never feels as lonely and as deserted as in this metropolitan crush of persons. For here, as elsewhere, it is by no means necessary that the freedom of man reflect itself in his emotional life only as a pleasant experience.

It is not only the immediate size of the area and population which, on the basis of world-historical correlation between the increase in the size of the social unit and the degree of personal inner and outer freedom, makes the metropolis the locus of this condition. It is rather in transcending this purely tangible extensiveness that the metropolis also becomes the seat of cosmopolitanism. Comparable with the form of the development of wealth—(beyond a certain point property increases in ever more rapid progression as out of its own inner being)—the individual's horizon is enlarged. In the same way, economic, personal and intellectual relations in the city (which are its ideal reflection), grow in a geometrical progression as soon as, for the first time, a certain limit has been passed. Every dynamic extension becomes a preparation not only for a similar extension but rather for a larger one and from every thread which is spun out of it there

continue, growing as out of themselves, an endless number of others. This may be illustrated by the fact that within the city the "unearned increment" of ground rent, through a mere increase in traffic, brings to the owner profits which are self-generating. At this point the quantitative aspects of life are transformed qualitatively. The sphere of life of the small town is, in the main, enclosed within itself. For the metropolis it is decisive that its inner life is extended in a wave-like motion over a broader national or international area. Weimar was no exception because its significance was dependent upon individual personalities and died with them, whereas the metropolis is characterised by its essential independence even of the most significant individual personalities; this is rather its antithesis and it is the price of independence which the individual living in it enjoys. The most significant aspect of the metropolis lies in this functional magnitude beyond its actual physical boundaries and this effectiveness reacts upon the latter and gives to it life, weight, importance and responsibility. A person does not end with limits of his physical body or with the area to which his physical activity is immediately confined but embraces, rather, the totality of meaningful effects which emanates from him temporally and spatially. In the same way the city exists only in the totality of the effects which transcend their immediate sphere. These really are the actual extent in which their existence is expressed. This is already expressed in the fact that individual freedom, which is the logical historical complement of such extension, is not only to be understood in the negative sense as mere freedom of movement and emancipation from prejudices and philistinism. Its essential characteristic is rather to be found in the fact that the particularity and incomparability which ultimately every person possesses in some way is actually expressed, giving form to life. That we follow the laws of our inner nature—and this is what freedom is—becomes perceptible and convincing to us and to others only when the expressions of this nature distinguish themselves from others; it is our irreplaceability by others which shows that our mode of existence is not imposed upon us from the outside.

Cities are above all the seat of the most advanced economic division of labor. They produce such extreme phenomena as the lucrative vocation of the *quatorzieme* in Paris. These are persons who may be recognized by shields on their houses and who hold themselves ready at the dinner hour in appropriate costumes so they can be called upon on short notice in case thirteen persons find themselves at the table. Exactly in the measure of its extension the city offers to an increasing degree the determining conditions for the division of labor. It is a unit which, because of its large size, is receptive to a highly diversified plurality of achievements while at the same time the agglomeration of individuals and their struggle for the customer forces the individual to a type of specialized accomplishment in which he cannot be so easily exterminated by the other. The decisive fact here is that in the life of a city, struggle with nature for the means of life is transformed into a conflict with human beings and the gain which is fought for is granted, not by nature, but by man. For here we find not only the previously mentioned source of specialization but rather the deeper one in which the seller must seek to produce in the person to whom he wishes to sell ever new and unique needs. The necessity to specialize one's product in order to find a source of income which is not yet exhausted and also to specialize a function which cannot be easily supplanted is conducive to differentiation, refinement and enrichment of the needs of the public which obviously must lead to increasing personal variation within this public.

All this leads to the narrower type of intellectual individuation of mental qualities to which the city gives rise in proportion to its size. There is a whole series of causes for this. First of all there is the difficulty of giving one's own personality a certain status within the framework of metropolitan life. Where quantitative increase of value and energy has reached its limits, one seizes on qualitative distinctions, so that, through taking advantage of the existing sensitivity to differences, the attention of the social world can, in some way, be won for oneself. This leads ultimately to the strangest eccentricities, to specifically metropolitan extravagances of self-distanciation, of caprice, of fastidiousness, the

meaning of which is no longer to be found in the content of such activity itself but rather in its being a form of "being different"—of making oneself noticeable. For many types of persons these are still the only means of saving for oneself, through the attention gained from others, some sort of self-esteem and the sense of filling a position. In the same sense there operates an apparently insignificant factor which in its effects however is perceptibly cumulative, namely, the brevity and rarity of meetings which are allotted to each individual as compared with social intercourse in a small city. For here we find the attempt to appear to-the-point, clear-cut and individual with extraordinarily greater frequency than where frequent and long association assures to each person an unambiguous conception of the other's personality.

This appears to me to be the most profound cause of the fact that the metropolis places emphasis on striving for the most individual forms of personal existence—regardless of whether it is always correct or always successful. The development of modern culture is characterised by the predominance of what one can call the objective spirit over the subjective; that is, in language as well as in law, in the technique of production as well as in art, in science as well as in the objects of domestic environment, there is embodied a sort of spirit [*Geist*], the daily growth of which is followed only imperfectly and with an even greater lag by the intellectual development of the individual. If we survey for instance the vast culture which during the last century has been embodied in things and in knowledge, in institutions and comforts, and if we compare them with the cultural progress of the individual during the same period—at least in the upper classes—we would see a frightful difference in rate of growth between the two which represents, in many points, rather a regression of the culture of the individual with reference to spirituality, delicacy and idealism. This discrepancy is in essence the result of the success of the growing division of labor. For it is this which requires from the individual an ever more one-sided type of achievement which, at its highest point, often permits his personality as a whole to fall into neglect. In any case this overgrowth of objective culture has been less and less satisfactory for the individual. Perhaps less conscious than in practical activity and in the obscure complex of feelings which flow from him, he is reduced to a negligible quantity. He becomes a single cog as over against the vast overwhelming organization of things and forces which gradually take out of his hands everything connected with progress, spirituality and value. The operation of these forces results in the transformation of the latter from a subjective form into one of purely objective existence. It need only be pointed out that the metropolis is the proper arena for this type of culture which has outgrown every personal element. Here in buildings and in educational institutions, in the wonders and comforts of space-conquering technique, in the formations of social life and in the concrete institutions of the State is to be found such a tremendous richness of crystallizing, depersonalized cultural accomplishments that the personality can, so to speak, scarcely maintain itself in the face of it. From one angle life is made infinitely more easy in the sense that stimulations, interests, and the taking up of time and attention, present themselves from all sides and carry it in a stream which scarcely requires any individual efforts for its ongoing. But from another angle, life is composed more and more of these impersonal cultural elements and existing goods and values which seek to suppress peculiar personal interests and incomparabilities. As a result, in order that this most personal element be saved, extremities and peculiarities and individualizations must be produced and they must be over-exaggerated merely to be brought into the awareness even of the individual himself. The atrophy of individual culture through the hypertrophy of objective culture lies at the root of the bitter hatred which the preachers of the most extreme individualism, in the footsteps of Nietzsche, directed against the metropolis. But it is also the explanation of why indeed they are so passionately loved in the metropolis and indeed appear to its residents as the saviors of their unsatisfied yearnings.

When both of these forms of individualism which are nourished by the quantitative relationships of the metropolis, i.e., individual independence and the elaboration of personal peculiarities, are examined with reference to their historical position, the metropolis attains an entirely new value and meaning in the world history of the spirit. The eighteenth century found the individual in the grip of powerful bonds which had become meaningless—bonds of a political, agrarian, guild and religious nature—delimitations which imposed upon the human being at the same time an unnatural form and for a long time an unjust inequality. In this situation arose the cry for freedom and equality—the belief in the full freedom of movement of the individual in all his social and intellectual relationships which would then permit the same noble essence to emerge equally from all individuals as Nature had placed it in them and as it had been distorted by social life and historical development. Alongside of this liberalistic ideal there grew up in the nineteenth century from Goethe and the Romantics, on the one hand, and from the economic division of labor on the other, the further tendency, namely, that individuals who had been liberated from their historical bonds sought now to distinguish themselves from one another. No longer was it the "general human quality" in every individual but rather his qualitative uniqueness and irreplaceability that now became the criteria of his value. In the conflict and shifting interpretations of these two ways of defining the position of the individual within the totality is to be found the external as well as the internal history of our time. It is the function of the metropolis to make a place for the conflict and for the attempts at unification of both of these in the sense that its own peculiar conditions have been revealed to us as the occasion and the stimulus for the development of both. Thereby they attain a quite unique place, fruitful with an inexhaustible richness of meaning in the development of the mental life. They reveal themselves as one of those great historical structures in which conflicting life-embracing currents find themselves with equal legitimacy. Because of this, however, regardless of whether we are sympathetic or antipathetic with their individual expressions, they transcend the sphere in which a judge-like attitude on our part is appropriate. To the extent that such forces have been integrated, with the fleeting existence of a single cell, into the root as well as the crown of the totality of historical life to which we belong—it is our task not to complain or to condone but only to understand.

Discussion Questions

1. With his concept of the tragedy of culture, Simmel explored how cultural objects, once they confront their users as external forces, often constrain our attempts to cultivate freedom and individuality. How do cars and computers allow us to fulfill some of our needs, while at the same time constraining our pursuit of independence and individual goals? What other cultural objects possess these opposing elements? How do Simmel's views on such issues compare with those offered by Marx's understanding of alienation and commodity fetishism?

2. In his essay "Exchange," Simmel explores the origins of economic value. According to his argument, what role does interaction play in establishing the value of goods or objects? What roles does the quality and scarcity of goods and objects play in establishing their value? How does Simmel's view on the source of value compare with that offered by Marx?

3. Simmel describes the stranger as someone who is near and far, or close and remote, at the same time. How might his analysis of the stranger shed light on current debates regarding immigration and the influx of "illegal aliens" into the United States?

4. According to Simmel, what effects does the metropolis have on the emotions and intellect of the individual? What are the causes responsible for producing these effects? In what ways is the metropolis both "freeing" and "dehumanizing" with regard to the cultivation of individuality?

5. How does Simmel's analysis of modern society outlined in "The Metropolis and Mental Life" compare to the views offered by Marx, Durkheim, and Weber? According to their respective arguments, where does the locus for the modern condition lie? What prospects, if any, does each hold out for change or further societal development?

6. Simmel described how changes in fashion occur as the lower classes adopt the styles originating in the upper classes. How might Simmel explain the practice of purchasing $300 jeans that are prefaded and preripped? How might race influence changes in fashion trends? More generally, how might class, gender, and racial forms of stratification affect changes in musical tastes or participation in sports as a player or spectator or even the everyday language a person speaks? To what degree are tennis and golf seen as "exclusive" sports and on what is their exclusivity based?

7 W. E. B. Du Bois (1868–1963)

Key Concepts

- The color line
- Double consciousness
- The veil

The problem of the 20th century is the problem of the color line—the relation of the darker to the lighter races of men in Asia and Africa, in America and the islands of the sea.

(Du Bois 1903/1989:13)

From Jim Crow, to the civil rights movement, to affirmative action; from Willie Horton, to Rodney King and O. J. Simpson, to the "Million Man March," it seems clear that W. E. B. Du Bois's famous prophecy as to the salience of the "color line" in the twentieth century has proven true. Yet in the above passage, Du Bois was not talking only about the United States. Du Bois saw that the forces of white oppression were "all but world-wide" (Du Bois 1920/2003d:66–67). He was perhaps the first social theorist to critically remark on the global racial order, and to understand not only the economic, but also the racial dimensions of the European colonization of Africa, Asia, India, and Latin America.

Ironically, Du Bois's own life is further testament to the existence and pervasiveness of the color line. Despite his exceptional work and provocative sociological insights, Du Bois was virtually ignored by fellow American sociologists in his day. This, despite the fact that he was the first African American to receive a Ph.D. from Harvard University, and despite the fact that his most famous book of essays, *The Souls of Black Folk*, which has been republished in no fewer than 119 editions since 1903, has been called one the most important books ever on the subject of "race."[1] At the turn of the twentieth century, though, Du Bois's work on African American communities was neither well known nor highly respected. Most white reviewers could not see past the "bitterness" and "hateful" stance that was said to characterize his works (Feagin 2003:22). What this demonstrates, of course, is that the canon of sociological theory is a product not only of intellectual accomplishments, but also of social, cultural, and historical dynamics. Many people wrote insightful sociological treatises at the same time as founding figures such as Durkheim and Weber, but were not widely recognized by their peers. One of the most important of these, however, is W. E. B. Du Bois.

At the turn of the twentieth century, racism precluded most social scientists from a cogent understanding of race. Indeed, during the nineteenth and early twentieth centuries, most scientists and essayists writing about "Negroes" denied the very humanity of African Americans. It was widely believed that Negro brains weighed less than those of whites and that "prefrontal deficiencies" caused inferior "psychic faculties, especially reason, judgment, self control or voluntary inhibitions" (Rudwick 1960/1982:123). Moreover, between the end of the Civil War and the dawn of the twentieth century, racial violence and tension, segregation, and vigilantism were not only widespread—but also increasing. In 1891, the Georgia legislature passed legislation segregating all public streetcars, and by 1900, most southern states had similar codes. Between 1882 and 1901, there were some 150 lynchings each year. Most appallingly, in the late nineteenth and early twentieth centuries, lynchings were not the domain of deviants and outlaws. Rather, they were communal events, with the

Photo 7.1   A crowd mills around the bodies of Abram South, 19, and Thomas Shipp, 18, who were taken from the Grand County Jail, in Marion, Indiana, and lynched in the public square in 1930.

---

[1]We use quotation marks around the concept of "race" to reflect that we are talking about the social perception and experience of race—not a biological concept of race. (See Edles 2002.) This is the meaning to be taken throughout the remainder of the chapter.

tacit if not explicit support of many "upstanding" white citizens, including religious clergy (Patterson 1998:173). As Du Bois maintained, it was the "nucleus of ordinary men that continually [gave] the mob its initial and awful impetus" (Du Bois 1935/1962:678).

Within this context of social as well as scientific ignorance and injustice, Du Bois stood virtually alone in his quest to sociologically explain the complex, intertwined dimensions of race and class. This is not to say that Du Bois was the only turn-of-the-century intellectual to astutely explore the race question. On the contrary, many activists and essayists spoke or wrote eloquently about the issue of race, among them Sojourner Truth (1797–1883), Frederick Douglass (1817–1895), Booker T. Washington (1856–1915), Anna Julia Cooper (1858–1964; see Significant Others Box on next page), and Ida B. Wells-Barnett (1862–1931). The point here, however, is that Du Bois sought to investigate and explain racial issues *sociologically*. He viewed sociological investigation as the solution to racism. As he maintained,

> [t]he Negro problem was in my mind a matter of systematic investigation and intelligent understanding. The world was thinking wrong about race because it did not know. The ultimate evil was stupidity. The cure for it was knowledge based on scientific investigation. (Du Bois 1940/1984:58)

Consequently, Du Bois set the theoretical as well as empirical parameters in the field of race relations. His recognition of the interconnections between race and class continues to steer sociological inquiry to this day.

While few sociologists deny the importance of Du Bois in the field of race relations, not all sociologists consider him to be a significant social theorist. This perception stems from two somewhat contradictory beliefs: (1) Du Bois's work is perceived as being too empirical and theoretically narrow to qualify him as a theorist, and (2) his work is deemed too subjective to merit the label of "theory." In fact, Du Bois did not write explicitly about the "big" theoretical questions such as the nature of action or social order. Instead, much of his work centered on describing and explaining the actual social conditions in which African Americans lived. Additionally, in contrast to the other founding figures of sociology, Du Bois not only wrote in the subjective form of essays, but also sometimes in the first person. As you will see, Du Bois uses poetry, slave songs, and his own autobiographical experience in his most widely read book of essays, *The Souls of Black Folk* (Du Bois 1903/1989).

But interestingly, Du Bois himself made no apologies for his style. He maintained that European thought is but one kind of thought; it does not represent a "universal," let alone superior, approach to the world. As Du Bois states,

> [t]he style is tropical—African. This needs no apology. The blood of my fathers spoke through me and cast off the English restraint of my training and surroundings. . . . One who is born with a cause is predestined to certain narrowness of view, and at the same time to some clearness of vision within his limits with which the world often finds it well to reckon. (Du Bois 1904, quoted in E. Griffin 2003:32)

Given his standing among his contemporaries and succeeding generations of sociologists, the question remains: why include Du Bois's work in a theory reader such as this? Our answer to this question is twofold: First, we include Du Bois because his work brings to the canon a "clearness of vision" regarding specific phenomena—most obviously race, but also epistemology, or "how we know what we know." Anyone writing on these issues today is in some fashion indebted to Du Bois's groundbreaking insights. Second, akin to the case of Charlotte Perkins Gilman (Chapter 5), we include Du Bois for historical reasons. In the nineteenth and early twentieth centuries, African Americans were marginalized, both explicitly and implicitly, from formal sociological theorizing. As a result, we must broaden our notion of "theory" if we want to uncover the ideas of African Americans such as Du Bois who lived during this patently racist period. Indeed, as the acclaimed contemporary sociologist Lawrence Bobo maintains,

> [w]ere it not for the deeply entrenched racism in the United States during his early professional years, Du Bois would be recognized alongside the likes of Albion Small, Edward A. Ross, Robert E. Park, Lewis With, and W. I. Thomas as one of the fountainheads of American sociology. Had not racism so thoroughly excluded him from placement in the center of the academy, he might arguably have come to rank with Max Weber or Émile Durkheim in stature. (Bobo 2000:187)

Because his work is so intimately tied to his personal life, in the next section we provide a rather lengthy biographical sketch of Du Bois, after which we turn to more theoretical concerns.

Significant Others

Anna Julia Cooper (1858–1964): A Voice From the South

Born in 1858 to a slave mother, Hannah Stanley Haywood, and her white master, George Washington Haywood, Anna Julia Cooper is one of the earliest individuals to offer what would today be recognized as a social theory of women of color. Cooper articulated her perspective in her major work, *A Voice from the South by a Black Woman of the South*, first published in 1892. This provocative and compelling book turned Cooper into a leading spokeswoman of her time. However, Cooper dedicated most of her long and extraordinary life (she died at the age of 105) to the education of others, and in particular "to the education of neglected people" (Washington 1988:xxviii). In addition to being a high school teacher and principal, she started a night school for working people who could not attend college during the day. Cooper maintained that the ideals and integrity of any group can be measured "by its treatment of those who suffer the greatest oppression" (ibid.:xxix).

Cooper's *A Voice From the South* is lauded as "the first *systematic* working out of the insistence that no one social category can capture the reality of black women" (Lemert 1998:15; emphasis in original). In this inimitable book, Cooper points out that categories such as race, gender, and class do not capture, by themselves, the situation of black women. Instead, the black woman "is confronted by both a woman question and a race problem," as Cooper bluntly states (Cooper 1892/1988:112, as cited by Lemert 1998:15). In addition, Cooper sought to counteract the prevailing assumptions about black women as immoral and ignorant. She criticizes black men for securing higher education for themselves while erecting roadblocks to deny women access to those same opportunities. Cooper herself had to fight against both black and white men in order to attain her own education. Nevertheless, she attained bachelors and masters degrees from Oberlin College in 1984 and 1987, respectively, and a Ph.D. from the Sorbonne in Paris in 1925, at 66 years of age. If black men are a "muffled chord," she maintained, then black women are "the mute and voiceless note" of the race, "with no language but a cry" (Washington 1988:xxix). Cooper was equally critical of the white women's movement for its elitism, and she challenged white women to link their cause with all of the "undefended" (ibid.).

Cooper was passionately committed to higher education as the key to ending women's physical, emotional, and economic dependence on men. Indeed, she was forced to resign from her job as principal of the only black high school in Washington, DC, when, in defiance of her white supervisor, who told her that black children should be taught trades, Cooper continued to prepare them for college (Washington 1988:xxxiv). (Several of her students attended prestigious universities such as Harvard, Brown, Oberlin, Yale, Amherst, Dartmouth, and Radcliffe.) Like Du Bois, Cooper opposed Booker T. Washington's prescriptions for advancing the cause of African Americans. Cooper and Du Bois also shared an emphasis on "voice": both sought to expose the standpoint of oppressed folks, to present submerged points of view. As Cooper (1892:60) states, "there is a feminine as well as a masculine side to truth; that these are related not as inferior and superior, not as better and worse, not as weaker and stronger, but as complements—complements in one necessary and symmetric whole" (ibid.:60). Ironically, however, Cooper was largely ignored by the leading black male intellectuals of her day, including Du Bois, though she did not blame him but retained hope:

> It is no fault of man's that he has not been able to see truth from her standpoint. It does credit both to his head and heart that no greater mistakes have been committed or even wrongs perpetrated while she sat making tatting and snipping paper flowers. Man's own innate chivalry and the mutual interdependence of their interests have insured his treating her causes, in the main at least, as his own. And he is pardonably surprised and even a little chagrined, perhaps to find his legislation not considered "perfectly lovely" in every respect." But in any case his work is only impoverished by her remaining dumb. The world has had to limp along with the wobbling gait and one-sided hesitancy of a man with one eye. Suddenly the bandage is removed from the other eye and the whole body is filled with light. It sees a circle where before it saw a segment. The darkened eye restored, every member rejoices with it. (ibid.:122)

A Biographical Sketch ▪▪

William Edward Burghardt (W. E. B.) Du Bois was born in Great Barrington, Massachusetts, in 1868.[2] Du Bois's father, Alfred, was a restless, unhappy, "just visibly colored" man, who tried a wide variety of occupations, including barbering and even preaching, before he abandoned his wife and child and drifted away, never to return. Du Bois's mother, Mary Burghardt, was a "silent, repressed," "dark shining bronze" woman who struggled to make ends meet by working as a domestic and by boarding a relative.[3]

---

[2]Great Barrington was a small town of around 5,000 mostly middle-class families of European descent. The wealthier farm and factory owners tended primarily to be of English and Dutch stock, while the small farmers, merchants, and skilled laborers tended to be mostly of German and Irish ancestry. There were a handful of black families in Great Barrington, and they (including the Burghardts) were among the town's oldest residents. They worked primarily as laborers and servants. According to Du Bois, "The color line was manifest and yet not absolutely drawn" in Great Barrington, allowing blacks to organize their own social life to an extent (Du Bois 1940/1984:10; Rudwick 1960/1982:16).

[3]These descriptions of his parents' skin color come from *The Autobiography of W. E. B. Du Bois* (Du Bois 1968). Du Bois described his own racial background as "a flood of Negro blood, a strain of French, a bit of Dutch but thank God! no 'Anglo-Saxon'" (Rudwick 1960/1982:15–16).

While Du Bois was still a child, his mother suffered a stroke and became partially paralyzed. Though she was able to continue working as a maid, William began to shoulder a share of the burden for maintaining their small family, seeking out odd jobs wherever he could. Despite his new obligations, William did exceptionally well in school.

Du Bois soon came to recognize that his race set him apart from his schoolmates. As he stated later, "In early youth a great bitterness entered my life and kindled a great ambition" (Marable 1986:6). He began to recognize not only the existence of the color line, but also the ways in which it was distinct from class-based inequality. Indeed, by age 16 Du Bois had formulated the basic premise of the social and political philosophy that would guide him the rest of his years: black Americans must organize themselves as a race-conscious bloc in order to win and exercise their freedom. They must "act in concert" if they want to become a "power not to be despised" (ibid.). Though such race consciousness would seem to run counter to the individualism and egalitarianism supposedly underlying American society, Du Bois maintained that in a racist society democracy could not be colorblind.

Du Bois was the first black graduate in his high school's history. Shortly after his graduation, his mother suddenly died. Du Bois's formal education might have ended right then were it not for the benevolence of four liberal white men who took great interest in Du Bois and agreed to pay for his college education. These community leaders (who included Du Bois's high school principal and several prominent pastors) decided that Du Bois should attend Fisk University—a historically black school in Nashville, Tennessee—rather than Harvard, as Du Bois initially desired.

In fall 1885, Du Bois arrived at Fisk University. In his autobiography, he described himself as

> thrilled to be for the first time among so many people of my own color or rather of such various and such extraordinary colors, which I had only glimpsed before, but who it seemed were bound to me by new and exciting and eternal ties. (Du Bois 1968:107)

For Du Bois, Fisk represented the tremendous yet unfulfilled potential of the entire black world that had been "held back by race prejudice and legal bonds, as well as by deep ignorance and dire poverty" (Marable 1986:9).

Du Bois witnessed "the real seat of slavery" not at Fisk, but in the east Tennessee countryside where he went to teach summer school in 1886 and 1887. There he taught the poorest of the black poor in a schoolhouse he described as a "log hut" with "neither door or furniture" (Du Bois 1903/1989:55–56). Du Bois eloquently describes this experience in an essay titled, "Of the Meaning of Progress" that would later appear in *The Souls of Black Folk* (Du Bois 1903/1989).

In 1888, Du Bois transferred to Harvard. In 1892, he received a master's degree in philosophy from Harvard, and set off to study further at the University of Berlin. Harvard would not offer a degree in sociology for another 20 years, and Germany was at the heart of the fledgling new discipline of sociology. Du Bois attended lectures and communicated with some of the great sociologists of the day, including Max Weber. In addition, Du Bois attended local meetings of the German Social Democratic party, at that time the largest socialist party in the world. Enthusiastic about this new intersection of philosophy and social reform, Du Bois sought to apply "philosophy to an historical interpretation of race relations," that is, "to take my first steps toward sociology as the science of human action" (Du Bois 1968:148).

In Europe, Du Bois's racial consciousness was transformed just as it had been seven years earlier when he moved to the American South. In Europe, Du Bois encountered white people who exhibited little or no racial prejudice:

> In Germany in 1892, I found myself on the outside of the American world, looking in. With me were white folk—students, acquaintances, teachers—who viewed the scene

with me. They did not always pause to regard me as a curiosity, or something sub-human; I was just a man of the somewhat privileged student rank, with whom they were glad to meet and talk over the world; particularly, the part of the world whence I came. (ibid.:157)

Correspondingly, Du Bois's view of white folks was itself transformed. White Europeans became "not white folks, but folks. The unity beneath all life clutched me" (Marable 1986:20). Nevertheless, Du Bois fully recognized that racism in the United States was only one virulent example of racial/ethnic/national subordination and that German anti-Semitism had "much in common with our own race question" (ibid.:17,18).

For two years, Du Bois "dreamed and loved and wandered and sang" in Europe, but in 1894, low on funds, he was forced to return to the United States. "Dropped suddenly back into 'nigger'-hating America" (Du Bois 1968:183), Du Bois began teaching at Wilberforce University, a small African Methodist Episcopal school in central Ohio. While there, Du Bois also finished his doctoral thesis (which he had begun at Harvard before he left for Europe) and turned it into his first book, *The Suppression of the African Slave-Trade to the United States of America, 1638–1870* (Du Bois 1896). This was the first major research study on the U.S. slave trade (Feagin 2003: 23). Only two years would pass before he published his second book, *The Philadelphia Negro: A Social Study* (1899), which was the first sociological text on an African American community published in the United States.

In 1896, Du Bois married Nina Gomer, and the following year their first child was born. Sadly, the child, named Burghardt, died 18 months later, allegedly a victim of sewage pollution in the city's water system (Marable 1986:30). Du Bois scholars such as Manning Marable suggest that it was perhaps Du Bois's deep sorrow that prompted the softer, more sympathetic tone apparent in his masterpiece, *The Souls of Black Folk*, published in 1903. This work would become not only Du Bois's most famous book, but also arguably one of the most important books on race and class ever written in the United States. (The significance of this book is discussed further below.)

In 1897, Du Bois began teaching at Atlanta University in Georgia, where he would remain until 1910. He wrote prolifically during this time, while receiving the recognition of a few leading sociologists, such as Max Weber, who hailed Du Bois's *The Souls of Black Folk* as a "splendid work" that "ought to be translated in German" (ibid.:63). Indeed, Europeans appreciated Du Bois's sociological research and writing much more than did his fellow (white) American intellectuals.

Soon after arriving in Atlanta, Du Bois turned his attention to political activism. In 1905, he helped found a militant civil rights organization, the Niagara Movement, devoted to "freedom of speech and criticism," "suffrage," "the abolition of all caste distinctions based simply on race or color," and "the principle of human brotherhood as a practical present creed" (ibid.:55). In 1910, Du Bois left his academic post at Atlanta University in order to work full time for the National Association for the Advancement of Colored People (NAACP) in New York City, an association that evolved out of the Niagara Movement. Against the advice of some NAACP board members (several of whom were white), Du Bois founded the journal *Crisis*, which from the beginning had a distinctly militant tone. The journal became a forum for Du Bois, who wrote dozens of articles and editorials on the profound discrepancy between American democratic principles and racism, beseeching the country to expand political and economic opportunities for blacks (Rudwick 1960/1982:151). He also challenged the laws against miscegenation, asserting that a black man should be able "to marry any sane grown person who wants to marry him" (Marable 1986:79).

During his tenure as the editor of *Crisis*, Du Bois managed to write several more books, including the highly acclaimed *Darkwater: Voices From Within the Veil* (1920), *The Gift of Black Folk* (1924), and *Dark Princess* (1928), as well as books on Africa: *The Negro* (1915),

Africa—Its Place in Modern History (1930a), and *Africa—Its Geography, People, and Products* (1930b). Du Bois, moreover, was sensitive to the dual oppression faced by black women, as is evident in his poem, "The Burden of Black Women" (first published as "The Riddle of the Sphinx in 1907), and his essays "The Black Mother" (1912), "Hail Columbia!" (1913), and "The Damnation of Women" (1920).[4]

Historians and biographers have held that Du Bois was not an easy man to work with. He was known to be stubborn, snobbish, arrogant, and demanding. His sharp tongue and harsh editorials created a series of problems for the NAACP board: over the years there were significant power struggles between Du Bois and the board members of the NAACP. In 1934, these struggles came to a head, and Du Bois took a leave of absence from the NAACP. He then returned to Atlanta University where he became professor and chair of the department of sociology. Meanwhile, Du Bois's political position shifted farther and farther to the left. Drawing on Marxist-Leninism and the program of the Communist Party, he portrayed slaves who abandoned plantations as proletariat engaged in "general strikes" and romantically deemed the Reconstruction governments a Marxist experiment (Rudwick 1960/1982:287).

From the 1940s until his death in 1963, Du Bois wrote eloquently and provocatively about the twin evils of colonialism and imperialism. He considered colonies "the slums of the world . . . centers of helplessness, of discouragement of initiative, or forced labor, and of legal suppression of all activities or thoughts which the master country fears or dislikes" (Du Bois 1945:17), and he assailed capitalist nations and corporations for profiting from war. Maintaining that class divisions within America's Negro communities had undermined the prospects for black unity and social advance, he advocated a racially segregated, socialized economy for the United States. As Du Bois became increasingly interested in socialist political alternatives, he called for the creation of a unified, socialist African state and began to devote himself almost exclusively to the Pan-African movement.

Given his outspoken views on racial injustice and stated sympathies with the Soviet Union, it is not surprising that in the early 1950s Du Bois became a target in the Cold War. The Justice Department ordered Du Bois to register as an "agent of a foreign principal." He refused and faced a possible fine of $10,000 and five years in federal prison (Marable 1986:182). At his arraignment, Du Bois coolly read the following statement:

> It is a sad commentary that we must enter a courtroom today to plead Not Guilty to something that cannot be a crime—advocating peace and friendship between the American people and the peoples of the world. . . . In a world which has barely emerged from the horrors of the Second World War and which trembles on the brink of atomic catastrophe, can it be criminal to hope and work for peace? (ibid.)

After a long and costly trial, Du Bois was acquitted, though his passport was retained by the Justice Department and he was refused the right to travel abroad. Du Bois refused to be intimidated by McCarthyism and the Red Scare, however. The following year, when the State Department asked Du Bois to make a sworn statement "as to whether you are now or ever have been [a] communist," Du Bois indignantly refused, stating to the press that his political beliefs "are none of your business" (ibid.:203).

Meanwhile, the topic of Africa continued to dominate much of Du Bois's intellectual thought. In 1960, President Kwame Nkrumah of Ghana invited Du Bois to direct

---

[4]As F. Griffin (2000:30,31) aptly notes, this does not mean that Du Bois completely escaped the sexism of his day. Rather, Du Bois can be considered sexist, elitist, and "color-struck," in that his image of the "Talented Tenth," or African American intellectual leadership, privileged "biracial" or light-skinned, "cultured" males.

a major scholarly project, an *Encyclopedia Africana*. Du Bois was reluctant to accept the offer, since he was already 92 years old. But President Nkrumah maintained that, as the father of Pan-Africanism and the leading advocate of socialism in the black world, Du Bois was the best possible person for the project, and in the end the president's pleas won out.

In Ghana, Du Bois officially joined the Communist Party of the United States and made clear his affinities with Marxist theory. Remarking on his commitment to communism, he stated,

> [c]apitalism cannot reform itself; it is doomed to self-destruction. No universal selfishness can bring social good to all. Communism—the effort to give all men what they need and to ask of each the best they can contribute—this is the only way of human life. . . . In the end communism will triumph. I want to help bring that day. The path of the American Communist Party is clear: It will provide the United States with a real Third party and thus restore democracy to this land. (ibid.:212)

Photo 7.2   Du Bois with Mao Tse-Tung at Mao's Villa in South-Central China, May 1959

Significant costs came with joining the Communist Party. Du Bois was prevented from returning to the United States, as the American Consulate refused to renew his passport. For that matter, it was illegal for members of the Communist Party to possess an American passport, leaving Du Bois subject to 10 years of imprisonment for his crime (ibid.:213). Du Bois and his wife had no choice but to renounce their American citizenship. In March 1963, they became citizens of Ghana. Less than six months later, at the age of 95, Du Bois died.

INTELLECTUAL INFLUENCES AND CORE IDEAS ▪▪

As indicated previously, Du Bois was primarily concerned with the nature and intersection of race and class—an issue virtually ignored by the discipline's "founding figures." Within this program, Du Bois conducted three types of research: (1) empirical studies illuminating the actual social conditions of African Americans (e.g., *The Philadelphia Negro* [1899]); (2) interpretive essays informed by careful historical research and personal experience, as well as keen observation (e.g., *The Souls of Black Folk* [1903]) that emphasized the subjective experience and sources of inequality; and (3) explicitly political essays focusing on Pan-Africanist and socialist solutions to inequality and racism (e.g., *Color and Democracy: Colonies and Peace* [1945]). To be sure, the three styles overlap, and elements of each are apparent in most of Du Bois's work.

On the surface, Du Bois's first book, *The Suppression of the African Slave-Trade to the United States of America, 1638–1870* appears to be a formal, legalistic study, but underneath it is "a sharp moral and ethical repudiation of chattel slavery" (Marable 1986:22). In it, Du Bois harshly condemns the United States for its moral and political hypocrisy and the inability of its leaders to check "this real, existent, growing evil" that led "straight to the

Civil War" (ibid.:23). As you will see, this same detachment coupled with an implicit moralistic tone is also evident in *The Philadelphia Negro* (1899).

At the same time, Du Bois did not hesitate to criticize the black community. He censured black parents for not sufficiently reinforcing the value of formal education, assailed the black church for not adequately combating social corruption and moral decay, and lamented the existence of "the usual substratum of loafers and semicriminals who will not work" (Marable 1986:28). Nevertheless, he insisted that the vast majority of African Americans suffered not from moral laxity, but from "irregular employment" (ibid.).

Du Bois studied less impoverished and oppressed African American communities, too. In "The College-Bred Negro" (1910), he identified and contacted thousands of African American college graduates. He also wrote on "The Negro Landholders of Georgia" (1901) and African American artisans and skilled workers (see Du Bois et al. 1980:146–8). Du Bois (1899/1996:392) maintained that "the better classes of Negroes should recognize their duty toward the masses," an idea that would become one of Du Bois's most famous. Du Bois believed that the burden for winning freedom and justice for all African Americans rested on the shoulders of those who were best prepared, educationally and economically. It would thus fall to the so-called "Talented Tenth" (Du Bois 1903/1989) to lead the fight against racial discrimination (Du Bois in Zuckerman 2004:51; see also Du Bois 1968:236).

In 1935, Du Bois published *Black Reconstruction in America: An Essay Toward a History of the Part Which Black Folk Played in the Attempt to Reconstruct Democracy in America 1860–1880*. This book, based on ignored data about blacks' role in Reconstruction, is most notable in that it challenged the elitism that pervaded historical studies. In contrast to an elitist emphasis on a few "great men" and magnificent military battles, Du Bois sought to write about "ordinary" people—"ordinary" black people—though it would be a full generation later before a "people's" social history would find its place in academia. Indeed, *Black Reconstruction* uncovered the latent elitism and ethnocentrism of the entire historical profession, which under the guise of scientific objectivity was actually, according to Du Bois, a font of racist oppression, as historians' monographs reflected their prejudices (Rudwick 1960/1982:302).

In 1939, Du Bois founded the journal *Phylon: The Atlanta University Review of Race and Culture* as a forum for passionate and scientific works dedicated to promoting black freedom and justice. He explicitly rejected the sterility of a cloistered research approach in favor of an interventionist social science, even though this kind of sociology would not be accepted (and in some circles, has yet to be accepted) for another 50 years. In both his history and his sociology, Du Bois refused to write from a position of value neutrality even as he strove for the truth—a stance that makes perfect sense in the contemporary context of humanistic and interpretive social science, but was quite unheard of in his day.

It's no wonder, then, that Du Bois has been proclaimed "an authentic American genius" not only ahead of *his* time, but also *"ahead of our time"'* (Feagin 2003:22,23; emphasis in original). No wonder Du Bois was nominated for *Time* magazine's "Person of the Twentieth Century" (Zuckerman 2000:1). Du Bois's sociological legacy is based not only on his intrepid exploration of the workings of race and class, but also on his interpretive commitment to truth and justice. Indeed, as Reverend Martin Luther King, Jr., stated in a tribute to him in 1968,

> Dr. Du Bois was a tireless explorer and a gifted discoverer of social truths. His singular greatness lay in his quest for truth about his own people. . . . [His] greatest virtue was his committed empathy with all the oppressed and his divine dissatisfaction with all forms of injustice. (King 1970:181–83)

Photo 7.3 Du Bois believed that higher education was essential to developing the leadership capacity of African Americans, and achieving political and civil equality. These graduates of Fisk University, circa 1906, would be considered part of "the Talented Tenth," that is, among the most able 10 percent of African Americans, who could dedicate themselves to "leavening the lump" and "inspiring the masses" (Du Bois 1903 in Zuckerman 2004:193).

DU BOIS'S THEORETICAL ORIENTATION ▦

Du Bois's work is exemplary in that it illuminates the intertwined structural and subjective causes and consequences of class, race, and racism. This multidimensionality is evident in the multiple approaches that Du Bois used. As discussed previously, Du Bois incorporated rich empirical data and rigorous historical information as well as autobiographical and literary styles into his work. In so doing, Du Bois addressed the distinct subjectivities that motivate action and shape our perception and experience of the world. At the same time, he underscored the economic, and more specifically capitalistic, social structures behind such social and cultural patterns.

As shown in Figure 7.1, in terms of order, Du Bois tended to be collectivist in orientation; he was primarily interested in examining how broad social and cultural patterns—most important, class and racial stratification systems—shaped individual behavior and perceptions. In terms of action, Du Bois continually emphasized both that race and racism are formed by social structural forces, and that they are lived and felt experiences (see Figure 7.2). However, here we consider Du Bois primarily nonrationalist in terms of action. Although Du Bois was, as you will see, quite theoretically multidimensional, it is for his groundbreaking work on racial *consciousness* (as motivating particular lines of conduct) that he is best known. Du Bois powerfully demonstrated how race and racism are rooted in taken-for-granted symbolic structures, thereby reflecting the nonrational realm.

Du Bois's multidimensional theoretical orientation is clearly evident in *The Philadelphia Negro* (Du Bois 1899; discussed further below), in which he explores both the collectivist and individualist dimensions and the rationalist and nonrationalist dimensions of race. On the one hand, Du Bois develops an empirical typology of four economic classes within the

Figure 7.1 Du Bois's Basic Theoretical Orientation

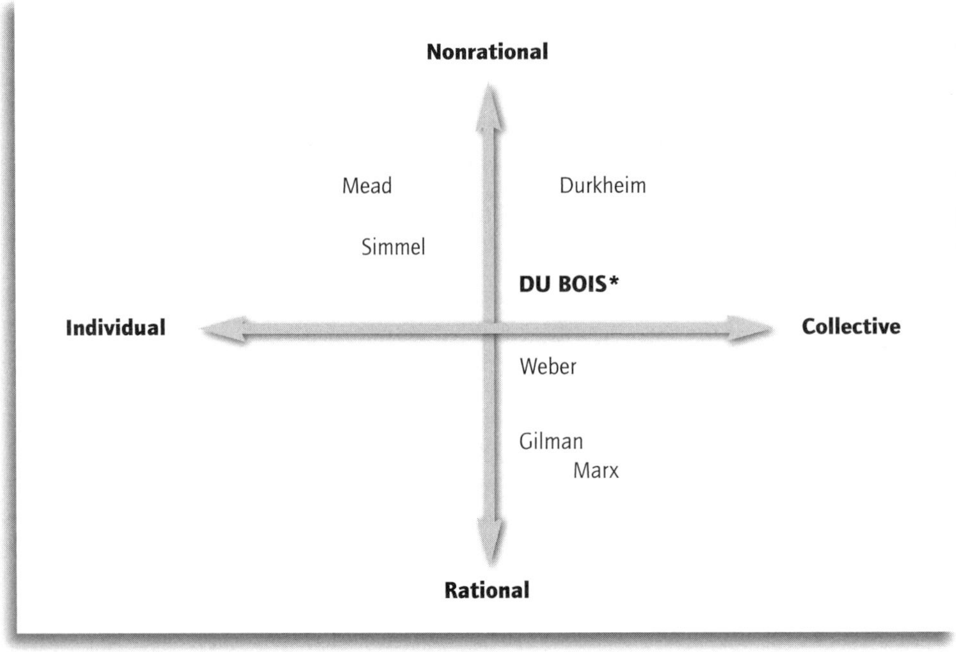

*Our placement of Du Bois on the nonrational side of the continuum reflects the excerpts in this volume that were chosen because of their theoretical significance. In our view, it is his understanding of racial consciousness that constitutes his single most important theoretical contribution. However, he continually underscored the intertwined, structural underpinnings of race and class that, in the latter part of his life, led him to adopt a predominantly rationalist, Marxist-inspired orientation. In our view, however, Du Bois's later work has more empirical than theoretical significance.

Philadelphia black community: the well-to-do; the decent hard workers, who were doing well; the "worthy poor," who were barely making ends meet; and the "submerged tenth," who were beneath the level of socioeconomic viability. Du Bois emphasizes that this stratification system resulted from the increasing industrialization of the time, reflecting a collectivist and rationalist approach to order and action (see Figure 7.2). In short, he contends that the economic system of capitalism led to industrialization and, hence, to the decay of the black community within Philadelphia.

Nevertheless, in *The Philadelphia Negro* Du Bois recognizes the power of nonrational factors too. For instance, Du Bois's typology depicts an upper-class "colortocracy" within the black community: light-skinned blacks were a caste apart and formed the bulk of Du Bois's so-called Talented Tenth (Anderson 2000:56). This "colortocracy" reflects the nonrational realm (at the collective level) in the sense that it represents a symbolic code or scheme in Durkheimian terms, a status system in Weber's terms, and an ideology in Marxist terms (see Figure 7.2). As Du Bois states,

> [i]ndividual members of the colortocracy at times developed a notorious but distinctive racial complex involving an ideology that set them apart from those they viewed as their inferiors. They would take excessive pride in their "white" features, including light skin, thin noses and lips, and "good" hair. Often "colorstruck," they mimicked and voiced the anti-black prejudices of whites, whose fears, concerns, and values they understood and partly shared. (Du Bois, quoted in Anderson 2000:57)

Figure 7.2 Du Bois's Multidimensional Approach to Race and Class

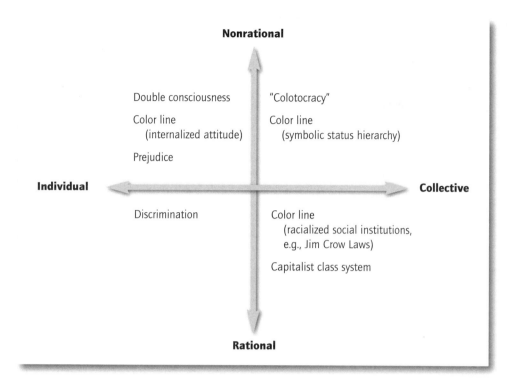

In addition, while Du Bois emphasized the structural or collectivist dimension of class and race, he also recognized that their institutional and symbolic features were perpetuated and disseminated in everyday interaction, at the level of the individual. Consider, for instance, the following passage about his experiences as a waiter at a summer resort hotel:

> I did not mind the actual work or the kind of work, but it was the dishonesty and deception, the flattery and cajolery, the unnatural assumption that worker and diner had no common humanity. It was uncanny. It was inherently and fundamentally wrong. I stood staring and thinking, while the other boys hustled about. Then I noticed one fat hog, feeding at a heavily gilded trough, who could not find his waiter. He beckoned me. It was not his voice, for his mouth was too full. It was his way, his air, his assumption. Thus Caesar ordered his legionnaires or Cleopatra her slaves. Dogs recognized the gesture. I did not. He may be beckoning yet for all I know, for something froze within me. I did not look his way again. Then and there I disowned menial service for me and my people. . . . When I finally walked out of that hotel and out of menial service forever, I felt as though, in a field of flowers, my nose had been unpleasantly long to the worms and manure at their roots. (Du Bois 1920/2003c:128–29)

On the one hand, in this passage, Du Bois describes how it *felt* to be treated as a second-class citizen, a "dog," a "slave." Speaking to the nonrational realm, he illuminates the profound dehumanization that is at the heart of racism (at the level of the individual). On the other hand, Du Bois also emphasizes here that the font of such dehumanization is a hierarchical racial and class *system* that enables and reaffirms both this symbolic code and the hierarchical social roles. The point is (and this is made more explicit in his later, more Marxist-oriented works) that it is the menial service role that makes for such degradation or alienation. This is why Du Bois followed Marx in suggesting that capitalism itself precluded equality and democracy, resting as it did on exploitation of labor. So too, this passage suggests that the sordid "assumptions" and "airs" of white elites stem from their privileged position at the top of the racial and class hierarchies. Du Bois's multidimensional approach to race is illustrated in Figure 7.2.

Readings

We begin this section with excerpts from *The Philadelphia Negro* (1899), which is the first major sociological study of an African American community ever published in the United States. We then turn to Du Bois's most well-known and highly acclaimed work, *The Souls of Black Folk* (1903). This is a book of essays that relies not only on "objective" "social facts," but also on subjective, interpretative experience. We conclude this chapter with "The Souls of White Folk," a provocative, albeit less well-known, essay which was first published in 1910, but was revised with references to World War I for republication in *Darkwater* in 1920. While at first glance each of these works exemplifies a different aspect of Du Bois's life and work—ranging from the more empirical, to the more interpretive, to the more militant—you will see that in fact all three of these aspects of Du Bois's thinking and writing are evident in all three works.

Introduction to *The Philadelphia Negro*

As indicated previously, *The Philadelphia Negro* is the first major sociological study of an African American community ever published in the United States. In this groundbreaking book, Du Bois presents a rich array of empirical information about the lives of Philadelphia African Americans, focusing on the "social condition of the Colored People of the Seventh Ward of Philadelphia," an impoverished area of about 10,000 African Americans (nearly one-fourth of the entire Philadelphia black population). Du Bois carried out some 5,000 surveys and interviews with African Americans in the Seventh Ward. Du Bois and his wife, Nina, also directly experienced many of the conditions he described so vividly, living in the Settlement House, in the heart of the Seventh Ward. In addition to his careful statistical and ethnographic research, Du Bois also paid particular attention to historical information because he believed that "one cannot study the Negro in freedom . . . without knowing his history in slavery" (Marable 1986:24,25). Du Bois adamantly believed that careful sociological documentation combined with social and cultural understanding of a social group (reminiscent of Weber's *Verstehen*—see Chapter 4) would make possible a cogent agenda of social reform.

Du Bois was commissioned to carry out his study of the "Philadelphia Negro" by the University of Pennsylvania. In several different places in his autobiographical writings, Du Bois comments on the circumstances behind which the study was commissioned:

It all happened this way: Philadelphia, then and still one of the worst governed of America's badly governed cities, was having one of its periodic spasms of reforms. A thorough study of causes was called for. . . . [T]he underlying cause was evident to most white Philadelphians: the corrupt, semi-criminal vote of the Negro Seventh Ward. Everyone agreed that here lay the cancer; but would it not be well to give sci-entific sanction to the known causes by an investigation, with imprimatur of the University? (Du Bois 1968:194)

Although in this passage Du Bois somewhat cynically describes the circumstances that led to the birth of the study, *The Philadelphia Negro* itself does not read cynically. Rather, as you will see, it forthrightly explains African Americans' lives in rich, historical, statistical, and ethnographic detail. Chapter by chapter, Du Bois systematically outlines relevant demo-graphic and other information about the black population, including the "size, age and sex of the Negro population"; family; marriage; "education and illiteracy"; housing; occupations; and institutions, most importantly the Black church. Du Bois points out that blacks received low wages for undesirable work, lived in "less pleasant quarters than most people," and paid for them "higher rents . . . [while] the Negro who ventures away from the mass of his people and their organized life, finds himself alone, shunned and taunted, stared at, and made uncomfortable" (Marable 1986:26). Du Bois does not lay all the problems in the Seventh Ward at the doorstep of the white population, however. Instead, he outlines black individuals' own role in creating pauperism, alcoholism, and criminality in the black population as well.

The selection below begins with two brief excerpts that give you the flavor of the over-all study. Next, we turn to a section from a chapter called "The Contact of the Races," in which Du Bois systematically sets out the specific social consequences of prejudice (as well as of discrimination). The final excerpt below, from the conclusion of the book, reflects the theoretically multidi-mensional nature of Du Bois's work. On the one hand, Du Bois outlines the severe social structural (collectivist) conditions that result in so many different problems for African Americans. On the other hand, Du Bois insists that the "vastest of the Negro problems" is that which he would later famously call the "problem of the Color Line," the relation of the "darker to the lighter races" (Du Bois 1903/1989:13). This problem reflects the power of the nonra-tional realm in that, as discussed previously, "color" is a status system, a symbolic code or scheme (see Figure 7.2). Meanwhile, Du Bois's specific recommendations at the end of the book reflect his understanding that even though "Negro problems" are a function of structural and historical conditions, social change will come only if and when *individual* whites and blacks change their attitudes as well as their behavior.

Photo 7.4   The Seventh Ward of Philadelphia, Circa 1900

From *The Philadelphia Negro* (1899)

W. E. B. Du Bois

THE SCOPE OF THIS STUDY

General Aim.—This study seeks to present the results of an inquiry undertaken by the University of Pennsylvania into the condition of the forty thousand or more people of Negro blood now living in the city of Philadelphia. This inquiry extended over a period of fifteen months and sought to ascertain something of the geographical distribution of this race, their occupations and daily life, their homes, their organizations, and, above all, their relation to their million white fellow-citizens. The final design of the work is to lay before the public such a body of information as may be a safe guide for all efforts toward the solution of the many Negro problems of a great American city.

The Methods of Inquiry.—The investigation began August the first, 1896, and, saving two months, continued until December the thirty-first, 1897. The work commenced with a house-to-house canvass of the Seventh Ward. This long narrow ward, extending from South Seventh street to the Schuylkill River and from Spruce street to South street, is an historic centre of Negro population, and contains to-day a fifth of all the Negroes in this city.[i] It was therefore thought best to make an intensive study of conditions in this district, and afterward to supplement and correct this information by general observation and inquiry in other parts of the city.

Six schedules were used among the nine thousand Negroes of this ward; a family schedule with the usual questions as to the number of members, their age and sex, their conjugal condition and birthplace, their ability to read and write, their occupation and earnings, etc.; an individual schedule with similar inquiries; a home schedule with questions as to the number of rooms, the rent, the lodgers, the conveniences, etc.; a street schedule to collect data as to the various small streets and alleys, and an institution schedule for organizations and institutions; finally a slight variation of the individual schedule was used for house-servants living at their places of employment.

This study of the central district of Negro settlement furnished a key to the situation in the city; in the other wards therefore a general survey was taken to note any striking differences of condition, to ascertain the general distribution of these people, and to collect information and statistics as to organizations, property, crime and pauperism, political activity, and the like. This general inquiry, while it lacked precise methods of measurement in most cases, served nevertheless to correct the errors and illustrate the meaning of the statistical material obtained in the house-to-house canvass.

Throughout the study such official statistics and historical matter as seemed reliable were used, and experienced persons, both white and colored, were freely consulted. . . .

THE PROBLEM

The Negro Problems of Philadelphia.— In Philadelphia, as elsewhere in the United States, the existence of certain peculiar social problems affecting the Negro people are plainly manifest. Here is a large group of people— perhaps forty-five thousand, a city within a city—who do not form an integral part of the larger social group. This in itself is not altogether unusual; there are other unassimilated groups: Jews, Italians, even Americans; and yet in the case of the Negroes the segregation is more conspicuous, more patent to the eye, and so intertwined with along historic evolution,

SOURCE: From *The Philadelphia Negro* by W. E. B. Du Bois, 1899.

[i] I shall throughout this study use the term "Negro," to designate all persons of Negro descent, although the appellation is to some extent illogical. I shall, moreover, capitalize the word, because I believe that eight million Americans are entitled to a capital letter.

with peculiarly pressing social problems of poverty, ignorance, crime and labor, that the Negro problem far surpasses in scientific interest and social gravity, most of the other race or class questions.

The student of these questions must first ask, What is the real condition of this group of human beings? Of whom is it composed, what sub-groups and classes exist, what sort of individuals are being considered? Further, the student must clearly recognize that a complete study must not confine itself to the group, but must specially notice the environment; the physical environment of city, sections and houses, the far mightier social environment—the surrounding world of custom, wish, whim, and thought which envelops this group and powerfully influences its social development.

Nor does the clear recognition of the field of investigation simplify the work of actual study; it rather increases it, by revealing lines of inquiry far broader in scope than first thought suggests. To the average Philadelphian the whole Negro question reduces itself to a study of certain slum districts. His mind reverts to Seventh and Lombard streets and to Twelfth and Kater streets of to-day, or to St. Mary's in the past. Continued and widely known charitable work in these sections make the problem of poverty familiar to him; bold and daring crime too often traced to these centres has called his attention to a problem of crime, while the scores of loafers, idlers and prostitutes who crowd the sidewalks here night and day remind him of a problem of work.

All this is true—all these problems are there and of threatening intricacy; unfortunately, however, the interest of the ordinary man of affairs is apt to stop here. Crime, poverty and idleness affect his interests unfavorably and he would have them stopped; he looks upon these slums and slum characters as unpleasant things which should in some way be removed for the best interests of all. The social student agrees with him so far, but must point out that the removal of unpleasant features from our complicated modern life is a delicate operation requiring knowledge and skill; that a slum is not a simple fact, it is a symptom, and that to know the removable

causes of the Negro slums of Philadelphia requires a study that takes one far beyond the slum districts. For few Philadelphians realize how the Negro population has grown and spread. There was a time in the memory of living men when a small district near Sixth and Lombard streets comprehended the great mass of the Negro population of the city. This is no longer so. Very early the stream of the black population started northward, but the increased foreign immigration of 1830 and later, turned it back. It started south also but was checked by poor houses and worse police protection. Finally with gathered momentum the emigration from the slums started west, rolling on slowly and surely taking Lombard street as its main thoroughfare, gaining early foothold in West Philadelphia, and turning at the Schuylkill River north and south to the newer portions of the city.

Thus to-day the Negroes are scattered in every ward of the city, and the great mass of them live far from the whilom centre of colored settlement. What, then, of this great mass of the population? Manifestly they form a class with social problems of their own—the problems of the Thirtieth Ward differ from the problems of the Fifth, as the black inhabitants differ. In the former ward we have represented the rank and file of Negro working-people; laborers and servants, porters and waiters. This is at present the great middle class of Negroes feeding the slums on the one hand and the upper class on the other. Here are social questions and conditions which must receive the most careful attention and patient interpretation.

Not even here, however, can the social investigator stop. He knows that every group has its upper class; it may be numerically small and socially of little weight, and yet its study is necessary to the comprehension of the whole—it forms the realized ideal of the group, and as it is true that a nation must to some extent be measured by its slums, it is also true that it can only be understood and finally judged by its upper class.

The best class of Philadelphia Negroes, though sometimes forgotten or ignored in discussing the Negro problems, is nevertheless known to many Philadelphians. Scattered throughout the better parts of the Seventh Ward, and on Twelfth, lower Seventeenth and Nineteenth streets, and here and there in the residence wards of the northern, southern, and western sections of the city is a class of caterers, clerks, teachers, professional men, small merchants, etc., who constitute the aristocracy of the Negroes. Many are well-to-do, some are wealthy, all are fairly educated, and some liberally trained. Here too are social problems—differing from those of the other classes, and differing too from those of the whites of a corresponding grade, because of the peculiar social environment in which the whole race finds itself, which the whole race feels, but which touches this highest class at most points and tells upon them most decisively.

Many are the misapprehensions and misstatements as to the social environment of Negroes in a great Northern city. Sometimes it is said, here they are free; they have the same chance as the Irishman, the Italian, or the Swede; at other times it is said, the environment is such that it is really more oppressive than the situation in Southern cities. The student must ignore both of these extreme statements and seek to extract from a complicated mass of facts the tangible evidence of a social atmosphere surrounding Negroes, which differs from that surrounding most whites; of a different mental attitude, moral standard, and economic judgment shown toward Negroes than toward most other folk. That such a difference exists and can now and then plainly be seen, few deny; but just how far it goes and how large a factor it is in the Negro problems, nothing but careful study and measurement can reveal.

Such then are the phenomena of social condition and environment which this study proposes to describe, analyze, and, so far as possible, interpret. . . .

The Contact of the Races

Color Prejudice. Incidentally throughout this study the prejudice against the Negro has been again and again mentioned. It is time now to reduce this somewhat indefinite term to something tangible. Everybody speaks of the matter, everybody knows that it exists, but in just

what form it shows itself or how influential it is few agree. In the Negro's mind, color prejudice in Philadelphia is that widespread feeling of dislike for his blood, which keeps him and his children out of decent employment, from certain public conveniences and amusements, from hiring houses in many sections, and in general, from being recognized as a man. Negroes regard this prejudice as the chief cause of their present unfortunate condition. On the other hand most white people are quite unconscious of any such powerful and vindictive feeling; they regard color prejudice as the easily explicable feeling that intimate social intercourse with a lower race is not only undesirable but impracticable if our present standards of culture are to be maintained; and although they are aware that some people feel the aversion more intensely than others, they cannot see how such a feeling has much influence on the real situation, or alters the social condition of the mass of Negroes.

As a matter of fact, color prejudice in this city is something between these two extreme views: it is not to-day responsible for all, or perhaps the greater part of the Negro problems, or of the disabilities under which the race labors; on the other hand it is a far more powerful social force than most Philadelphians realize. The practical results of the attitude of most of the inhabitants of Philadelphia toward persons of Negro descent are as follows:

1. As to getting work:
No matter how well trained a Negro may be, or how fitted for work of any kind, he cannot in the ordinary course of competition hope to be much more than a menial servant.

He cannot get clerical or supervisory work to do save in exceptional cases.

He cannot teach save in a few of the remaining Negro schools.

He cannot become a mechanic except for small transient jobs, and cannot join a trades union.

A Negro woman has but three careers open to her in this city: domestic service, sewing, or married life.

2. As to keeping work:
The Negro suffers in competition more severely than white men.

Change in fashion is causing him to be replaced by whites in the better paid positions of domestic service.

Whim and accident will cause him to lose a hard-earned place more quickly than the same things would affect a white man.

Being few in number compared with the whites the crime or carelessness of a few of his race is easily imputed to all, and the reputation of the good, industrious and reliable suffer thereby.

Because Negro workmen may not often work side by side with white workmen, the individual black workman is rated not by his own efficiency, but by the efficiency of a whole group of black fellow workmen which may often be low.

Because of these difficulties which virtually increase competition in his case, he is forced to take lower wages for the same work than white workmen.

3. As to entering new lines of work:

Men are used to seeing Negroes in inferior positions; when, therefore, by any chance a Negro gets in a better position, most men immediately conclude that he is not fitted for it, even before he has a chance to show his fitness.

If, therefore, he set up a store, men will not patronize him.

If he is put into public position men will complain.

If he gain a position in the commercial world, men will quietly secure his dismissal or see that a white man succeeds him.

4. As to his expenditure:

The comparative smallness of the patronage of the Negro, and the dislike of other customers makes it usual to increase the charges or difficulties in certain directions in which a Negro must spend money.

He must pay more house-rent for worse houses than most white people pay.

He is sometimes liable to insult or reluctant service in some restaurants, hotels and stores, at public resorts, theatres and places of recreation; and at nearly all barbershops.

5. As to his children:

The Negro finds it extremely difficult to rear children in such an atmosphere and not have them either cringing or impudent: if he impresses upon them patience with their lot, they may grow up satisfied with their condition; if he inspires them with ambition to rise, they may grow to despise their own people, hate the whites and become embittered with the world.

His children are discriminated against, often in public schools.

They are advised when seeking employment to become waiters and maids.

They are liable to species of insult and temptation peculiarly trying to children.

6. As to social intercourse:

In all walks of life the Negro is liable to meet some objection to his presence or some discourteous treatment; and the ties of friendship or memory seldom are strong enough to hold across the color line.

If an invitation is issued to the public for any occasion, the Negro can never know whether he would be welcomed or not; if he goes he is liable to have his feelings hurt and get into unpleasant altercation; if he stays away, he is blamed for indifference.

If he meet a lifelong white friend on the street, he is in a dilemma; if he does not greet the friend he is put down as boorish and impolite; if he does greet the friend he is liable to be flatly snubbed.

If by chance he is introduced to a white woman or man, he expects to be ignored on the next meeting, and usually is.

White friends may call on him, but he is scarcely expected to call on them, save for strictly business matters.

If he gain the affections of a white woman and marry her he may invariably expect that slurs will be thrown on her reputation and on his, and that both his and her race will shun their company.

When he dies he cannot be buried beside white corpses.

7. The result:

Any one of these things happening now and then would not be remarkable or call for especial comment; but when one group of people suffer all these little differences of treatment and discriminations and insults continually, the result is either discouragement, or bitterness, or over-sensitiveness, or recklessness. And a people feeling thus cannot do their best.

Presumably the first impulse of the average Philadelphian would be emphatically to deny any such marked and blighting discrimination as the above against a group of citizens in this metropolis. Every one knows that in the past color prejudice in the city was deep and passionate; living men can remember when a Negro could not sit in a street car or walk many streets in peace. These times have passed, however, and many imagine that active discrimination against the Negro has passed with them. Careful inquiry will convince any such one of his error. To be sure a colored man to-day can walk the streets of Philadelphia without personal insult; he can go to theatres, parks and some places of amusement without meeting more than stares and discourtesy; he can be accommodated at most hotels and restaurants, although his treatment in some would not be pleasant. All this is a vast advance and augurs much for the future. And yet all that has been said of the remaining discrimination is but too true. . . .

A FINAL WORD

The Meaning of All This.—Two sorts of answers are usually returned to the bewildered American who asks seriously: What is the Negro problem? The one is straightforward and clear: it is simply this, or simply that, and one simple remedy long enough applied will in time cause it to disappear. The other answer is apt to be hopelessly involved and complex—to indicate no simple panacea, and to end in a somewhat hopeless—There it is; what can we do? Both of these sorts of answers have something of truth in them: the Negro problem looked at in one way is but the old world questions of ignorance, poverty, crime, and the dislike of the stranger. On the other hand it is a mistake to think that attacking each of these questions single-handed without reference to the others will settle the matter: a combination of social problems is far more than a matter of mere addition,—the combination itself is a problem. Nevertheless the Negro problems are not more hopelessly complex than many others have been. Their elements despite their bewildering complication can be kept clearly in view: they

are after all the same difficulties over which the world has grown gray: the question as to how far human intelligence can be trusted and trained; as to whether we must always have the poor with us; as to whether it is possible for the mass of men to attain righteousness on earth; and then to this is added that question of questions: after all who are Men? Is every featherless biped to be counted a man and brother? Are all races and types to be joint heirs of the new earth that men have striven to raise in thirty centuries and more? Shall we not swamp civilization in barbarism and drown genius in indulgence if we seek a mythical Humanity which shall shadow all men? The answer of the early centuries to this puzzle was clear: those of any nation who can be called Men and endowed with rights are few: they are the privileged classes—the well-born and the accidents of low-birth called up by the King. The rest, the mass of the nation, the *pöbel,* the mob, are fit to follow, to obey, to dig and delve, but not to think or rule or play the gentleman. We who were born to another philosophy hardly realize how deep-seated and plausible this view of human capabilities and powers once was; how utterly incomprehensible this republic would have been to Charlemagne or Charles V or Charles I. We rather hasten to forget that once the courtiers of English kings looked upon the ancestors of most Americans with far greater contempt than these Americans look upon Negroes—and perhaps, indeed, had more cause. We forget that once French peasants were the "Niggers" of France, and that German princelings once discussed with doubt the brains and humanity of the *bauer.*

Much of this—or at least some of it—has passed and the world has glided by blood and iron into a wider humanity, a wider respect for simple manhood unadorned by ancestors or privilege. Not that we have discovered, as some hoped and some feared, that all men were created free and equal, but rather that the differences in men are not so vast as we had assumed. We still yield the well-born the advantages of birth, we still see that each nation has its dangerous flock of fools and rascals; but we also find most men have brains to be cultivated and souls to be saved.

And still this widening of the idea of common Humanity is of slow growth and to-day but dimly realized. We grant full citizenship in the World-Commonwealth to the "Anglo-Saxon" (whatever that may mean), the Teuton and the Latin; then with just a shade of reluctance we extend it to the Celt and Slav. We half deny it to the yellow races of Asia, admit the brown Indians to an ante-room only on the strength of an undeniable past; but with the Negroes of Africa we come to a full stop, and in its heart the civilized world with one accord denies that these come within the pale of nineteenth century Humanity. This feeling, widespread and deep-seated, is, in America, the vastest of the Negro problems; we have, to be sure, a threatening problem of ignorance but the ancestors of most Americans were far more ignorant than the freedmen's sons; these ex-slaves are poor but not as poor as the Irish peasants used to be; crime is rampant but not more so, if as much, as in Italy; but the difference is that the ancestors of the English and the Irish and the Italians were felt to be worth educating, helping and guiding because they were men and brothers, while in America a census which gives a slight indication of the utter disappearance of the American Negro from the earth is greeted with ill-concealed delight.

Other centuries looking back upon the culture of the nineteenth would have a right to suppose that if, in a land of freemen, eight millions of human beings were found to be dying of disease, the nation would cry with one voice, "Heal them!" If they were staggering on in ignorance, it would cry, "Train them!" If they were harming themselves and others by crime, it would cry, "Guide them!" And such cries are heard and have been heard in the land; but it was not one voice and its volume has been ever broken by counter-cries and echoes, "Let them die!" "Train them like slaves!" "Let them stagger downward!"

This is the spirit that enters in and complicates all Negro social problems and this is a problem which only civilization and humanity can successfully solve. Meantime we have the other problems before us—we have the problems arising from the uniting of so many social questions about one centre. In such a situation we need only to avoid underestimating the difficulties on the one hand and overestimating them on the other. The problems are difficult, extremely difficult,

but they are such as the world has conquered before and can conquer again. Moreover the battle involves more than a mere altruistic interest in an alien people. It is a battle for humanity and human culture. If in the hey-dey of the greatest of the world's civilizations, it is possible for one people ruthlessly to steal another, drag them helpless across the water, enslave them, debauch them, and then slowly murder them by economic and social exclusion until they disappear from the face of the earth—if the consummation of such a crime be possible in the twentieth century, then our civilization is vain and the republic is a mockery and a farce.

But this will not be; first, even with the terribly adverse circumstances under which Negroes live, there is not the slightest likelihood of their dying out; a nation that has endured the slave-trade, slavery, reconstruction, and present prejudice three hundred years, and under it increased in numbers and efficiency, is not in any immediate danger of extinction. Nor is the thought of voluntary or involuntary emigration more than a dream of men who forget that there are half as many Negroes in the United States as Spaniards in Spain. If this be so then a few plain propositions may be laid down as axiomatic:

1. The Negro is here to stay.

2. It is to the advantage of all, both black and white, that every Negro should make the best of himself.

3. It is the duty of the Negro to raise himself by every effort to the standards of modern civilization and not to lower those standards in any degree.

4. It is the duty of the white people to guard their civilization against debauchment by themselves or others; but in order to do this it is not necessary to hinder and retard the efforts of an earnest people to rise, simply because they lack faith in the ability of that people.

5. With these duties in mind and with a spirit of self-help, mutual aid and co-operation, the two races should strive side by side to realize the ideals of the republic and make this truly a land of equal opportunity for all men.

Introduction to *The Souls of Black Folk*

Du Bois's most famous work, *The Souls of Black Folk* (1903/1989), is a compilation of 14 essays that Du Bois had published previously in the *Atlantic Monthly*, the *New World*, and other journals. As Marable (1986:47) quite rightly observes, "Its grace and power is still overwhelming." *The Souls of Black Folk* is notable not only for its grace, though, but also for at least three distinct, albeit interrelated, characteristics.

First, historically, *The Souls of Black Folk* is important because it explicitly exposed an important intellectual and political schism in the black community between the more moderate Booker T. Washington; and the more radical Du Bois (as well as radical political activists, such as the journalist Ida B. Wells-Barnett, a feminist and antilynching activist). In his essay "Of Mr. Booker T. Washington and Others" (below), Du Bois pointedly attacks Washington for his "capitulating" agenda. Du Bois was enraged by Washington's "Atlanta Compromise" (favored by many whites) in which blacks would disavow open agitation in exchange for developing their own segregated institutions—which Du Bois interpreted as nothing more than "the old attitude of adjustment and submission" (Du Bois 1903/1989:43). Du Bois also abhorred Washington's "gradualist" solution to the issue of suffrage and his notion that Negro education should primarily be in the form of trade schools (rather than "abstract" knowledge).

Second, from a social science standpoint, *The Souls of Black Folk* is significant because Du Bois writes in a new "voice." Disgusted by the failure of sound empirical research to lead to desperately needed social change for the African American community and having done the empirical work himself, Du Bois became convinced that empirical data alone would never convince white Americans of the true workings of racial discrimination and prejudice. As a result, he turned away from the more empirical, strictly scientific accounting of race after writing *The Philadelphia Negro*, convinced that facts alone did nothing to influence people toward improving conditions for blacks. He began to write with a more "soulful" voice, because, quite rightly (in our view), he recognized that race does not work or exist solely at the "rational" level.[1]

As Du Bois states,

> For many years it was the theory of most Negro leaders that . . . white America did not know or realize the continuing plight of the Negro. Accordingly, for the last two decades, we have striven by book and periodical, by speech and appeal, by various dramatic methods of agitation to put the essential facts before the American people. Today there can be no doubt that Americans know the facts; and yet they remain for the most part indifferent and unmoved. (Du Bois 1898, quoted in Berry 2000:106)

Thus, Du Bois himself "stepped within the Veil, raising it that you [the presumably white reader] may view faintly its deeper recesses,—the meaning of its religion, the passion of its human sorrow, and the struggle of its greater souls" (Du Bois 1903/1989:1,2). Adding, "Need I add that I who speak here am bone of the bone and flesh of the flesh of them that live within the Veil?," Du Bois pointedly used his own biographical experience to illuminate the reality of race in the United States (ibid.:2).

Relatedly, although Du Bois was intensely critical of the black church for failing to address the real economic needs of the black community, in *The Souls of Black Folk*,

---

[1]Nevertheless, Du Bois did not completely abandon a positivist stance; he continued to seek "hard data," be they from official censuses, government documents, or field notes.

Du Bois turns his attention to the essence and power of black spirituality. He begins each chapter with "a bar of the Sorrow Songs,—some echo of haunting melody from the only American music which welled up from black souls in the dark past" to convey the essence of blackness and "our spiritual strivings" (ibid.:2–3). As Du Bois eloquently points out in "Of the Faith of the Fathers" (below), "Under the stress of law and whip, it [the Music of the Negro] became the one true expression of a people's sorrow, despair, and hope" (ibid.:156). In addition, Du Bois underscores the tremendous social significance of the Negro church. He calls the church the "social center of Negro life in the United States" (ibid.:157) and notes that "in the South, at least, practically every American Negro is a church member" (ibid.:158). Of course, this is precisely why Du Bois was critical of the Negro church as well; given its social and institutional vitality, Du Bois believed that the Black church could and should do more for its people.

Social scientifically, then, *The Souls of Black Folk* contains a crucial methodological lesson: the workings of such complex phenomena as race and class cannot be fully understood using only "scientific" means. Du Bois explored subjectivity because he believed that race and racism did not work at a strictly rational level. Interestingly, today this position is not only fully embraced by postmodernists, but also is taken to a radical extreme. Whereas Du Bois sought to combine empirical, scientific, and historical data with more subjective, intuitive understandings, postmodernists deny the existence of "social scientific" ways of knowing altogether, saying that all knowledge, in the end, is subjective and idiosyncratic.

Put in another way, Du Bois's approach was a precursor to the interpretive shift that began to emerge in sociology (and the social sciences in general) in the 1980s. Interpretive social scientists emphasize that complete objectivity is neither possible nor desirable. So-called value-free, positivist sociology is disingenuous, if not impossible to practice; a sociologist who does not acknowledge his or her own subjectivity runs the risk of reifying existing prejudices and beliefs. From this point of view, the notion that truth is arrived at through objectivity is unfounded. We cannot adequately capture the richness and complexity of reality using only objectivist tools (e.g., historical documents, surveys, statistics). We can and must use our own subjective experience and humanistic understanding in order to explore and explain the complexities of reality.

Third, *The Souls of Black Folk* is important theoretically because it contains three interrelated concepts for which Du Bois is now famous: the **color line, double consciousness,** and the **veil.** All of these concepts reflect the intertwined dimensions of race discussed previously (see Figure 7.2). Consider, first, one of Du Bois's most oft-cited sentences, which appears in that work: "The problem of the twentieth century is the problem of the color line—the relation of the darker to the lighter races of men in Asia and Africa, in America and the islands of the sea" (ibid.:13). Du Bois's concept of the "color line" is most intriguing, in part, because it speaks to the collective/rational, collective/nonrational, and individual/nonrational realms. As shown in Figure 7.2, the color line is both a preexisting social and cultural structure, and an internalized attitude.

Specifically, on the one hand, the color line addresses the historical and institutional (particularly colonial) dimensions of race. For instance, Du Bois maintains that once "the 'Color Line' began to pay dividends" through the colonization and exploitation of Africa and Africans beginning in the fifteenth century, race became central to world history (Monteiro 2000:229). Africa's poverty is inexorably linked to colonialism and imperial domination; the wealth of the colonial empires of England, France, Germany, and the United States "comes directly from the darker races of the world" (Du Bois 1947/1995:645).

At the same time, the color line has important subjective or nonrational dimensions (see Figure 7.2). Du Bois not only examines race in its objective, demographic, and historical

Photo 7.5 Segregated drinking fountains labeled "white" and "colored" in the Dougherty County Courthouse, Albany, Georgia, circa 1963. Note the unequal size of the fountains.

aspects, but also addresses race as a symbolic and experiential reality. This emphasis on the nonrational workings of the color line is highlighted in his question, "How does it feel to be a problem?" found at the very beginning of *The Souls of Black Folk* (Du Bois 1903/1989:3).

This focus on the subjectivity integral to race and racism is also readily apparent in Du Bois's concepts of the veil and double consciousness:

> The Negro is a sort of seventh son, born with a veil, and gifted with second sight in this American world—a world which yields him no self-consciousness, but only lets him see himself through the revelation of the other world. It is a peculiar sensation, this double-consciousness, this sense of always looking at one's self through the eyes of others, of measuring one's soul by the tape of a world that looks on in amused contempt and pity. One ever feels his two-ness—an American, a Negro; two souls, two thoughts, two unreconciled strivings; two warring ideals in one dark body, whose dogged strength alone keeps it from being torn asunder. (ibid.:5)

Here Du Bois reflects not on the structural causes, but rather on the subjective (or nonrational) consequences of being black in America. He explores the "peculiar *sensation*" (emphasis added) of being black in America. He wonders, "Why did God make me an outcast and a stranger in mine own house?" (ibid.:5). Nevertheless, as discussed previously, in exploring the subjective dimensions of racism Du Bois never lost sight of its "objective," historical origins; the subjective experience of race— "double consciousness"—is rooted ultimately in the marginal structural position of "blacks." In this way, Du Bois's argument can be said to parallel that of Marx. As you will recall (see Chapter 2), Marx always tied the subjective experience of "alienation" and "estrangement" to workers' relationship to the means of production, that is, to class structure.

Today, in the context of the recent election of Barack Obama as president of the United States, new questions about "the color line" abound. On the one hand, clearly the structural and symbolic situation of African Americans has dramatically changed with the election of Barack Obama. As holder of the most powerful position in the nation (if not the world), with his picture adorning the walls of the White House as well as public schools and government offices across the land (and around the globe), it seems absurd that President Obama might consider himself "an outcast and a stranger in mine own house" as did Du Bois and African Americans of his generation. Some would even argue that the election of Obama signifies that the "problem of the color line" has been "resolved": since African Americans make up only some 13 percent of the American population, his electoral victory *must* be evidence of a fundamental reduction in prejudice and racism in the United States.

Certainly, the election of Barack Obama reflects a tremendous sea change in the history of the United States. For instance, children today (both white and nonwhite) will be less likely to take white supremacy for granted, simply because their biographical experience will include living in a moment in which educated, highly esteemed African Americans are president and first lady. Yet, despite this phenomenal turn of events, huge reservoirs of inequality in the United States persist and important issues regarding race and class remain. Based on current rates of first incarceration, an estimated 32 percent of black males will enter state or federal prison during their lifetime, compared to 17 percent of Hispanic males and 5.9 percent of white males (http://www.ojp.usdoj.gov/bjs/crimoff.htm#inmates). So, too, white households continue to be far wealthier than black or Hispanic households, and educational differences extend directly from those differentials in class and wealth. For instance, students with family incomes of more than $200,000 had an average SAT math score of 570, those in the $80,000–$100,000 cohort had an average of 525, and those with family income up to $20,000 had an average of 456.

Photo 7.6 While the election of Barack Obama to the presidency of the United States in 2008 reflects significant shifts in the "color line," racialization continues to be a powerful force in the United States.

Getting back to Du Bois, what the election of Obama makes clear is that the "place" of African Americans in the United States' system of stratification is complex. As Du Bois readily recognized when he coined the term "Talented Tenth," even before legal emancipation in 1865, when "race" was understood as literally an issue of "blood," the binary symbolic system of "black" and "white" oversimplified a far more nuanced scheme of racial classification. In the past two centuries, we have witnessed a proliferation of even more multifarious and contradictory systems of meaning. On the one hand, Obama's remarkable journey parallels that of Du Bois himself; both are the crème de la crème of the "Talented Tenth" of their respective generations. In Du Bois's day, the divide was between "good" (subservient, often mulatto) "house" blacks who "knew their place," and ("uppity") "city blacks" or (unrefined) "field" blacks who were deemed arrogant or ignorant, respectively. Today, the divide between "good" (upper-middle class, upstanding) citizens who are black—for example, Obama, as well as Oprah Winfrey, Bill Cosby, Tiger Woods, and Condoleezza Rice—and "bad" blacks (inner-city thugs and gangsters) remains—although African Americans are vulnerable to either symbolic classificatory scheme. (Thus, for instance, Obama was charged with being "uppity" and "white-washed" because of his degree from Harvard.) The media continuously juxtaposes regal images of Barack and Michelle Obama with threatening images of inner-city African American thugs and gangsters (e.g., Rodney King and Willie Horton). But even this dual imagery is more complex than this narrative lets on: "gangstas" might be glorified or vilified, epitomizing celebrity idols or abject poverty and bad behavior. And of course even the most esteemed blacks can instantaneously spiral downward (like O.J. Simpson—see Edles 2002:121–27).

This brings us to the concept of double consciousness. Interestingly, in some ways Du Bois's notion of double consciousness parallels Simmel's discussion of the stranger discussed in the previous chapter. Both theorists emphasized the sense of otherness that not only inhibits social solidarity, but also prevents the formation of a unified sense of self (Calhoun et al. 2002:237,238). Moreover, both Simmel and Du Bois used the metaphor of a veil to describe the social distance between people. Of course, the theoretical power of these terms is that they can be used to discuss an array of social groups—from the disabled, to the transgendered, to the "undocumented" populations. Today these concepts are especially familiar to feminists. In the 1970s, Dorothy Smith coined the term "bifurcation of consciousness" to refer to how women are forced to live a double existence as they seek to reconcile the male-dominated, abstract "governing consciousness" with their own practical consciousness rooted in the everyday world.

While up until the 1970s the reality of double consciousness was submerged in the context of white male hegemony, today it is more transparent, and the advantages of it are more clear. The same double consciousness that enabled Du Bois to banter with both Harvard intellectuals and Philadelphians from the Seventh Ward—while being oblivious to whites— is overt in the person of Barack Obama. Obama's dual Kansan and Kenyan background is considered an asset rather than scandalous or exotic. From this point of view, Obama's phenomenal success reflects not only his personal charisma and intelligence, but also his immersion in and ability to capitalize on multiple symbolic worlds. Working-class tropes are prevalent today as well. White, upper-class intellectual elites and "bluebloods" are often considered suspect and "out of touch" (this was the symbolic problem that plagued both Al Gore and John Kerry in their presidential bids). Instead, it is the "realness" of "Joe the Plumber" that commands attention and offers credibility. To be sure, such shifts in meanings do not necessarily directly reflect or impact institutionalized systems of social stratification. The point is that multiculturalism is our new reality, a critical component of our globalized and hypermediated world.

From *The Souls of Black Folk* (1903)

W. E. B. Du Bois

THE FORETHOUGHT

Herein lie buried many things which if read with patience may show the strange meaning of being black here in the dawning of the Twentieth Century. This meaning is not without interest to you, Gentle Reader; for the problem of the Twentieth Century is the problem of the color-line.

I pray you, then, receive my little book in all charity, studying my words with me, forgiving mistake and foible for sake of the faith and passion that is in me, and seeking the grain of truth hidden there.

I have sought here to sketch, in vague, uncertain outline, the spiritual world in which ten thousand thousand Americans live and strive. First, in two chapters I have tried to show what Emancipation meant to them, and what was its aftermath.

In a third chapter I have pointed out the slow rise of personal leadership, and criticised candidly the leader who bears the chief burden of his race today. Then, in two other chapters I have sketched in swift outline the two worlds within and without the Veil, and thus have come to the central problem of training men for life. Venturing now into deeper detail, I have in two chapters studied the struggles of the massed millions of the black peasantry, and in another have sought to make clear the present relations of the sons of master and man.

Leaving, then, the world of the white man, I have stepped within the Veil, raising it that you may view faintly its deeper recesses,—the meaning of its religion, the passion of its human sorrow, and the struggle of its greater souls. All this I have ended with a tale twice told but seldom written. . . .

Before each chapter, as now printed, stands a bar of the Sorrow Songs,—some echo of haunting melody from the only American music which welled up from black souls in the dark past. And, finally, need I add that I who speak here am bone of the bone and flesh of the flesh of them that live within the Veil?

OF OUR SPIRITUAL STRIVINGS

Between me and the other world there is ever an unasked question: unasked by some through feelings of delicacy; by others through the difficulty of rightly framing it. All, nevertheless, flutter round it. They approach me in a half-hesitant sort of way, eye me curiously or compassionately, and then, instead of saying directly, How does it feel to be a problem? they say, I know an excellent colored man in my town; or, I fought at Mechanicsville; or, Do not these Southern outrages make your blood boil? At these I smile, or am interested, or reduce the boiling to a simmer, as the occasion may require. To the real question, How does it feel to be a problem? I answer seldom a word.

And yet, being a problem is a strange experience,—peculiar even for one who has never been anything else, save perhaps in babyhood and in Europe. It is in the early days of rollicking boyhood that the revelation first bursts upon one, all in a day, as it were. I remember well when the shadow swept across me. I was a little thing, away up in the hills of New England, where the dark Housatonic winds between Hoosac and Taghkanic to the sea. In a wee wooden schoolhouse, something put it into the boys' and girls' heads to buy gorgeous visiting-cards—ten cents a package—and exchange. The exchange was merry, till one girl, a tall newcomer, refused my card,—refused it peremptorily, with a glance. Then it dawned upon me with a certain suddenness that I was different from the others; or like, mayhap, in heart and life and longing, but shut out from their world by a vast veil. I had thereafter no desire to tear down that veil, to creep through; I held all beyond it in common contempt, and lived above it in a region of blue sky and great wandering shadows. That sky was bluest when I could beat my mates at examination-time, or beat them at a foot-race, or even beat their stringy heads. Alas, with the years all this fine contempt began to fade; for the worlds I longed for, and all their dazzling opportunities, were theirs, not mine. But they should not keep these prizes, I said; some, all, I would wrest from them. Just how I would do it I could never decide: by reading law, by healing the sick, by telling the wonderful tales that swam in my head,—some way. With other black boys the strife was not so fiercely sunny: their youth shrunk into tasteless sycophancy, or into silent hatred of the pale world about them and mocking distrust of everything white; or wasted itself in a bitter cry, Why did God make me an outcast and a stranger in mine own house? The shades of the prison-house closed round about us all: walls strait and stubborn to the whitest, but relentlessly narrow, tall, and unscalable to sons of night who must plod darkly on in resignation, or beat unavailing palms against the stone, or steadily, half hopelessly, watch the streak of blue above.

After the Egyptian and Indian, the Greek and Roman, the Teuton and Mongolian, the Negro is a sort of seventh son, born with a veil, and gifted with second-sight in this American world,—a world which yields him no true self-consciousness, but only lets him see himself through the revelation of the other world. It is a peculiar sensation, this double-consciousness, this sense of always looking at one's self through the eyes of others, of measuring one's soul by the tape of a world that looks on in amused contempt and pity. One ever feels his two-ness,—an American, a Negro; two souls, two thoughts, two unreconciled strivings; two warring ideals in one dark body, whose dogged strength alone keeps it from being torn asunder.

The history of the American Negro is the history of this strife—this longing to attain self-conscious manhood, to merge his double self into a better and truer self. In this merging he wishes neither of the older selves to be lost. He would not Africanize America, for America has too much to teach the world and Africa. He would not bleach his Negro soul in a flood of white Americanism, for he knows that Negro blood has a message for the world. He simply wishes to make it possible for a man to be both a Negro and an American, without being cursed and spit upon by his fellows, without having the doors of Opportunity closed roughly in his face.

This, then, is the end of his striving: to be a co-worker in the kingdom of culture, to escape both death and isolation, to husband and use his best powers and his latent genius. These powers of body and mind have in the past been strangely wasted, dispersed, or forgotten. The shadow of a mighty Negro past flits through the tale of Ethiopia the Shadowy and of Egypt the Sphinx. Throughout history, the powers of single black men flash here and there like falling stars, and die sometimes before the world has rightly gauged their brightness. Here in America, in the few days since Emancipation, the black man's turning hither and thither in hesitant and doubtful striving has often made his very strength to lose effectiveness, to seem like absence of power, like weakness. And yet it is not weakness,—it is the contradiction of double aims. The double-aimed struggle of the black artisan—on the one hand to escape white contempt for a nation of mere hewers of wood and drawers of water, and on the other hand to plough and nail and dig for a poverty-stricken horde—could only result in making him a poor craftsman, for he had but half a heart in either cause. By the poverty and ignorance of his people, the Negro minister or doctor was tempted toward quackery and demagogy; and by the criticism of the other world, toward ideals that made him ashamed of his lowly tasks. The would-be black *savant* was confronted by the paradox that the knowledge his people needed was a twice-told tale to his white neighbors, while the knowledge which would teach the white world was Greek to his own flesh and blood. The innate love of harmony and beauty that set the ruder souls of his people a-dancing and a-singing raised but confusion and doubt in the soul of the black artist; for the beauty revealed to him was the soul-beauty of a race which his larger audience despised, and he could not articulate the message of another people. This waste of double aims, this seeking to satisfy two unreconciled ideals, has wrought sad havoc with the courage and faith and deeds of ten thousand thousand people,—has sent them often wooing false gods and invoking false means of salvation, and at times has even seemed about to make them ashamed of themselves.

Away back in the days of bondage they thought to see in one divine event the end of all doubt and disappointment; few men ever worshipped Freedom with half such unquestioning faith as did the American Negro for two centuries. To him, so far as he thought and dreamed, slavery was indeed the sum of all villainies, the cause of all sorrow, the root of all prejudice; Emancipation was the key to a promised land of sweeter beauty than ever stretched before the eyes of wearied Israelites. In song and exhortation swelled one refrain—Liberty; in his tears and curses the God he implored had Freedom in his right hand. At last it came,—suddenly, fearfully, like a dream. With one wild carnival of blood and passion came the message in his own plaintive cadences:—

> "Shout, O children!
> Shout, you're free!
> For God has bought your liberty!"

Years have passed away since then,—ten, twenty, forty; forty years of national life, forty years of renewal and development, and yet the swarthy spectre sits in its accustomed seat at the Nation's feast. In vain do we cry to this our vastest social problem:—

> "Take any shape but that, and my firm nerves
> Shall never tremble!"

The Nation has not yet found peace from its sins; the freedman has not yet found in freedom his promised land. Whatever of good may have come in these years of change, the shadow of a deep disappointment rests upon the Negro people,—a disappointment all the more bitter because the unattained ideal was unbounded save by the simple ignorance of a lowly people.

The first decade was merely a prolongation of the vain search for freedom, the boon that seemed ever barely to elude their grasp,—like a tantalizing will-o'-the-wisp, maddening and misleading the headless host. The holocaust of war, the terrors of the Ku-Klux Klan, the lies of carpet- baggers, the disorganization of industry, and the contradictory advice of friends and foes, left the bewildered serf with no new watch-word beyond the old cry for freedom. As the time flew, however, he began to grasp a new idea. The ideal of liberty demanded for its attainment powerful means, and these the Fifteenth Amendment gave him. The ballot, which before he had looked upon as a visible sign of freedom, he now regarded as the chief means of gaining and perfecting the liberty with which war had partially endowed him. And why not? Had not votes made war and emancipated millions? Had not votes enfranchised the freedmen? Was anything impossible to a power that had done all this? A million black men started with renewed zeal to vote themselves into the kingdom. So the decade flew away, the revolution of 1876 came, and left the half-free serf weary, wondering, but still inspired. Slowly but steadily, in the following years, a new vision began gradually to replace the dream of political power,—a powerful movement, the rise of another ideal to guide the unguided, another pillar of fire by night after a clouded day. It was the ideal of "book-learning"; the curiosity, born of compulsory ignorance, to know and test the power of the cabalistic letters of the white man, the longing to know. Here at last seemed to have been discovered the mountain path to Canaan; longer than the highway of Emancipation and law, steep and rugged, but straight, leading to heights high enough to overlook life.

Up the new path the advance guard toiled, slowly, heavily, doggedly; only those who have watched and guided the faltering feet, the misty minds, the dull understandings, of the dark pupils of these schools know how faithfully, how piteously, this people strove to learn. It was weary work. The cold statistician wrote down the inches of progress here and there, noted also where here and there a foot had slipped or some one had fallen. To the tired climbers, the horizon was ever dark, the mists were often cold, the Canaan was always dim and far away. If, however, the vistas disclosed as yet no goal, no resting-place, little but flattery and criticism, the journey at least gave

leisure for reflection and self-examination; it changed the child of Emancipation to the youth with dawning self-consciousness, self-realization, self-respect. In those sombre forests of his striving his own soul rose before him, and he saw himself,—darkly as through a veil; and yet he saw in himself some faint revelation of his power, of his mission. He began to have a dim feeling that, to attain his place in the world, he must be himself, and not another. For the first time he sought to analyze the burden he bore upon his back, that dead-weight of social degradation partially masked behind a half-named Negro problem. He felt his poverty; without a cent, without a home, without land, tools, or savings, he had entered into competition with rich, landed, skilled neighbors. To be a poor man is hard, but to be a poor race in a land of dollars is the very bottom of hardships. He felt the weight of his ignorance,—not simply of letters, but of life, of business, of the humanities; the accumulated sloth and shirking and awkwardness of decades and centuries shackled his hands and feet. Nor was his burden all poverty and ignorance. The red stain of bastardy, which two centuries of systematic legal defilement of Negro women had stamped upon his race, meant not only the loss of ancient African chastity, but also the hereditary weight of a mass of corruption from white adulterers, threatening almost the obliteration of the Negro home.

A people thus handicapped ought not to be asked to race with the world, but rather allowed to give all its time and thought to its own social problems. But alas! while sociologists gleefully count his bastards and his prostitutes, the very soul of the toiling, sweating black man is darkened by the shadow of a vast despair. Men call the shadow prejudice, and learnedly explain it as the natural defence of culture against barbarism, learning against ignorance, purity against crime, the "higher" against the "lower" races. To which the Negro cries Amen! and swears that to so much of this strange prejudice as is founded on just homage to civilization, culture, righteousness, and progress, he humbly bows and meekly does obeisance. But before that nameless prejudice that leaps beyond all this he stands helpless, dismayed, and well-nigh speechless; before that personal disrespect and mockery, the ridicule and systematic humiliation, the distortion of fact and wanton license of fancy, the cynical ignoring of the better and the boisterous welcoming of the worse, the all-pervading desire to inculcate disdain for everything black, from Toussaint to the devil,—before this there rises a sickening despair that would disarm and discourage any nation save that black host to whom "discouragement" is an unwritten word.

But the facing of so vast a prejudice could not but bring the inevitable self-questioning, self-disparagement, and lowering of ideals which ever accompany repression and breed in an atmosphere of contempt and hate. Whisperings and portents came borne upon the four winds: Lo! we are diseased and dying, cried the dark hosts; we cannot write, our voting is vain; what need of education, since we must always cook and serve? And the Nation echoed and enforced this self-criticism, saying: Be content to be servants, and nothing more; what need of higher culture for half-men? Away with the black man's ballot, by force or fraud,—and behold the suicide of a race! Nevertheless, out of the evil came something of good,—the more careful adjustment of education to real life, the clearer perception of the Negroes' social responsibilities, and the sobering realization of the meaning of progress.

So dawned the time of *Sturm und Drang:* storm and stress today rocks our little boat on the mad waters of the world-sea; there is within and without the sound of conflict, the burning of body and rending of soul; inspiration strives with doubt, and faith with vain questionings. The bright ideals of the past,—physical freedom, political power, the training of brains and the training of hands,—all these in turn have waxed and waned, until even the last grows dim and overcast. Are they all wrong,—all false? No, not that, but each alone was oversimple and incomplete,—the dreams of a credulous race-childhood, or the fond imaginings of the other world which does not know and does not want to know our power. To be really true, all these ideals must be melted and welded into one. The training of the schools we need to-day more than ever,—the training of deft hands, quick eyes and ears, and above all the broader,

deeper, higher culture of gifted minds and pure hearts. The power of the ballot we need in sheer self-defence,—else what shall save us from a second slavery? Freedom, too, the long-sought, we still seek,—the freedom of life and limb, the freedom to work and think, the freedom to love and aspire. Work, culture, liberty,—all these we need, not singly but together, not successively but together, each growing and aiding each, and all striving toward that vaster ideal that swims before the Negro people, the ideal of human brotherhood, gained through the unifying ideal of Race; the ideal of fostering and developing the traits and talents of the Negro, not in opposition to or contempt for other races, but rather in large conformity to the greater ideals of the American Republic, in order that some day on American soil two world-races may give each to each those characteristics both so sadly lack. We the darker ones come even now not altogether empty-handed: there are to-day no truer exponents of the pure human spirit of the Declaration of Independence than the American Negroes; there is no true American music but the wild sweet melodies of the Negro slave; the American fairy tales and folk-lore are Indian and African; and, all in all, we black men seem the sole oasis of simple faith and reverence in a dusty desert of dollars and smartness. Will America be poorer if she replace her brutal dyspeptic blundering with light-hearted but determined Negro humility? or her coarse and cruel wit with loving jovial good-humor? or her vulgar music with the soul of the Sorrow Songs?

Merely a concrete test of the underlying principles of the great republic is the Negro Problem, and the spiritual striving of the freedmen's sons is the travail of souls whose burden is almost beyond the measure of their strength, but who bear it in the name of an historic race, in the name of this the land of their fathers' fathers, and in the name of human opportunity.

And now what I have briefly sketched in large outline let me on coming pages tell again in many ways, with loving emphasis and deeper detail, that men may listen to the striving in the souls of black folk.

OF THE DAWN OF FREEDOM

The problem of the twentieth century is the problem of the color-line,—the relation of the darker to the lighter races of men in Asia and Africa, in America and the islands of the sea. It was a phase of this problem that caused the Civil War; and however much they who marched South and North in 1861 may have fixed on the technical points of union and local autonomy as a shibboleth, all nevertheless knew, as we know, that the question of Negro slavery was the real cause of the conflict. Curious it was, too, how this deeper question ever forced itself to the surface despite effort and disclaimer. No sooner had Northern armies touched Southern soil than this old question, newly guised, sprang from the earth,—What shall be done with Negroes? Peremptory military commands, this way and that, could not answer the query; the Emancipation Proclamation seemed but to broaden and intensify the difficulties; and the War Amendments made the Negro problems of to-day. . . .

The passing of a great human institution before its work is done, like the untimely passing of a single soul, but leaves a legacy of striving for other men. The legacy of the Freedmen's Bureau is the heavy heritage of this generation. To-day, when new and vaster problems are destined to strain every fibre of the national mind and soul, would it not be well to count this legacy honestly and carefully? For this much all men know: despite compromise, war, and struggle, the Negro is not free. In the backwoods of the Gulf States, for miles and miles, he may not leave the plantation of his birth; in well-nigh the whole rural South the black farmers are peons, bound by law and custom to an economic slavery, from which the only escape is death or the penitentiary. In the most cultured sections and cities of the South the Negroes are a segregated servile caste, with restricted rights and privileges. Before the courts, both in law and custom, they stand on a different and peculiar basis. Taxation without representation is the rule of their political life. And the result of all this, and in nature must have been, lawlessness and crime. That is the large legacy of the Freedmen's Bureau, the work it did not do because it could not.

I have seen a land right merry with the sun, where children sing, and rolling hills lie like passioned women wanton with harvest. And there in the King's Highway sat and sits a figure veiled and bowed, by which the traveller's footsteps hasten as they go. On the tainted air broods fear. Three centuries' thought has been the raising and unveiling of that bowed human heart, and now behold a century new for the duty and the deed. The problem of the Twentieth Century is the problem of the color-line.

OF MR. BOOKER T. WASHINGTON AND OTHERS

Easily the most striking thing in the history of the American Negro since 1876 is the ascendancy of Mr. Booker T. Washington. It began at the time when war memories and ideals were rapidly passing; a day of astonishing commercial development was dawning; a sense of doubt and hesitation overtook the freedmen's sons,—then it was that his leading began. Mr. Washington came, with a simple definite programme, at the psychological moment when the nation was a little ashamed of having bestowed so much sentiment on Negroes, and was concentrating its energies on Dollars. His programme of industrial education, conciliation of the South, and submission and silence as to civil and political rights, was not wholly original; the Free Negroes from 1830 up to wartime had striven to build industrial schools, and the American Missionary Association had from the first taught various trades; and Price and others had sought a way of honorable alliance with the best of the Southerners. But Mr. Washington first indissolubly linked these things; he put enthusiasm, unlimited energy, and perfect faith intro this programme, and changed it from a by-path into a veritable Way of Life. And the ale of the methods by which he did this is a fascinating study of human life.

It startled the nation to hear a Negro advocating such a programme after many decades of bitter complaint; it startled and won the applause of the South, it interested and won the admiration of the North; and after a confused murmur of protest, it silenced if it did not convert the Negroes themselves.

To gain the sympathy and cooperation of the various elements comprising the white South was Mr. Washington's first task; and this, at the time Tuskegee was founded, seemed, for a black man, well-nigh impossible. And yet ten years later it was done in the word spoken at Atlanta: "In all things purely social we can be as separate as the five fingers, and yet one as the hand in all things essential to mutual progress." This "Atlanta Compromise" is by all odds the most notable thing in Mr. Washington's career. The South interpreted it in different ways: The radicals received it as a complete surrender of the demand for civil and political equality; the conservatives, as a generously conceived working basis for mutual understanding. So both approved it, and to-day its author is certainly the most distinguished Southerner since Jefferson Davis, and the one with the largest personal following.

Next to this achievement comes Mr. Washington's work in gaining place and consideration in the North. Others less shrewd and tactful had formerly essayed to sit on these two stools and had fallen between them; but as Mr. Washington knew the heart of the South from birth and training, so by singular insight he intuitively grasped the spirit of the age which was dominating the North. And so thoroughly did he learn the speech and thought of triumphant commercialism, and the ideals of material prosperity, that the picture of a lone black boy poring over a French grammar amid the weeds and dirt of a neglected home soon seemed to him the acme of absurdities. One wonders what Socrates and St. Francis of Assisi would say to this.

And yet this very singleness of vision and thorough oneness with his age is a mark of the successful man. It is as though Nature must needs make men narrow in order to give them force. So Mr. Washington's cult has gained unquestioning followers, his work has wonderfully prospered, his friends are legion, and his enemies are confounded. To-day he stands as the one recognized spokesman of his ten million fellows, and one of the most notable figures in a nation of seventy millions. One hesitates, therefore, to criticise a life which, beginning with so little, has done so much. And yet the time is come when one may speak in all sincerity and utter courtesy of the mistakes and shortcomings of Mr. Washington's career, as well as of his triumphs, without being thought captious or envious, and without forgetting that it is easier to do ill than well in the world.

The criticism that has hitherto met Mr. Washington has not always been of this broad character. In the South especially has he had to walk warily to avoid the harshest judgments,—and naturally so, for he is dealing with the one subject of deepest sensitiveness to that section. Twice—once when at the Chicago celebration of the Spanish-American War he alluded to the color-prejudice that is "eating away the vitals of the South," and once when he dined with President Roosevelt—has the resulting Southern criticism been violent enough to threaten seriously his popularity. In the North the feeling has several times forced itself into words, that Mr. Washington's counsels of submission overlooked certain elements of true manhood, and that his educational programme was unnecessarily narrow. Usually, however, such criticism has not found open expression, although, too, the spiritual sons of the Abolitionists have not been prepared to acknowledge that the schools founded before Tuskegee, by men of broad ideals and self-sacrificing spirit, were wholly failures or worthy of ridicule. While, then, criticism has not failed to follow Mr. Washington, yet the prevailing public opinion of the land has been but too willing to deliver the solution of a wearisome problem into his hands, and say, "If that is all you and your race ask, take it."

Among his own people, however, Mr. Washington has encountered the strongest and most lasting opposition, amounting at times to bitterness, and even to-day continuing strong and insistent even though largely silenced in outward expression by the public opinion of the nation. Some of this opposition is, of course, mere envy; the disappointment of displaced demagogues and the spite of narrow minds. But aside from this, there is among educated and thoughtful colored men in all parts of the land a feeling of deep regret, sorrow, and apprehension at the wide currency and ascendancy which some of Mr. Washington's theories have gained. These same men admire his sincerity of purpose, and are willing to forgive much to honest endeavor which is doing something worth the doing. They coöperate with Mr. Washington as far as they conscientiously can; and, indeed, it is no ordinary tribute to this man's tact and power that, steering as he must between so many diverse interests and opinions, he so largely retains the respect of all. . . .

Mr. Washington represents in Negro thought the old attitude of adjustment and submission; but adjustment at such a peculiar time as to make his programme unique. This is an age of unusual economic development, and Mr. Washington's programme naturally takes an economic cast, becoming a gospel of Work and Money to such an extent as apparently almost completely to overshadow the higher aims of life. Moreover, this is an age when the more advanced races are coming in closer contact with the less developed races, and the race-feeling is therefore intensified; and Mr. Washington's programme practically accepts the alleged inferiority of the Negro races. Again, in our own land, the reaction from the sentiment of war time has given impetus to race-prejudice against Negroes, and Mr. Washington withdraws many of the high demands of Negroes as men and American citizens. In other periods of intensified prejudice all the Negro's tendency to self-assertion has been called forth; at this period a policy of submission is advocated. In the history of nearly all other races and peoples the doctrine preached at such crises has been that manly self-respect is worth more than lands and houses, and that a people who voluntarily surrender such respect, or cease striving for it, are not worth civilizing.

In answer to this, it has been claimed that the Negro can survive only through submission. Mr. Washington distinctly asks that black people give up, at least for the present, three things,—First, political power,

Second, insistence on civil rights,

Third, higher education of Negro youth,—

And concentrate all their energies on industrial education, the accumulation of wealth, and the conciliation of the South. This policy has been courageously and insistently advocated for over fifteen years, and has been triumphant for perhaps ten years. As a result of this tender of the palm-branch, what has been the return? In these years there have occurred:

1. The disfranchisement of the Negro.

2. The legal creation of a distinct status of civil inferiority for the Negro.

3. The steady withdrawal of aid from institutions for the higher training of the Negro.

These movements are not, to be sure, direct results of Mr. Washington's teachings; but his propaganda has, without a shadow of doubt, helped their speedier accomplishment. The question then comes: Is it possible, and probable, that nine millions of men can make effective progress in economic lines if they are deprived of political rights, made a servile caste, and allowed only the most meagre chance for developing their exceptional men? If history and reason give any distinct answer to these questions, it is an emphatic *No*. And Mr. Washington thus faces the triple paradox of his career:

1. He is striving nobly to make Negro artisans business men and property-owners; but it is utterly impossible, under modern competitive methods, for workingmen and property-owners to defend their rights and exist without the right of suffrage.

2. He insists on thrift and self-respect, but at the same time counsels a silent submission to civic inferiority such as is bound to sap the manhood of any race in the long run.

3. He advocates common-school and industrial training, and depreciates institutions of higher learning; but neither the Negro common-schools, nor Tuskegee itself, could remain open a day were it not for teachers trained in Negro colleges, or trained by their graduates. . . .

To-day even the attitude of the Southern whites toward the blacks is not, as so many assume, in all cases the same; the ignorant Southerner hates the Negro, the workingmen fear his competition, the money-makers wish to use him as a laborer, some of the educated see a menace in his upward development, while others—usually the sons of the masters—wish to help him to rise. National opinion has enabled this last class to maintain the Negro common schools, and to protect the Negro partially in property, life, and limb. Through the pressure of the money-makers, the Negro is in danger of being reduced to semi-slavery, especially in the country districts; the workingmen, and those of the educated who fear the Negro, have united to disfranchise him, and some have urged his deportation; while the passions of the ignorant are easily aroused to lynch and abuse any black man . . .

The South ought to be led, by candid and honest criticism, to assert her better self and do her full duty to the race she has cruelly wronged and is still wronging. The North— her co-partner in guilt—cannot salve her conscience by plastering it with gold. We cannot settle this problem by diplomacy and suaveness, by "policy" alone. If worse comes to worst, can the moral fibre of this country survive the slow throttling and murder of nine millions of men?

The black men of America have a duty to perform, a duty stern and delicate,—a forward movement to oppose a part of the work of their greatest leader. So far as Mr. Washington preaches Thrift, Patience, and Industrial Training for the masses, we must hold up his hands and strive with him, rejoicing in his honors and glorying in the strength of this Joshua called of God and of man to lead the headless host. But so far as Mr. Washington apologizes for injustice, North or South, does not rightly value the prvilege and duty of voting, belittles the emasculating effects of caste distinctions, and opposes the higher training and ambition of our brighter minds,—so far as he, the South, or the Nation, does this,—we must unceasingly and firmly oppose them. By every civilized and peaceful method we must strive for the rights which the world accords to men, clinging unwaveringly to those great words which the sons of the Fathers would fain forget: "We hold these truths to be self-evident: That all men are created equal; that they are endowed by their Creator with certain unalienable rights; that among these are life, liberty, and the pursuit of happiness."

Of the Faith of the Fathers

It was out in the country, far from home, far from my foster home, on a dark Sunday night. The road wandered from our rambling log-house up the stony bed of a creek, past wheat and corn, until we could hear dimly across the fields a rhythmic cadence of song,—soft, thrilling, powerful, that swelled and died sorrowfully in our ears. I was a country schoolteacher then, fresh from the East, and had never seen a Southern Negro revival. To be sure, we in Berkshire were not perhaps as stiff and formal as they in Suffolk of olden time; yet we were very quiet and subdued, and I know not what would have happened those clear Sabbath mornings had some one punctuated the sermon with a wild scream, or interrupted the long prayer with a loud Amen! And so most striking to me, as I approached the village and the little plain church perched aloft, was the air of intense excitement that possessed that mass of black folk. A sort of suppressed terror hung in the air and seemed to seize us,—a pythian madness, a demoniac possession, that lent terrible reality to song and word. The black and massive form of the preacher swayed and quivered as the words crowded to his lips and flew at us in singular eloquence. The people moaned and fluttered, and then the gaunt-cheeked brown woman beside me suddenly leaped straight into the air and shrieked like a lost soul, while round about came wail and groan and outcry, and a scene of human passion such as I had never conceived before.

Those who have not thus witnessed the frenzy of a Negro revival in the untouched backwoods of the South can but dimly realize the religious feeling of the slave; as described, such scenes appear grotesque and funny, but as seen they are awful. Three things characterized this religion of the slave,—the Preacher, the Music, and the Frenzy. The preacher is the most unique personality developed by the Negro on American soil. A leader, a politician, an orator, a "boss," an intriguer, an idealist,— all these he is, and ever, too, the centre of a group of men, now twenty, now a thousand in number. The combination of a certain adroitness with deep-seated earnestness, of tact with consummate ability, gave him his preeminence, and helps him maintain it. The type, of course, varies according to time and place, from the West Indies in the sixteenth century to New England in the nineteenth, and from the Mississippi bottoms to cities like New Orleans or New York.

The Music of Negro religion is that plaintive rhythmic melody, with its touching minor cadences, which, despite caricature and defilement, still remains the most original and beautiful expression of human life and longing yet born on American soil. Sprung from the African forests, where its counterpart can still be heard, it was adapted, changed, and intensified by the tragic soul-life of the slave, until, under the stress of law and whip, it became the one true expression of a people's sorrow, despair, and hope.

Finally the Frenzy of "Shouting," when the Spirit of the Lord passed by, and, seizing the devotee, made him mad with supernatural joy, was the last essential of Negro religion and the one more devoutly believed in than all the rest. It varied in expression from the silent rapt countenance or the low murmur and moan to the mad abandon of physical fervor,—the stamping, shrieking, and shouting, the rushing to and fro and wild waving of arms, the weeping and laughing, the vision and the trance. All this is nothing new in the world, but old as religion, as Delphi and Endor. And so firm a hold did it have on the Negro, that many generations firmly believed that without this visible manifestation of the God there could be no true communion with the Invisible.

These were the characteristics of Negro religious life as developed up to the time of Emancipation. Since under the peculiar circumstances of the black man's environment they were the one expression of his higher life, they are of deep interest to the student of his development, both socially and psychologically. Numerous are the attractive lines of inquiry that here group themselves. What did slavery mean to the African savage? What was his attitude toward the World and Life? What seemed to him good and evil,—God and Devil? Whither went his longings and strivings, and wherefore were his heart-burnings and disappointments? Answers to such questions can come only from

a study of Negro religion as a development, through its gradual changes from the heathenism of the Gold Coast to the institutional Negro church of Chicago.

Moreover, the religious growth of millions of men, even though they be slaves, cannot be without potent influence upon their contemporaries. The Methodists and Baptists of America owe much of their condition to the silent but potent influence of their millions of Negro converts. Especially is this noticeable in the South, where theology and religious philosophy are on this account a long way behind the North, and where the religion of the poor whites is a plain copy of Negro thought and methods. The mass of "gospel" hymns which has swept through American churches and well-nigh ruined our sense of song consists largely of debased imitations of Negro melodies made by ears that caught the jingle but not the music, the body but not the soul, of the Jubilee songs. It is thus clear that the study of Negro religion is not only a vital part of the history of the Negro in America, but no uninteresting part of American history.

The Negro church of to-day is the social centre of Negro life in the United States, and the most characteristic expression of African character. Take a typical church in a small Virginia town: it is the "First Baptist"—a roomy brick edifice seating five hundred or more persons, tastefully finished in Georgia pine, with a carpet, a small organ, and stained-glass windows. Underneath is a large assembly room with benches. This building is the central club-house of a community of a thousand or more Negroes. Various organizations meet here,—the church proper, the Sunday-school, two or three insurance societies, women's societies, secret societies, and mass meetings of various kinds. Entertainments, suppers, and lectures are held beside the five or six regular weekly religious services. Considerable sums of money are collected and expended here, employment is found for the idle, strangers are introduced, news is disseminated and charity distributed. At the same time this social, intellectual, and economic centre is a religious centre of great power. Depravity, Sin, Redemption, Heaven, Hell, and Damnation are preached twice a Sunday after the crops are laid by; and few indeed of the community have the hardihood to withstand conversion. Back of this more formal religion, the Church often stands as a real conserver of morals, a strengthener of family life, and the final authority on what is Good and Right.

Thus one can see in the Negro church to-day, reproduced in microcosm, all the great world from which the Negro is cut off by color-prejudice and social condition. In the great city churches the same tendency is noticeable and in many respects emphasized. A great church like the Bethel of Philadelphia has over eleven hundred members, an edifice seating fifteen hundred persons and valued at one hundred thousand dollars, an annual budget of five thousand dollars, and a government consisting of a pastor with several assisting local preachers, an executive and legislative board, financial boards and tax collectors; general church meetings for making laws; sub-divided groups led by class leaders, a company of militia, and twenty-four auxiliary societies. The activity of a church like this is immense and far-reaching, and the bishops who preside over these organizations throughout the land are among the most powerful Negro rulers in the world.

Such churches are really governments of men, and consequently a little investigation reveals the curious fact that, in the South, at least, practically every American Negro is a church member. Some, to be sure, are not regularly enrolled, and a few do not habitually attend services; but, practically, a proscribed people must have a social centre and that centre for this people is the Negro church. The census of 1890 showed nearly twenty-four thousand Negro churches in the country, with a total enrolled membership of over two and a half millions, or ten actual church members to every twenty eight persons, and in some Southern states one in every two persons. Besides these there is the large number who, while not enrolled as members, attend and take part in many of the activities of the church. There is an organized Negro church for every sixty black families in the nation, and in some States for every forty families, owning, on an average, a thousand dollars' worth of property each, or nearly twenty-six million dollars in all.

Such, then, is the large development of the Negro church since Emancipation. The question now is, What have been the successive steps of this social history and what are the present tendencies? First, we must realize that no such institution as the Negro church could rear itself without definite historical foundations. These foundations we can find if we remember that the social history of the Negro did not start in America. He was from a definite social environment—the polygamous clan life under the headship of the chief and the potent influence of the priest. His religion was nature-worship, with profound belief in invisible surrounding influences, good and bad, and his worship was through incantation and sacrifice. The first rude change in this life was the slave ship and the West Indian sugar-fields. The plantation organization replaced the clan and tribe, and the white master replaced the chief with far greater and more despotic powers. Forced and long-continued toil became the rule of life, the old ties of blood relationship and kinship disappeared, and instead of the family appeared a new polygamy and polyandry, which, in some cases, almost reached promiscuity. It was a terrific social revolution, and yet some traces were retained of the former group life, and the chief remaining institution was the Priest or Medicine-man. He early appeared on the plantation and found his function as the healer of the sick, the interpreter of the Unknown, the comforter of the sorrowing, the supernatural avenger of wrong, and the one who rudely but picturesquely expressed the longing, disappointment, and resentment of a stolen and oppressed people. Thus, as bard, physician, judge, and priest, within the narrow limits allowed by the slave system, rose the Negro preacher, and under him the first church was not at first by any means Christian nor definitely organized; rather it was an adaptation and mingling of heathen rites among the members of each plantation, and roughly designated as Voodooism. Association with the masters, missionary effort and motive of expediency gave these rites an early veneer of Christianity, and after the lapse of many generations the Negro church became Christian.

Two characteristic things must be noticed in regard to the church. First, it became almost entirely Baptist and Methodist in faith; secondly, as a social institution it antedated by many decades the monogamic Negro home. From the very circumstances of its beginning, the church was confined to the plantation, and consisted primarily of a series of disconnected units; although, later on, some freedom of movement was allowed, still this geographical limitation was always important and was one cause of the spread of the decentralized and democratic Baptist faith among the slaves. At the same time, the visible rite of baptism appealed strongly to their mystic temperament. To-day the Baptist Church is still largest in membership among Negroes, and has a million and a half communicants. Next in popularity came the churches organized in connection with the white neighboring churches, chiefly Baptist and Methodist, with a few Episcopalian and others. The Methodists still form the second greatest denomination, with nearly a million members. The faith of these two leading denominations was more suited to the slave church from the prominence they gave to religious feeling and fervor. The Negro membership in other denominations has always been small and relatively unimportant, although the Episcopalian and Presbyterians are gaining among the more intelligent classes to-day, and the Catholic Church is making headway in certain sections. After Emancipation, and still earlier in the North, the Negro churches largely severed such affiliations as they had had with the white churches, either by choice or by compulsion. The Baptist churches became independent, but the Methodists were compelled early to unite for purposes of episcopal government. This gave rise to the great African Methodist Church, the greatest Negro organization in the world, to the Zion Church and the Colored Methodist, and to the black conferences and churches in this and other denominations.

The second fact noted, namely, that the Negro church antedates the Negro home, leads to an explanation of much that is paradoxical in this communistic institution and in the morals of its members. But especially it leads us to regard this institution as peculiarly the expression of the inner ethical life of a people in a sense seldom true elsewhere. Let us turn, then, from the outer physical development of the church to the more important inner ethical life of the people who compose it. The Negro has already been pointed out many times as a religious animal—a being of that deep emotional nature which turns instinctively toward the supernatural. Endowed with a rich tropical imagination and a keen, delicate appreciation of Nature, the transplanted African lived in a world animate with gods and devils, elves and witches; full of strange influences,—of Good to be implored, of Evil to be propitiated. Slavery, then, was to him the dark triumph of Evil over him. All the hateful powers of the Under-world were striving against him, and a spirit of revolt and revenge filled his heart. He called up all the resources of heathenism to aid,—exorcism and witch-craft, the mysterious Obi worship with its barbarious rites, spells, and blood-sacrifice even, now and then, of human victims. Weird midnight orgies and mystic conjurations were invoked, the witch-woman and the voodoo-priest became the centre of Negro group life, and that vein of vague superstition which characterizes the unlettered Negro even to-day was deepened and strengthened.

In spite, however, of such success as that of the fierce Maroons, the Danish blacks, and others, the spirit of revolt gradually died away under the untiring energy and superior strength of the slave masters. By the middle of the eighteenth century the black slave had sunk, with hushed murmurs, to his place at the bottom of a new economic system, and was unconsciously ripe for a new philosophy of life. Nothing suited his condition then better than the doctrines of passive submission embodied in the new newly learned Christianity. Slave masters early realized this, and cheerfully aided religious propaganda within certain bounds. The long system of repression and degradation of the Negro tended to emphasize the elements of his character which made him a valuable chattel: courtesy became humility, moral strength degenerated into submission, and the exquisite native appreciation of the beautiful became an infinite capacity for dumb suffering. The Negro, losing the joy of this world, eagerly seized upon the offered conceptions of the next; the avenging Spirit of the Lord enjoining patience in this world, under sorrow and tribulation until the Great Day when He should lead His dark children home,—this became his comforting dream. His preacher repeated the prophecy, and his bards sang,—

"Children, we all shall be free
When the Lord shall appear!"

This deep religious fatalism, painted so beautifully in "Uncle Tom," came soon to breed, as all fatalistic faiths will, the sensualist side by side with the martyr. Under the lax moral life of the plantation, where marriage was a farce, laziness a virtue, and property a theft, a religion of resignation and submission degenerated easily, in less strenuous minds, into a philosophy of indulgence and crime. Many of the worst characteristics of the Negro masses of to-day had their seed in this period of the slave's ethical growth. Here it was that the Home was ruined under the very shadow of the Church, white and black; here habits of shiftlessness took root, and sullen hopelessness replaced hopeful strife.

With the beginning of the abolition movement and the gradual growth of a class of free Negroes came a change. We often neglect the influence of the freedman before the war, because of the paucity of his numbers and the small weight he had in the history of the nation. But we must not forget that his chief influence was internal,—was exerted on the black world; and that there he was the ethical and social leader. Huddled as he was in a few centres like Philadelphia, New York, and New Orleans, the masses of the freedmen sank into poverty and listlessness; but not all of them. The free Negro leader early arose and his chief characteristic was intense earnestness and deep feeling on the slavery question. Freedom became to him a real thing and not a dream. His religion became darker and more intense, and into his ethics crept a note of revenge, into his songs a day of reckoning close at hand. The "Coming of the Lord" swept this side of Death, and came to be a thing to be hoped for in this day. Through fugitive slaves and irrepressible discussion this desire for freedom seized the black millions still in bondage, and became their one ideal of life. The black bards caught new notes, and sometimes even dared to sing,—

"O Freedom, O Freedom,
O Freedom over me!
Before I'll be a slave
I'll be buried in my grave,
And go home to my Lord
And be free."

For fifty years Negro religion thus transformed itself and identified itself with the dream of Abolition, until that which was a radical fad in the white North and an anarchistic plot in the white South had become a religion to the black world.

Thus, when Emancipation finally came, it seemed to the freedman a literal Coming of the Lord. His fervid imagination was stirred as never before, by the tramp of armies, the blood and dust of battle, and the wail and whirl of social upheaval. He stood dumb and motionless before the whirlwind: what had he to do with it? Was it not the Lord's doing, and marvellous in his eyes? Joyed and bewildered with what came, he stood awaiting new wonders till the inevitable Age of Reaction swept over the nation and brought the crisis of to-day.

It is difficult to explain clearly the present critical stage of Negro religion. First, we must remember that living as the blacks do in close contact with a great modern nation, and sharing, although imperfectly, the soul-life of that nation, they must necessarily be affected more or less directly by all the religious and ethical forces that are to-day moving the United States. These questions and movements are, however, overshadowed and dwarfed by the (to them) all-important question of their civil, political, and economic status. They must perpetually discuss the "Negro Problem,"—must live, move, and have their being in it, and interpret all else in its light or darkness. With this come, too, peculiar problems of their inner life,—of the status of women, the maintenance of Home, the training of children, the accumulation of wealth, and the prevention of crime. All this must mean a time of intense ethical ferment, of religious heart-searching and intellectual unrest. From the double life every American Negro must live, as a Negro and as an American, as swept on by the current of the nineteenth while yet struggling in the eddies of the fifteenth century,—from this must arise a painful self-consciousness, an almost morbid sense of personality and a moral hesitancy which is fatal to self-confidence. The worlds within and without the Veil of Color are changing, and changing rapidly, but not at the same rate, not in the same way; and this must produce a peculiar wrenching of the soul, a peculiar sense of doubt and bewilderment. Such a double life, with double thoughts, double duties, and double social classes, must give rise to double words and double ideals, and tempt the mind to pretence or revolt, to hypocrisy or radicalism.

In some such doubtful words and phrases can one perhaps most clearly picture the peculiar ethical paradox that faces the Negro of to-day and is tingeing and changing his religious life. Feeling that his rights and his dearest ideals are being trampled upon, that the public conscience is ever more deaf to his righteous appeal, and that all the reactionary forces of prejudice, greed, and revenge are daily gaining new strength and fresh allies, the Negro faces no enviable dilemma. Conscious of his impotence, and pessimistic, he often becomes bitter and vindictive; and his religion, instead of a worship, is a complaint and a curse, a wail rather than a hope, a sneer rather than a faith. On the other hand, another type of mind, shrewder and keener and more tortuous too, sees in the very strength of the anti-Negro movement its patent weaknesses, and with Jesuitic casuistry is deterred by no ethical considerations in the endeavor to turn this weakness to the black man's strength. Thus we have two great and hardly reconcilable streams of thought and ethical strivings; the danger of the one lies in anarchy, that of the other in hypocrisy. The one type of Negro stands almost ready to curse God and die, and the other is too often found a traitor to right and a coward before force; the one is wedded to ideals remote, whimsical, perhaps impossible of realization; the other forgets that life is more than meat and the body more than raiment. But, after all, is not this simply the writhing of the age translated into black, the triumph of the Lie which today, with its false culture, faces the hideousness of the anarchist assassin?

To-day the two groups of Negroes, the one in the North, the other in the South, represent these divergent ethical tendencies, the first tending toward radicalism, the other toward hypocritical compromise. It is no idle regret with which the white South mourns the loss of the old-time Negro,—the frank, honest, simple old servant who stood for the earlier religious age of submission and humility. With all his laziness and lack of many elements of true manhood, he was at least open-hearted, faithful, and sincere. To-day he is gone, but who is to blame for his going? Is it not those very persons who mourn for him? Is it not the tendency, born of Reconstruction and Reaction, to found a society

on lawlessness and deception, to tamper with the moral fibre of a naturally honest and straight-forward people until the whites threaten to become ungovernable tyrants and the blacks criminals and hypocrites? Deception is the natural defence of the weak against the strong, and the South used it for many years against its conquerors; to-day it must be prepared to see its black proletariat turn that same two-edged weapon against itself. And how natural this is! The death of Denmark Vesey and Nat Turner proved long since to the Negro the present hopelessness of physical defence. Political defence is becoming less and less available, and economic defence is still only partially effective. But there is a patent defence at hand,—the defence of deception and flattery, of cajoling and lying. It is the same defence which peasants of the Middle Age used and which left its stamp on their character for centuries. To-day the young Negro of the South who would succeed cannot be frank and outspoken, honest and self-assertive, but rather he is daily tempted to be silent and wary, politic and sly; he must flatter and be pleasant, endure petty insults with a smile, shut his eyes to wrong; in too many cases he sees positive personal advantage in deception and lying. His real thoughts, his real aspirations, must be guarded in whispers; he must not criticise, he must not complain. Patience, humility, and adroitness must, in these growing black youth, replace impulse, manliness, and courage. With this sacrifice there is an economic opening, and perhaps peace and some prosperity. Without this there is riot, migration, or crime. Nor is this situation peculiar to the Southern United States, is it not rather the only method by which undeveloped races have gained the right to share modern culture? The price of culture is a Lie.

On the other hand, in the North the tendency is to emphasize the radicalism of the Negro. Driven from his birthright in the South by a situation at which every fibre of his more outspoken and assertive nature revolts, he finds himself in a land where he can scarcely earn a decent living amid the harsh competition and the color discrimination. At the same time, through schools and periodicals, discussions and lectures, he is intellectually quickened and awakened. The soul, long pent up and dwarfed, suddenly expands in new-found freedom. What wonder that every tendency is to excess, radical complaint, radical remedies, bitter denunciation or angry silence. Some sink, some rise. The criminal and the sensualist leave the church for the gambling-hell and the brothel, and fill the slums of Chicago and Baltimore; the better classes segregate themselves from the group-life of both white and black, and form an aristocracy, cultured but pessimistic, whose bitter criticism stings while it points out no way of escape. They despise the submission and subserviency of the Southern Negroes, but offer no other means by which a poor and oppressed minority can exist side by side with its masters. Feeling deeply and keenly the tendencies and opportunities of the age in which they live, their souls are bitter at the fate which drops the Veil between; and the very fact that this bitterness is natural and justifiable only serves to intensify it and make it more maddening.

Between the two extreme types of ethical attitude which I have thus sought to make clear wavers the mass of the millions of Negroes, North and South; and their religious life and activity partake of this social conflict within their ranks. Their churches are differentiating,—now into groups of cold, fashionable devotees, in no way distinguishable from similar white groups save in color of skin; now into large social and business institutions catering to the desire for information and amusement of their members, warily avoiding unpleasant questions both within and without the black world, and preaching in effect if not in word: *Dum vivimus, vivamus.*

But back of this still broods silently the deep religious feeling of the real Negro heart, the stirring, unguided might of powerful human souls who have lost the guiding star of the past and seek in the great night a new religious ideal. Some day the Awakening will come, when the pent-up vigor of ten million souls shall sweep irresistibly toward the Goal, out of the Valley of the Shadow of Death, where all that makes life worth living—Liberty, Justice, and Right—is marked "For White People Only."

Introduction to "The Souls of White Folk"

Throughout his long life, Du Bois's basic argument remained unchanged. However, after World War I, Du Bois stepped up his demands for emancipation, and his work acquired a more aggressive tone. African Americans coming home from battle in Europe were incensed to find that despite their wartime contributions, they were treated, at best, as second-class citizens. In addition, between 1910 and 1920, black literacy and militancy had been increasing, and a "New Negro" movement was taking hold. Membership in the NAACP nearly doubled between 1918 and 1919, and there were dozens of riots in scores of American cities. In a single riot in Chicago, 15 whites and 23 blacks were killed, and more than 500 were injured (Lewis 2000:8; Rudwick 1960/1982:237).

"The Souls of White Folk" was originally published in 1910, but it was revised with references to World War I for republication in *Darkwater* in 1920. *Darkwater* set off a tremendous public storm, in large part because Du Bois asserted, "the dark was preparing to meet the white world in battle" (Rudwick 1960/1982:242). Of course, Du Bois had already been stirring up controversy with his provocative editorials in the *Crisis*. For instance, in the May 1919 issue of that journal (which sold approximately 100,000 copies), Du Bois maintained,

> By the God of Heaven, we are cowards and jackasses if now that the war is over, we do not marshal every ounce of our brain and brawn to fight a sterner, longer, more unbending battle against the forces of hell in our own land. (Du Bois, quoted in Rudwick 1960/1982:238)

Sadly (albeit predictably), many whites reacted hysterically to the "New Negro" movement. In 1919, 77 African Americans were lynched in various parts of the country (ibid.:237). Racial tensions completely exploded in rural Arkansas in October 1919 after gunfire from black sharecroppers who had been meeting in a church left a white deputy sheriff dead and several white citizens wounded. In response, white planters and farmers instigated a seven-day frenzy of violence, during which time nearly 200 people were killed. Within two days of the bloodbath, more than 1,000 African Americans were rounded up; 12 were hanged for "conspiracy to seize control of the county," and 67 received prison terms of 5 to 25 years. Meanwhile, because of the five white deaths, the whites' actions were deemed legal (Lewis 2000:8).

Thus, although you will see similarities between "The Souls of White Folk" and *The Souls of Black Folk* (most obviously, both explore the dual workings of race and class), "The Souls of White Folk" reads far more militantly. Another notable feature of this essay is that Du Bois reverses the gaze of racial domination. Here it is *white* consciousness (or the lack thereof) that is being explored. Indeed, "The Souls of White Folk" has been called "the first major analysis in Western intellectual history to probe deeply White identity and the meaning of Whiteness" (Feagin 2003:11).

There are several important points in this essay worth highlighting. First, Du Bois suggests that white privilege, ironically, is invisible to whites. Indeed, Du Bois contends that while African Americans have a "double consciousness," whites have no racial consciousness at all. In other words, just as the fish in the fishbowl is the last to know that it lives in water, the white man does not realize what it is, and what it entails to be white. However, Du Bois suggests that, unbeknownst to whites, African Americans *can* see what it means to be white. Blacks' "clairvoyance" comes from their servile position. As servants in one form or another, blacks are exposed to the intimate details of whites' lives, hence they see whites as they really are.

Most important, what blacks see is that whites typically practice the very opposite of what they preach. Du Bois condemns whites not only for their hypocrisy, but also for their delusion. Thus, Du Bois declares white Christianity "a miserable failure" because the number of whites who actually practice "the democracy and unselfishness of Jesus Christ" is so small as to be farcical. So too, Du Bois finds the United States' call to make "the World Safe for Democracy" ludicrous. Asks Du Bois, "How could America condemn in Germany that which she commits, just as brutally, within her own borders?" (Du Bois 1920/2003d:59).

Of course, this is the very same message preached by Dr. Martin Luther King, Jr.—albeit in a more positive way—30 years later. Dr. King challenged America to live up to its own moral and democratic code. He shamed America by showing that it did not live up to its own sacred ideals—most important, "freedom and justice for all." Indeed, in his final tribute to Du Bois, Dr. King still challenged his audience "to be dissatisfied":

Let us be dissatisfied until every man can have food and material necessities for his body, culture, and education for his mind, freedom and human dignity for his spirit. Let us be dissatisfied until rat-infested, vermin-filled slums will be a thing of a dark past and every family will have a decent sanitary house in which to live. Let us be dissatisfied until the empty stomachs of Mississippi are filled and the idle industries of Appalachia are revitalized. Let us be dissatisfied until brotherhood is no longer a meaningless word at the end of a prayer but the first order of business on every legislative agenda. Let us be dissatisfied until our brother of the Third World—Asia, Africa, and Latin America—will no longer be the victim of imperialist exploitation but will be lifted from the long night of poverty, illiteracy, and disease. Let us be dissatisfied until this pending cosmic elegy will be transformed into a creative psalm of peace and "justice will roll down like waters from a mighty stream." (King 1970:183)

From "The Souls of White Folk" (1920)

W. E. B. Du Bois

High in the tower, where I sit above the loud complaining of the human sea, I know many souls that toss and whirl and pass, but none there are that intrigue me more than the Souls of White Folk.

Of them I am singularly clairvoyant. I see in and through them. I view them from unusual points of vantage. Not as a foreigner do I come, for I am native, not foreign, bone of their thought and flesh of their language. Mine is not the knowledge of the traveler or the colonial composite of dear memories, words and wonder. Nor yet is my knowledge that which servants have of masters, or mass of class, or capitalist of artisan. Rather I see these souls undressed and from the back and side. I see the working of their entrails. I know their thoughts and they know that I know. This knowledge makes them now embarrassed, now furious! They deny my right to live and be and call me misbirth! My word is to them mere bitterness and my soul, pessimism. And yet as they preach and strut and shout and threaten, crouching as they clutch at rags of facts and fancies to hide their nakedness, they go twisting, flying by my tired eyes and I see them ever stripped,—ugly, human.

The discovery of personal whiteness among the world's peoples is a very modern thing,—a nineteenth and twentieth century matter, indeed. The ancient world would have laughed at such a distinction. The Middle Age regarded skin color with mild curiosity; and even up into the eighteenth century we were hammering our national manikins into one, great, Universal Man, with fine frenzy which ignored color and race even more than birth. Today we have changed all that, and the world in a sudden, emotional conversion has discovered that it is white and by that token, wonderful!

This assumption that of all the hues of God whiteness alone is inherently and obviously better than brownness or tan leads to curious acts; even the sweeter souls of the dominant world as they discourse with me on weather, weal, and woe are continually playing above their actual words an obligato of tune and tone, saying:

"My poor, un-white thing! Weep not nor rage. I know, too well, that the curse of God lies heavy on you. Why? That is not for me to say, but be brave! Do your work in your lowly sphere, praying the good Lord that into heaven above, where all is love, you may, one day, be born—white!"

I do not laugh. I am quite straight-faced as I ask soberly:

"But what on earth is whiteness that one should so desire it?" Then always, somehow, some way, silently but clearly, I am given to understand that whiteness is the ownership of the earth forever and ever, Amen!

Now what is the effect on a man or a nation when it comes passionately to believe such an extraordinary dictum as this? That nations are coming to believe it is manifest daily. Wave on wave, each with increasing virulence, is dashing this new religion of whiteness on the shores of our time. Its first effects are funny: the strut of the Southerner, the arrogance of the Englishman amuck, the whoop of the hoodlum who vicariously leads your mob. Next it appears dampening generous enthusiasm in what we once counted glorious; to free the slave is discovered to be tolerable only in so far as it freed his master! Do we sense somnolent writhings in black Africa or angry groans in India or triumphant banzais in Japan? "To your tents, O Israel!" These nations are not white!

After the more comic manifestations and the chilling of generous enthusiasm come subtler, darker deeds. Everything considered, the title to the universe claimed by White Folk is faulty. It ought, at least, to look plausible. How easy, then, by emphasis and omission to make children believe that every great soul the world ever saw was a white man's soul; that every

SOURCE: From *Dark Water: Voices from Within the Veil* by W. E. B. Du Bois, pp. 55–60 and 65–70 (Amherst, NY: Prometheus Books). Published in 2003 (Classics in Black Studies Series).

great thought the world ever knew was a white man's thought; that every great deed the world ever did was a white man's deed; that every great dream the world ever sang was a white man's dream. In fine, that if from the world were dropped everything that could not fairly be attributed to White Folk, the world would, if anything, be even greater, truer, better than now. And if all this be a lie, is it not a lie in a great cause?

Here it is that the comedy verges to tragedy. The first minor note is struck, all unconsciously, by those worthy souls in whom consciousness of high descent brings burning desire to spread the gift abroad,—the obligation of nobility to the ignoble. Such sense of duty assumes two things: a real possession of the heritage and its frank appreciation by the humble-born. So long, then, as humble black folk, voluble with thanks, receive barrels of old clothes from lordly and generous whites, there is much mental peace and moral satisfaction. But when the black man begins to dispute the white man's title to certain alleged bequests of the Fathers in wage and position, authority and training; and when his attitude toward charity is sullen anger rather than humble jollity; when he insists on his human right to swagger and swear and waste,— then the spell is suddenly broken and the phil-anthropist is ready to believe that Negroes are impudent, that the South is right, and that Japan wants to fight America.

After this the descent to Hell is easy. On the pale, white faces which the great billows whirl upward to my tower I see again and again, often and still more often, a writing of human hatred, a deep and passionate hatred, vast by the very vagueness of its expressions. Down through the green waters, on the bottom of the world, where men move to and fro, I have seen a man—an educated gentleman—grow livid with anger because a little, silent, black woman was sitting by herself in a Pullman car. He was a white man. I have seen a great, grown man curse a little child, who had wandered into the wrong wait-ing-room, searching for its mother: "Here, you damned black—" He was white. In Central Park I have seen the upper lip of a quiet, peaceful man curl back in a tigerish snarl of rage because black folk rode by in a motor car. He was a white man. We have seen, you and I, city after

city drunk and furious with ungovernable lust of blood; mad with murder, destroying, killing, and cursing; torturing human victims because some-body accused of crime happened to be of the same color as the mob's innocent victims and because that color was not white! We have seen,—Merciful God! in these wild days and in the name of Civilization, Justice, and Motherhood,—what have we not seen, right here in America, of orgy, cruelty, barbarism, and murder done to men and women of Negro descent.

Up through the foam of green and weltering waters wells this great mass of hatred, in wilder, fiercer violence, until I look down and know that today to the millions of my people no mis-fortune could happen,—of death and pestilence, failure and defeat—that would not make the hearts of millions of their fellows beat with fierce, vindictive joy! Do you doubt it? Ask your own soul what it would say if the next census were to report that half of black America was dead and the other half dying.

Unfortunate? Unfortunate. But where is the misfortune? Mine? Am I, in my blackness, the sole sufferer? I suffer. And yet, somehow, above the suffering, above the shackled anger that beats the bars, above the hurt that crazes there surges in me a vast pity,—pity for a people imprisoned and enthralled, hampered and made miserable for such a cause, for such a phantasy!

Conceive this nation, of all human peoples, engaged in a crusade to make the "World Safe for Democracy"! Can you imagine the United States protesting against Turkish atrocities in Armenia, while the Turks are silent about mobs in Chicago and St. Louis; what is Louvain com-pared with Memphis, Waco, Washington, Dyersburg, and Estill Springs? In short, what is the black man but America's Belgium, and how could America condemn in Germany that which she commits, just as brutally, within her own borders?

A true and worthy ideal frees and uplifts a people; a false ideal imprisons and lowers. Say to men, earnestly and repeatedly: "Honesty is best, knowledge is power; do unto others as you would be done by." Say this and act it and the nation must move toward it, if not to it. But say to a people: "The one virtue is to be white," and the people rush to the inevitable conclusion, "Kill the 'nigger'!"

Is not this the record of present America? Is not this its headlong progress? Are we not coming more and more, day by day, to making the statement "I am white," the one fundamental tenet of our practical morality? Only when this basic, iron rule is involved is our defense of right nationwide and prompt. Murder may swagger, theft may rule and prostitution may flourish and the nation gives but spasmodic, intermittent and lukewarm attention. But let the murderer be black or the thief brown or the violator of womanhood have a drop of Negro blood, and the righteousness of the indignation sweeps the world. Nor would this fact make the indignation less justifiable did not we all know that it was blackness that was condemned and not crime.

In the awful cataclysm of World War, where from beating, slandering, and murdering us the white world turned temporarily aside to kill each other, we of the Darker Peoples looked on in mild amaze. . . .

The European world is using black and brown men for all the uses which men know. Slowly but surely white culture is evolving the theory that "darkies" are born beasts of burden for white folk. It were silly to think otherwise, cries the cultured world, with stronger and shriller accord. The supporting arguments grow and twist themselves in the mouths of merchant, scientist, soldier, traveler, writer, and missionary: Darker peoples are dark in mind as well as in body; of dark, uncertain, and imperfect descent; of frailer, cheaper stuff; they are cowards in the face of mausers and maxims; they have no feelings, aspirations, and loves; they are fools, illogical idiots,—"half-devil and half-child."

Such as they are civilization must, naturally, raise them, but soberly and in limited ways. They are not simply dark white men. They are not "men" in the sense that Europeans are men. To the very limited extent of their shallow capacities lift them to be useful to whites, to raise cotton, gather rubber, fetch ivory, dig diamonds,—and let them be paid what men think they are worth—white men who know them to be well-nigh worthless.

Such degrading of men by men is as old as mankind and the invention of no one race or people. Ever have men striven to conceive of their victims as different from the victors, endlessly different, in soul and blood, strength and cunning, race and lineage. It has been left, however, to Europe and to modern days to discover the eternal worldwide mark of meanness,—color!

Such is the silent revolution that has gripped modern European culture in the later nineteenth and twentieth centuries. Its zenith came in Boxer times: White supremacy was all but world-wide, Africa was dead, India conquered, Japan isolated, and China prostrate, while white America whetted her sword for mongrel Mexico and mulatto South America, lynching her own Negroes the while. Temporary halt in this program was made by little Japan and the white world immediately sensed the peril of such "yellow" presumption! What sort of a world would this be if yellow men must be treated "white"? Immediately the eventual overthrow of Japan became a subject of deep thought and intrigue, from St. Petersburg to San Francisco, from the Key of Heaven to the Little Brother of the Poor.

The using of men for the benefit of masters is no new invention of modern Europe. It is quite as old as the world. But Europe proposed to apply it on a scale and with an elaborateness of detail of which no former world ever dreamed. The imperial width of the thing,—the heaven-defying audacity—makes its modern newness.

The scheme of Europe was no sudden invention, but a way out of long-pressing difficulties. It is plain to modern white civilization that the subjection of the white working classes cannot much longer be maintained. Education, political power, and increased knowledge of the technique and meaning of the industrial process are destined to make a more and more equitable distribution of wealth in the near future. The day of the very rich is drawing to a close, so far as individual white nations are concerned. But there is a loophole. There is a chance for exploitation on an immense scale for inordinate profit, not simply to the very rich, but to the middle class and to the laborers. This chance lies in the exploitation of darker peoples. It is here that the golden hand beckons. Here are no labor unions or votes or questioning onlookers or inconvenient consciences. These men may be used down to the very bone, and shot and maimed in "punitive" expeditions when they revolt. In these dark lands "industrial development" may repeat in exaggerated form every horror of the industrial history of Europe, from slavery and rape to disease and maiming, with only one test of success,—dividends!

This theory of human culture and its aims has worked itself through warp and woof of our daily thought with a thoroughness that few realize. Everything great, good, efficient, fair, and honorable is "white"; everything mean, bad, blundering, cheating, and dishonorable is "yellow"; a bad taste is "brown"; and the devil is "black." The changes of this theme are continually rung in picture and story, in newspaper heading and moving-picture, in sermon and school book, until, of course, the King can do no wrong,—a White Man is always right and a Black Man has no rights which a white man is bound to respect.

There must come the necessary despisings and hatreds of these savage half-men, this unclean *canaille* of the world—these dogs of men. All through the world this gospel is preaching. It has its literature, it has its priests, it has its secret propaganda and above all—it pays!

There's the rub,—it pays. Rubber, ivory, and palm-oil; tea, coffee, and cocoa; bananas, oranges, and other fruit; cotton, gold, and copper—they, and a hundred other things which dark and sweating bodies hand up to the white world from their pits of slime, pay and pay well, but of all that the world gets the black world gets only the pittance that the white world throws it disdainfully.

Small wonder, then, that in the practical world of things-that-be there is jealousy and strife for the possession of the labor of dark millions, for the right to bleed and exploit the colonies of the world where this golden stream may be had, not always for the asking, but surely for the whipping and shooting. It was this competition for the labor of yellow, brown, and black folks that was the cause of the World War. Other causes have been glibly given and other contributing causes there doubtless were, but

they were subsidiary and subordinate to this vast quest of the dark world's wealth and toil.

Colonies, we call them, these places where "niggers" are cheap and the earth is rich; they are those outlands where like a swarm of hungry locusts white masters may settle to be served as kings, wield the lash of slave-drivers, rape girls and wives, grow as rich as Croesus and send homeward a golden stream. They belt the earth, these places, but they cluster in the tropics, with its darkened peoples: in Hong Kong and Anam, in Borneo and Rhodesia, in Sierra Leone and Nigeria, in Panama and Havana—these are the El Dorados toward which the world powers stretch itching palms.

Germany, at last one and united and secure on land, looked across the seas and seeing England with sources of wealth insuring a luxury and power which Germany could not hope to rival by the slower processes of exploiting her own peasants and workingmen, especially with these workers half in revolt, immediately built her navy and entered into a desperate competition for possession of colonies of darker peoples. To South America, to China, to Africa, to Asia Minor, she turned like a hound quivering on the leash, impatient, suspicious, irritable, with blood-shot eyes and dripping fangs, ready for the awful word. England and France crouched watchfully over their bones, growling and wary, but gnawing industriously, while the blood of the dark world whetted their greedy appetites. In the background, shut out from the highway to the seven seas, sat Russia and Austria, snarling and snapping at each other and at the last Mediterranean gate to the El Dorado, where the Sick Man enjoyed bad health, and where millions of serfs in the Balkans, Russia, and Asia offered a feast to greed wellnigh as great as Africa.

The fateful day came. It had to come. The cause of war is preparation for war; and of all that Europe has done in a century there is nothing that has equaled in energy, thought, and time her preparation for wholesale murder. The only adequate cause of this preparation was conquest and conquest, not in Europe, but primarily among the darker peoples of Asia and Africa; conquest, not for assimilation and uplift, but for commerce and degradation. For this, and this mainly, did Europe gird herself at frightful cost for war.

Discussion Questions

1. Explain Du Bois's concept of the color line. Using specific examples from Du Bois's work as well as real life, discuss how this concept encompasses both nonrational and rational motivation for action, at the individual as well as collective levels.

2. Compare and contrast Simmel's notion of the stranger with Du Bois's notion of double consciousness. What similarities do these concepts share? How are they different?

3. Discuss Weber's distinction between status and class in the context of Du Bois's work. Is race *merely* a "status"? How so or why not?

4. Compare and contrast Du Bois's theory as to the oppression of African Americans with Gilman's theory as to the oppression of women. What similarities do you see in their arguments? What are the differences in these two theories of oppression?

5. Compare and contrast the Marxist notion of false consciousness with the delusion of whites, as described by Du Bois. Who are affected by each type of delusion, and how? Do you see any similarities in these two types of delusion? How so or why not? Discuss the Marxist dimensions of Du Bois's work in general.

6. Discuss multiracialism and multiculturalism in the context of Du Bois's theories of race and class. To what extent do you think that the election of Barack Obama signifies a new era in the racial history of the United States?

8 GEORGE HERBERT MEAD (1863–1931)

Key Concepts

- "I"
- "Me"
- Generalized other
- Meaning
- Significant symbols
- Taking the attitude of the other
- Play stage
- Game stage

The individual experiences himself as such, not directly, but only indirectly, from the particular standpoints of other individual members of the same social group, or from the generalized standpoint of the social group as a whole to which he belongs. . . and he becomes an object to himself only by taking the attitudes of other individuals toward himself.

(Mead 1934/1962:138)

I t's time to get ready for tonight's date and you're trying on different clothes. As you put them on, what are you doing? More than likely, you're looking in a mirror and *think-ing*—"How do I look in this? What impression about myself am I giving?" That is, in deciding what to wear, you're talking to yourself. But this conversation involves someone other than you. It's a dialogue between you and an *imagined* other, your date. Through the use of language, you are able to answer your own questions as if you were another person looking at yourself. Through language, you are able to see yourself as an object, just as your

371

date does, and react to your appearance as she or he does. In an important sense, then, you experience your "self" socially, through taking the presumed attitudes of others toward your own behaviors, and on the basis of these attitudes, you control your own conduct.

What is the link between thinking and behavior? When you are thinking, what are you doing? What role does an individual's thinking play in the evolution of society? These are the questions that fascinated George Herbert Mead, the pragmatist philosopher whose views laid the foundation for the theory of symbolic interactionism, one of the major theoretical paradigms within sociology. In contrast to theorists who define society as a system of inter-related parts that promote consensus and stability or conflict and domination, Mead defined society as "generalized social attitudes" that continually *emerge* through coordinated inter-action between individuals and groups. In this chapter, we examine Mead's perspective on the relationship between the individual and society through selections from his posthu-mously published book, *Mind, Self, and Society* (Mead 1934/1962). Before turning to the primary texts, however, we first provide a brief sketch of Mead's biography and then an overview of the intellectual currents that informed his philosophy.

▚ A Biographical Sketch

Mead was born in 1863 in South Hadley, Massachusetts. At the age of seven, Mead moved to Ohio when his father, a Congregationalist minister, had accepted a position at Oberlin Theological Seminary. After graduating from Oberlin College in 1883, Mead left for the Northwest, where he worked for three years as a tutor and as a railroad surveyor. Undecided on his career options, Mead's next move took him to Harvard University, where he pursued graduate studies in philosophy and psychology. During this period, Mead encountered some of the thinkers who would prove pivotal in formulating his own philosophy of the mind and self—perhaps most important, the American pragmatist William James and the German philosopher Wilhelm Wundt.

Upon completing his master's degree at Harvard in 1888, Mead was hired as a philoso-phy instructor at the University of Michigan. Here, Mead kindled important intellectual rela-tionships with eminent American scholars John Dewey and Charles Horton Cooley. Mead's stay at Michigan was short-lived, however. Two years after arriving, Mead left for the recently founded University of Chicago, where Dewey had been appointed to head the Department of Philosophy. Mead remained at Chicago until his death in 1931.

Mead's influence on sociology stemmed largely from his graduate course in social psy-chology. A dynamic speaker, Mead's lectures attracted students from an array of disciplines including, of course, sociology. Recognizing the significance of the ideas Mead presented in the classroom, some of his students had his lectures transcribed. Shortly after Mead's death, the transcriptions were edited and published in *Mind, Self, and Society* (Mead 1934/1962), a text that has since earned a central position in social theory literature.

Mead's academic career was complemented by his ongoing commitment to progressive social causes. Not unlike Marx, Mead envisioned the creation of a social utopia. Mead's adherence to liberal ideals (e.g., individualism, volunteerism), however, led him to advo-cate reform over revolution as the means for realizing a more just and democratic society. For Mead, such a society would be based on expanding universal rights that enabled indi-viduals to pursue their own interests, while at the same time creating a more cooperative, united democratic order. At the root of this evolutionary process stood individuals coordi-nating their actions with one another through sympathetic understanding and the self-inhibition of behaviors detrimental to social cohesion. Rejecting a social gospel based on religious doctrine, Mead nevertheless was active in the reform movement of the early twen-tieth century. His optimistic faith in science and reason as means of creating a full-fledged, participatory democracy spurred his efforts in educational reform and feminist causes as well as his work with Jane Addams (1860–1935), the pioneer activist, reformer, and architect of professional social work.[1]

INTELLECTUAL INFLUENCES AND CORE IDEAS ∷

Mead's philosophy was informed by a number of intellectual currents. In this section, we briefly discuss three of the most significant theoretical influences on his work: pragmatism, behaviorism, and evolutionism.

Pragmatism

Pragmatism is a uniquely American philosophical doctrine developed primarily through the work of Charles S. Peirce (1839–1914), William James (1842–1910), and John Dewey (1859–1952). Unlike most European philosophical schools, pragmatism was not oriented toward uncovering general "Truths" or formal principles of human behaviors or desires. Instead, it was offered as a method or instrument for studying the meaning of behavior. More specifically, the early pragmatists argued that the meaning of objects and actions lie in their practical aspects, that is, how they allow individuals to adapt to and solve the problems they confront. "Truth," or meaning, is thus found in what "works." It is a result, not a fixed ideal or an intrinsic feature of a given object, event, or situation.

For instance, consider something as ordinary as water. What *is* water? Is it fundamentally only a combination of hydrogen and oxygen? From the point of view of pragmatism, what water is—its meaning—depends on the situation in which it is encountered and thus the use to which it will be put. For the chemist, water *is* a solvent used to dissolve substances. For the athlete, water *is* a liquid used for hydration or as a means for exercise or sport. For the firefighter, water *is* a tool used for extinguishing fires, or even perhaps a potential danger, depending on the type of fire. For the gardener, water *is* a necessary ingredient that must be used in order to grow healthy plants. The point is that the meaning of objects or social interaction is rooted in action, in everyday practical conduct, that is, the uses that are made of them as individuals go about the business of constructing their behavior.

Although such a position may not appear to be particularly controversial, it is based on a host of assumptions regarding the nature of the individual and his relationship to the external world. Particularly important in this regard is the indebtedness of Mead (and pragmatists more generally) to German idealism and the philosophy of Immanuel Kant (1724–1804) and Georg Hegel (1770–1831). It was Kant who proposed the existence of a "twofold self or "double I": in being conscious of one's self, the individual "splits" into "the I as subject and the I as object" (Wrong 1994:61,62). Indeed, for Kant and many other philosophers, it is this capacity for "reflexivity," or the ability to experience one's self simultaneously as a thinking, perceiving subject, *and* as a perceived object, that separates humans from all other animal forms. The influence of these ideas on Mead is most apparent in his conversion of Kant's twofold self into a distinction between the "I" and the "me" (see a discussion of Mead's distinction below).

Equally important, from a Kantian view, there is no reality separable from the perceiving subject. Instead, the external world of objects and events exists only through the conscious

---

apprehension of them. The world is thus not something "out there" to be experienced by the subject, but rather is "a task to be accomplished" (Coser 1977:350). Moreover, consciousness itself cannot be said to exist unless there is some object in the external world that one notes or becomes aware of. The act of knowing (subject) and the known (object) are thus intimately connected.[2]

Behaviorism

The pragmatist view of consciousness and its importance for guiding our actions is not without its detractors. Of particular relevance here is the behaviorist branch of psychology, in response to which Mead fashioned his own theoretical framework that he labeled "social behaviorism." Psychological behaviorism is a resolutely empirical branch of psychology that focuses solely on observable actions. Indeed, its proponents argue that *only* overt behaviors are open to scientific investigation. From this it follows that because states of mind, feelings, desires, and thinking cannot be observed, they cannot be studied scientifically. As a result, behaviorism is confined to studying the links between visible stimuli and the learned responses that are associated with them. As you will read below, some of Mead's most significant contributions stem from his counterargument that thinking is indeed a behavior and thus available for analysis.

During Mead's time, the leading figure in behaviorism was the psychologist **John B. Watson** (1878–1958). Watson argued that human behavior differed little in principle from animal behavior—both could be explained and predicted on the basis of laws that govern the association of behavioral responses to external stimuli. Thus, as a *learned* response to a specific stimulus, a rabbit's retreat from the path of a snake is no different in kind from a person's efforts to perform well in her job. Watson's conviction regarding the power of behaviorism to unlock the secrets of human development and action is perhaps nowhere more plainly asserted than in his claim:

> Give me a dozen healthy infants, well-formed, and my own specified world to bring them up in and I'll guarantee to take any one at random and train him to become any type of specialist I might select—doctor, lawyer, artist, merchant-chief, and yes, even beggarman and thief, regardless of his talents, penchants, tendencies, abilities, vocations, and race of his ancestors. (Watson 1924/1966:104)

It is against this picture of passive, nonreflexive individuals that Mead fashioned his social behaviorism. For Mead, the mind is not an ephemeral "black box" that is inaccessible to investigation. Instead, Mead viewed the mind as a *behavioral process* that entails a "conversation of significant gestures." In this internal conversation, an individual **takes the attitude of the other**, arousing in his own mind the same responses to his potential action that are aroused in the other person's.[3] Individuals, then, shape their actions on the basis of the imagined responses they attribute to others. Self-control of what we say and do is thus in actuality a form of social control as we check our behaviors—discarding some options while pursuing others—against the responses that we anticipate will be elicited from others.

---

[2]Although Mead borrowed from Kant's philosophy of knowledge, he did not adopt some of its fundamental principles. Perhaps most important is that Mead, unlike Kant, did not maintain that the mental categories (e.g., our experience of time, space, or beauty) through which our perceptions are organized and given meaning exist independently of the knowing subject. Instead, Mead argued that such categories are learned and emerge through the process of interaction with others.

[3]Depending on the situation, "others" might be a specific person (a sibling or neighbor), an identifiable group (classmates or coworkers), or the community at large, which Mead referred to as the "generalized other" (e.g., Southerners, Americans).

Significant Others

Charles Horton Cooley (1864–1929): The "Looking-Glass Self"

Born in Ann Arbor, Michigan, most of Charles Cooley's life revolved around that city's hub, the University of Michigan. It was here that his father served on the faculty of the Michigan Law School and where he would attend school, earning a degree in engineering and later a Ph.D. in political economy and sociology. While pursuing his graduate degree, Cooley began teaching at the University of Michigan, where he remained on the faculty for his entire career. Mead's notion of "taking the attitude of the other" and the internal conversation that it entails shares much with Cooley's discussion of the "looking-glass self." Cooley argues that an individual's sense of self is developed through interaction with others. More specifically, there are three facets to the looking-glass self: "the imagination of our appearance to the other person, the imagination of his judgment of that appearance, and some sort of self-feeling, such as pride or mortification" (Cooley 1902/1964:184). His underscoring of the role of imaginations in the construction of self-identity likewise informed his general perspective on society and sociology. Indeed, Cooley maintained that the "imaginations which people have of one another are the *solid* facts of society, and . . . to observe and interpret these must be the chief aim of sociology" (ibid.:121, emphasis in the original). It is this overly "mentalistic" understanding of society and of the task of sociology that has been the primary cause of the marginalization of Cooley's ideas in the discipline. Perhaps, in some ways, his focus on mental processes stemmed from his personal disposition. Cooley was a shy, semi-invalid whose list of ailments included a speech impediment. As a result, he led a quiet social life, preferring books and his own imaginative introspections to the company of others.

SOURCE: Coser 1977:314–18.

It is important to note that Mead's view of the relationship between self-control and social control is not based on a deterministic understanding of individual action. A deterministic view contends that social structures or divisions within society determine behaviors or attitudes, leaving little possibility for individuals to shape their life courses and counteract the destinies supposedly derived from broad social forces. Instead, Mead emphasized that the individual and the society to which she belongs are mutually dependent; each requires the other for its progressive evolution. Thus, for Mead, "social control, so far from tending to crush out the human individual or to obliterate his self-conscious individuality, is, on the contrary, actually constitutive of and inextricably associated with that individuality" (Mead 1934/1962:255).

This view of the mind as a process of thinking that entails both self- and social control can be illustrated through any number of commonplace examples, for it is something we all continually experience. Consider, for instance, the internal conversation you engage in before asking someone out for a date, going on a job interview, determining how you will resolve an argument with a friend, or deciding whether to ask a question in class. In each case, you take the attitude of the other, viewing yourself as an object as other individuals do during interaction. So, you may ask yourself, "What type of movie should I take him or her to?" or "What music should we listen to when we're driving in the car?" Similarly, before raising your hand you may think, "Will I seem stupid if I ask the professor this question? Will it will seem like I'm trying to score 'brownie' points?" The answers to such questions, and thus the behavior you intend to undertake, are shaped by the responses evoked in your mind. The responses are not entirely your own, however; instead, they reflect the assumed attitude that others take toward your behavior.

Evolutionism

A third major stream in Mead's framework is found in Darwinian evolutionism. The impact of evolutionism on Mead's thought comes in two forms. First, Mead, like Darwin, saw humans as the most advanced species. Humankind's superiority lies in the capacity to communicate symbolically. Through the use of language, we are able to take the attitude of the other and respond to our behavior in the same way as those to whom it is directed. Language thus allows us to see ourselves as an object. Symbolically casting off one's self as an object is the essence of self-consciousness, something other animals do not possess. Instead, other animals react to each other and their environment on the basis of instinct without anticipating the possible effects of their behavior.

The human capacity for self-consciousness, then, allows us to temporarily suspend our behaviors as we symbolically test different solutions to the situations we face. The ability to interact symbolically with others, ourselves, and the environment not only provides for an individual's adjustment to problems that may arise; it also creates the mechanism for social evolution. The second evolutionary theme in Mead's work is thus found in his view of social progress. Mead outlined both a practical and a utopian version of progress. His practical view highlighted the role of science as an instrument for addressing the problems that we confront in the natural and social environments. For Mead, science is "the evolutionary process grown self-conscious" (Mead 1936/1964:23). While the process of natural selection may take centuries to enable a species of plant or animal to adapt better to its environment, science is a conscious endeavor to identify and solve problems. Whether it is the problem of crime, pollution, or illness, the scientific method generates alternative solutions that can then be tested. When the problem has been fixed (a vaccine is invented) evolution is furthered. Science thus allows for an orderly adaptation to and increasing advantage over our environment.

Mead's utopian view of social progress rests with his emphasis on language and the eventual creation of a democratic "universal society." For Mead (and other early social theorists), the problem to be solved in modern, differentiated societies lies in fashioning an orderly coordination of social activities. Yet, how can such coordination be achieved without stifling innovation and adjustment to a changing environment? How can social stability be maintained without diminishing individual expression and freedom? In short, how can a "true" democracy be attained in light of the opposing interests and unequal access to resources that make up a highly diverse society?

Mead found his answer in the ever-widening circle of social contact experienced by individuals living in modern, heterogeneous societies. Such contact expands the vocabulary of significant symbols (shared meanings), allowing individuals to take the attitude of a broader range of others.

A more highly evolved empathetic understanding would then breed a "universal discourse" of shared meanings and with it an ideal democracy based on cooperation and a full recognition of the rights and duties to be freely and equally exercised by all. Mead states the matter thusly:

> The ideal of human society is one which brings people so closely together in their interrelationships, so fully develops the necessary system of communication [i.e., significant symbols], that the individuals who exercise their peculiar functions can take the attitude of those whom they affect. . . . If that system of communication could be made theoretically perfect, the individual would affect himself as he affects others in every way. That would be the ideal of communication. . . . The meaning of that which is said is here the same to one as it is to everybody else. Universal discourse is then the formal ideal of communication. If communication can be carried through and made perfect, then there would exist the kind of democracy . . . in which each individual would carry just the response in himself that he knows he calls out in the community. (Mead 1934/1962:327)

Significant Others

William James (1842–1910): Consciousness and the Self

William James was one of America's leading scholars at the turn of the twentieth century. Born into one of the most illustrious families in American intellectual life, his father was an independently wealthy eccentric and amateur theologian, and his younger brother, Henry James, Jr., is one of the great names in American literature. Though James's formal academic degree was in medicine (Harvard, 1869), his work has been most influential in the fields of psychology and philosophy, two disciplines that better fit his quest to resolve his own emotional, spiritual, and intellectual dilemmas. Often house-bound due to physical illnesses such as smallpox, as well as serious bouts of depression and a mental condition he described as "soul sickness," James nevertheless published a number of major works, most importantly his two-volume, 1,200-page opus, *The Principles of Psychology* (James 1890), and *The Varieties of Religious Experience* (1902). Through his books, essays, and lectures, he tackled such topics as the nature of truth (it is "good" because it is "useful"), the connection between physiological and emotional responses to stimuli (the James-Lange theory), the constitution and experience of the self, religious experiences of the "divine," and the nature of consciousness and its relationship to reality.

For James, consciousness does not exist as a sort of substance or entity that is brought *to* experiences; rather, it is a continuous series of changing relations *within* experiences that perform the "function of knowing." Life flows along an ongoing "stream of consciousness" that selectively attends to and emphasizes some features of our outer world while ignoring most of what transpires around us. "Now we are seeing, now hearing; now reasoning, now willing; now recollecting, now expecting; now loving, now hating. . . . Like a bird's life, [consciousness] seems to be an alternation of flights and perchings" in which our every thought, our every state of mind, is always unique (James 1892/2001:27). Out of the indistinguishable, chaotic swirl of movements that make up our environment, our attention is focused only on those things that happen to be of interest to us and, which, having singled out its independent existence, we thus name.

Of all the things that flow through our stream of consciousness, one holds our interest unlike any other: the "Empirical Self" or "me"—that "empirical aggregate of things objectively known" (James 1890/2007:215) by the individual. James divides the self or "me" into three parts: its "constituents" (which he further divides into the "material Self," the "social Self," the "spiritual Self," and the "pure Ego"), the feelings and emotions aroused by the various constituents of the self ("Self-feelings"), and the actions prompted by the self ("Self-seeking" and "Self-preservation"). Taken together, *"a man's Self is the sum total of all that he CAN call his"* (ibid.:291; emphasis in original). Thus, the self is composed of one's physical body and mental powers, material possessions and social relationships, reputation and deeds, and all the feelings and actions to which they give rise. These "MUST *be the supremely interesting* OBJECTS *for each human mind*" (ibid.:324; emphasis in original).

:: MEAD'S THEORETICAL ORIENTATION

With regard to our metatheoretical framework, Mead's work is predominantly individualistic and nonrationalist in orientation. (See Figure 8.1.) His vision is one where the social order is continually emerging through the ongoing activities of individuals (individualistic) who are attempting to navigate or make sense of the situations in which they find themselves (nonrationalist).

To explicate this understanding of Mead's perspective, consider first the individualistic dimension of his theory. As we discussed above, Mead saw society as the product of coordinated activities undertaken by *individuals* reflexively taking the attitude of the other(s) with whom they are interacting. Society is neither a "thing," an overarching system of pre-existing structures, nor a form of relationship that connects individuals with one another. Instead, Mead envisioned society as shared attitudes that consciously shape individuals' behaviors. Even institutions, such as legal, educational, or kinship systems, were defined by Mead as "social habits" or a "common response on the part of all members of the community to a particular situation" (Mead 1934/1962:261). Hence, they affect our behavior insofar as we adopt the attitude of others with whom we interact in recurring situations. Institutions (e.g., courts, schools, marriage) continually *emerge* as patterns of action are reenacted by participants in the encounter.

Unlike Marx or Weber, then, Mead granted little attention to *categories* such as institutions, class, or status, because he did not believe that these collectivist or structural forces determined one's consciousness or behavior. Certainly, he did not deny society a role in shaping an individual's attitudes and behaviors. Indeed, our self is essentially a social construct, given its basis in perceptions of how *others* will respond to our behaviors. In important aspects, then, Mead viewed the individual and society as dialectically intertwined—each one shapes and cannot exist without the other. Nevertheless, for Mead, the central elements of social interaction take place in one's imagination. In turn, he

Figure 8.1   Mead's Basic Theoretical Orientation

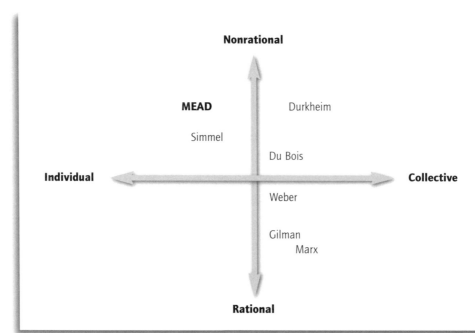

emphasized the importance of the "I," "me," "generalized other," and the role of thinking and language in social life. Importantly, Mead viewed language as a neutral means for symbolically communicating with others. Language, in other words, was not seen as reflecting relations of domination and subordination based, for example, on class or gender.[4] One's mind or thinking, which for Mead is intimately tied to language, is therefore not a product of one's station in society. Compare this view with that expressed by Marx, who argued that an individual's consciousness and view of the world are determined by her class position.

But what of the nonrationalist presuppositions underlying Mead's theory? Here, his position is perhaps more difficult to glean. Dennis Wrong points to the source of this difficulty when he remarks that Mead was "only casually and tangentially interested in the self as a source of motivational energy, or as an object of affective attachment" (Wrong 1994:65). Because our distinction between rationalist and nonrationalist refers precisely to a theorist's view of the motivational foundations of action, Mead's "casual interest" on this issue leaves us with less to work with than we may like. Nevertheless, we can point out the general tendencies in Mead's perspective.

First, Mead's account of the self and interaction posits individuals as essentially cognitive. In other words, he offers much insight into the mental behaviors that we undertake in preparing to act with others and in assigning meaning to conduct and events. Hence, Mead details the role of language and self-objectification in *thinking*. This focus should not, however, be confused with a rationalist orientation to motivation. As we discussed in Chapter 1, rationalist presuppositions portray actors and groups as *motivated* by their self-interested maximization of rewards or pleasures and minimizing of costs. Nowhere does Mead make such claims regarding the motivational source of behavior.

Based on our definition of rationalism, we are left with Mead as a theorist given to a nonrationalist orientation. However, such a description is based on more than a simple process of elimination. For instance, in describing institutions as "social habits," Mead suggests that actions are motivated out of just that—habit. Also, in arguing that we approach situations pragmatically, Mead suggests that our actions are guided by an attempt to solve problems. Thus, we pursue behaviors that work and are less preoccupied with fashioning the optimal line of action with regard to any particular payoff. Last, Mead occasionally notes the role of disapproval in shaping our behaviors. Here, he argues that in taking the attitude of the other and responding to our gestures as the other might, we suppress actions that would elicit disapproval from others. To the extent that experiencing the disapproval of others arouses negative emotions or affects, we can say it is a nonrationalist motivating force.[5] As central points of debate, it is, of course, up to the reader to determine where Mead's work falls in the metatheoretical framework.

---

[4]Mead's assumptions regarding language are by no means unchallenged. For instance, many observers argue that the use of "he" as a universal pronoun renders women invisible. In nullifying women's experiences, the reality of patriarchal relations is obscured and thus further reinforced. Similarly, some argue that the wording of SATs reflects class, race, and gender biases. To the extent that occupation and wage are tied to educational attainment, the ostensibly "objective" test therefore perpetuates structurally based patterns of economic inequality.

[5]One could argue that it is rational to avoid disapproval because failing to do so might compromise our attempts to gain rewards. However, this leads to tautological or circular reasoning, where it is impossible for motivations of any kind to fall outside rational, cost-benefit calculations. In turn, this logic would suggest that it is rational to act on the basis of tradition or habit, rational to be in love or desire companionship, or rational to be fearful in certain situations.

Figure 8.2 Mead's Core Concepts

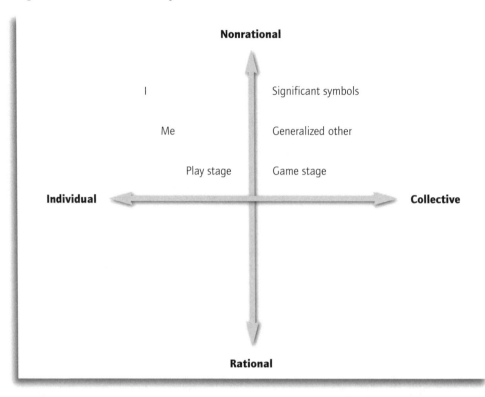

Readings

In what follows, you will read selections from Mead's *Mind, Self, and Society*. The first excerpts are from the portion titled "Mind," in which Mead discusses the social and symbolic bases of thinking and of the construction of meaning. We then move to excerpts taken from the second portion of the book, titled "Self." Here, Mead explores, among other issues, the "phases" of the self and their development through social experiences. In the excerpts from "Society," the third section of the book, Mead extends his discussion of the self as he outlines the conditions necessary for creating a truly democratic society.

Introduction to "Mind"

In these first selections, drawn from *Mind, Self, and Society*, you will encounter three central themes in Mead's work: (1) mind, (2) symbols and language, and (3) the essence of meaning. Here, you will read that, for Mead, "mind" is a process or behavior that allows for the conscious control of one's actions. More specifically, the mind involves an internal conversation of gestures that makes possible the imagined testing of alternative lines of conduct. In the delay of responses produced by such testing lies the crux of intelligent behavior: controlling one's present action with reference to *ideas* about possible future consequences.

It is through symbols, or language, that we point out objects to ourselves and orient our behavior. For instance, when looking for a place to sit at the movies, you probably will have an image of an empty seat in your mind. You know what it is you are looking for even if you do not actually see an empty seat because thinking is carried out through symbols. "Symbols stand for the meanings of those things or objects which have meanings" (Mead 1934/1962:122). Moreover, it is our capacity to think symbolically—to hold within our minds the meanings of things—that allows us to mentally rehearse lines of action without actually performing them. Herein lies the locus of behavioral control, yet controlling one's behavior through thinking is a social process. This is because the mind "emerges" as we point out to others and to ourselves the meaning of things.

Our awareness of the meaning of an object, gesture, or event makes it possible for us to form meaningful responses—that is, responses that indicate to others and ourselves how we are going to act in reference to the situation at hand. And, in indicating our forthcoming responses, we are at the same time indicating the meaning of the object, gesture, or event. This viewpoint, however, does not indicate *where* meaning is produced. Again, we find Mead emphasizing the social nature of individual experience.

Mead defines **meaning** as a "threefold relationship" between (1) an individual's gesture, (2) the adjustive response by another to that gesture, and (3) the completion of the social act initiated by the gesture of the first individual. A gesture thus signifies (1) what the individual making it intends to do, (2) what the individual to whom it is directed is expected to do, and (3) the coordinated activity that is to take place. Meaning, in turn, is not an *idea*, but a *response* to a gesture developed within a social act. In other words, meaning is not intrinsic to a given object or action, nor does it exist within one's own consciousness independently of interaction, as the earlier German idealist philosophers argued. One's gesture to another (for example, asking someone, "May I borrow a pencil?") refers to a desired result (getting a pencil) of the interaction. But an individual's gesture is socially meaningful, or significant, *only* if it elicits the desired response from the person to whom it is directed. That is, your gesture is understood only if the other person responds to your words or actions as you responded in your own mind. If a person hands you a book instead of the pencil you asked for, your gesture lacked meaning, as the other's behavior did not lead to the successful completion of the act. (This point underscores Mead's roots in pragmatic philosophy.) How many times have you said to someone, "You don't understand what I'm saying," when they did not respond as you had imagined they would? Such instances illustrate how meaning develops through a social process.

If the meaning of one's gesture is the response to that gesture by another, how then are you able to ensure the proper interpretation of your actions? How do you make your intentions known such that you are able to bring out the desired responses from others? For Mead, such coordinated activity becomes possible only with the development of language in the form of significant symbols. **Significant symbols** are words and gestures that have the same meaning for all those involved in a social act. They call out the same response in the person who initiates the gesture as they do in those to whom the gesture is directed. Shared meanings provide the basis from which actions can be planned and carried out. Without such a consensus, whether it is preexisting or created by the participants, social interaction would proceed haltingly, as we would be unable to indicate our intentions to others in light of their presumed interpretation of our actions.

"Mind" (1934)

George Herbert Mead

MIND AND THE SYMBOL

The mechanism of the central nervous system enables us to have now present, in terms of attitudes or implicit responses, the alternative possible overt completions of any given act in which we are involved; and this fact must be realized and recognized, in virtue of the obvious control which later phases of any given act exert over its earlier phases. More specifically, the central nervous system provides a mechanism of implicit response which enables the individual to test out implicitly the various possible completions of an already initiated act in advance of the actual completion of the act—and thus to choose for himself, on the basis of this testing, the one which it is most desirable to perform explicitly or carry into overt effect. The central nervous system, in short, enables the individual to exercise conscious control over his behavior. It is the possibility of delayed response which principally differentiates reflective conduct from non-reflective conduct in which the response is always immediate. The higher centers of the central nervous system are involved in the former type of behavior by making possible the interposition, between stimulus and response in the simple stimulus-response arc, of a process of selecting one or another of a whole set of possible responses and combinations of responses to the given stimulus.

Mental processes take place in this field of attitudes as expressed by the central nervous system; and this field is hence the field of ideas: the field of the control of present behavior in terms of its future consequences, or in terms of future behavior; the field of that type of intelligent conduct which is peculiarly characteristic of the higher forms of life, and especially of human beings. The various attitudes expressible through the central nervous system can be organized into different types of subsequent acts; and the delayed reactions or responses thus made possible by the central nervous system are the distinctive feature of mentally controlled or intelligent behavior.[i]

What is the mind as such, if we are to think in behavioristic terms? Mind, of course, is a very ambiguous term, and I want to avoid ambiguities. What I suggested as characteristic of the mind is the reflective intelligence of the human animal which can be distinguished from the intelligence of lower forms. If we should try to regard reason as a specific faculty which deals with that which is universal we should find responses in lower forms which are universal. We can also point out that their conduct is purposive, and that types of conduct which do not lead up to certain ends are eliminated. This would seem to answer to what we term "mind" when we talk about the animal mind, but what we refer to as reflective intelligence we generally recognize as belonging only to the human organism. The non-human animal acts with reference to a future in the sense that it has impulses which are seeking expression that can only be satisfied in later experience, and however this is to be explained, this later experience does determine what the present experience shall be. If one accepts a Darwinian explanation he says that only those forms survive whose conduct has a certain relationship to a specific future, such as belongs to the environment of the specific form. The forms whose conduct does insure the future will naturally survive. In such a

[i]In considering the rôle or function of the central nervous system—important though it is—in intelligent human behavior, we must nevertheless keep in mind the fact that such behavior is essentially and fundamentally social; that it involves and presupposes an ever ongoing social life-process; and that the unity of this ongoing social process—or of any one of its component acts—is irreducible, and in particular cannot be adequately analyzed simply into a number of discrete nerve elements. This fact must be recognized by the social psychologist. These discrete nerve elements lie within the unity of this ongoing social process, or within the unity of any one of the social acts in which this process is expressed or embodied; and the analysis which isolates them—the analysis of which they are the results or end-products—does not and cannot destroy that unity.

statement, indirectly at least, one is making the future determine the conduct of the form through the structure of things as they now exist as a result of past happenings.

When, on the other hand, we speak of reflective conduct we very definitely refer to the presence of the future in terms of ideas. The intelligent man as distinguished from the intelligent animal presents to himself what is going to happen. The animal may act in such a way as to insure its food tomorrow. A squirrel hides nuts, but we do not hold that the squirrel has a picture of what is going to happen. The young squirrel is born in the summer time, and has no directions from other forms, but it will start off hiding nuts as well as the older ones. Such action shows that experience could not direct the activity of the specific form. The provident man, however, does definitely pursue a certain course, pictures a certain situation, and directs his own conduct with reference to it. The squirrel follows certain blind impulses, and the carrying-out of its impulses leads to the same result that the storing of grain does for the provident man. It is this picture, however, of what the future is to be as determining our present conduct that is the characteristic of human intelligence—the future as present in terms of ideas.

When we present such a picture it is in terms of our reactions, in terms of what we are going to do. There is some sort of a problem before us, and our statement of the problem is in terms of a future situation which will enable us to meet it by our present reactions. That sort of thinking characterizes the human form and we have endeavored to isolate its mechanism. What is essential to this mechanism is a way of indicating characters of things which control responses, and which have various values to the form itself, so that such characters will engage the attention of the organism and bring about a desired result. The odor of the victim engages the attention of the beast of prey, and by attention to that odor he does satisfy his hunger and insure his future. What is the difference between such a situation and the conduct of the man who acts, as we say, rationally? The fundamental difference is that the latter individual in some way indicates this character, whatever it may be, to another person and to himself; and the symbolization of it by

means of this indicative gesture is what constitutes the mechanism that gives the implements, at least, for intelligent conduct. Thus, one points to a certain footprint, and says that it means bear. Now to identify that sort of a trace by means of some symbol so that it can be utilized by the different members of the group, but particularly by the individual himself later, is the characteristic thing about human intelligence. To be able to identify "this as leading to that," and to get some sort of a gesture, vocal or otherwise, which can be used to indicate the implication to others and to himself so as to make possible the control of conduct with reference to it, is the distinctive thing in human intelligence which is not found in animal intelligence.

What such symbols do is to pick out particular characteristics of the situation so that the response to them can be present in the experience of the individual. We may say they are present in ideal form, as in a tendency to run away, in a sinking of the stomach when we come on the fresh footprints of a bear. The indication that this is a bear calls out the response of avoiding the bear, or if one is on a bear hunt, it indicates the further progress of the hunt. One gets the response into experience before that response is overtly carried out through indicating and emphasizing the stimulus that instigates it. When this symbol is utilized for the thing itself one is, in Watson's terms, conditioning a reflex. The sight of the bear would lead one to run away, the footprint conditioned that reflex, and the word "bear" spoken by one's self or a friend can also condition the reflex, so that the sign comes to stand for the thing so far as action is concerned.

What I have been trying to bring out is the difference between the foregoing type of conduct and the type which I have illustrated by the experiment on the baby with the white rat and the noise behind its head. In the latter situation there is a conditioning of the reflex in which there is no holding apart of the different elements. But when there is a conditioning of the reflex which involves the word "bear," or the sight of the footprint, there is in the experience of the individual the separation of the stimulus and the response. Here the symbol means bear, and that in turn means getting out of the way, or furthering the hunt. Under those circumstances the person who stumbles on the footprints of the bear is not afraid of the footprints—he is afraid of the bear. The footprint means a bear. The child is afraid of the rat, so that the response of fear is to the sight of the white rat; the man is not afraid of the footprint, but of the bear. The footprint and the symbol which refers to the bear in some sense may be said to condition or set off the response, but the bear and not the sign is the object of the fear. The isolation of the symbol, as such, enables one to hold on to these given characters and to isolate them in their relationship to the object, and consequently in their relation to the response. It is that, I think, which characterizes our human intelligence to a peculiar degree. We have a set of symbols by means of which we indicate certain characters, and in indicating those characters hold them apart from their immediate environment, and keep simply one relationship clear. We isolate the footprint of the bear and keep only that relationship to the animal that made it. We are reacting to that, nothing else. One holds on to it as an indication of the bear and of the value that object has in experience as something to be avoided or to be hunted. The ability to isolate these important characters in their relationship to the object and to the response which belongs to the object is, I think, what we generally mean when we speak of a human being thinking a thing out, or having a mind. Such ability makes the world-wide difference between the conditioning of reflexes in the case of the white rat and the human process of thinking by means of symbols.[ii]

What is there in conduct that makes this level of experience possible, this selection of certain characters with their relationship to other characters and to the responses which these call out? My own answer, it is clear, is in terms of such a set of symbols as arise in our social conduct, in the conversation of gestures—in a word, in terms of language. When we get into conduct these symbols which indicate certain characters and their relationship to things and to responses, they enable us to pick out these characters and hold them in so far as they determine our conduct.

A man walking across country comes upon a chasm which he cannot jump. He wants to go ahead but the chasm prevents this tendency from being carried out. In that kind of a situation there arises a sensitivity to all sorts of characters which he has not noticed before. When he stops, mind, we say, is freed. He does not simply look for the indication of the path going ahead. The dog and the man would both try to find a point where they could cross. But what the man could do that the dog could not would be to note that the sides of the chasm seem to be approaching each other in one direction. He picks out the best places to try, and that approach which he indicates to himself determines the way in which he is going to go. If the dog saw at a distance a narrow place he would run to it, but probably he would not be affected by the gradual approach which the human individual symbolically could indicate to himself.

The human individual would see other objects about him, and have other images appear in his experience. He sees a tree which might serve as a bridge across the space ahead of him. He might try various sorts of possible actions which would be suggested to him in such a situation, and present them to himself by means of the symbols he uses. He has not simply conditioned certain

[ii]The meanings of things or objects are actual inherent properties or qualities of them; the locus of any given meaning is in the thing which, as we say, "has it." We refer to the meaning of a thing when we make use of the symbol. Symbols stand for the meanings of those things or objects which have meanings; they are given portions of experience which point to, indicate, or represent other portions of experience not directly present or given at the time when, and in the situation in which, any one of them is thus present (or is immediately experienced). The symbol is thus more than a mere substitute stimulus—more than a mere stimulus for a conditioned response or reflex. For the conditioned reflex—the response to a mere substitute stimulus—does not or need not involve consciousness; whereas the response to a symbol does and must involve consciousness. Conditioned reflexes plus consciousness of the attitudes and meanings they involve are what constitute language, and hence lay the basis, or comprise the mechanism for, thought and intelligent conduct. Language is the means whereby individuals can indicate to one another what their responses to objects will be, and hence what the meanings of objects are; it is not a mere system of conditioned reflexes. Rational conduct always involves a reflexive reference to self, that is, an indication to the individual of the significances which his actions or gestures have for other individuals. And the experiential or behavioristic basis for such conduct—the neuro-physiological mechanism of thinking—is to be found, as we have seen, in the central nervous system.

responses by certain stimuli. If he had, he would be bound to those. What he does do by means of these symbols is to indicate certain characters which are present, so that he can have these responses there all ready to go off. He looks down the chasm and thinks he sees the edges drawing together, and he may run toward that point. Or he may stop and ask if there is not some other way in which he can hasten his crossing. What stops him is a variety of other things he may do. He notes all the possibilities of getting across. He can hold on to them by means of symbols, and relate them to each other so that he can get a final action. The beginning of the act is there in his experience. He already has a tendency to go in a certain direction and what he would do is already there determining him. And not only is that determination there in his attitude but he has that which is picked out by means of the term "that is narrow, I can jump it." He is ready to jump, and that reflex is ready to determine what he is doing. These symbols, instead of being a mere conditioning of reflexes, are ways of picking out the stimuli so that the various responses can organize themselves into a form of action.[iii]

The situation in which one seeks conditioning responses is, I think, as far as effective intelligence is concerned, always present in the form of a problem. When a man is just going ahead he seeks the indications of the path but he does it unconsciously. He just sees the path ahead of him; he is not aware of looking for it under those conditions. But when he reaches the chasm, this onward movement is stopped by the very process of drawing back from the chasm. That conflict, so to speak, sets him free to see a whole set of other things. Now,

the sort of things he will see will be the characters which represent various possibilities of action under the circumstances. The man holds on to these different possibilities of response in terms of the different stimuli which present themselves, and it is his ability to hold them there that constitutes his mind.

We have no evidence of such a situation in the case of the lower animals, as is made fairly clear by the fact that we do not find in any animal behavior that we can work out in detail any symbol, any method of communication, anything that will answer to these different responses so that they can all be held there in the experience of the individual. It is that which differentiates the action of the reflectively intelligent being from the conduct of the lower forms; and the mechanism that makes that possible is language. We have to recognize that language is a part of conduct. Mind involves, however, a relationship to the characters of things. Those characters are in the things, and while the stimuli call out the response which is in one sense present in the organism, the responses are to things out there. The whole process is not a mental product and you cannot put it inside of the brain. Mentality is that relationship of the organism to the situation which is mediated by sets of symbols. . . .

Mentality on our approach simply comes in when the organism is able to point out meanings to others and to himself. This is the point at which mind appears, or if you like, emerges. What we need to recognize is that we are dealing with the relationship of the organism to the environment selected by its own sensitivity. The psychologist is interested in the mechanism which the human species has evolved to get control over these relationships. The relationships

---

[iii]The reflective act consists in a reconstruction of the perceptual field so that it becomes possible for impulses which were in conflict to inhibit action no longer. This may take place by such a temporal readjustment that one of the conflicting impulses finds a later expression. In this case there has entered into the perceptual field other impulses which postpone the expression of that which had inhibited action. Thus, the width of the ditch inhibits the impulse to jump. There enters into the perceptual field the image of a narrower stretch and the impulse to go ahead finds its place in a combination of impulses, including that of movement toward the narrower stretch.

The reconstruction may take place through the appearance of other sensory characters in the field ignored before. A board long enough to bridge the ditch is recognized. Because the individual has already the complex of impulses which lead to lifting it and placing it across the ditch it becomes a part of the organized group of impulses that carry the man along toward his destination. In neither case would he be ready to respond to the stimulus (in the one case the image of the narrower stretch of the ditch, in the other the sight of the board) if he had not reactions in his nature answering to these objects, nor would these tendencies to response sensitize him to their stimuli if they were not freed from firmly organized habits. It is this freedom, then, that is the prerequisite of reflection, and it is our social self-reflective conduct that gives this freedom to human individuals in their group life (MS).

have been there before the indications are made, but the organism has not in its own conduct controlled that relationship. It originally has no mechanism by means of which it can control it. The human animal, however, has worked out a mechanism of language communication by means of which it can get this control. Now, it is evident that much of that mechanism does not lie in the central nervous system, but in the relation of things to the organism. The ability to pick these meanings out and to indicate them to others and to the organism is an ability which gives peculiar power to the human individual. The control has been made possible by language. It is that mechanism of control over meaning in this sense which has, I say, constituted what we term "mind." The mental processes do not, however, lie in words any more than the intelligence of the organism lies in the elements of the central nervous system. Both are part of a process that is going on between organism and environment. The symbols serve their part in this process, and it is that which makes communication so important. Out of language emerges the field of mind.

It is absurd to look at the mind simply from the standpoint of the individual human organism; for, although it has its focus there, it is essentially a social phenomenon; even its biological functions are primarily social. The subjective experience of the individual must be brought into relation with the natural, socio-biological activities of the brain in order to render an acceptable account of mind possible at all; and this can be done only if the social nature of mind is recognized. The meagerness of individual experience in isolation from the processes of social experience—in isolation from its social environment—should, moreover, be apparent. We must regard mind, then, as arising and developing within the social process, within the empirical matrix of social interactions. We must, that is, get an inner individual experience from the standpoint of social acts which include the experiences of separate individuals in a social context wherein those individuals interact. The processes of experience which the human brain makes possible are made possible only for a group of interacting individuals: only for individual organisms which are members of a society; not for the individual organism in isolation from other individual organisms.

Mind arises in the social process only when that process as a whole enters into, or is present in, the experience of any one of the given individuals involved in that process. When this occurs the individual becomes self-conscious and has a mind; he becomes aware of his relations to that process as a whole, and to the other individuals participating in it with him; he becomes aware of that process as modified by the reactions and interactions of the individuals—including himself—who are carrying it on. The evolutionary appearance of mind or intelligence takes place when the whole social process of experience and behavior is brought within the experience of any one of the separate individuals implicated therein, and when the individual's adjustment to the process is modified and refined by the awareness or consciousness which he thus has of it. It is by means of reflexiveness—the turning-back of the experience of the individual upon himself—that the whole social process is thus brought into the experience of the individuals involved in it; it is by such means, which enable the individual to take the attitude of the other toward himself, that the individual is able consciously to adjust himself to that process, and to modify the resultant of that process in any given social act in terms of his adjustment to it. Reflexiveness, then, is the essential condition, within the social process, for the development of mind.

Meaning

We are particularly concerned with intelligence on the human level, that is, with the adjustment to one another of the acts of different human individuals within the human social process; an adjustment which takes place through communication: by gestures on the lower planes of human evolution, and by significant symbols (gestures which posses meanings and are hence more than mere substitute stimuli) on the higher planes of human evolution.

The central factor in such adjustment is "meaning." Meaning arises and lies within the field of the relation between the gesture of a given human organism and the subsequent behavior of this organism as indicated to another human organism by that gesture. If that gesture does so indicate to another organism the subsequent (or resultant) behavior of the given organism, then it has meaning. In other words, the relationship between a given stimulus—as a gesture—and the later phases of the social act of which it is an early (if not the initial) phase constitutes the field within which meaning originates and exists. Meaning is thus a development of something objectively there as a relation between certain phases of the social act; it is not a psychical addition to that act and it is not an "idea" as traditionally conceived. A gesture by one organism, the resultant of the social act in which the gesture is an early phase, and the response of another organism to the gesture, are the relata in a triple or threefold relationship of gesture to first organism, of gesture to second organism, and of gesture to subsequent phases of the given social act; and this threefold relationship constitutes the matrix within which meaning arises, or which develops into the field of meaning. The gesture stands for a certain resultant of the social act, a resultant to which there is a definite response on the part of the individuals involved therein; so that meaning is given or stated in terms of response. Meaning is implicit—if not always explicit—in the relationship among the various phases of the social act to which it refers, and out of which it develops. And its development takes place in terms of symbolization at the human evolutionary level.

We have been concerning ourselves, in general, with the social process of experience and behavior as it appears in the calling out by the act of one organism of an adjustment to that act in the responsive act of another organism. We have seen that the nature of meaning is intimately associated with the social process as it thus appears, that meaning involves this threefold relation among phases of the social act as the context in which it arises and develops: this relation of the gesture of one organism to the adjustive response of another organism (also implicated in the given act), and to the completion of the given act—a relation such that the second organism responds to the gesture of the first as indicating or referring to the completion of the given act. For example, the chick's response to the cluck of the mother hen is a response to the meaning of the cluck; the cluck refers to danger or to food, as the case may be, and has this meaning or connotation for the chick.

The social process, as involving communication, is in a sense responsible for the appearance of new objects in the field of experience of the individual organisms implicated in that process.

Organic processes or responses in a sense constitute the objects to which they are responses; that is to say, any given biological organism is in a way responsible for the existence (in the sense of the meanings they have for it) of the objects to which it physiologically and chemically responds. There would, for example, be no food—no edible objects—if there were no organisms which could digest it. And similarly, the social process in a sense constitutes the objects to which it responds, or to which it is an adjustment. That is to say, objects are constituted in terms of meanings within the social process of experience and behavior through the mutual adjustment to one another of the responses or actions of the various individual organisms involved in that process, an adjustment made possible by means of a communication which takes the form of a conversation of gestures in the earlier evolutionary stages of that process, and of language in its later stages.

Awareness or consciousness is not necessary to the presence of meaning in the process of social experience. A gesture on the part of one organism in any given social act calls out a response on the part of another organism which is directly related to the action of the first organism and its outcome; and a gesture is a symbol of the result of the given social act of one organism (the organism making it) in so far as it is responded to by another organism (thereby also involved in that act) as indicating that result. The mechanism of meaning is thus present in the social act before the emergence of consciousness or awareness of meaning occurs. The act or adjustive response of the second organism gives to the gesture of the first organism the meaning which it has.

Symbolization constitutes objects not constituted before, objects which would not exist except for the context of social relationships wherein symbolization occurs. Language does not simply symbolize a situation or object which is already there in advance; it makes possible the existence or the appearance of that situation or object, for it is a part of the mechanism whereby that situation or object is created. The social process relates the responses of one individual to the gestures of another, as the meanings of the latter, and is thus responsible for the rise and existence of new objects in the social situation, objects dependent upon or constituted by these meanings. Meaning is thus not to be conceived, fundamentally, as a state of consciousness, or as a set of organized relations existing or subsisting mentally outside the field of experience into which they enter; on the contrary, it should be conceived objectively, as having its existence entirely within this field itself.[iv] The response of one organism to the gesture of another in any given social act is the meaning of that gesture, and also is in a sense responsible for the appearance or coming into being of the new object—or new content of an old object—to which that gesture refers through the outcome of the given social act in which it is an early phase. For, to repeat, objects are in a genuine sense constituted within the social process of experience, by the communication and mutual adjustment of behavior among the individual organisms which are involved in that process and which carry it on. Just as in fencing the parry is an interpretation of the thrust, so, in the social act, the adjustive response of one organism to the gesture of another is the interpretation of that gesture by that organism—it is the meaning of that gesture.

At the level of self-consciousness such a gesture becomes a symbol, a significant symbol. But the interpretation of gestures is not, basically, a process going on in a mind as such, or one necessarily involving a mind; it is an external, overt, physical, or physiological process going on in the actual field of social experience. Meaning can be described, accounted for, or stated in terms of symbols or language at its highest and most complex stage of development (the stage it reaches in human experience), but language simply lifts out of the social process a situation which is logically or implicitlythere already. The language symbol is simply a significant or conscious gesture.

Two main points are being made here: (1) that the social process, through the communication which it makes possible among the individuals implicated in it, is responsible for the appearance of a whole set of new objects in nature, which exist in relation to it (objects, namely, of "common sense"); and (2) that the gesture of one organism and the adjustive response of another organism to that gesture within any given social act bring out the relationship that exists between the gesture as the beginning of the given act and the completion

or resultant of the given act, to which the gesture refers. These are the two basic and complementary logical aspects of the social process.

The result of any given social act is definitely separated from the gesture indicating it by the response of another organism to that gesture, a response which points to the result of that act as indicated by that gesture. This situation is all there—is completely given—on the non-mental, non-conscious level, before the analysis of it on the mental or conscious level. Dewey says that meaning arises through communication.[v] It isto the content to which the social process gives rise that this statement refers; not to bare ideas or printed words as such, but to the social process which has been so largely responsible for the objects constituting the daily environment in which we live: a process in which communication plays the main part. That process can give rise to these new objects in nature only in so far as it makes possible communication among the individual organisms involved in it. And the sense in which it is responsible for their existence—indeed for the existence of the whole world of common-sense objects—is the sense in which it determines, conditions, and makes possible their abstraction from the total structure of events, as identities which are relevant for everyday social behavior; and in that sense, or as having that meaning, they are existent only relative to that behavior. In the same way, at a later, more advanced stage of its development, communication is responsible for the existence of the whole realm of scientific objects as well as identities abstracted from the total structure of events by virtue of their relevance for scientific purposes.

The logical structure of meaning, we have seen, is to be found in the threefold relationship of gesture to adjustive response and to the resultant of the given social act. Response on the part of the second organism to the gesture of the first is the interpretation—and brings out the meaning—of that gesture, as indicating the resultant of the social act which it initiates, and in which both organisms are thus involved. This threefold or triadic relation between gesture, adjustive response, and resultant of the social act which the gesture initiates is the basis of meaning; for the existence of meaning depends upon the fact

---

[iv]Nature has meaning and implication but not indication by symbols. The symbol is distinguishable from the meaning it refers to. Meanings are in nature, but symbols are the heritage of man.

[v]See *Experience and Nature*, Chap. V.

that the adjustive response of the second organism is directed toward the resultant of the given social act as initiated and indicated by the gesture of the first organism. The basis of meaning is thus objectively there in social conduct, or in nature in its relation to such conduct. Meaning is a content of an object which is dependent upon the relation of an organism or group of organisms to it. It is not essentially or primarily a psychical content (a content of mind or consciousness), for it need not be conscious at all, and is not in fact until significant symbols are evolved in the process of human social experience. Only when it becomes identified with such symbols does meaning become conscious. The meaning of a gesture on the part of one organism is the adjustive response of another organism to it, as indicating the resultant of the social act it initiates, the adjustive response of the second organism being itself directed toward or related to the completion of that act. In other words, meaning involves a reference of the gesture of one organism to the resultant of the social act it indicates or initiates, as adjustively responded to in this reference by another organism; and the adjustive response of the other organism is the meaning of the gesture.

Gestures may be either conscious (significant) or unconscious (non-significant). The conversation of gestures is not significant below the human level, because it is not conscious, that is, not self-conscious (though it is conscious in the sense of involving feelings or sensations). An animal as opposed to a human form, in indicating something to, or bringing out a meaning for, another form, is not at the same time indicating or bringing out the same thing or meaning to or for himself; for he has no mind, no thought, and hence there is no meaning here in the significant or self-conscious sense. A gesture is not significant when the response of another organism to it does not indicate to the organism making it what the other organism is responding to.[vi]

Much subtlety has been wasted on the problem of the meaning of meaning. It is not necessary, in attempting to solve this problem, to have recourse to psychical states, for the nature of meaning, as we have seen, is found to be implicit in the structure of the social act, implicit in the relations among its three basic individual components: namely, in the triadic relation of a gesture of one individual, a response to that gesture by a second individual, and completion of the given social act initiated by the gesture of the first individual. And the fact that the nature of meaning is thus found to be implicit in the structure of the social act provides additional emphasis upon the necessity, in social psychology, of starting off with the initial assumption of an ongoing social process of experience and behavior in which any given group of human individuals is involved, and upon which the existence and development of their minds, selves, and self-consciousness depend.

:: ::

Introduction to "Self"

In the selections taken from the chapters on the self, Mead presents a number of concepts central to his theory. Here, we present an overview of Mead's notion of the "I," "me," and "generalized other," and their realization through the "play" and "game" stages of the development of self-consciousness. Moreover, in his discussion of the self, you will again encounter two themes that form the core of Mead's program: (1) the interconnectedness between the self and social experiences, and (2) language as the tool that mediates this relationship.

[vi]There are two characters which belong to that which we term "meanings," one is participation and the other is communicability. Meaning can arise only in so far as some phase of the act which the individual is arousing in the other can be aroused in himself. There is always to this extent participation. And the result of this participation is communicability, i.e., the individual can indicate to himself what he indicates to others. There is communication without significance where the gesture of the individual calls out the response in the other without calling out or tending to call out the same response in the individual himself. Significance from the standpoint of the observer may be said to be present in the gesture which calls out the appropriate response in the other or others within a cooperative act, but it does not become significant to the individuals who are involved in the act unless the tendency to the act is aroused within the individual who makes it, and unless the individual who is directly affected by the gesture puts himself in the attitude of the individual who makes the gesture (MS).

For Mead, the self does not exist as a "personality" or an "identity" that answers the question, "Who am I?" Instead, the self exists as self-consciousness, that is, the capacity to be both subject and object to one's self that is made possible solely through interaction. As we project the possible implications of courses of action and attempt to elicit the desired responses from others, we become an "object" to ourselves. However, an individual is aware of his self as an object, "not directly, but only indirectly, from the particular standpoints of other individual [s] . . . or from the generalized standpoint of the social group as a whole to which he belongs" (Mead 1934/1962:138). In turn, the experience of our self as an object becomes possible only by taking the attitudes of others toward our self.

Viewing ourselves as an object is the phase of the self Mead termed the "**me**."[1] This phase represents a sense of who we are that is created, sustained, and modified through our interaction with others. (In the opening to this chapter, "me" is who you see in the mirror.) It is the "organized set of attitudes of others which one himself assumes" (ibid.:175); the self one is aware of when thinking, or taking the attitude of the other. Thus, the self is a social product that is rooted in our perceptions of how others interpret our behaviors.

That we see ourselves and respond to our conduct as others would is not the only way in which the self is a reflection of social interaction. According to Mead, "We divide ourselves up in all sorts of different selves with reference to our acquaintances" (ibid.:142). In other words, it is interaction and the context within which it takes place that determines who we are for the moment—the side of our self that we show to others. The self you experience as a sibling is different from the one you experience as a coworker or a student. We all have, in a sense, multiple social selves or personalities, as assuming the attitudes of a particular other or community shapes our behavior.

While the "me" is involved in thinking or reflexive role taking, the self enters into another phase during moments of interaction. Here, Mead argues that the "**I**" is in the foreground. The "**I**" reacts or answers to the "me," the phase of the self that one is conscious of. (The "I" is the phase of the self that *looks* in the mirror in our opening example.) It is "the response of the [individual] to the attitudes of the others" (ibid.:175). The individual is aware of the "I" of the present moment only as it passes into memory, into the "me." It represents the here-and-now, creative aspect of one's self, for no matter how much you rehearse your behaviors or try to anticipate the reactions of others, interaction may take an unplanned

Photo 8.1   The Play Stage: A "Mommy" Feeding Her "Baby"

---

[1]In developing his view of the self, Mead borrowed much from William James, including the latter's idea of the "me" and the notion of possessing multiple social selves. See the Significant Others box that briefly summarizes James's work.

course. You never know for certain what you will say or how others will interpret your behaviors until after you have spoken and a response has been made. For Mead, it is this spontaneous, unpredictable aspect of the self and social interaction that sparks personal and cultural innovation. Without the "I," social life would be static and relentlessly conformist.

According to Mead, the self (that is, self-consciousness) develops through two cognitive stages: "play" and the "game."[2] The **play stage** is marked by the ability to assume the attitude

Photo 8.2a   Six-Year-Old Children Playing Soccer

Photo 8.2b   Twelve-Year-Old Children Playing Soccer. Among other things, notice the difference in spacing of players in the age groups.

---

[2]Contemporary theorists are not in agreement as to how many stages of self-development Mead proposes. Some identify three distinct stages (play, game, and generalized other), while others read Mead as outlining four stages (preplay, play, game, and generalized other). We present a two-stage model, conceiving of the generalized other not as a distinct phase in the development of the self, but as a "community attitude" that is assumed in the game stage. We believe this is a more accurate reading of Mead's ideas.

of only one particular individual at a time. This is the stage of self-consciousness that we find in children until around the age of eight. Here, children perhaps are pretending to be a parent, a superhero, or a princess, moving from one role to the next in an unconnected fashion. In play, the child is able only to switch successively between discrete roles while taking the attitude of the specific other toward her self.

The next phase of development is the **game stage.** Here, the child is able to move beyond simply taking the role of particular others and assume the roles of multiple others simultaneously. Moreover, the child has the ability to control his actions on the basis of abstract "rules of the game." This configuration of "roles organized according to rules" brings the attitudes of all participants together in an abstract unity called the "generalized other." The **generalized other** is the organized community or group to which an individual belongs. When taking the attitude of the generalized other towards one's self, one assumes the attitudes that are common to the whole group. (It is the vantage point from which one sees, and thus gives meaning to, the "who" that is reflected in the mirror from the opening to this chapter.) Responding to ourselves from the point of view of the whole community is the mechanism through which that community controls the behaviors of its members, making possible the coordination of diverse activities in large groups or institutions. Moreover, by assuming the attitude of the generalized other, we are able to orient our behavior according to abstract ideals and principles such as freedom, individual rights, and fairness.

The notion of the generalized other clearly illustrates Mead's view of the dialectical relationship between self and society (a theme echoed in the work of Georg Simmel). As the generalized other develops from its rudimentary form found in the games children play to ever-widening, increasingly abstract social circles, self-consciousness likewise develops more fully as the individual begins to take the imagined attitudes of others toward himself that he does not—and never will—know. As Mead remarks, "We cannot be ourselves unless we are also members in whom there is a community of attitudes which control the attitudes of all. . . . The individual possesses a self only in relation to the selves of the other members of his social group; and the structure of his self expresses or reflects the general behavior pattern of this social group to which he belongs" (ibid.:164).[3] Yet, organized group processes and activities—that is, society itself—are made possible "only in so far as every individual involved in them or belonging to that society can take the general attitudes of all such other individuals with reference to these processes and activities" (ibid.:155).

Early in the game stage, children are preoccupied with rules, often becoming quite literal and rigid in following them. For instance, a batter might not be willing to "take one" for the team, allowing himself to be hit by a pitch, or a player might not move outside her usual position even if it means enhancing the team's chances for success. At this point, children are still unable to abstract themselves from the game as such; they are unable to understand the constructed nature, and thus flexibility, of the rules that organize interaction. With the development of the ability to take the attitude of the generalized other, however, one is not simply following the rules; instead, the player has become conscious of her part in the realization of the rules. No longer is she merely "in" the game; now the player is "above" it, as the rules become subject to self-conscious interpretation and, possibly, to change.

Similarly, consider how a six-year-old might decide whether or not to steal. The child's internal conversation or thinking would entail taking the attitude of a particular other toward his potential lines of conduct. His decision is then based on the imagined response of that other, a parent perhaps, to the action in question. Internally responding to the contemplated action as his parent might, the child's decision to not steal is based on his reaction to the

---

[3]Again, our self is realized only in relation to a wider community of others. One is a teacher only if she has students, or an artist or actor only if she has an audience.

anticipated disapproval of that other. In contrast, an adult's decision would be based also on the imagined response of the generalized other. Here, the same decision is not derived solely from a reaction to the anticipated disapproval of a specific other, but also from an internal response to the community's shared *principle* that it is wrong to steal.

The difference between the play and the game stages is readily apparent if we look at how a soccer game is played by six-year-olds as opposed to a game played by 15-year-olds. In the former, children play as if they are all magnetically attracted to the ball. They have no sense of position or organized, structured participation. There is little if any passing or cooperation between the players. The game consists mainly in a flock of children running wildly after the ball (though it is not uncommon for some players to lose interest in the game completely—e.g., to stop to pick dandelions, wave at their parents, gaze into space, or examine their socks).

Compare this scenario to a game of soccer played by tenth graders. Here, there is a sense of position and an understanding of when to pass the ball and who to pass it to. Teenagers do not selfishly follow the ball, because *it* no longer is the object of their attention. Instead, the *game* is now the focus of the players. The ability to cooperate as a team produces a kind of rudimentary society where each player is able simultaneously to take the attitude of all the other players, anticipate their reactions, and adjust her own line of action accordingly.

In taking the attitude of the other (whether a particular individual or generalized other), we, of course, are not capable of getting inside another person's head, nor do we experience our self as a physical, tangible object. Instead, when determining what path of behaviors we should pursue we think and develop ourselves in terms of language. Language is thus the essential medium for the genesis of the mind and self. Moreover, as the medium through which we control our actions and coordinate them with those of others, language makes society itself possible.

"Self" (1934)

George Herbert Mead

THE SELF AND THE ORGANISM

In our statement of the development of intelligence we have already suggested that the language process is essential for the development of the self. The self has a character which is different from that of the physiological organism proper. The self is something which has a development; it is not initially there, at birth, but arises in the process of social experience and activity, that is, develops in the given individual as a result of his relations to that process as a whole and to other individuals within that process. The intelligence of the lower forms of animal life, like a great deal of human intelligence, does not involve a self. In our habitual actions, for example, in our moving about in a world that is simply there and to which we are so adjusted that no thinking is involved, there is a certain amount of sensuous experience such as persons have when they are just waking up, a bare thereness of the world. Such characters about us may exist in experience without taking their place in relationship to the self. One must, of course, under those conditions, distinguish between the experience that immediately takes place and our own organization of it into the experience of the self. One says upon analysis that a certain item had its place in his experience, in the experience of his self. We do inevitably tend at a certain level of sophistication to organize all experience into that of a self. We do so intimately identify our experiences, especially our affective experiences, with the self that it takes a moment's abstraction to realize that pain and pleasure can be there without being the experience of the self. Similarly, we normally organize our memories upon the string of our self. If we date things we always date them from the point of view of our past experiences. We frequently have memories that we cannot date, that we cannot place. A picture comes before us suddenly and we are at a loss to explain when that experience originally took place. We remember perfectly distinctly the picture, but we do not have it definitely placed, and until we can place it in terms of our past experience we are not satisfied. Nevertheless, I think it is obvious when one comes to consider it that the self is not necessarily involved in the life of the organism, nor involved in what we term our sensuous experience, that is, experience in a world about us for which we have habitual reactions.

We can distinguish very definitely between the self and the body. The body can be there and can operate in a very intelligent fashion without there being a self involved in the experience. The self has the characteristic that it is an object to itself, and that characteristic distinguishes it from other objects and from the body. It is perfectly true that the eye can see the foot, but it does not see the body as a whole. We cannot see our backs; we can feel certain portions of them, if we are agile, but we cannot get an experience of our whole body. There are, of course, experiences which are somewhat vague and difficult of location, but the bodily experiences are for us organized about a self. The foot and hand belong to the self. We can see our feet, especially if we look at them from the wrong end of an opera glass, as strange things which we have difficulty in recognizing as our own. The parts of the body are quite distinguishable from the self. We can lose parts of the body without any serious invasion of the self. The mere ability to experience different parts of the body is not different from the experience of a table. The table presents a different feel from what the hand does when one hand feels another, but it is an experience of something with which we come definitely into contact. The body does not experience itself as a whole, in the sense in which the self in some way enters into the experience of the self.

It is the characteristic of the self as an object to itself that I want to bring out. This characteristic is represented in the word "self," which is a reflexive, and indicates that which can be both subject and object. This type of object is essentially different from other objects, and in the past it has been distinguished as conscious, a term which indicates an experience with, an experience of, one's self. It was assumed that consciousness in some way carried this capacity of being an object to itself. In giving a behavioristic statement of consciousness we have to look for some sort of experience in which the physical organism can become an object to itself.[i]

When one is running to get away from someone who is chasing him, he is entirely occupied in this action, and his experience may be swallowed up in the objects about him, so that he has, at the time being, no consciousness of self at all. We must be, of course, very completely occupied to have that take place, but we can, I think, recognize that sort of a possible experience in which the self does not enter. . . . In such instances there is a contrast between an experience that is absolutely wound up in outside activity in which the self as an object does not enter, and an activity of memory and imagination in which the self is the principal object. The self is then entirely distinguishable from an organism that is surrounded by things and acts with reference to things, including parts of its own body. These latter may be objects like other objects, but they are just objects out there in the field, and they do not involve a self that is an object to the organism. This is, I think, frequently overlooked. It is that fact which makes our anthropomorphic reconstructions of animal life so fallacious. How can an individual get outside himself (experientially) in such a way as to become an object to himself? This is the essential psychological

---

[i] Man's behavior is such in his social group that he is able to become an object to himself, a fact which constitutes him a more advanced product of evolutionary development than are the lower animals. Fundamentally it is this social fact—and not his alleged possession of a soul or mind with which he, as an individual, has been mysteriously and supernaturally endowed, and with which the lower animals have not been endowed—that differentiates him from them.

problem of selfhood or of self-consciousness; and its solution is to be found by referring to the process of social conduct or activity in which the given person or individual is implicated. The apparatus of reason would not be complete unless it swept itself into its own analysis of the field of experience; or unless the individual brought himself into the same experiential field as that of the other individual selves in relation to whom he acts in any given social situation. Reason cannot become impersonal unless it takes an objective, non-affective attitude toward itself; otherwise we have just consciousness, not *self*-consciousness. And it is necessary to rational conduct that the individual should thus take an objective, impersonal attitude toward himself, that he should become an object to himself. For the individual organism is obviously an essential and important fact or constituent element of the empirical situation in which it acts; and without taking objective account of itself as such, it cannot act intelligently, or rationally.

The individual experiences himself as such, not directly, but only indirectly, from the particular standpoints of other individual members of the same social group, or from the generalized standpoint of the social group as a whole to which he belongs. For he enters his own experience as a self or individual, not directly or immediately, not by becoming a subject to himself, but only in so far as he first becomes an object to himself just as other individuals are objects to him or in his experience; and he becomes an object to himself only by taking the attitudes of other individuals toward himself within a social environment or context of experience and behavior in which both he and they are involved.

The importance of what we term "communication" lies in the fact that it provides a form of behavior in which the organism or the individual may become an object to himself. It is that sort of communication which we have been discussing—not communication in the sense of the cluck of the hen to the chickens, or the bark of a wolf to the pack, or the lowing of a cow, but communication in the sense of significant symbols, communication which is directed not only to others but also to the individual himself. So far as that type of communication is a part of behavior it at least introduces a self. Of course,

one may hear without listening; one may see things that he does not realize; do things that he is not really aware of. But it is where one does respond to that which he addresses to another and where that response of his own becomes a part of his conduct, where he not only hears himself but responds to himself, talks and replies to himself as truly as the other person replies to him, that we have behavior in which the individuals become objects to themselves. . . .

The self, as that which can be an object to itself, is essentially a social structure, and it arises in social experience. After a self has arisen, it in a certain sense provides for itself its social experiences, and so we can conceive of an absolutely solitary self. But it is impossible to conceive of a self arising outside of social experience. When it has arisen we can think of a person in solitary confinement for the rest of his life, but who still has himself as a companion, and is able to think and to converse with himself as he had communicated with others. That process to which I have just referred, of responding to one's self as another responds to it, taking part in one's own conversation with others, being aware of what one is saying and using that awareness of what one is saying to determine what one is going to say thereafter—that is a process with which we are all familiar. We are continually following up our own address to other persons by an understanding of what we are saying, and using that understanding in the direction of our continued speech. We are finding out what we are going to say, what we are going to do, by saying and doing, and in the process we are continually controlling the process itself. In the conversation of gestures what we say calls out a certain response in another and that in turn changes our own action, so that we shift from what we started to do because of the reply the other makes. The conversation of gestures is the beginning of communication. The individual comes to carry on a conversation of gestures with himself. He says something, and that calls out a certain reply in himself which makes him change what he was going to say. One starts to say something, we will presume an unpleasant something, but when he starts to say it he realizes it is cruel. The effect on himself of what he is saying checks him; there is here a conversation of gestures between the individual and himself. We mean by significant speech that the action is one that affects the individual himself, and that the effect upon the individual himself is part of the intelligent carrying-out of the conversation with others. Now we, so to speak, amputate that social phase and dispense with it for the time being, so that one is talking to one's self as one would talk to another person.[ii]

This process of abstraction cannot be carried on indefinitely. One inevitably seeks an audience, has to pour himself out to somebody. In reflective intelligence one thinks to act, and to act solely so that this action remains a part of a social process. Thinking becomes preparatory to social action. The very process of thinking is, of course, simply an inner conversation that goes on, but it is a conversation of gestures which in its completion implies the expression of that which one thinks to an audience. One separates the significance of what he is saying to others from the actual speech and gets it ready before saying it. He thinks it out, and perhaps writes it in the form of a book; but it is still a part of social intercourse in which one is addressing other persons and at the same time addressing one's self, and in which one controls the address to other persons by the response made to one's own gesture. That the person should be responding to himself is necessary to the self, and it is this sort of social conduct which provides behavior within which that self appears. I know of no other form of behavior than the linguistic in which the individual is an object to himself, and, so far as I can see, the individual is not a self in the reflexive sense unless he is an object to himself. It is this fact that gives a critical importance to communication, since this is a type of behavior in which the individual does so respond to himself.

We realize in everyday conduct and experience that an individual does not mean a great deal of what he is doing and saying. We frequently say that such an individual is not himself. We come away from an interview with a realization that we have left out important things, that there are parts of the self that did not get into what was said. What determines the amount of the self that gets into communication is the social experience itself. Of course, a good deal of the self does not need to get expression. We carry on a whole series of different relationships to different people. We are one thing to one man and another thing to another. There are parts of the self which exist only for the self in relationship to itself. We divide ourselves up in all sorts of different selves with reference to our acquaintances. We discuss politics with one and religion with another. There are all sorts of different selves answering to all sorts of different social reactions. It is the social process itself that is responsible for the appearance of the self; it is not there as a self apart from this type of experience. . . .

The unity and structure of the complete self reflects the unity and structure of the social process as a whole; and each of the elementary selves of which it is composed reflects the unity and structure of one of the various aspects of that process in which the individual is implicated. In other words, the various elementary selves which constitute, or are organized into, a complete self are the various aspects of the structure of that complete self answering to the various aspects of the structure of the social process as a whole; the structure of the complete self is thus a reflection of the complete social process. The organization and unification of a social group is identical with the organization and unification of any one of the selves arising within the social

---

[ii]It is generally recognized that the specifically social expressions of intelligence, or the exercise of what is often called "social intelligence," depend upon the given individual's ability to take the rôles of, or "put himself in the place of," the other individuals implicated with him in given social situations; and upon his consequent sensitivity to their attitudes toward himself and toward one another. These specifically social expressions of intelligence, of course, acquire unique significance in terms of our view that the whole nature of intelligence is social to the very core—that this putting of one's self in the places of others, this taking by one's self of their roles or attitudes, is not merely one of the various aspects or expressions of intelligence or of intelligent behavior, but is the very essence of its character. . . .

process in which that group is engaged, or which it is carrying on. . . .[iii]

The Background of the Genesis of the Self

The problem now presents itself as to how, in detail, a self arises. We have to note something of the background of its genesis. First of all there is the conversation of gestures between animals involving some sort of co-operative activity. There the beginning of the act of one is a stimulus to the other to respond in a certain way, while the beginning of this response becomes again a stimulus to the first to adjust his action to the oncoming response. Such is the preparation for the completed act, and ultimately it leads up to the conduct which is the outcome of this preparation. The conversation of gestures, however, does not carry with it the reference of the individual, the animal, the organism, to itself. It is not acting in a fashion which calls for a response from the form itself, although it is conduct with reference to the conduct of others. We have seen, however, that there are certain gestures that do affect the organism as they affect other organisms and may, therefore, arouse in the organism responses of the same character as aroused in the other. Here, then, we have a situation in which the individual may at least arouse responses in himself and reply to these responses, the condition being that the social stimuli have an effect on the individual which is like that which they have on the other. That, for example, is what is implied in language; otherwise language as significant symbol would disappear, since the individual would not get the meaning of that which he says.

The peculiar character possessed by our human social environment belongs to it by virtue of the peculiar character of human social activity; and that character, as we have seen, is to be found in the process of communication, and more particularly in the triadic relation on which the existence of meaning is based: the relation of the gesture of one organism to the adjustive response made to it by another organism, in its indicative capacity as pointing to the completion or resultant of the act it initiates (the meaning of the gesture being thus the response of the second organism to it as such, or as a gesture). What, as it were, takes the gesture out of the social act and isolates it as such—what makes it something more than just an early phase of an individual act—is the response of another organism, or of other organisms, to it. Such a response is its meaning, or gives it its meaning. The social situation and process of behavior are here presupposed by the acts of the individual organisms implicated therein. The gesture arises as a separable element in the social act, by virtue of the fact that it is selected out by the sensitivities of other organisms to it; it does not exist as a gesture merely in the experience of the single individual. The meaning of a gesture by one organism, to repeat, is found in the response of another organism to what would be the completion of the act of the first organism which that gesture initiates and indicates.

We sometimes speak as if a person could build up an entire argument in his mind, and then put it into words to convey it to someone else. Actually, our thinking always takes place by means of some sort of symbols. It is possible that one could have the meaning of "chair" in his experience without there being a symbol, but we would not be thinking about it in that case. We may sit down in a chair without thinking about what we are doing, that is, the approach to the chair is presumably already aroused in our experience, so that the meaning is there. But if one is thinking about the chair he must have some sort of a symbol for it. It may be the form of the chair, it may be the attitude that somebody else takes in sitting down, but it is more apt to be some language symbol that arouses this response. In a thought process there has to be some sort of a symbol that can refer to this meaning, that is, tend to call out this response, and also serve this purpose for other persons as well. It would not be a thought process if that were not the case.

[iii]The unity of the mind is not identical with the unity of the self. The unity of the self is constituted by the unity of the entire relational pattern of social behavior and experience in which the individual is implicated, and which is reflected in the structure of the self; but many of the aspects or features of this entire pattern do not enter into consciousness, so that the unity of the mind is in a sense an abstraction from the more inclusive unity of the self.

Our symbols are all universal.[iv] You cannot say anything that is absolutely particular; anything you say that has any meaning at all is universal. You are saying something that calls out a specific response in anybody else provided that the symbol exists for him in this experience as it does for you. There is the language of speech and the language of hands, and there may be the language of the expression of the countenance. One can register grief or joy and call out certain responses. There are primitive people who can carry on elaborate conversations just by expressions of the countenance. Even in these cases the person who communicates is affected by that expression just as the expects somebody else to be affected. Thinking always implies a symbol which will call out the same response in another that it calls out in the thinker. Such a symbol is a universal of discourse; it is universal in its character. We always assume that the symbol we use is one which will call out in the other person the same response, provided it is a part of his mechanism of conduct. A person who is saying something is saying to himself what he says to others; otherwise he does not know what he is talking about.

There is, of course, a great deal in one's conversation with others that does not arouse in one's self the same response it arouses in others. That is particularly true in the case of emotional attitudes. One tries to bully somebody else; he is not trying to bully himself. There is, further, a whole set of values given in speech which are not of a symbolic character. The actor is conscious of these values; that is, if he assumes a certain attitude he is, as we say, aware that this attitude represents grief. If it does he is able to respond to his own gesture in some sense as his audience does. It is not a natural situation; one is not an actor all of the time. We do at times act and consider just what the effect of our attitude is going to be, and we may deliberately use a certain tone of voice to bring about a certain result. Such a tone arouses the same response in ourselves that we want to arouse in somebody else. But a very large part of what goes on in speech has not this symbolic status. . . .

We do not normally use language stimuli to call out in ourselves the emotional response which we are calling out in others. One does, of course, have sympathy in emotional situations; but what one is seeking for there is something which is, after all, that in the other which supports the individual in his own experience. In the case of the poet and actor, the stimulus calls out in the artist that which it calls out in the other, but this is not the natural function of language; we do not assume that the person who is angry is calling out the fear in himself that he is calling out in someone else. The emotional part of our act does not directly call out in us the response it calls out in the other. If a person is hostile the attitude of the other that he is interested in, an attitude which flows naturally from his angered tones, is not one that he definitely recognizes in himself. We are not frightened by a tone which we may use to frighten somebody else. On the emotional side, which is a very large part of the vocal gesture, we do not call out in ourselves in any such degree the response we call out in others as we do in the case of significant speech. Here we should call out in ourselves the type of response we are calling out in others; we must know what we are saying, and the attitude of the other which we arouse in ourselves should control what we do say. Rationality means that the type of the response which we call out in others should be so called out in ourselves, and that this response should in turn take its place in determining what further thing we are going to say and do.

What is essential to communication is that the symbol should arouse in one's self what it arouses in the other individual. It must have that sort of universality to any person who finds

---

[iv]Thinking proceeds in terms of or by means of universals. A universal may be interpreted behavioristically as simply the social act as a whole, involving the organization and interrelation of the attitudes of all the individuals implicated in the act, as controlling their overt responses. This organization of the different individual attitudes and interactions in a given social act, with reference to their interrelations as realized by the individuals themselves, is what we mean by a universal; and it determines what the actual overt responses of the individuals involved in the given social act will be, whether that act be concerned with a concrete project of some sort (such as the relation of physical and social means to ends desired) or with some purely abstract discussion, say the theory of relativity or the Platonic ideas.

himself in the same situation. There is a possibility of language whenever a stimulus can affect the individual as it affects the other. With a blind person such as Helen Keller, it is a contact experience that could be given to another as it is given to herself. It is out of that sort of language that the mind of Helen Keller was built up. As she has recognized, it was not until she could get into communication with other persons through symbols which could arouse in herself the responses they arouse in other people that she could get what we term a mental content, or a self.

Another set of background factors in the genesis of the self is represented in the activities of play and the game.

Among primitive people, as I have said, the necessity of distinguishing the self and the organism was recognized in what we term the "double": the individual has a thing-like self that is affected by the individual as it affects other people and which is distinguished from the immediate organism in that it can leave the body and come back to it. This is the basis for the concept of the soul as a separate entity.

We find in children something that answers to this double, namely, the invisible, imaginary companions which a good many children produce in their own experience. They organize in this way the responses which they call out in other persons and call out also in themselves. Of course, this playing with an imaginary companion is only a peculiarly interesting phase of ordinary play. Play in this sense, especially the stage which precedes the organized games, is a play at something. A child plays at being a mother, at being a teacher, at being a policeman; that is, it is taking different rôles, as we say. We have something that suggests this in what we call the play of animals: a cat will play with her kittens, and dogs play with each other. Two dogs playing with each other will attack and defend, in a process which if carried through would amount to an actual fight. There is a combination of responses which checks the depth of the bite. But we do not have in such a situation the dogs taking a definite rôle in the sense that a child deliberately takes the rôle of another. This tendency on the part of the children is what we are working with in the kindergarten where the rôles which the children assume are made the basis for training. When a child does assume a rôle he has in himself the stimuli which call out that particular response or group of responses. He may, of course, run away when he is chased, as the dog does, or he may turn around and strike back just as the dog does in his play. But that is not the same as playing at something. Children get together to "play Indian." This means that the child has a certain set of stimuli which call out in itself the responses that they would call out in others, and which answer to an Indian. In the play period the child utilizes his own responses to these stimuli which he makes use of in building a self. The response which he has a tendency to make to these stimuli organizes them. He plays that he is, for instance, offering himself something, and he buys it; he gives a letter to himself and takes it away; he addresses himself as a parent, as a teacher; he arrests himself as a policeman. He has a set of stimuli which call out in himself the sort of responses they call out in others. He takes this group of responses and organizes them into a certain whole. Such is the simplest form of being another to one's self. It involves a temporal situation. The child says something in one character and responds in another character, and then his responding in another character is a stimulus to himself in the first character, and so the conversation goes on. A certain organized structure arises in him and in his other which replies to it, and these carry on the conversation of gestures between themselves.

If we contrast play with the situation in an organized game, we note the essential difference that the child who plays in a game must be ready to take the attitude of everyone else involved in that game, and that these different rôles must have a definite relationship to each other. Taking a very simple game such as hide-and-seek, everyone with the exception of the one who is hiding is a person who is hunting. A child does not require more than the person who is hunted and the one who is hunting. If a child is playing in the first sense he just goes on playing, but there is no basic organization gained. In that early stage he passes from one rôle to another just as a whim takes him. But in a game where a number of individuals are involved, then the child taking one rôle must be ready to take the rôle of everyone else. If he gets in a ball nine he must have the responses of each position involved in his own position. He must know what everyone else is going to do in order to carry out his own play. He has to take all of these rôles. They do not all have to be present in consciousness at the same time, but at some moments he has to have three or four individuals present in his own attitude, such as the one who is going to throw the ball, the one who is going to catch it, and so on. These responses must be, in some degree, present in his own make-up. In the game, then, there is a set of responses of such others so organized that the attitude of one calls out the appropriate attitudes of the other. . . .

PLAY, THE GAME, AND THE GENERALIZED OTHER

The fundamental difference between the game and play is that in the latter the child must have the attitude of all the others involved in that game. The attitudes of the other players which the participant assumes organize into a sort of unit, and it is that organization which controls the response of the individual. The illustration used was of a person playing baseball. Each one of his own acts is determined by his assumption of the action of the others who are playing the game. What he does is controlled by his being everyone else on that team, at least in so far as those attitudes affect his own particular response. We get then an "other" which is an organization of the attitudes of those involved in the same process.

The organized community or social group which gives to the individual his unity of self may be called "the generalized other." The attitude of the generalized other is the attitude of the whole community.[v] Thus, for example, in the case of such a social group as a ball team, the team is the generalized other in so far as it enters—as an organized process or social activity—into the experience of any one of the individual members of it.

If the given human individual is to develop a self in the fullest sense, it is not sufficient for him merely to take the attitudes of other human individuals toward himself and toward one another within the human social process, and to bring that social process as a whole into his individual experience merely in these terms: he must also, in the same way that he takes the attitudes of other individuals toward himself and toward one another, take their attitudes toward the various phases or aspects of the common social activity or set of social undertakings in which, as members of an organized society or social group, they are all engaged; and he must then, by generalizing these individual attitudes of that organized society or social group itself, as a whole, act toward different social projects which at any given time it is carrying out, or toward the various larger phases of the general social process which constitutes its life and of which these projects are specific manifestations. This getting of the broad activities of any given social whole or organized society as such within the experiential field of any one of the individuals involved or included in that whole is, in other words, the essential basis and prerequisite of the fullest development of that individual's self: only in so far as he takes the attitudes of the organized social group to which he belongs toward the organized, co-operative social activity or set of such activities in which that group as such is

---

[v] It is possible for inanimate objects, no less than for other human organisms, to form parts of the generalized and organized—the completely socialized—other for any given human individual, in so far as he responds to such objects socially or in a social fashion (by means of the mechanism of thought, the internalized conversation of gestures). Any thing—any object or set of objects, whether animate or inanimate, human or animal, or merely physical—toward which he acts, or to which he responds, socially, is an element in what for him is the generalized other; by taking the attitudes of which toward himself he becomes conscious of himself as an object or individual, and thus develops a self or personality. Thus, for example, the cult, in its primitive form, is merely the social embodiment of the relation between the given social group or community and its physical environment—an organized social means, adopted by the individual members of that group or community, of entering into social relations with that environment, or (in a sense) of carrying on conversations with it; and in this way that environment becomes part of the total generalized other for each of the individual members of the given social group or community.

engaged, does he develop a complete self or possess the sort of complete self he has developed. And on the other hand, the complex co-operative processes and activities and institutional functionings of organized human society are also possible only in so far as every individual involved in them or belonging to that society can take the general attitudes of all other such individuals with reference to these processes and activities and institutional functionings, and to the organized social whole of experiential relations and interactions thereby constituted—and can direct his own behavior accordingly.

It is in the form of the generalized other that the social process influences the behavior of the individuals involved in it and carrying it on, i.e., that the community exercises control over the conduct of its individual members; for it is in this form that the social process or community enters as a determining factor into the individual's thinking. In abstract thought the individual takes the attitude of the generalized other[vi] toward himself, without reference to its expression in any particular other individuals; and in concrete thought he takes that attitude in so far as it is expressed in the attitudes toward his behavior of those other individuals with whom he is involved in the given social situation or act. But only by taking the attitude of the generalized other toward himself, in one or another of these ways, can he think at all; for only thus can thinking—or the internalized conversation of gestures which constitutes thinking—occur. And only through the taking by individuals of the attitude or attitudes of the generalized other toward themselves is the existence of a universe of discourse, as that system of common or social meanings which thinking presupposes at its context, rendered possible.

The self-conscious human individual, then, takes or assumes the organized social attitudes of the given social group or community (or of some one section thereof) to which he belongs, toward the social problems of various kinds which confront that group or community at any given time, and which arise in connection with the correspondingly different social projects or organized co-operative enterprises in which that group or community as such is engaged; and as an individual participant in these social projects or co-operative enterprises, he governs his own conduct accordingly. In politics, for example, the individual identifies himself with an entire political party and takes the organized attitudes of that entire party toward the rest of the given social community and toward the problems which confront the party within the given social situation; and he consequently reacts or responds in terms of the organized attitudes of the party as a whole. He thus enters into a special set of social relations with all the other individuals who belong to that political party; and in the same way he enters into various other special sets of social relations, with various other classes of individuals respectively, the individuals of each of these classes being the other members of some one of the particular organized subgroups (determined in socially functional terms) of which he himself is a member within the entire given society or social community. In the most highly developed, organized, and complicated human social communities—those evolved by civilized man—these various socially functional classes or subgroups of individuals to which any given individual belongs (and with the other individual members of which he thus enters into a special set of social relations) are of two kinds. Some of them are concrete social classes or subgroups, such as political parties, clubs, corporations, which are all actually functional social units, in terms of which their individual members are directly

---

[vi]We have said that the internal conversation of the individual with himself in terms of words or significant gestures—the conversation which constitutes the process or activity of thinking—is carried on by the individual from the standpoint of the "generalized other." And the more abstract that conversation is, the more abstract thinking happens to be, the further removed is the generalized other from any connection with particular individuals. It is especially in abstract thinking, that is to say, that the conversation involved is carried on by the individual with the generalized other, rather than with any particular individuals. Thus it is, for example, that abstract concepts are concepts stated in terms of the attitudes of the entire social group or community; they are stated on the basis of the individual's consciousness of the attitudes of the generalized other toward them, as a result of his taking these attitudes of the generalized other and then responding to them. And thus it is also that abstract propositions are stated in a form which anyone—any other intelligent individual—will accept.

related to one another. The others are abstract social classes or subgroups, such as the class of debtors and the class of creditors, in terms of which their individual members are related to one another only more or less indirectly, and which only more or less indirectly function as social units, but which afford or represent unlimited possibilities for the widening and ramifying and enriching of the social relations among all the individual members of the given society as an organized and unified whole. The given individual's membership in several of these abstract social classes or subgroups makes possible his entrance into definite social relations (however indirect) with an almost infinite number of other individuals who also belong to or are included within one or another of these abstract social classes or subgroups cutting across functional lines of demarcation which divide different human social communities from one another, and including individual members from several (in some cases from all) such communities. Of these abstract social classes or subgroups of human individuals the one which is most inclusive and extensive is, of course, the one defined by the logical universe of discourse (or system of universally significant symbols) determined by the participation and communicative interaction of individuals; for of all such classes or subgroups, it is the one which claims the largest number of individual members, and which enables the largest conceivable number of human individuals to enter into some sort of social relation, however indirect or abstract it may be, with one another—a relation arising from the universal functioning of gestures as significant symbols in the general human social process of communication. . . .

The game has a logic, so that such an organization of the self is rendered possible: there is a definite end to be obtained; the actions of the different individuals are all related to each other with reference to that end so that they do not conflict; one is not in conflict with himself in the attitude of another man on the team. If one has the attitude of the person throwing the ball he can also have the response of catching the ball. The two are related so that they further the purpose of the game itself. They are interrelated in a unitary, organic fashion. There is a definite unity,

then, which is introduced into the organization of other selves when we reach such a stage as that of the game, as over against the situation of play where there is a simple succession of one rôle after another, a situation which is, of course, characteristic of the child's own personality. The child is one thing at one time and another at another, and what he is at one moment does not determine what he is at another. That is both the charm of childhood as well as its inadequacy. You cannot count on the child; you cannot assume that all the things he does are going to determine what he will do at any moment. He is not organized into a whole. The child has no definite character, no definite personality.

The game is then an illustration of the situation out of which an organized personality arises. In so far as the child does take the attitude of the other and allows that attitude of the other to determine the thing he is going to do with reference to a common end, he is becoming an organic member of society. He is taking over the morale of that society and is becoming an essential member of it. He belongs to it in so far as he does allow the attitude of the other that he takes to control his own immediate expression. What is involved here is some sort of an organized process. That which is expressed in terms of the game is, of course, being continually expressed in the social life of the child, but this wider process goes beyond the immediate experience of the child himself. The importance of the game is that it lies entirely inside of the child's own experience, and the importance of our modern type of education is that it is brought as far as possible within this realm. The different attitudes that a child assumes are so organized that they exercise a definite control over his response, as the attitudes in a game control his own immediate response. In the game we get an organized other, a generalized other, which is found in the nature of the child itself, and finds its expression in the immediate experience of the child. And it is that organized activity in the child's own nature controlling the particular response which gives unity, and which builds up his own self. . . .

We may illustrate our basic concept by a reference to the notion of property. If we say "This is my property, I shall control it," that affirmation calls out a certain set of responses which must be the same in any community in which property exists. It involves an organized attitude with reference to property which is common to all the members of the community. One must have a definite attitude of control of his own property and respect for the property of others. Those attitudes (as organized sets of responses) must be there on the part of all, so that when one says such a thing he calls out in himself the response of the others. He is calling out the response of what I have called a generalized other. That which makes society possible is such common responses, such organized attitudes, with reference to what we term property, the cults of religion, the process of education, and the relations of the family. Of course, the wider the society the more definitely universal these objects must be. In any case there must be a definite set of responses, which we may speak of as abstract, and which can belong to a very large group. Property is in itself a very abstract concept. It is that which the individual himself can control and nobody else can control. The attitude is different from that of a dog toward a bone. A dog will fight any other dog trying to take the bone. The dog is not taking the attitude of the other dog. A man who says "This is my property" is taking an attitude of the other person. The man is appealing to his rights because he is able to take the attitude which everybody else in the group has with reference to property, thus arousing in himself the attitude of others.

What goes to make up the organized self is the organization of the attitudes which are common to the group. A person is a personality because he belongs to a community, because he takes over the institutions of that community into his own conduct. He takes its language as a medium by which he gets his personality, and then through a process of taking the different rôles that all the others furnish he comes to get the attitude of the members of the community. Such, in a certain sense, is the structure of a man's personality. There are certain common responses which each individual has toward certain common things, and in so far as those common responses are awakened in the individual when he is affecting other persons he arouses his own self. The structure, then, on which the self is built is this response which is common to all, for one has to be a member of a community to be a self. Such responses are abstract attitudes, but they constitute just what we term a man's character. They give him what we term his principles, the acknowledged attitudes of all members of the community toward what are the values of that community. He is putting himself in the place of the generalized other, which represents the organized responses of all the members of the group. It is that which guides conduct controlled by principles, and a person who has such an organized group of responses is a man whom we say has character, in the moral sense. . . .

I have so far emphasized what I have called the structures upon which the self is constructed, the framework of the self, as it were. Of course we are not only what is common to all: each one of the selves is different from everyone else; but there has to be such a common structure as I have sketched in order that we may be members of a community at all. We cannot be ourselves unless we are also members in whom there is a community of attitudes which control the attitudes of all. We cannot have rights unless we have common attitudes. That which we have acquired as self-conscious persons makes us such members of society and gives us selves. Selves can only exist in definite relationships to other selves. No hard-and-fast line can be drawn between our own selves and the selves of others, since our own selves exist and enter as such into our experience only in so far as the selves of others exist and enter as such into our experience also. The individual possesses a self only in relation to the selves of the other members of his social group; and the structure of his self expresses or reflects the general behavior pattern of this social group to which he belongs, just as does the structure of the self of every other individual belonging to this social group. . . .

THE SELF AND THE SUBJECTIVE

There is one other matter which I wish briefly to refer to now. The only way in which we can react against the disapproval of the entire community is by setting up a higher sort of community which in a certain sense out-votes the one we find. A person may reach a point of going against the whole world about him; he may stand out by himself over against it. But to do that he has to speak with the voice of reason to himself. He has to comprehend the voices of the past and of the future. That is the only way in which the self can get a voice which is more than the voice of the community. As a rule we assume that this general voice of the community is identical with the larger community of the past and the future; we assume that an organized custom represents what we call morality. The things one cannot do are those which everybody would condemn. If we take the attitude of the community over against our own responses, that is a true statement, but we must not forget this other capacity, that of replying to the community and insisting on the gesture of the community changing. We can reform the order of things; we can insist on making the community standards better standards. We are not simply bound by the community. We are engaged in a conversation in which what we say is listened to by the community and its response is one which is affected by what we have to say. This is especially true in critical situations. A man rises up and defends himself for what he does; he has his "day in court"; he can present his views. He can perhaps change the attitude of the community toward himself. The process of conversation is one in which the individual has not only the right but the duty of talking to the community of which he is a part, and bringing about those changes which take place through the interaction of individuals. That is the way, of course, in which society gets ahead, by just such interactions as those in which some person thinks a thing out. We are continually changing our social system in some respects, and we are able to do that intelligently because we can think. . . .

The "I" and the "Me"

The "I" is the response of the organism to the attitudes of the others; the "me" is the organized set of attitudes of others which one himself assumes. The attitudes of the others constitute the organized "me," and then one reacts toward that as an "I." I now wish to examine these concepts in greater detail.

There is neither "I" nor "me" in the conversation of gestures; the whole act is not yet carried out, but the preparation takes place in this field of gesture. Now, in so far as the individual arouses in himself the attitudes of the others, there arises an organized group of responses. And it is due to the individual's ability to take the attitudes of these others in so far as they can be organized that he gets self-consciousness. The taking of all of those organized sets of attitudes gives him his "me"; that is the self he is aware of. He can throw the ball to some other member because of the demand made upon him from other members of the team. That is the self that immediately exists for him in his consciousness. He has their attitudes, knows what they want and what the consequence of any act of his will be, and he has assumed responsibility for the situation. Now, it is the presence of those organized sets of attitudes that constitutes that "me" to which he as an "I" is responding. But what that response will be he does not know and nobody else knows. Perhaps he will make a brilliant play or an error. The response to that situation as it appears in his immediate experience is uncertain, and it is that which constitutes the "I."

The "I" is his action over against that social situation within his own conduct, and it gets into his experience only after he has carried out the act. Then he is aware of it. He had to do such a thing and he did it. He fulfils his duty and he may look with pride at the throw which he made. The "me" arises to do that duty—that is the way in which it arises in his experience. He had in him all the attitudes of others, calling for a certain response; that was the "me" of that situation, and his response is the "I."

I want to call attention particularly to the fact that this response of the "I" is something that is more or less uncertain. The attitudes of others which one assumes as affecting his own conduct constitute the "me," and that is something that is there, but the response to it is as yet not given. When one sits down to think anything out, he has certain data that are there. Suppose that it is a social situation which he has to straighten out. He sees himself from the point of view of one individual or another in the group. These individuals, related all together, give him a certain self. Well, what is he going to do? He does not know and nobody else knows. He can get the situation into his experience because he can assume the attitudes of the various individuals involved in it. He knows how they feel about it by the assumption of their attitudes. He says, in effect, "I have done certain things that seem to commit me to a certain course of conduct." Perhaps if he does so act it will place him in a false position with another group. The "I" as a response to this situation, in contrast to the "me" which is involved in the attitudes which he takes, is uncertain. And when the response takes place, then it appears in the field of experience largely as a memory image. . . .

The "I," then, in this relation of the "I" and the "me," is something that is, so to speak, responding to a social situation which is within the experience of the individual. It is the answer which the individual makes to the attitude which others take toward him when he assumes an attitude toward them. Now, the attitudes he is taking toward them are present in his own experience, but his response to them will contain a novel element. The "I" gives the sense of freedom, of initiative. The situation is there for us to act in a self-conscious fashion. We are aware of ourselves, and of what the situation is, but exactly how we will act never gets into experience until after the action takes place.

Such is the basis for the fact that the "I" does not appear in the same sense in experience as does the "me." The "me" represents a definite organization of the community there in our own attitudes, and calling for a response, but the response that takes place is something that just happens. There is no certainty in regard to it. There is a moral necessity but no mechanical necessity for the act. When it does take place then we find what has been done. The above account gives us, I think, the relative position of the "I" and "me" in the situation, and the grounds for the separation of the two in behavior. The two are separated in the process but they belong together in the sense of being parts of a whole. They are separated and yet they belong together. The separation of the "I" and the "me" is not fictitious. They are not identical, for, as I have said, the "I" is something that is never entirely calculable. The "me" does call for a certain sort of an "I" in so far as we meet the obligations that are given in conduct itself, but the "I" is always something different from what the situation itself calls for. So there is always that distinction, if you like, between the "I" and the "me." The "I" both calls out the "me" and responds to it. Taken together they constitute a personality as it appears in social experience. The self is essentially a social process going on with these two distinguishable phases. If it did not have these two phases there could not be conscious responsibility, and there would be nothing novel in experience.

Introduction to "Society"

In this last selection of readings, Mead offers his vision of a possible future, an ideal society where individuals are able to realize their full potential while promoting the advancement of the broader community. Although the blueprint he provides for attaining a "universal society" is perhaps simplistic, it is instructive on three counts. First, it further crystallizes Mead's core ideas concerning the importance of significant symbols in taking the attitudes of others and the fundamental interconnection between the individual self and society. Second, it provides an important counterpoint to the evolutionary schemes expressed by other classical social theorists (including those presented in this volume). Third, Mead's notion of the basis for and realization of democracy shares much with the conventional wisdom informing American political culture. Mead's liberalism is the liberalism of America. Thus, studying these selections provides an opportunity for investigating the liberal mythology of American political culture and for shedding light on its presuppositions.

For Mead, the evolution of separate societies into a universal whole mirrors the process that the individual undergoes in relation to the society to which he belongs. In brief, Mead argues, like Durkheim, that modern societies are characterized by a greater degree of complexity and functional interdependence. Unlike in premodern societies, modern individuals rely for their survival on expanding networks of increasingly distant others who perform specialized or differentiated functions. A similar process occurs not only within societies, but also across them, as whole societies become more and more dependent on each other for their continued survival and development. Both instances promote everwidening circles of social contacts and opportunities not only for cooperation, but also for domination.

The key to creating a universal democracy lies in how the twin processes of social control and individual creativity evolve. In order for the expanding networks of social relations (both individual and societal) to produce the freedom and equality associated with democracy, interaction must be based on taking the attitude of the others involved. Only through such self-conscious control of action can a cooperative social unity be achieved wherein the future consequences of one's conduct shapes present lines of action. Without anticipating the responses of others to one's actions, whether it be the actions and responses of an individual, a community, an entire nation, or a group of nations, the exercise of one's own freedom may prevent others from doing the very same.

Photo 8.3 Today's Democratic, Universal Community? Delegates to the United Nations participate in international policy making.

However, social control through taking the attitude of the other, while necessary for sustaining a democratic society, must not impede the development of individual expression and the opportunity to realize one's potential. The solution to this problem emerges when taking the attitude of the other involves an assertion of an individual's or group's "functional superiority." In other words, an ideal democracy is one in which an individual or group is able to assert its superior skills without seeking to dominate others who in turn are able to benefit from such superior contributions. The individual or group is able to develop its "superior" sense of self while recognizing that one's unique contribution is made possible only through willing relations with others. In turn, a more highly evolved community based on cooperative social relations is created. Indeed, this more highly developed, "global" ability to take the attitude of the other found expression in the creation of the League of Nations shortly after World War I as countries realized their growing mutual dependence on one another for their own continued survival and development. For Mead, the establishment of this international organization, "where every community recognizes every other community in the very process of asserting itself" (Mead 1934/1962:287), marked the beginning stages of a truly democratic, universal society.[1]

"Society" (1934)

George Herbert Mead

CONFLICT AND INTEGRATION

I have been emphasizing the continued integration of the social process, and the psychology of the self which underlies and makes possible this process. A word now as to the factors of conflict and disintegration. In the baseball game there are competing individuals who want to get into the limelight, but this can only be attained by playing the game. Those conditions do make a certain sort of action necessary, but inside of them there can be all sorts of jealously competing individuals who may wreck the team. There seems to be abundant opportunity for disorganization in the organization essential to the team. This is so to a much larger degree in the economic process. There has to be distribution, markets, mediums of exchange; but within that field all kinds of competition and disorganizations are possible, since there is an "I" as well as a "me" in every case.

Historical conflicts start, as a rule, with a community which is socially pretty highly organized. Such conflicts have to arise between different groups where there is an attitude of hostility to others involved. But even here a wider social organization is usually the result; there is, for instance, an appearance of the tribe over against the clan. It is a larger, vaguer organization, but still it is there. This is the sort of situation we have at the present time; over against the potential hostility of nations to each other, they recognize themselves as forming some sort of community, as in the League of Nations.

The fundamental socio-physiological impulses or behavior tendencies which are common to all human individuals, which lead those individuals collectively to enter or form themselves into organized societies or social communities, and which constitute the ultimate basis of those societies or social communities, fall, from the social point of view, into two main classes: those which lead to social co-operation, and those which lead to social antagonism among individuals; those which give rise to friendly attitudes and relations, and those which give rise to hostile attitudes and relations, among the human individuals implicated in the social situations. We have used the term "social" in its broadest and strictest sense; but in that quite common narrower sense, in which it bears an ethical connotation, only the fundamental physiological human impulses or behavior

---

[1]Unfortunately, modern history has rendered Mead's vision of a utopian democracy and its potential realization in organizations such as the League of Nations more and more a fantasy. The League dissolved in 1946, in no small measure due to the United States' unwillingness to join. While the establishment of the United Nations in 1945 perhaps has allowed for a lingering hope for the creation of a universal democracy, it, too, has been beset by internal conflicts. While by no means the only country that has obstructed the efforts of the UN, many of the conflicts have been sparked by the United States' blocking the passage of international policy or refusing to abide by existing regulations. Such a "might makes right" stance was demonstrated in the war with Iraq as American political leaders avowed to "go it alone" without passage of a UN-backed resolution.

tendencies of the former class (those which are friendly, or which make for friendliness and co-operation among the individuals motivated by them) are "social" or lead to "social" conduct; whereas those impulses or behavior tendencies of the latter class (those which are hostile, or which make for hostility and antagonism among the individuals motivated by them) are "anti-social" or lead to "anti-social" conduct. Now it is true that the latter class of fundamental impulses or behavior tendencies in human beings are "anti-social" in so far as they would, by themselves, be destructive of all human social organization, or could not, alone, constitute the basis of any organized human society; yet in the broadest and strictest non-ethical sense they are obviously no less social than are the former class of such impulses or behavior tendencies. They are equally common to, or universal among, all human individuals, and, if anything, are more easily and immediately aroused by the appropriate social stimuli; and as combined or fused with, and in a sense controlled by, the former impulses or behavior tendencies, they are just as basic to all human social organization as are the former, and play a hardly less necessary and significant part in that social organization itself and in the determination of its general character. Consider, for example, from among these "hostile" human impulses or attitudes, the functioning or expression or operation of those of self-protection and self-preservation in the organization and organized activities of any given human society or social community, let us say, of a modern state or nation. Human individuals realize or become aware of themselves as such, almost more easily and readily in terms of the social attitudes connected or associated with these two "hostile" impulses (or in terms of these two impulses as expressed in these attitudes) than they do in terms of any other social attitudes or behavior tendencies as expressed by those attitudes. Within the social organization of a state or nation the "anti-social" effects of these two impulses are curbed and kept under control by the legal system which is one aspect of that organization; these two impulses are made to constitute the fundamental principles in terms of which the economic system, which is another aspect of that organization, operates; as

combined and fused with, and organized by means of the "friendly" human impulses—the impulses leading to social co-operation among the individuals involved in that organization—they are prevented from giving rise to the friction and enmity among those individuals which would otherwise be their natural consequence, and which would be fatally detrimental to the existence and well-being of that organization; and having thus been made to enter as integral elements into the foundations of that organization, they are utilized by that organization as fundamental impulsive forces in its own further development, or they serve as a basis for social progress within its relational framework. Ordinarily, their most obvious and concrete expression or manifestation in that organization lies in the attitudes of rivalry and competition which they generate inside the state or nation as a whole, among different socially functional subgroups of individuals—subgroups determined (and especially economically determined) by that organization; and these attitudes serve definite social ends or purposes presupposed by that organization, and constitute the motives of functionally necessary social activities within that organization. But self-protective and self-preservational human impulses also express or manifest themselves indirectly in that organization, by giving rise through their association in that organization with the "friendly" human impulses, to one of the primary constitutive ideals or principles or motives of that organization—namely, the affording of social protection, and the lending of social assistance, to the individual by the state in the conduct of his life; and by enhancing the efficacy, for the purposes of that organization, of the "friendly" human impulses with a sense or realization of the possibility and desirability of such organized social protection and assistance to the individual. Moreover, in any special circumstances in which the state or nation is, as a whole, confronted by some danger common to all its individual members, they become fused with the "friendly" human impulses in those individuals, in such a way as to strengthen and intensify in those individuals the sense of organized social union and co-operative social interrelationship among them in terms of the state; in such circumstances, so far from constituting forces of disintegration or destruction within the social organization of the state or nation, they become, indirectly, the principles of increased social unity, coherence, and co-ordination within that organization. In time of war, for example, the self-protective impulse in all the individual members of the state is unitedly directed against their common enemy and ceases, for the time being, to be directed among themselves; the attitudes of rivalry and competition which that impulse ordinarily generates between the different smaller, socially functional groups of those individuals within the state are temporarily broken down; the usual social barriers between these groups are likewise removed; and the state presents a united front to the given common danger, or is fused into a single unity in terms of the common end shared by, or reflected in, the respective consciousnesses of all its individual members. It is upon these war-time expressions of the self-protective impulse in all the individual members of the state or nation that the general efficacy of national appeals to patriotism is chiefly based.

Further, in those social situations in which the individual self feels dependent for his continuation or continued existence upon the rest of the members of the given social group to which he belongs, it is true that no feeling of superiority on his part toward those other members of that group is necessary to his continuation or continued existence. But in those social situations in which he cannot, for the time being, integrate his social relations with other individual selves into a common, unitary pattern (i.e., into the behavior pattern of the organized society or social community to which he belongs, the social behavior pattern that he reflects in his self-structure and that constitutes this structure), there ensues, temporarily (i.e., until he can so integrate his social relations with other individual selves), an attitude of hostility, of "latent opposition," on his part toward the organized society or social community of which he is a member; and during that time the given individual self must "call in" or rely upon the feeling of superiority toward that society or social community, or toward its other individual members, in order to buoy himself up and "keep himself going" as such. We always present

ourselves to ourselves in the most favorable light possible; but since we all have the job of keeping ourselves going, it is quite necessary that if we are to keep ourselves going we should thus present ourselves to ourselves.

A highly developed and organized human society is one in which the individual members are interrelated in a multiplicity of different intricate and complicated ways whereby they all share a number of common social interests,—interests in, or for the betterment of, the society—and yet, on the other hand, are more or less in conflict relative to numerous other interests which they posses only individually, or else share with one another only in small and limited groups. Conflicts among individuals in a highly developed and organized human society are not mere conflicts among their respective primitive impulses but are conflicts among their respective selves or personalities, each with its definite social structure—highly complex and organized and unified—and each with a number of different social facets or aspects, a number of different sets of social attitudes constituting it. Thus, within such a society, conflicts arise between different aspects or phases of the same individual self (conflicts leading to cases of split personality when they are extreme or violent enough to be psychopathological), as well as between different individual selves. And both these types of individual conflict are settled or terminated by reconstructions of the particular social situations, and modifications of the given framework of social relationships, wherein they arise or occur in the general human social life-process—these reconstructions and modifications being performed, as we have said, by the minds of the individuals in whose experience or between whose selves these conflicts take place.

Mind, as constructive or reflective or problem-solving thinking, is the socially acquired means or mechanism or apparatus whereby the human individual solves the various problems of environmental adjustment which arise to confront him in the course of his experience, and which prevent his conduct from proceeding harmoniously on its way, until they have thus been dealt with. And mind or thinking is also—as possessed by the individual members of human society—the means or mechanism or apparatus whereby social reconstruction is effected or accomplished by these individuals. For it is their possession of minds or powers of thinking which enables human individuals to turn back critically, as it were, upon the organized social structure of the society to which they belong (and from their relations to which their minds are in the first instance derived), and to reorganize or reconstruct or modify that social structure to a greater or less degree, as the exigencies of social evolution from time to time require. Any such social reconstruction, if it is to be at all far-reaching, presupposes a basis of common social interests shared by all the individual members of the given human society in which that reconstruction occurs; shared, that is, by all the individuals whose minds must participate in, or whose minds bring about, that reconstruction. And the way in which any such social reconstruction is actually effected by the minds of the individuals involved is by a more or less abstract intellectual extension of the boundaries of the given society to which these individuals all belong, and which is undergoing the reconstruction—an extension resulting in a larger social whole in terms of which the social conflicts that necessitate the reconstruction of the given society are harmonized or reconciled, and by reference to which, accordingly, these conflicts can be solved or eliminated.[i]

The changes that we make in the social order in which we are implicated necessarily involve our also making changes in ourselves. The social conflicts among the individual members of a given organized human society, which, for their removal, necessitate conscious or intelligent reconstructions and modifications of that society by those individuals, also and equally necessitate such reconstructions or modifications by those individuals of their own selves or personalities.

Thus the relations between social reconstruction and self or personality reconstruction are reciprocal and internal or organic; social reconstruction by the individual members of any organized human society entails self or personality reconstruction in some degree or other by each of these individuals, and vice versa, for, since their selves or personalities are constituted by their organized social relations to one another, they cannot reconstruct those selves or personalities without also reconstructing, to some extent, the given social order, which is, of course, likewise constituted by their organized social relations to one another. In both types of reconstruction the same fundamental material of organized social relations among human individuals is involved, and is simply treated in different ways, or from different angles or points of view, in the two cases, respectively; or in short, social reconstruction and self or personality reconstruction are the two sides of a single process—the process of human social evolution. Human social progress involves the use by human individuals of their socially derived mechanism of self-consciousness, both in the effecting of such progressive social changes, and also in the development of their individual selves or personalities in such a way as adaptively to keep pace with such social reconstruction.

Ultimately and fundamentally societies develop in complexity of organization only by means of the progressive achievement of greater and greater degrees of functional, behavioristic differentiation among the individuals who constitute them; these functional, behavioristic differentiations among the individual members implying or presupposing initial oppositions among them of individual needs and ends, oppositions which in terms of social organization, however, are or have been transformed into these differentiations, or

---

[i]The reflexive character of self-consciousness enables the individual to contemplate himself as a whole; his ability to take the social attitudes of other individuals and also of the generalized other toward himself, within the given organized society of which he is a member, makes possible his bringing himself, as an objective whole, within his own experiential purview; and thus he can consciously integrate and unify the various aspects of his self, to form a single consistent and coherent and organized personality. Moreover, by the same means, he can undertake and effect intelligent reconstructions of that self or personality in terms of its relations to the given social order, whenever the exigencies of adaptation to his social environment demand such reconstructions.

into mere specializations of socially functional individual behavior.

The human social ideal—the ideal or ultimate goal of human social progress—is the attainment of a universal human society in which all human individuals would possess a perfected social intelligence, such that all social meanings would each be similarly reflected in their respective individual consciousness—such that the meanings of any one individual's acts or gestures (as realized by him and expressed in the structure of his self, through his ability to take the social attitudes of other individuals toward himself and toward their common social ends or purposes) would be the same for any other individual whatever who responded to them. . . .

OBSTACLES AND PROMISES IN THE DEVELOPMENT OF THE IDEAL SOCIETY

Ethical ideas, within any given human society, arise in the consciousness of the individual members of that society from the fact of the common social dependence of all these individuals upon one another (or from the fact of the common social dependence of each one of them upon that society as a whole or upon all the rest of them), and from their awareness or sensing or conscious realization of this fact. But ethical problems arise for individual members of any given human society whenever they are individually confronted with a social situation to which they cannot readily adjust and adapt themselves, or in which they cannot easily realize themselves, or with which they cannot immediately integrate their own behavior; and the feeling in them which is concomitant with their facing and solution of such problems (which are essentially problems of social adjustment and adaptation to the interests and conduct of other individuals) is that of self-superiority and temporary opposition to other individuals. In the case of ethical problems, our social relationships with other individual members of the given human society to which we belong depend upon our opposition to them, rather than, as in the case of the development or formulation of ethical ideals, upon our unity, co-operation, and identification with

them. Every human individual must, to behave ethically, integrate himself with the pattern of organized social behavior which, as reflected or prehended in the structure of his self, makes him a self-conscious personality. Wrong, evil, or sinful conduct on the part of the individual runs counter to this pattern of organized social behavior which makes him, as a self, what he is, just as right, good, or virtuous behavior accords with this pattern; and this fact is the basis of the profound ethical feeling of conscience—of "ought" and "ought not"—which we all have, in varying degrees, respecting our conduct in given social situations. The sense which the individual self has of his dependence upon the organized society or social community to which he belongs is the basis and origin, in short, of his sense of duty (and in general of his ethical consciousness); and ethical and unethical behavior can be defined essentially in social terms: the former as behavior which is socially beneficial or conducive to the well-being of society, the latter as behavior which is socially harmful or conductive to the disruption of society. From another point of view, ethical ideals and ethical problems may be considered in terms of the conflict between the social and the asocial (the impersonal and the personal) sides or aspects of the individual self. The social or impersonal aspect of the self integrates it with the social group to which it belongs and to which it owes its existence; and this side of the self is characterized by the individual's feeling of cooperation and equality with the other members of that social group. The asocial or personal aspect of the self (which, nevertheless, is also and equally social, fundamentally in the sense of being socially derived or originated and of existentially involving social relations with other individuals, as much as the impersonal aspect of the self is and does), on the other hand, differentiates it from, or sets it in distinctive and unique opposition to, the other members of the social group to which it belongs; and this side of the self is characterized by the individual's feeling of superiority toward the other members of that group. The "social" aspect of human society—which is simply the social aspect of the selves of all individual members taken collectively—with its concomitant feelings on the parts of all these individuals of co-operation and social interdependence, is the basis for the development and existence of ethical ideals in that society; whereas the "asocial" aspect of human society—which is simply the asocial aspect of the selves of all individual members taken collectively—with its concomitant feelings on the parts of all these individuals of individuality, self-superiority to other individual selves, and social independence, is responsible for the rise of ethical problems in that society. These two basic aspects of each single individual self are, of course, responsible in the same way or at the same time for the development of ethical ideals and the rise of ethical problems in the individual's own experience as opposed to the experience of human society as a whole, which is obviously nothing but the sum-total of the social experiences of all its individual members.

Those social situations in which the individual finds it easiest to integrate his own behavior with the behavior of the other individual selves are those in which all the individual participants are members of some one of the numerous socially functional groups of individuals (groups organized, respectively, for various special social ends and purposes) within the given human society as a whole; and in which he and they are acting in their respective capacities as members of this particular group. (Every individual member of any given human society, of course, belongs to a large number of such different functional groups.) On the other hand, those social situations in which the individual finds it most difficult to integrate his own behavior with the behavior of others are those in which he and they are acting as members, respectively, of two or more different socially functional groups: groups whose respective social purposes or interests are antagonistic or conflicting or widely separated. In social situations of the former general type each individual's attitude toward the other individuals is essentially social; and the combination of all these social attitudes toward one another of the individuals represents, or tends to realize more or less completely, the ideal of any social situation respecting organization, unification, co-operation, and integration of the behavior of the several individuals involved. In any social situation of this general type the individual realizes himself as such in his relation to all the other members of the given socially functional group and realizes his own particular social function in its relations to the respective functions of all other individuals. He takes or assumes the social attitudes of all these other individuals toward himself and toward one another, and integrates himself with that situation or group by controlling his own behavior or conduct accordingly; so that there is nothing in the least competitive or hostile in his relations with these other individuals. In social situations of the latter general type on the other hand, each individual's attitude toward the other individuals is essentially asocial or hostile (though these attitudes are of course social in the fundamental non-ethical sense, and are socially derived); such situations are so complex that the various individuals involved in any one of them either cannot be brought into common social relations with one another at all or else can be brought into such relations only with great difficulty, after long and tortuous processes of mutual social adjustment; for any such situation lacks a common group or social interest shared by all the individuals—it has no one common social end or purpose characterizing it and serving to unite and co-ordinate and harmoniously interrelate the actions of all those individuals; instead, those individuals are motivated, in that situation, by several different and more or less conflicting social interests or purposes. Examples of social situations of this general type are those involving interactions or relations between capital and labor, i.e., those in which some of the individuals are acting in their socially functional capacity as members of the capitalistic class, which is one economic aspect of modern human social organization; whereas the other individuals are acting in their socially functional capacity as members of the laboring class, which is another (and in social interests directly opposed) economic aspect of that social organization. Other examples of social situations of this general type are those in which the individuals involved stand in the economic relations to each other of producers and consumers, or buyers and sellers, and are acting in their respective socially functional capacities as such. But even the social situations of this general type (involving complex social antagonisms and diversities of social interests among the individuals implicated in any one of them, and respectively lacking the co-ordinating, integrating, unifying influence of common social ends and motives shared by those individuals), even these social situations, as occurring within the general human social process of experience and behavior, are definite aspects of or ingredients in the general relational pattern of that process as a whole.

What is essential to the order of society in its fullest expression on the basis of the theory of the self that we have been discussing is, then, an organization of common attitudes which shall be found in all individuals. It might be supposed that such an organization of attitudes would refer only to that abstract human being which could be found as identical in all members of society, and that that which is peculiar to the personality of the individual would disappear. The term "personality" implies that the individual has certain common rights and values obtained in him and through him; but over and above that sort of social endowment of the individual, there is that which distinguishes him from anybody else, makes him what he is. It is the most precious part of the individual. The question is whether that can be carried over into the social self or whether the social self shall simply embody those reactions which can be common to him in a great community. On the account we have given we are not forced to accept the latter alternative.

When one realizes himself, in that he distinguishes himself, he asserts himself over others in some peculiar situation which justifies him in maintaining himself over against them. If he could not bring that peculiarity of himself into the common community, if it could not be recognized, if others could not take his attitude in some sense, he could not have appreciation in emotional terms, he could not be the very self he is trying to be. The author, the artist, must have his audience; it may be an audience that belongs to posterity, but there must be an audience. One has to find one's self in his own individual creation as appreciated by others; what the individual accomplishes must be something that is in itself social. So far as he is a self, he must be an organic part of the life of the community, and his contribution has to be something that is social. It may be an ideal which he has discovered, but it has its value in the fact that it belongs

to society. One may be somewhat ahead of his time, but that which he brings forward must belong to the life of the community to which he belongs. There is, then, a functional difference, but it must be a functional difference which can be entered into in some real sense by the rest of the community. Of course, there are contributions which some make that others cannot make, and there may be contributions which people cannot enter into; but those that go to make up the self are only those which can be shared. To do justice to the recognition of the uniqueness of an individual in social terms, there must be not only the differentiation which we do have in a highly organized society but a differentiation in which the attitudes involved can be taken by other members of the group.

Take, for example, the labor movement. It is essential that the other members of the community shall be able to enter into the attitude of the laborer in his functions. It is the caste organization, of course, which makes it impossible; and the development of the modern labor movement not only brought the situation actually involved before the community but inevitably helped to break down the caste organization itself. The caste organization tended to separate in the selves the essential functions of the individuals so that one could not enter into the other. This does not, of course, shut out the possibility of some sort of social relationship; but any such relationship involves the possibility of the individual's taking the attitude of the other individuals, and functional differentiation does not make that impossible. A member of the community is not necessarily like other individuals because he is able to identify himself with them. He may be different. There can be a common content, common experience, without there being an identity of function. A difference of functions does not preclude a common experience; it is possible for the individual to put himself in the place of the other although his function is different from the other. It is that sort of functionally differentiated personality that I wanted to refer to as over against that which is simply common to all members of a community.

There is, of course, a certain common set of reactions which belong to all, which are not differentiated on the social side but which get their expression in rights, uniformities, the common methods of action which characterize members of different communities, manners of speech, and so on. Distinguishable from those is the identity which is compatible with the difference of social functions of the individuals, illustrated by the capacity of the individual to take the part of the others whom he is affecting, the warrior putting himself in the place of those whom he is proceeding against, the teacher putting himself in the position of the child whom he is undertaking to instruct. That capacity allows for exhibiting one's own peculiarities, and at the same time taking the attitude of the others whom he is himself affecting. It is possible for the individual to develop his own peculiarities, that which individualizes him, and still be a member of a community, provided that he is able to take the attitude of those whom he affects. Of course, the degree to which that takes place varies tremendously, but a certain amount of it is essential to citizenship in the community.

One may say that the attainment of that functional differentiation and social participation in the full degree is a sort of ideal which lies before the human community. The present stage of it is presented in the ideal of democracy. It is often assumed that democracy is an order of society in which those personalities which are sharply differentiated will be eliminated, that everything will be ironed down to a situation where everyone will be, as far as possible, like everyone else. But of course that is not the implication of democracy: the implication of democracy is rather that the individual can be as highly developed as lies within the possibilities of his own inheritance, and still can enter into the attitudes of the others whom he affects. There can still be leaders, and the community can rejoice in their attitudes just in so far as these superior individuals can themselves enter into the attitudes of the community which they undertake to lead.

How far individuals can take the rôles of other individuals in the community is dependent upon a number of factors. The community may in its size transcend the social organization, may go beyond the social organization which makes such identification possible. The most striking illustration of that is the economic community.

This includes everybody with whom one can trade in any circumstances, but it represents a whole in which it would be next to impossible for all to enter into the attitudes of the others. The ideal communities of the universal religions are communities which to some extent may be said to exist, but they imply a degree of identification which the actual organization of the community cannot realize. We often find the existence of castes in a community which make it impossible for persons to enter into the attitude of other people although they are actually affecting and are affected by these other people. The ideal of human society is one which does bring people so closely together in their interrelationships, so fully develops the necessary system of communication, that the individuals who exercise their own peculiar functions can take the attitude of those whom they affect. The development of communication is not simply a matter of abstract ideas, but is a process of putting one's self in the place of the other person's attitude, communicating through significant symbols. Remember that what is essential to a significant symbol is that the gesture which affects others should affect the individual himself in the same way. It is only when the stimulus which one gives another arouses in himself the same or like response that the symbol is a significant symbol. Human communication takes place through such significant symbols, and the problem is one of organizing a community which makes this possible. If that system of communication could be made theoretically perfect, the individual would affect himself as he affects others in every way. That would be the ideal of communication, an ideal attained in logical discourse wherever it is understood. The meaning of that which is said is here the same to one as it is to everybody else. Universal discourse is then the formal ideal of communication. If communication can be carried through and made perfect, then there would exist the kind of democracy to which we have referred, in which each individual would carry just the response in himself that he knows he calls out in the community. That is what makes communication in the significant sense the organizing process in the community. It is not simply a process of transferring abstract symbols; it is always a gesture in a social act which calls out in the individual himself the tendency to the same act that is called out in others.

What we call the ideal of a human society is approached in some sense by the economic society on the one side and by the universal religions on the other side, but it is not by any means fully realized. Those abstractions can be put together in a single community of the democratic type. As democracy now exists, there is not this development of communication so that individuals can put themselves into the attitudes of those whom they affect. There is a consequent leveling-down, and an undue recognition of that which is not only common but identical. The ideal of human society cannot exist as long as it is impossible for individuals to enter into the attitudes of those whom they are affecting in the performance of their own peculiar functions.

Discussion Questions

1. How does Mead define self-consciousness? What role does language play in the development of self-consciousness? What role does self-consciousness play in social interaction?

2. Compare and contrast Mead's concept of the "I," the "me," and the "generalized other" with Simmel's views on the relationship between the self and group affiliations. How might Simmel's concept of the "stranger" be used to critique Mead's understanding of the "generalized other"?

3. Can television, and mass media more generally, affect the development of the self? If so, how?

4. Given his emphasis on the self, language, thinking, and meaning, what factors might be missing from Mead's view of social interaction?

5. What is Mead's vision of a "universal society" based on? What obstacles might prevent its realization? How does Mead's view differ from those expressed by Marx and Weber?

6. What are some of the implications of Du Bois's discussion on race and Gilman's discussion on gender for Mead's theory of "taking the attitude of the other"?

REFERENCES

Alexander, Jeffrey C. 1987. *Twenty Lectures.* New York: Columbia University Press.

Anderson, Elijah. 2000. "The Emerging Philadelphia African American Class Structure." *The Annals of the American Academy of Political and Social Science* 568 (March):54–77.

Antonio, Robert J. 1995. "Nietzsche's Antisociology: Subjectified Culture and the End of History." *American Journal of Sociology* 101:1–43.

Aron, Raymond. 1965/1970. *Main Currents in Sociological Thought*, vols. 1 and 2, translated by Richard Howard and Helen Weaver. New York: Anchor Books.

Ashe, Fidelma. 1999. "The Subject." Pp. 88–110 in *Contemporary Social and Political Theory*, edited by Alan Finlayson Ashe, Moya Lloyd, Iain MacKenzie, James Martin, and Shane O'Neill. Philadelphia: Open University Press.

Ashley, David and David Michael Orenstein. 1998. *Sociological Theory: Classical Statements.* Boston: Allyn and Bacon.

Bellah, Robert. 1973. "Introduction." Pp. ix–lv in *Émile Durkheim on Morality and Society*, edited by Robert Bellah. Chicago: University of Chicago Press.

Bendix, Richard. 1977. *Max Weber: An Intellectual Portrait.* Berkeley: University of California Press.

Berry, Mary Frances. 2000. "Du Bois as Social Activist: Why We Are Not Saved." *The Annals of the American Academy of Political and Social Science* 568 (March):100–110.

Bierstedt, Robert. 1981. *American Sociological Theory: A Critical History.* New York: Academic Press.

Blum, Deborah. 1997. *Sex on the Brain: The Biological Differences Between Men and Women.* New York: Penguin.

Bobo, Lawrence. 2000. "Reclaiming a Du Boisian Perspective on Racial Attitudes." *The Annals of the American Academy of Political and Social Science* 568 (March):186–202.

Calhoun, Craig J., Joseph Gerteis, James Moody, Steven Pfaff, Kathryn Schmidt, and Indermohan Virk, eds. 2002. *Classical Sociological Theory.* Malden, MA: Blackwell.

Carroll, Lewis. 1865/1960. *Alice's Adventures in Wonderland.* New York: New American Library of World Literature.

Collins, Patricia Hill. 1987. "The Meaning of Motherhood in Black Culture and Black Mother/ Daughter Relationships." *Sage* 4(2):3–10.

Collins, Randall and Michael Makowsky. 1998. *The Discovery of Society.* Boston: McGraw-Hill.

Cooley, Charles Horton. 1902/1964. *Human Nature and the Social Order.* New York: Schocken Books.

Cooper, Anna Julia. 1892/1988. *A Voice From the South by a Black Woman of the South.* New York: Oxford University Press.

Comte, Auguste. 1830–42/1974. *The Positive Philosophy*, edited and translated by Harriet Martineau. 6 volumes. New York: AMS Press.

Coser, Lewis A. 1977. *Masters of Sociological Thought.* New York: Harcourt, Brace, Jovanovich.

Deegan, Mary Jo. 1997. "Introduction: Gilman's Sociological Journey from *Herland* to *Ourland.*" Pp. 1–57 in Charlotte Perkins Gilman, *With Her in Ourland*, edited by Deegan and Michael Hill. Westport, CT: Greenwood Press.

Degler, Carl. 1966. "Introduction." Pp. vi–xxxv in Charlotte Perkins Gilman, *Women and Economics,* edited by Degler. New York: Harper and Row.

Dickens, Charles. 1854. *Hard Times.* London: Bradbury and Evans.

Du Bois, W. E. B. 1896. *The Suppression of the African Slave-Trade to the United States of America, 1638–1870.* New York: Longmans and Green.

———. 1898/2000. "The Study of Negro Problems." *The Annals of the American Academy of Political and Social Science* 568 (March):13–27.

———. 1899/1996. *Philadelphia Negro: A Social Study.* Philadelphia: University of Pennsylvania Press.

———. 1903/1989. *The Souls of Black Folk.* New York: Penguin Books.

———. 1903/2004. "The Talented Tenth." Pp. 185–95 in *The Social Theory of W.E.B. Du Bois*, edited by Philip Zuckerman. Thousand Oaks, CA: Pine Forge.

———. 1914/2000. "The Burden of Black Women." Pp. 102–04 in *Du Bois on Religion*, edited by Phil Zuckerman. Walnut Creek, CA: AltaMira.

————. 1912/2000. "The Black Mother." P. 294 in *W.E.B. Du Bois: A Reader*, edited by David Levering Lewis. New York: Henry Holt.

————. 1913/2000. "Hail Columbia!" Pp. 295–96 in *W.E.B. Du Bois: A Reader*, edited by David Levering Lewis. New York: Henry Holt.

————. 1915. *The Negro*. New York: Henry Holt.

————. 1920/2003a. "The Damnation of Women." Pp. 171–91 in *Darkwater: Voices from Within the Veil*. (With an introduction by Joe Feagin.) Amherst, NY: Humanity Books.

————. 1920/2003b. *Darkwater: Voices From Within the Veil*. (With an introduction by Joe Feagin.) Amherst, NY: Humanity Books.

————. 1920/2003c. "The Servant in the House." Pp. 125–35 in *Darkwater: Voices from Within the Veil*. (With an introduction by Joe Feagin.) Amherst, NY: Humanity Books.

————. 1920/2003d. "The Souls of White Folk." Pp. 55–75 in *Darkwater: Voices from Within the Veil*. (With an introduction by Joe Feagin.) Amherst, NY: Humanity Books.

————. 1924/2007. *The Gift of Black Folk: The Negroes in the Making of America*. Oxford University Press.

————. 1928/1995. *Dark Princess*. Jackson: University Press of Mississippi.

————. 1930a/2007. *Africa: Its Place in Modern History*. Oxford: Oxford University Press.

————. 1930b/2007. *Africa: Its Geography, People, and Products*. Oxford: Oxford University Press.

————. 1935/1962. *Black Reconstruction in America: An Essay Toward a History of the Part Which Black Folk Played in the Attempt to Reconstruct Democracy in America, 1860–1880*. New York: Atheneum.

————. 1940/1984. *Dusk of Dawn: An Essay Toward an Autobiography of the Race Concept*. New Brunswick, NJ: Transaction.

————. 1945. *Color and Democracy: Colonies and Peace*. New York: Harcourt Brace.

————. 1947/1995. *The World and Africa*. New York: Viking.

————. 1968. *The Autobiography of W. E. B. Du Bois*. New York: International Publishers.

Du Bois, W. E. B., Dan S. Green, Edwin D. Driver, 1980. *W. E. B. Du Bois On Sociology and the Black Community*. Chicago: University of Chicago Press.

Durkheim, Émile. 1893/1984. *The Division of Labor in Society*. New York: Free Press.

————. 1895/1966. *The Rules of Sociological Method*. New York: Free Press.

————. 1897/1951. *Suicide: A Study in Sociology*. New York: Free Press.

————. 1898/1973. "Individualism and the Intellectuals." Pp. 43-57 in Émile Durkheim, *On Morality and Society*, edited by Robert Bellah. Chicago: University of Chicago Press.

————. 1912/1995. *The Elementary Forms of Religious Life*, translated by Karen Fields. New York: Free Press.

Edles, Laura Desfor. 2002. *Cultural Sociology in Practice*. Malden, MA: Blackwell.

Eliade, Mircea and Ioan P. Couliano. 1991. *The Eliade Guide to World Religions*. New York: Harper Collins.

Engels, Friedrich. 1845/1987. *The Condition of the Working Class in England*. New York: Penguin.

————. 1884/1942. *The Origin of the Family, Private Property and the State*. New York: International Publishers.

Feagin, Joe. 2003. "Introduction." Pp. 9–24 in W. E. B. Du Bois, *Darkwater: Voices from Within the Veil*. New York: Humanity Books.

Fish, M. Steven. 2003. "Muslim World: Repressing Women, Repressing Democracy." *Los Angeles Times*, October 12, M3.

Fletcher, Ronald. 1966. "Auguste Comte and the Making of Sociology." Speech delivered November 4, 1965, at the London School of Economics and Political Science. Athlone Press, London.

Ford, Daniel. 2002. "The Horror of the Human Bomb-Delivery System." Review of Albert Axell and Hideaki Kase, *Kamikaze: Japan's Suicide Gods*. *Wall Street Journal*, September 10, D8.

Frisby, David. 1984. *Georg Simmel*. London: Tavistock.

Gamson, Joshua. 1998. *Freaks Talk Back: Tabloid Talk Shows and Sexual Nonconformity*. Chicago: University of Chicago Press.

Gilman, Charlotte Perkins. 1892/1973. "The Yellow Wallpaper." Originally published as a short story in *New England Magazine* (January). Printed in book form in 1899 by the Feminist Press: *The Yellow Wallpaper*. New York: City University of New York Feminist Press.

————. 1893/1895. *In This Our World*. Oakland: McCombs & Vaughn (London: T. Fisher Unwin, 1895. 2nd ed.; San Francisco: Press of James H. Barry.)

————. 1898/1998. *Women and Economics: A Study of the Economic Relation Between Men and Women as a Factor in Social Evolution*. Mineola, NY: Dover.

————. 1900. Concerning Children. Boston: Small, Maynard & Co.

————. 1903. *The Home: Its Work and Influence*. New York: McClure, Phillips & Co.

————. 1904. *Human Work*. New York: McClure, Phillips & Co.

———. 1911/1970. *The Man-Made World, or Our Androcentric Culture.* New York: Charlton Co.; NY: Source Book Press.

———. 1923. "Is America Too Hospitable?" *Forum,* 70 (October) pp. 1983–89.

———. 1935/1972. *The Living of Charlotte Perkins Gilman: An Autobiography.* New York: D. Appleton-Century.

Golden, Catherine and Joanna Schneider Zangrando. 2000. "Introduction." Pp. 11–22 in *The Mixed Legacy of Charlotte Perkins Gilman,* edited by Golden and Zangrando. Newark, NJ: University of Delaware Press.

Gramsci, Antonio. 1971. *Prison Notebooks.* Edited and translated by Quentin Hoare and Geoffrey Nowell Smith. New York: International Publishers.

Griffin, Erica, compiler. 2003. "Reviews of *The Souls of Black Folk.*" Pp. 18–33 in *The Souls of Black Folk: One Hundred Years Later,* edited by Dolan Hubbard. Columbia: University of Missouri Press.

Griffin, Farah Jasmine. 2000. "Black Feminists and Du Bois: Respectability, Protection and Beyond." *The Annals of the American Academy of Political and Social Science* 568 (March):28–40.

Hedges, Elaine. 1973. "Afterword." Pp. 37–65 in Charlotte Perkins Gilman, *The Yellow Wallpaper.* New York: City University of New York Feminist Press.

Hill, Mary. 1989. "Charlotte Perkins Gilman: A Feminist's Struggle with Womanhood." Pp. 31–50 in *Charlotte Perkins Gilman: The Woman and her Work,* edited by Sheryl L. Meryering. Ann Arbor, MI: UMI Research Press.

Hugo, Victor. 1862/1879. *Les Miserables.* New York: Carleton Publishing Company.

International Labor Organization. 2006. "ILO Annual Jobs Report Says Global Unemployment Continues to Grow, Youth Now Make Up Half Those Out of Work." Retrieved July 27, 2009 (http://www.ilo.org/global/About_the_ILO/Media_ and_public_information/Press_releases/lang-- en/WCMS_065176/index.htm).

James, William. 1890/2007. *The Principles of Psychology.* New York: Cosimo Classics.

———. 1892/2001. "The Stream of Consciousness." Pp. 18–43 in *Psychology.* Mineola, NY: Dover.

———. 1902/1958. *The Varieties of Religious Experience.* New York: New American Library.

Jaspers, Karl. 1957/1962. *Socrates, Buddha, Confucius, Jesus.* (The Great Philosophers, vol. 1). New York: Harcourt Brace.

King, Martin Luther, Jr. 1970. "Honoring Dr. Du Bois." Pp. 176–85 in *Black Titan: W. E. B. Du Bois,* edited by John Henrik Clarke, Esther Jackson, Ernest Kaiser, and J. H. O'Dell. Boston: Beacon Press.

Lane, Ann J. 1990. *To Herland and Beyond: The Life and Work of Charlotte Perkins Gilman.* New York: Pantheon.

Lemert, Charles, ed. 1993. *Social Theory.* Boulder, CO: Westview Press.

———. 1998. "Anna Julia Cooper: The Colored Woman's Office." In Charles Lemert and Esme Bhand, eds., *The Voice of Anna Julia Cooper.* Lanham, MD: Rowman and Littlefield Publishers.

Levine, Donald. 1971. "Introduction." Pp. ix–lxv in *Georg Simmel: On Individuality and Social Forms.* Chicago: University of Chicago Press.

Lewis, David Levering. 2000. *W. E. B. Du Bois.* New York: Henry Holt.

National Assembly of France. 1789. "La Déclaration des Droits de l'Homme et du Citoyen" ("The Declaration of Rights of Man and of the Citizen.") Retrieved July 27, 2009 (http://www .aidh.org/Biblio/Text_fondat/FR_02.htm, https:// www.college.columbia.edu/core/students/cc/ settexts/nafman89.pdf).

Lukes, Steven. 1985. *Émile Durkheim: His Life and Work.* London: Allen Lane, Penguin Press.

Marable, Manning. 1986. *W. E. B. Du Bois: Black Radical Democrat.* Boston: Twayne.

Martin, James. 1999. "The Social and the Political." Pp. 155–77 in *Contemporary Social and Political Theory,* edited by Fidelma Ashe, Alan Finlayson, Moya Lloyd, Iain MacKenzie, James Martin, and Shane O'Neill. Buckingham, UK: Open University Press.

Martindale, Don. 1981. *The Nature and Types of Sociological Theory.* 2nd ed. Boston: Houghton Mifflin.

Martineau, Harriet. 1837. *Society in America.* New York: Saunders and Otley.

———. 1853. *The Positive Philosophy of Auguste Comte.* New York: Calvin Blanchard.

Marx, Karl. 1844/1978. *The Economic and Philosophic Manuscripts of 1844,* edited by Dirk J. Struik and translated by Martin Milligan. New York: International.

———. 1852/1978. "The Eighteenth Brumaire of Louis Bonaparte." Pp. 594–617 in *The Marx-Engels Reader,* edited by Robert C. Tucker. New York: W. W. Norton.

———. 1859/1978. "Preface to A Contribution to the Critique of Political Economy." Pp. 3–6 in *The Marx-Engels Reader,* edited by Robert C. Tucker. New York: W. W. Norton.

———. 1867/1978. *Capital,* vol. 1, translated by S. Moore and E. Aveling. New York: International.

Marx, Karl and Friedrich Engels. 1846/1977. *The German Ideology*, edited by C. J. Arthur. New York: International.

———. 1848/1978. "The Communist Manifesto." Pp. 469–500 in *The Marx-Engels Reader*, edited by Robert C. Tucker. New York: W. W. Norton.

McGuire, Meredith. 1997. *Religion: The Social Context*. 4th ed. Belmont, CA: Wadsworth.

McMahon, Darrin. 2001. *Enemies of the Enlightenment*. Oxford, UK: Oxford University Press.

Mead, George Herbert. 1934/1962. *Mind, Self, and Society*, edited by Charles W. Morris. Chicago: University of Chicago Press.

———. 1936/1964. "Mind Approached Through Behavior" Pp. 65–82 in *George Herbert Mead on Social Psychology*, edited by Anselm Strauss. Chicago: University of Chicago Press.

Michels, Robert. 1911/1958. *Political Parties*, translated by Eden and Cedar Paul. Glencoe, IL: The Free Press.

Monteiro, Anthony. 2000. "Being an African in the World of Du Boisian Epistemology." *The Annals of the American Academy of Political and Social Science*, 568 (March):220–34.

Morgan, Lewis. 1877/2000. *Ancient Society*. Piscataway, NJ: Transaction Publishers

National Association of Home Builders. 2003. *Los Angeles Times*, September 14, K4.

Nietzsche, Friedrich. 1866/1966. *Beyond Good and Evil*, translated by Walter Kaufmann. New York: Vintage Books.

———. 1882/1974. *The Joyful Wisdom*, translated by Walter Kaufmann. New York: Vintage Books.

———. 1883/1978. *Thus Spoke Zarathustra*, translated by Walter Kaufmann. New York: Penguin Books.

Patterson, Orlando. 1998. *Rituals of Blood*. Washington, DC: Civitas/Counterpoint.

Pichanick, Valerie K. 1980. *Harriet Martineau: The Woman and Her Work, 1802–76*. Ann Arbor: University of Michigan Press.

Putnam, Robert. 2001. *Bowling Alone: The Collapse and Revival of American Community*. New York: Simon & Schuster.

Ritzer, George. 2000. *Classical Sociological Theory*. 3rd ed. Boston: McGraw-Hill.

Robertson, Roland. 1970. *The Sociological Interpretation of Religion*. New York: Schocken.

Rothman, Paul. 1995. *Full Circle* [film]. New York: Filmakers Library.

Rousseau, Jean-Jacques. 1762. *The Social Contract*. Translated by G. D. H. Cole, public domain. Retrieved July 27, 2009 (http://www.constitution.org/jjr/socon.htm).

Rudwick, Elliot. 1960/1982. *W. E. B. Du Bois: Voice of the Black Protest Movement*. Urbana: University of Illinois Press.

Scharnhorst, Gary. 2000. "Historicizing Gilman: A Bibliographer's View." Pp. 65–73 in *The Mixed Legacy of Charlotte Perkins Gilman*, edited by C. Golden and J. Zangrando. Newark, NJ: University of Delaware Press.

Seidman, Steven. 1994. *Contested Knowledge*. Malden, MA: Blackwell.

Simmel, Georg. 1900a/1978. *The Philosophy of Money*, translated by Tom Bottomore and David Frisby. London: Routledge & Kegan Paul.

———. 1900b/1971. "Exchange." Pp. 43–69 in *On Individuality and Social Forms*, edited by Donald N. Levine. Chicago: University of Chicago Press.

———. 1903/1971. "The Metropolis and Mental Life." Pp. 324–339 in *On Individuality and Social Forms*, edited by Donald N. Levine. Chicago: University of Chicago Press.

———. 1904/1971. "Fashion." Pp. 294–323 in *On Individuality and Social Forms*, edited by Donald N. Levine. Chicago: University of Chicago Press.

———. 1908a/1955. *Conflict and the Web of Group Affiliations*, translated by Kurt H. Wolff and Reinhard Bendix. New York: Free Press.

———. 1908b/1971. "Conflict." Pp. 70–95 in *On Individuality and Social Forms*, edited by Donald N. Levine. Chicago: University of Chicago Press.

———. 1908c/1971. "The Stranger." Pp. 143–149 in *On Individuality and Social Forms*, edited by Donald N. Levine. Chicago: University of Chicago Press.

———. 1908d/1971. "The Problem of Sociology." Pp. 23–35 in *On Individuality and Social Forms*, edited by Donald N. Levine. Chicago: University of Chicago Press.

———. 1908e/1971. "How is Society Possible?" Pp. 6–22 in *On Individuality and Social Forms*, edited by Donald N. Levine. Chicago: University of Chicago Press.

———. 1908f/1971. "Subjective Culture." Pp. 227–234 in *On Individuality and Social Forms*, edited by Donald N. Levine. Chicago: University of Chicago Press.

———. 1910/1971. "Sociability." Pp. 127–140 in *On Individuality and Social Forms*, edited by Donald N. Levine. Chicago: University of Chicago Press.

———. 1917/1950. "Fundamental Problems of Sociology (Individual and Society)." Pp. 3–84 in *The Sociology of Georg Simmel*, edited and translated by Kurt Wolff. New York: Free Press.

Smith, Adam. 1776/1990. *An Inquiry into the Nature and Causes of the Wealth of Nations*. Chicago: Encyclopedia Britannica.

————. 1842/1993. "The Proper Sphere of Government." Originally printed in *The Nonconformist*; republished in *Spencer: Political Writings,* by Herbert Spencer, Author; John Offer, Editor. Cambridge: Cambridge University Press.

Stack, Carol. 1974. *All Our Kin: Strategies for Survival in a Black Community.* New York: Harper and Row.

Tönnies, Ferdinand. 1887/1957. *Community and Society (Gemeinschaft und Gesellschaft)*, translated by Charles P. Loomis. East Lansing, MI: Michigan State University Press.

Veblen, Thorstein. 1899/1934. *The Theory of the Leisure Class*, edited by Stuart Chase. New York: Macmillan.

————. 1904/1965. *The Theory of Business Enterprise.* New York: A. M. Kelley.

Wagner-Martin, Linda. 1989. "Gilman's 'The Yellow Wallpaper': A Centenary." Pp. 51–64 in *Charlotte Perkins Gilman: The Woman and her Work*, edited by Sheryl L. Meryering. Ann Arbor, MI: UMI Research Press.

Washington, Mary Helen. 1988. "Introduction." In Anna Julia Cooper, *A Voice From the South.* New York: Oxford University Press.

Watson, John B. 1924/1966. *Behaviorism.* Chicago: University of Chicago Press.

Weber, Marianne. 1907. *Marriage and Motherhood in the Development of Law. Ehefrau und Mutter in der Rechtensentwicklung.* Tübingen: J. C. B. Mohr.

————. 1926/1975. *Max Weber: A Biography.* New York: Wiley.

————. 1935. *Women and Love. Frauen und Liebe.* Koonigestein in Taunus: K. B. Langewissche.

Weber, Max. 1903–17/1949. *The Methodology of the Social Sciences*, edited and translated by Edward A. Shils and Henry A. Finch. New York: Free Press.

————. 1904–05/1958. *The Protestant Ethic and the Spirit of Capitalism*, translated by Talcott Parsons. New York: Charles Scribner's Sons.

————. 1915/1958. "The Social Psychology of the World Religions." Pp. 267–301 in *From Max Weber: Essays in Sociology*, edited and translated by H. H. Gerth and C. Wright Mills. New York: Oxford University Press.

————. 1919/1958. "Science as a Vocation." Pp. 129–156 in *From Max Weber: Essays in Sociology*, edited and translated by H. H. Gerth and C. Wright Mills. New York: Oxford University Press.

————. 1925a/1978. *Economy and Society*, vols. 1 and 2, edited by Guenther Roth and Claus Wittich. Berkeley: University California Press.

————. 1925b/1978. "Status Groups and Classes." Pp. 926–939 in *Economy and Society*, vols. 1 and 2, edited by Guenther Roth and Claus Wittich. Berkeley: University California Press.

————. 1925c/1978. "The Types of Legitimate Domination." Pp. 212–301 in *Economy and Society*, vols. 1 and 2, edited by Guenther Roth and Claus Wittich. Berkeley: University California Press.

————. 1925d/1978. "Bureaucracy." Pp. 956–1005 in *Economy and Society*, vols. 1 and 2, edited by Guenther Roth and Claus Wittich. Berkeley: University California Press.

————. 1947. *The Theory of Social and Economic Organization*, edited by Talcott Parsons and translated by A. M. Henderson and Parsons. New York: Oxford University Press.

————. 1958. *From Max Weber: Essays in Sociology*, edited and translated by H. H. Gerth and C. Wright Mills. New York: Oxford University Press.

Wrong, Dennis H. 1994. *The Problem of Order.* Cambridge: Harvard University Press.

Zuckerman, Phil, ed. 2000. *Du Bois on Religion.* Walnut Creek, CA: AltaMira.

————, ed. 2004. *The Social Theory of W. E. B. Du Bois.* Thousand Oaks, CA: Pine Forge.